I0065693

Signal Processing: Theory, Design and Algorithms

Signal Processing: Theory, Design and Algorithms

Edited by **Graham Eanes**

Willford Press

New York

Published by Willford Press,
118-35 Queens Blvd., Suite 400,
Forest Hills, NY 11375, USA
www.willfordpress.com

Signal Processing: Theory, Design and Algorithms
Edited by Graham Eanes

© 2016 Willford Press

International Standard Book Number: 978-1-68285-059-6 (Hardback)

This book contains information obtained from authentic and highly regarded sources. Copyright for all individual chapters remain with the respective authors as indicated. All chapters are published with permission under the Creative Commons Attribution License or equivalent. A wide variety of references are listed. Permission and sources are indicated; for detailed attributions, please refer to the permissions page and list of contributors. Reasonable efforts have been made to publish reliable data and information, but the authors, editors and publisher cannot assume any responsibility for the validity of all materials or the consequences of their use.

The publisher's policy is to use permanent paper from mills that operate a sustainable forestry policy. Furthermore, the publisher ensures that the text paper and cover boards used have met acceptable environmental accreditation standards.

Trademark Notice: Registered trademark of products or corporate names are used only for explanation and identification without intent to infringe.

Printed in the United States of America.

Contents

Preface

Every book is initially just a concept; it takes months of research and hard work to give it the final shape in which the readers receive it. In its early stages, this book also went through rigorous reviewing. The notable contributions made by experts from across the globe were first molded into patterned chapters and then arranged in a sensibly sequential manner to bring out the best results.

Signal processing is an ever evolving field. Its scope of applications is also rapidly expanding. In the modern times, it is being utilized in audio, speech, image and video processing. This comprehensive and innovative book covers the diverse aspects and applications of signal processing. Some interesting topics covered in this text revolve around signal processing theory, algorithms, architecture, design and implementation, signal processing design tools, signal processing for sensors, etc. This book is an essential guide for all associated with the discipline of signal processing, electronics and communication.

It has been my immense pleasure to be a part of this project and to contribute my years of learning in such a meaningful form. I would like to take this opportunity to thank all the people who have been associated with the completion of this book at any step.

Editor

Optimal blockwise subcarrier allocation policies in single-carrier FDMA uplink systems

Antonia Maria Masucci[1,2*], Elena Veronica Belmega[1] and Inbar Fijalkow[1]

Abstract

In this paper, we analyze the optimal (blockwise) subcarrier allocation schemes in single-carrier frequency division multiple access (SC-FDMA) uplink systems without channel state information at the transmitter side. The presence of the discrete Fourier transform (DFT) in SC-FDMA/orthogonal frequency division multiple access OFDMA systems induces correlation between subcarriers which degrades the transmission performance, and thus, only some of the possible subcarrier allocation schemes achieve better performance. We propose as a performance metric a novel sum-correlation metric which is shown to exhibit interesting properties and a close link with the outage probability. We provide the set of optimal block-sizes achieving the maximum diversity and minimizing the inter-carrier sum-correlation function. We derive the analytical closed-form expression of the largest optimal block-size as a function of the system's parameters: number of subcarriers, number of users, and the cyclic prefix length. The minimum value of sum-correlation depends only on the number of subcarriers, number of users and on the variance of the channel impulse response. Moreover, we observe numerically a close strong connection between the proposed metric and diversity: the optimal block-size is also optimal in terms of outage probability. Also, when the considered system undergoes carrier frequency offset (CFO), we observe the robustness of the proposed blockwise allocation policy to the CFO effects. Numerical Monte Carlo simulations which validate our analysis are illustrated.

Keywords: SC-FDMA/OFDMA; Subcarriers allocation; Channel frequency diversity; Cyclic prefix induced; Correlation; Carrier frequency offsets

1 Introduction

Due to its simplicity and flexibility to subcarrier allocation policies, single-carrier frequency division multiple access (SC-FDMA) has been proposed as the uplink transmission scheme for wireless standard of 4G technology such as 3GPP long-term evolution (LTE) [1-3]. SC-FDMA is a technique with similar performance and essentially the same general structure as an orthogonal frequency division multiple access (OFDMA) system. A remarkable advantage of SC-FDMA over OFDMA is that the signal has lower peak-to-average power ratio (PAPR) that guarantees the transmit power efficiency at the mobile terminal level [4]. However, similarly to OFDMA, SC-FDMA shows sensitivity to small values of carrier frequency offsets (CFOs) generated by the frequency misalignment between the mobile users' oscillators and the base station [5-7]. CFO is responsible for the loss of orthogonality among subcarriers by producing a shift of the received signals causing inter-carrier interferences (ICI).

In this work, we show that the SC-FDMA uplink systems without CFO and with imposed independent subcarriers attain the same channel diversity gain for any subcarrier allocation scheme. However, due to the discrete Fourier transform (DFT) of the channel, correlation between the subcarriers is induced, and thus, a degradation of the transmission performance occurs. Therefore, there exist some allocation schemes that are able to achieve an increased diversity gain when choosing the appropriate subcarrier allocation block-size.

In the uplink SC-FDMA transmissions, users spread their information across the set of available subcarriers. Subcarrier allocation techniques are used to split the available bandwidth between the users. In the case in which no channel state information (CSI) is available at the transmitter side, the most popular allocation scheme is the

*Correspondence: antonia.masucci@inria.fr
[1] ETIS/ENSEA - University of Cergy Pontoise - CNRS, 6 Avenue de Ponceau, 95014 Cergy, France
[2] INRIA Paris-Rocquencourt, Le Chesnay Cedex, France

blockwise allocation in [8]. In this *blockwise allocation scheme*, subsets of adjacent subcarriers, called blocks, are allocated to each user, (see Figures 1, 2, and 3). In particular, we call *mono-block allocation* the scheme with the maximum block-size given by the ratio between the number of subcarriers and the number of users, illustrated in Figure 1. The *interleaved allocation scheme* is a special case in which subcarriers are uniformly spaced at a distance equal to the number of users (block-size b is equal to one), as shown in Figure 3. The interleaved allocation is usually considered to benefit from frequency diversity (IEEE 802.16) [9]. However, robustness to CFO can be improved by choosing large block-sizes since they better combat the ICI. In the case of full CSI, an optimal block-size has been proposed for OFDMA systems in [10] as a good balance between the frequency diversity gain and robustness against CFO. In this paper, we study the optimal block-size allocation schemes, in the case of an uplink SC-FDMA without CSI.

To the best of our knowledge, the closest works to ours are references [11,12]. A subcarrier allocation scheme with respect to the user's outage probability has been proposed in [11] for OFDMA/SC-FDMA systems with and without CFO. In particular, the authors of [11] propose a semi-interleaved subcarrier allocation scheme capable of achieving the diversity gain with minimum CFO interference. However, the authors analyze only the case in which every user in the system transmits one symbol spread to all subcarriers, and as a consequence, their diversity results are restricted to the considered model and with a low data rate. We point out that our main contributions with respect to [11] consist in the following: we consider a more general model; we analyze all possible subcarrier allocation block-sizes; and we find the analytical expressions of the optimal blockwise allocation schemes that achieve maximum diversity. Moreover, we provide an analytical expression of the correlation between subcarriers and we analyze its effects on the system transmission's performance.

More precisely, in this work, we propose a new allocation policy based on the minimization of the correlation between subcarriers. In particular, in order to optimize the block-size subcarrier allocation, we propose a new performance metric, i.e., the sum-correlation function that we define as the sum of correlations of each subcarrier with respect to the others in the same allocation scheme. The introduction of the sum-correlation function as a performance metric is motivated by the fact that the correlation generated by the DFT implies that some allocation schemes achieve a higher diversity gain than others. The interest of the proposed approach is due to the fact that it allows us to find the exact expression of the block-sizes that achieve a higher diversity gain. It turns out that the minimum sum-correlation is achieved by block-size allocation policies that lie in a set composed of all block-sizes that are inferior or equal to a given threshold depending explicitly on the system's parameters: the number of subcarriers, the number of users, and the cyclic prefix length. Furthermore, we find the minimum value of the sum-correlation function. This value guarantees to achieve the maximum diversity gain, and what is more remarkable, it depends only on the number of subcarriers, number of users, and the variance of the channel impulse response. We also provide interesting properties of the individual sum-correlation terms: the auto-correlation term (i.e., the correlation between the subcarrier of reference and itself) depends on the length of cyclic prefix; the correlations between the subcarrier of reference and the ones that are spaced from it of a distance equal to a multiple of the ratio between the number of subcarriers, and the cyclic prefixes are equal to zero. The most interesting property of the proposed sum-correlation function is the close link to the outage probability and thus to the diversity gain. Numerically, we observe that the maximum diversity or the minimum outage allocation coincides with the one minimizing our sum-correlation function.

Moreover, we observe that when the SC-FDMA system undergoes CFO, we have the robustness to CFO for practical values of CFO. This means that when the CFO goes to zero, the CFO sum-correlation can be approximated with the sum-correlation defined in the case without CFO. This analysis has been done similarly to [12] in which coded

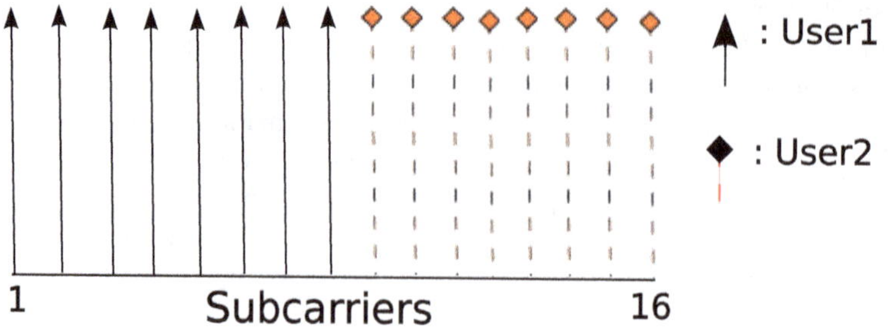

Figure 1 Mono-block allocation scheme. Block-size $b = 8$, $N_p = 16$ subcarriers, and $N_u = 2$ users (subcarriers allocated to user u_1 denoted by arrow markers and to user u_2 by diamond markers).

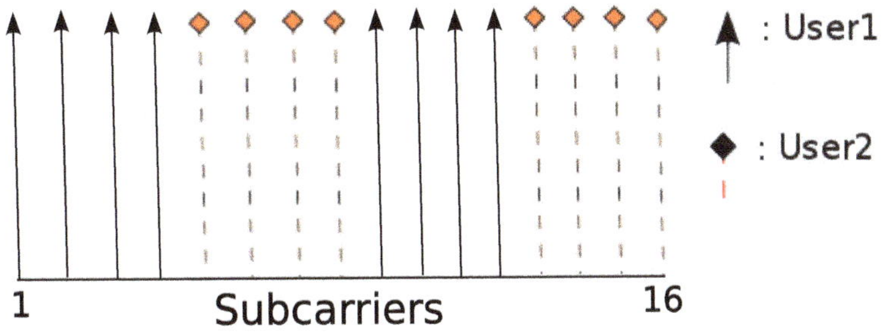

Figure 2 Blockwise allocation scheme. Block-size $b = 4$, $N_p = 16$ subcarriers, $N_u = 2$ users (subcarriers allocated to user u_1 denoted by arrow markers and to user u_2 by diamond markers).

OFDMA systems are analyzed. Uncoded OFDMA cannot exploit the frequency diversity of the channel; therefore, the use of channel coding with OFDMA in [12] reduces the errors resulting from the multipath fading environment recovering the diversity gain. Coding is not needed in SC-FDMA since it can be interpreted as a linearly precoded OFDMA system [4].

We underline that, with respect to [12] in which the results have been briefly announced, in this paper: 1) we provide a deeper and a more detailed theoretical analysis; 2) we consider a new performance metric, not identical to the one analyzed in [12], which takes into account the length of the channel impulse response $L_h \leq L$ and a general power delay profile which allow us to generalize our previous results in both cases, with and without CFO; 3) novel simulation results are presented in order to validate these new results.

The difficulty of our analytical study is related to the discrete feasible set of allocation block-sizes and also to the objective function (i.e., the sum-correlation function we propose) which is closely linked with the outage probability whose minimization is still an open issue in most non-trivial cases [13]. However, we provide extensive numerical Monte Carlo simulations that validate our analysis and all of our claims.

The sequel of our paper is organized as follows. In Section 2, we present the analytical model of the SC-FDMA system without CFO. In Section 3, we define a novel sum-correlation function and its properties; moreover, we find the optimal block-sizes for a subcarrier allocation scheme minimizing the subcarrier correlation function and we show the numerical results that validate our analysis. We present the SC-FDMA system with CFO in Section 4. We define the corresponding sum-correlation function and we observe its robustness against CFO. Numerical results that validate this analysis are also presented. At last, in Section 5 we conclude the paper.

2 System model without CFO

We consider a SC-FDMA uplink system where N_u mobile users communicate with a base station (BS) or access point. In the case in which the system is not affected by CFOs, the users are synchronized to the BS in time and frequency domains. No CSI is available at the transmitter side. The total bandwidth B is divided into N_p subcarriers and we denote by $M = \lfloor \frac{N_p}{N_u} \rfloor$ (where $\lfloor x \rfloor$ is the integer part of x) the number of subcarriers per user. Notice that we choose N_p as an integer power of two in order to optimize the DFT processing. To provide a fair allocation of the spectrum among the users (fair in the sense that the number of allocated subcarriers is the same for all users), notice that the number of not-allocated carriers is $N_p - N_u M < N_u << N_p$ which is a negligible fraction of the total available spectrum. Without loss of

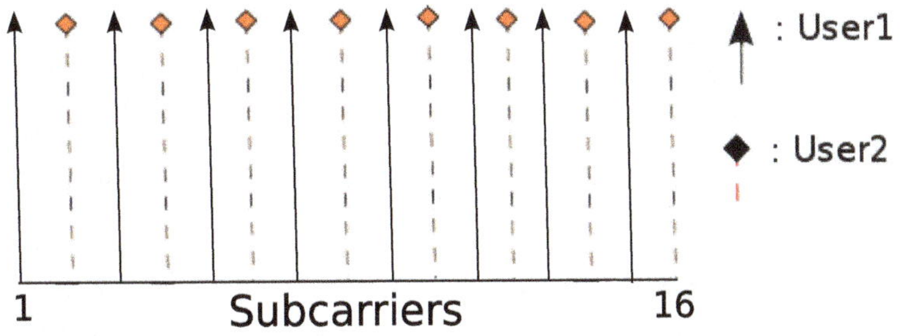

Figure 3 Interleaved allocation scheme. Block-size $b = 1$, $N_p = 16$ subcarriers, $N_u = 2$ users (subcarriers allocated to user u_1 denoted by arrow markers and to user u_2 by diamond markers).

generality and also to avoid complex notations[a], we will assume in the following that N_u is also a power of two and that $M = \frac{N_p}{N_u}$.

The signal at the input of the receiver DFT was expressed in [10] as follows:

$$\begin{pmatrix} y_{N_p-1} \\ \vdots \\ y_0 \\ \vdots \\ y_{-L} \end{pmatrix} = \sum_{u=1}^{N_u} \begin{pmatrix} h_0^{(u)} & \cdots & h_{L_h-1}^{(u)} & & 0 \\ & \ddots & & \ddots & \\ 0 & & h_0^{(u)} & \cdots & h_{L_h-1}^{(u)} \end{pmatrix} \times \begin{pmatrix} a_{N_p-1}^{(u)} \\ \vdots \\ a_0^{(u)} \\ \vdots \\ a_{-L}^{(u)} \end{pmatrix} + \begin{pmatrix} n_{N_p-1} \\ \vdots \\ n_0 \\ \vdots \\ n_{-L} \end{pmatrix} \tag{1}$$

where L is the length of the cyclic prefix. The vector $\mathbf{h}^{(u)} = \left[h_0^{(u)}, \ldots, h_{L_h-1}^{(u)}\right]$ is the channel impulse response whose dimension L_h is lower than or equal to L. The elements $a_k^{(u)}$ are the symbols at the output of the inverse discrete Fourier transform (IDTF) given by

$$\mathbf{a}^{(u)} = \begin{pmatrix} a_{N_p-1}^{(u)} \\ \vdots \\ a_0^{(u)} \end{pmatrix} = \mathbf{F}^{-1}\mathbf{\Pi}_b^{(u)}\mathbf{x}^{(u)} \tag{2}$$

with $\mathbf{F}^{-1}$ the N_p-size inverse DFT matrix, $\mathbf{x}^{(u)} = \mathbf{F}_{\frac{N_p}{N_u}}\tilde{\mathbf{x}}^{(u)}$ where $\mathbf{F}_{\frac{N_p}{N_u}}$ is the $\frac{N_p}{N_u}$-size DFT matrix, and $\tilde{\mathbf{x}}^{(u)}$ is the vector of the M-ary symbols transmitted by user u. The vector $\tilde{\mathbf{x}}^{(u)}$ does not have a particular structure, contrary to [11] where it is assumed to be equal to $\mathbf{1}_{\frac{N_p}{N_u}\times 1}\tilde{x}$ which means that one symbol is spread to all subcarriers. The symbol $\mathbf{\Pi}_b^{(u)}$ is the $N_p \times \frac{N_p}{N_u}$ subcarrier allocation matrix with only one element equal to 1 in each column which occurs at rows that represent the carriers allocated to user u according to the considered block-size $b \in \beta = \left\{1, \ldots, \frac{N_p}{N_u}\right\}$. The set β is composed of all divisors of $\frac{N_p}{N_u}$, this guarantees a fully utilized spectrum. The SC-FDMA can be viewed as a pre-coded version of OFDMA since the $\frac{N_p}{N_u}$-size DFT matrix does not affect the channel diversity.

Discarding in the signal at the input of the receiver DFT the L components corresponding to the cyclic prefix and rearranging the terms, we get

$$\underbrace{\begin{pmatrix} y_{N_p-1} \\ \vdots \\ y_0 \end{pmatrix}}_{\mathbf{y}} = \sum_{u=1}^{N_u} \underbrace{\begin{pmatrix} h_0^{(u)} & \cdots & h_{L_h-1}^{(u)} & & 0 \\ & \ddots & & \ddots & \\ & & & & h_{L_h-1}^{(u)} \\ h_{L_h-1}^{(u)} & 0 & & & \\ \vdots & \ddots & & \ddots & \\ h_1^{(u)} & \cdots & h_{L_h-1}^{(u)} & & h_0^{(u)} \end{pmatrix}}_{\mathbf{h}_{\text{circ}}^{(u)}} \times \begin{pmatrix} a_{N_p-1}^{(u)} \\ \vdots \\ a_0^{(u)} \end{pmatrix} + \begin{pmatrix} n_{N_p-1} \\ \vdots \\ n_0 \end{pmatrix} \tag{3}$$

where $\mathbf{h}_{\text{circ}}^{(u)}$ is a $N_p \times N_p$ circulant matrix. Denoting $\mathbf{r} = \mathbf{F}\mathbf{y}$, we have found that the received signal at the BS after the N_p-size DFT is given by:

$$\begin{aligned} \mathbf{r} &= \sum_{u=1}^{N_u} \mathbf{F}\mathbf{h}_{\text{circ}}^{(u)}\mathbf{F}^{-1}\mathbf{\Pi}_b^{(u)}\mathbf{x}^{(u)} + \mathbf{F}\mathbf{n} \\ &= \sum_{u=1}^{N_u} \mathbf{H}^{(u)}\mathbf{\Pi}_b^{(u)}\mathbf{x}^{(u)} + \tilde{\mathbf{n}} \end{aligned} \tag{4}$$

where $\mathbf{H}^{(u)} = \mathbf{F}\,\mathbf{h}_{\text{circ}}^{(u)}\mathbf{F}^{-1}$ is the diagonal channel matrix of user u with the diagonal (k,k)-entry given by

$$H_k^{(u)} = \frac{1}{\sqrt{N_p}} \sum_{m=0}^{L_h-1} h_m^{(u)} e^{-j2\pi mk/N_p}, \tag{5}$$

and $\tilde{\mathbf{n}} = \mathbf{F}\mathbf{n}$ is the $N_p \times 1$ additive Gaussian noise with variance $\sigma_n^2\mathbf{I}$. Therefore, over each subcarrier $k = 0, \ldots, N_p - 1$, we have

$$r_k = \sum_{u=1}^{N_u} \frac{1}{\sqrt{N_p}} \sum_{m=0}^{L_h-1} h_m^{(u)} e^{-j2\pi mk/N_p}\Pi_{k,k}^{(u)}x_k^{(u)} + n_k.$$

Note then that $\mathbf{H}^{(u)}$ is diagonal thanks to the assumption on the channel impulse response length being shorter than the cyclic prefix [10]. However, the diagonal entries (5) are correlated with each other.

3 Minimization of the subcarriers sum-correlation

Frequency diversity occurs in OFDMA systems by sending multiple replicas of the transmitted signal at different carrier frequencies. The idea behind diversity is to provide independent replicas of the same transmitted signal at the receiver and appropriately process them to make the detection more reliable. Different copies of the signal should be transmitted in different frequency bands, a condition which guarantees their independence. The subchannels given in (5) are correlated and no more than a L_h-order frequency diversity gain can be possible since there are only L_h independent channel coefficients, i.e.,

$h_0^{(u)}, \ldots, h_{L_h-1}^{(u)}$ [14]. Intuitively, we can imagine that there are L_h groups of $\frac{N_p}{L_h}$ frequencies which are 'identical'. Each user retrieves the maximal diversity if it has at least L_h blocks of size b which implies $b \leq \frac{N_p}{L_h N_u}$. Therefore, users can achieve full diversity when the block-size is within the coherent bandwidth $\frac{N_p}{L_h N_u}$.

In this work, we propose an original approach to find the block-size that guarantees the maximum diversity. This approach is based on the minimization of the subchannels/subcarrier sum-correlation function.

In the sequel, we define a measure of correlation between subcarriers, that we call sum-correlation function, and we derive its properties. Moreover, we find the set of optimal block-sizes $b^* \in \beta$ which minimizes this correlation in order to minimize the effects that it produces.

3.1 Properties of the sum-correlation function

Assuming that the channel impulse responses are independent but distributed accordingly to the complex Gaussian distribution $\mathcal{CN}(0, \sigma_h^{(u)2})$ [14], we define for each user the sum of correlations of each subcarrier with respect to the others in the same allocation scheme as follows:

$$\begin{aligned}
\Gamma_{u,m}(b) &= \sum_{c_u \in \mathcal{C}_u} \mathbb{E}\left[H_m^{(u)} H_{c_u}^{(u)*}\right] \\
&= \mathbb{E}\left[|H_m^{(u)}|^2\right] + \sum_{\substack{c_u \in \mathcal{C}_u \\ c_u \neq m}} \mathbb{E}\left[H_m^{(u)} H_{c_u}^{(u)*}\right] \\
&= \frac{\sigma_h^{(u)2}}{N_p} L_h + \frac{\sigma_h^{(u)2}}{N_p} \sum_{c_u \in \mathcal{C}_u} \frac{\sin\left[\frac{\pi L_h}{N_p}(m - c_u)\right]}{\sin\left[\frac{\pi}{N_p}(m - c_u)\right]} \\
&\quad \times e^{-\pi j \frac{(L_h - 1)}{N_p}(m - c_u)}
\end{aligned} \tag{6}$$

where $m \in \mathcal{C}_u$ is the reference subcarrier and

$$\mathcal{C}_u = \bigcup_{k \in \left\{1,2,3,\ldots,\frac{N_p}{bN_u}\right\}} \{(k-1)bN_u + (u-1)b + i, \forall i \in \{1, \ldots, b\}\}. \tag{7}$$

The set $\mathcal{C}_u$ is composed of all indices of subcarriers allocated to user u given a block-size b allocation scheme. The ratio $\frac{N_p}{bN_u}$ represents the number of blocks that can be allocated to each user given a block-size b. We consider, therefore, that the total N_p subcarriers are divided into $\frac{N_p}{bN_u}$ large-blocks that contain bN_u subcarriers corresponding to the N_u blocks, one for each user, of size b. The set $\mathcal{C}_u$ is the union of indices of subcarriers allocated to user u in all these large-blocks. Inside of the large-block of index k, the indices of the b subcarriers allocated to user u are $(k-1)bN_u + (u-1)b + i$, $\forall i \in \{1, \ldots, b\}$, where $(k-1)bN_u$ corresponds to the previous $k-1$ large-blocks and $(u-1)b$ corresponds to the previous allocated users $(1, 2, \ldots, u-1)$, see Figure 4. The function $\Gamma_{u,m}(b)$ is the sum of the correlations between subcarriers that are in the same allocation scheme.

Considering the subcarriers m and c_u in $\mathcal{C}_u$, we denote the distance between them by

$$\begin{aligned}
d &:= m - c_u \\
&= (k' - k'')bN_u + i' - i''
\end{aligned} \tag{8}$$

with $k', k'' \in \left\{1, 2, 3, \ldots, \frac{N_p}{bN_u}\right\}$ and $i', i'' \in \{1, \ldots, b\}$. We define the function

$$f(d) \triangleq \begin{cases} L_h & \text{if } d = 0 \\ \frac{\sin\left[\pi \frac{L_h}{N_p} d\right]}{\sin\left[\frac{\pi}{N_p} d\right]} e^{-\pi j \frac{(L_h - 1)}{N_p} d} & \text{otherwise}. \end{cases} \tag{9}$$

The next result guarantees that the function $f(d)$ is independent of the user index.

Proposition 1. *Given the parameters N_p, N_u, and L_h, the value $f(d)$ of the function in (9) for any $d = m - c_u$ in (8) is independent of the user index u.*

Proof: The dependence of f on the user index u is expressed by the term $(m - c_u)$, representing the distance between two subcarriers in the same allocation scheme, where $m, c_u \in \mathcal{C}_u$. If m and c_u are in two different blocks there exist two indices k' and k'' in $\left\{1, 2, 3, \ldots, \frac{N_p}{bN_u}\right\}$ such that

$$m = (k' - 1)bN_u + (u-1)b + i' \tag{10}$$

$$c_u = (k'' - 1)bN_u + (u-1)b + i'' \tag{11}$$

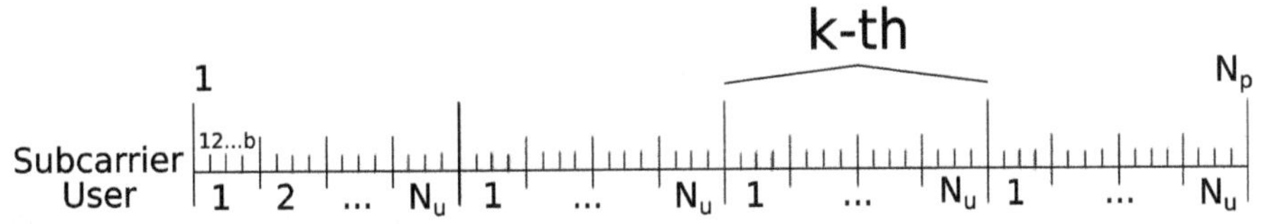

Figure 4 Set of subcarriers. The total N_p subcarriers are divided into $\frac{N_p}{bN_u}$ large blocks. The kth large block contains bN_u subcarriers, which are divided into N_u block of size b, one for each user.

with i' and i'' in $\{1, \ldots, b\}$. Therefore, the distance

$$m - c_u = \left(k' - k''\right) bN_u + i' - i'' \quad (12)$$

does not depend on the user index u. If m and c_u are in the same block, $k' = k''$ and the same reasoning holds. This guarantees the independence of the function $f(d)$ on the particular user u. ■

The independence between $f(d)$ and user index u comes from the fact that given a block-size b, the set of distances between subcarriers are the same for all users.

We observe that the function $f(d)$ has the following circularity property.

Proposition 2 (Circularity of f(d)). *Given N_p the number of subcarriers and for any distance d between subcarriers given in (8), we have*

$$f(d + N_p) = f(d). \quad (13)$$

Proof: From (9),

$$\begin{aligned} f\left(d + N_p\right) &= \frac{\sin\left[\pi \frac{L_h}{N_p}(d + N_p)\right]}{\sin\left[\frac{\pi}{N_p}(d + N_p)\right]} e^{-\pi j \frac{(L_h-1)}{N_p}(d+N_p)} \\ &= \frac{e^{j\pi \frac{L_h}{N_p}d} e^{j\pi L_h} - e^{-j\pi \frac{L_h}{N_p}d} e^{-j\pi L_h}}{e^{j\pi \frac{1}{N_p}d} e^{j\pi} - e^{-j\pi \frac{1}{N_p}d} e^{-j\pi}} \frac{e^{-j\pi \frac{L_h}{N_p}d} e^{-j\pi L_h}}{e^{-j\pi \frac{1}{N_p}d} e^{-j\pi}} \\ &= \frac{1 - e^{-j2\pi \frac{L_h}{N_p}d} e^{-j2\pi L_h}}{1 - e^{-j2\pi \frac{1}{N_p}d} e^{-j2\pi}} \\ &= \frac{e^{j\pi \frac{L_h}{N_p}d} - e^{-j\pi \frac{L_h}{N_p}d}}{e^{j\pi \frac{1}{N_p}d} - e^{-j\pi \frac{1}{N_p}d}} e^{-j\pi \frac{(L_h-1)}{N_p}d} \\ &= \frac{\sin\left[\pi \frac{L_h}{N_p}d\right]}{\sin\left[\frac{\pi}{N_p}d\right]} e^{-\pi j \frac{(L_h-1)}{N_p}d} \\ &= f(d). \end{aligned}$$

■

We consider the following sum-correlation metric:

$$\Gamma(b) = \sum_{u=1}^{N_u} \sum_{m \in C_u} \Gamma_{u,m}(b). \quad (14)$$

Thanks to Proposition 1 and Proposition 2, it can be expressed as follows:

$$\Gamma(b) = \sum_{u=1}^{N_u} |C_u| \frac{\sigma_h^{(u)2}}{N_p} \left(\sum_{d \in \mathcal{D}_b} f(d)\right) \quad (15)$$

$$= \sum_{u=1}^{N_u} \frac{N_p}{N_u} \frac{\sigma_h^{(u)2}}{N_p} \left(L_h + \sum_{\substack{d \in \mathcal{D}_b \\ d \neq 0}} f(d)\right) \quad (16)$$

where

$$\mathcal{D}_b = \left\{ \begin{array}{l} d = kbN_u + i,\ i \in \{0, \ldots, b-1\} \\ k \in \left\{0, \ldots, \left(\frac{N_p}{bN_u} - 1\right)\right\} \end{array} \right\}. \quad (17)$$

Given a subcarrier of reference, without loss of generality, the set $\mathcal{D}_b$ represents the set of the distances[b] between the subcarrier of reference and all the other subcarriers in the same allocation scheme (the k factor represents here the distance between the large-blocks). This definition is consistent since the function $f(d)$ has the circularity property with respect to N_p. This means that it does not matter which subcarrier of reference we consider. Therefore, the sum-correlation function defined in (15) is independent on the reference subcarrier. This guarantees that the next results hold for each user in the system.

We observe that there is no correlation between the subcarrier of reference and other carriers which are spaced from it at a distance equal to a multiple of $\frac{N_p}{L_h}$. It is obvious from the definition of the function $f(d)$ that it is equal to zero when the distance d is a multiple of the ratio $\frac{N_p}{L_h}$:

$$f\left(r\frac{N_p}{L_h}\right) = 0, \quad \forall r \in \mathbb{N}^*. \quad (18)$$

This is what we observe in Figure 5, in which we plot the absolute value of the function $\left|f(d)\right|$, with $m = 1$, $N_p = 32$, $N_u = 2$ and $L_h = 4$. We observe that this function is equal to zero for all the multiples of $\frac{N_p}{L_h} = 8$.

We have seen that the function $f(d)$ equals zero for all multiples of $\frac{N_p}{L_h}$ and that the distance d can be expressed in function of the block-size b (see (8) and (17)). In the following, we provide the expression of the block-size b such that $d = \frac{N_p}{L_h}$.

Proposition 3. *Given a fixed distance $d = \frac{N_p}{L_h}$ between subcarriers, the corresponding block-size is equal to $b = \frac{N_p}{L_h N_u}$.*

Proof: Notice that, given our allocation policy in Figure 4, not all the distances can be achieved for any possible block-size. We consider an arbitrary distance d in $\mathcal{D}_b$

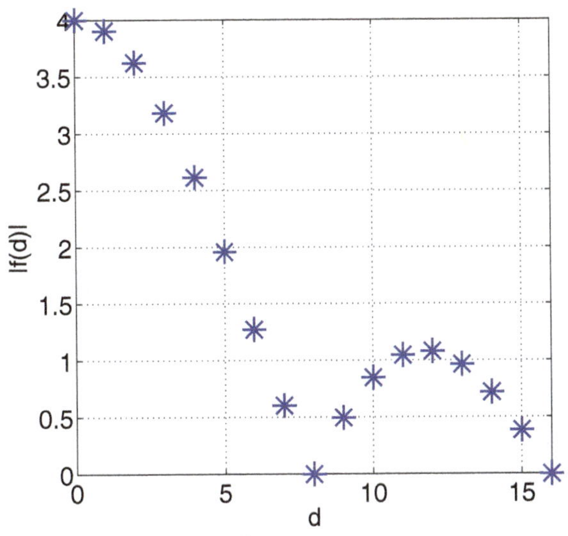

Figure 5 The function $\left|f(d)\right|$, with $m = 1, N_p = 32, N_u = 2$, and $L_h = 4$. In this case, $\frac{N_p}{L_h} = 8$.

and we are interested to find the block-size that ensures $d = \frac{N_p}{L_h}$:

$$\begin{aligned} d &= kbN_u + i \text{ with } i \in \{0, \ldots, b-1\}, \\ &\quad \text{and } k \in \left\{0, \ldots, \frac{N_p}{bN_u} - 1\right\} \\ &= r\frac{N_p}{L_h} \quad \text{with } r \leq 1. \end{aligned}$$

First, we analyze the case when $b \geq \frac{N_p}{L_h}$. In this case, we observe that the condition $d = \frac{N_p}{L_h}$ is satisfied if $k = 0$ and $i = \frac{N_p}{L_h}$. This means that for all $b \geq \frac{N_p}{L_h}$ there are two subcarriers in the same block such that their distance is $\frac{N_p}{L_h}$. More interesting is the case when $b < \frac{N_p}{L_h}$. From the fact that $i \in \{0, \ldots, b-1\}$ and $b < \frac{N_p}{L_h}$, in order to have the distance $d = \frac{N_p}{L_h}$, we have just to consider the case $i = 0$. Then, it is obvious that we have d equal to $\frac{N_p}{L_h}$ when $bN_u = \frac{N_p}{L_h}$. This means that the block-size is $b = \frac{N_p}{L_h N_u}$. ■

3.2 Novel blockwise allocation scheme

In the next Theorem, we give the set of optimal block-sizes that minimize the sum-correlation function $\Gamma(b)$. This set is given by $\beta^* = \left\{1, ..., \frac{N_p}{N_u L}, \frac{N_p}{N_u L_h}\right\} \subseteq \beta$.

Theorem 1. *We consider our uplink system with N_p subcarriers, N_u users and a channel impulse response length L_h. Given $\beta^* \triangleq \left\{1, \ldots, \frac{N_p}{N_u L}, \frac{N_p}{N_u L_h}\right\}$, we have*

1. *The elements in the set β^* minimize the sum-correlation function $\Gamma(b)$:*

$$\beta^* = \arg\min_{b \in \beta} \Gamma(b) \tag{19}$$

2. *The optimal value of the sum-correlation function depends only on the system parameters.*

$$\Gamma(b^*) = \sum_{u=1}^{N_u} \sigma_h^{(u)2} \frac{N_p}{N_u^2}, \quad \forall b^* \in \beta^*. \tag{20}$$

Proof: The proof is given in the Appendix 5. ■

Proposition 4. *In the case without CSI, assuming that L_h is not known at the transmitter side, we propose to use β^* restricted to $\left\{1, \ldots, \frac{N_p}{N_u L}\right\}$.*

Proof: We observe that $\left\{1, \ldots, \frac{N_p}{N_u L}\right\}$ is included in $\left\{1, \ldots, \frac{N_p}{N_u L}, \ldots, \frac{N_p}{N_u L_h}\right\}$. Intuitively, this means that, in a more realistic scenario in which only the knowledge of L and not of the channel length L_h is available, we can still provide the subset of optimal block-sizes. ■

We observe that, in the case without CSI, the minimum value of the sum-correlation function depends only on the number of subcarriers, number of users, and the variance of the channel impulse response and that the largest optimal block-size is given by

$$b^*_{\max} = \frac{N_p}{N_u L}, \tag{21}$$

which is a function of system's parameters: number of subcarriers, number of users, and cyclic prefix length.

In a more general scenario in which the channel length is different for each user, i.e., $L_h^{(u)} \neq L_h$, the optimal block-size maximizing the sum-correlation function is a difficult problem and an open issue. However, in a realistic scenario in which these parameters $L_h^{(u)}$ are not known, the system planner would assume the worse case scenario and approximate them with the length of the cyclic prefix L. Since the length of the cyclic prefix L is bigger, the chosen block-length is suboptimal and given by (21).

3.3 Numerical results: diversity and sum-correlation

In this section, the aim is to highlight the close relationship between diversity gain, outage probability and the sum-correlation function. We define the outage probability of the system under consideration as the maximum of the outage probabilities of the users:

$$P_{\text{out}} = \max_{1 \leq u \leq N_u} P_{\text{out}}^{(u)} \tag{22}$$

where $P_{\text{out}}^{(u)} = Pr\left\{C_b^{(u)} < R\right\}$ with R is a fixed target transmission rate and $C_b^{(u)}$ is the instantaneous mutual information of the user u defined in the following. We consider the transmitted symbols in (2) distributed accordingly to the Gaussian distribution such that $\mathbb{E}\left[\mathbf{x}^{(u)}\mathbf{x}^{(u)H}\right] = \mathbf{I}$. The user instantaneous achievable spectral efficiency assuming single-user decoding at the BS [15] in the case without CFO is as follows:

$$\begin{aligned} C_b^{(u)} &= \frac{N_u}{B} \log_2 \det \left[\mathbf{I} + \mathbf{H}^{(u)} \mathbf{\Pi}_b^{(u)} \mathbf{H}^{(u)\dagger} \right. \\ &\quad \left. \times \left(\mathbf{I}\sigma_n^2 + \sum_{\substack{v=1 \\ v \neq u}}^{N_u} \mathbf{H}^{(v)} \mathbf{\Pi}_b^{(v)} \mathbf{H}^{(v)\dagger} \right)^{-1} \right] \\ &= \frac{N_u}{B} \sum_{m \in \mathcal{C}_u} \log_2 \left(1 + \frac{1}{\sigma_n^2} |H_m^{(u)}|^2 \right). \end{aligned} \tag{23}$$

An explicit analytical relation between the sum-correlation function and the outage probability is still an open problem. The major issue is the fact that the distribution of the mutual information is very complex and closed-form expressions for the outage probability are not available in general. For example, Emre Telatar's conjecture on the optimal covariance matrix minimizing the outage probability in the single-user MIMO channels [13] is yet to be proven. We propose a new metric, the sum-correlation function, and show by simulations that there is an underlying relation between the sum-correlation function and the outage probability. Indeed, it is intuitive that, in SC-FDMA systems, correlation among the subcarriers decreases the diversity gain and, thus, the transmission reliability decreases [16,17]. This explains that the outage probability increases when the correlation among subcarriers is increasing. This connection has been validated via extensive numerical simulations. The interest behind this connection is that the sum-correlation function has a closed-form expression allowing us to perform a rigorous analysis and to find the blockwise subcarrier allocation minimizing the sum-correlation which is consistent with the optimal blockwise subcarrier allocation minimizing the outage probability. The following results illustrate numerically this connection.

3.3.1 *Uncorrelated subcarriers*

We consider the case of a SC-FDMA system with independent subcarriers. Since the subcarriers are independent the correlation between them is zero, which means that the sum-correlation $\Gamma(b)$ is equal to zero for any block-size b. In the next simulation, we observe that we obtain the same performance in terms of the outage probability regardless of the particular allocation scheme and the block-size, see Figure 6. Although this scenario is unrealistic from a practical standpoint, it is important to notice that, in this case, there are no privileged block-sizes to achieve better diversity gain.

In Figure 6, we plot the outage probability in the SC-FDMA system with independent subcarriers (subchannels) generated by complex Gaussian distribution with respect to SNR for the scenario $N_p = 64, N_u = 2$, and fixed rate $R = 1$ bits/s/Hz. In particular, in this case with independent subcarriers, we consider the matrix $\mathbf{H}^{(u)}$ in (4) to be diagonal with entries $H_k^{(u)}$ i.i.d $\sim \mathcal{CN}(0, \sigma^2)$. It is clear that for any block-size (hence, for any subcarrier allocation scheme) we obtain the same performance in terms of outage probability. Therefore, there are not any privileged block-size allocations to achieve better diversity gain. This motivates and strengthens our observation that the subcarrier correlation has a direct impact on the outage probability.

3.3.2 *Correlated subcarriers*

In this section, we consider a more interesting and realistic SC-FDMA system given in (4). For simplicity and

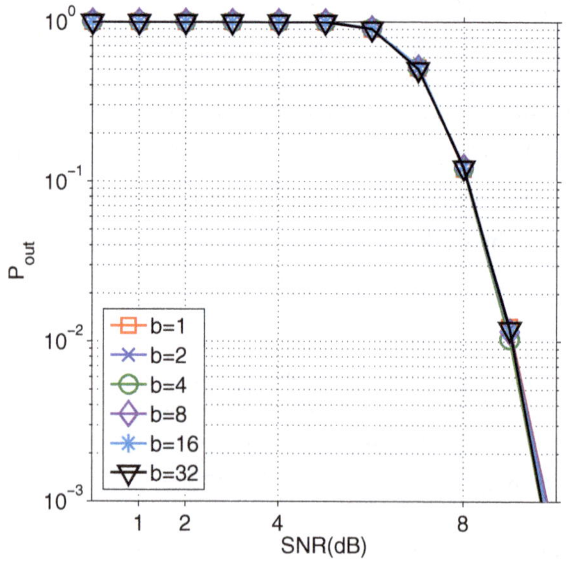

Figure 6 Outage probability for a SC-FDMA with independent channels with $N_p = 64$ and $N_u = 2$.

lack of space-related reasons, the simulations presented here have been done for the particular case $L = L_h$. Numerous other simulations were performed in the general case $L_h \leq L$, which confirm the theoretical result of Theorem 1.

In Figure 7, we have plotted with respect to the block-size b, the sum-correlation function $\Gamma(b)$ in the SC-FDMA system without CFO for the scenario $N_u = 4$, $L = 4$ and $\sigma^2(1) = 0.25, \sigma^2(2) = 0.5, \sigma^2(3) = 0.125, \sigma^2(4) = 0.3$. The illustrated markers represent the values of the function $\Gamma(b)$ for the given choice of the parameters of the system. We observe that the minimal values of $\Gamma(b)$ are obtained for the block-sizes $b^* \in \beta^* = \{1, 2, 4\}$. In particular, $\forall\, b^* \in \beta^* = \{1, 2, 4\}$ we have $\Gamma(b^*) = \sum_{u=1}^{N_u} \frac{\sigma_h^{(u)2} N_p}{N_u^2} = 4.7$.

In Figure 8, we use Binary Phase Shift Keying (BPSK) modulation in the following scenario: $N_p = 64$, $N_u = 2$, $L = 8$, and $\sigma_h^{(1)2} = \sigma_h^{(2)2}$. We observe that the optimal block-sizes are in $\beta^* = \{1, 2, 4\}$ (but here, we just plot the smallest and the biggest values) for the BER which confirms that these block-sizes optimize also the sum-correlation function we have proposed.

In Figure 9, we use BPSK modulation, $N_p = 64$, $N_u = 2$, $L = 4$, and $\sigma_h^{(2)2} = 2\sigma_h^{(1)2}$. The optimal block-sizes are given in the set $\beta^* = \{1, 2, 4, 8\}$.

In Figure 10, we use BPSK modulation in the following scenario: $N_p = 128$, $N_u = 2$, and $L = 8$. In this case, we consider an exponential power delay profile which means that $\sigma_h^{(u)2} = \frac{e^{-\tau/L}}{\sum_{\tau=0}^{L-1} e^{-\tau/L}}$ with $\tau \in \{0, 1, \ldots, L-1\}$.

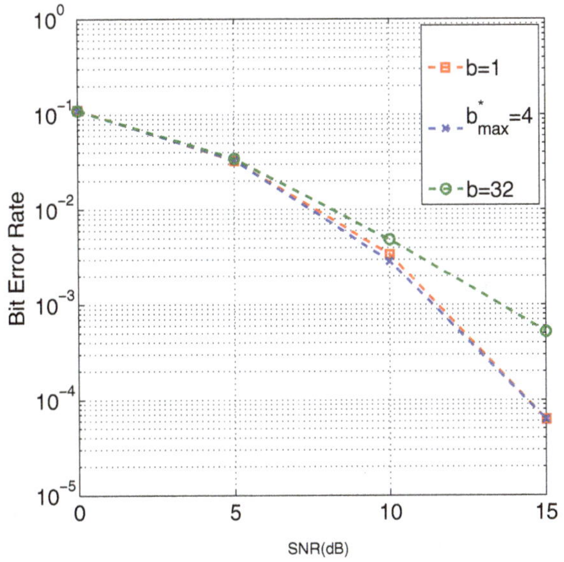

Figure 8 **Bit error rate for a SC-FDMA system without CFO for the scenario $N_p = 64$, $N_u = 2$, and $L = 8$, and $\sigma_h^{(1)2} = \sigma_h^{(2)2}$.** The optimal block-sizes are $b^* \in \beta^* = \{1, 2, 4\}$.

The theoretical results are confirmed since the optimal block-sizes are in $\beta^* = \{1, 2, 4, 8\}$.

In Figure 11, we evaluate the outage probability P_{out} in the SC-FDMA system without CFO for the scenario $N_u = 4$, $L = 4$, and $R = 1$ bits/s/Hz. We observe that the optimal block-sizes are the ones that correspond to the outage probabilities which have a higher

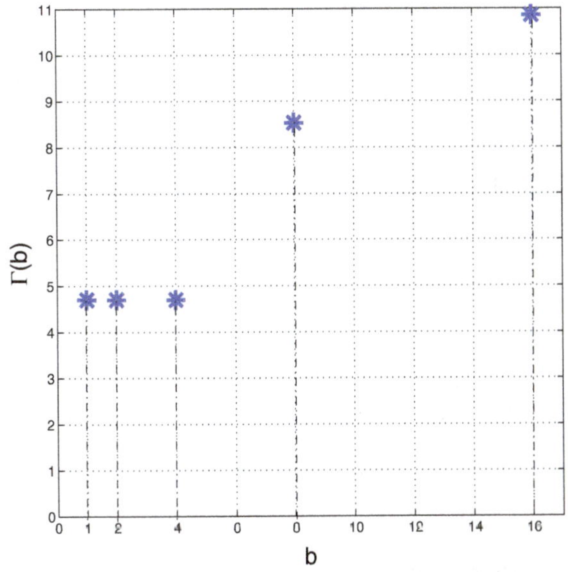

Figure 7 **Function $\Gamma(b)$ with $N_p = 64$, $N_u = 4$, $L = 4$, and $\sigma^2(1) = 0.25, \sigma^2(2) = 0.5, \sigma^2(3) = 0.125, \sigma^2(4) = 0.3$.** For any $b^* \in \beta^* = \{1, 2, 4\}$, we have $\Gamma(b^*) = (0.25 + 0.5 + 0.125 + 0.3)\frac{64}{4^2} = 4.7$.

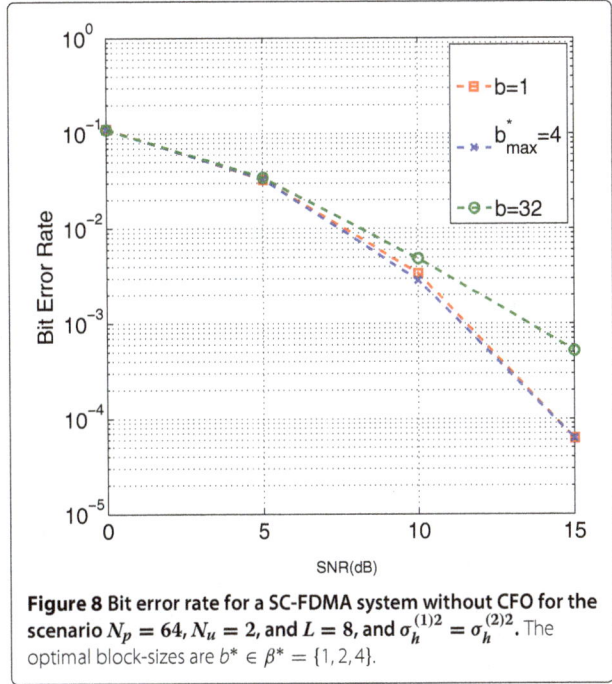

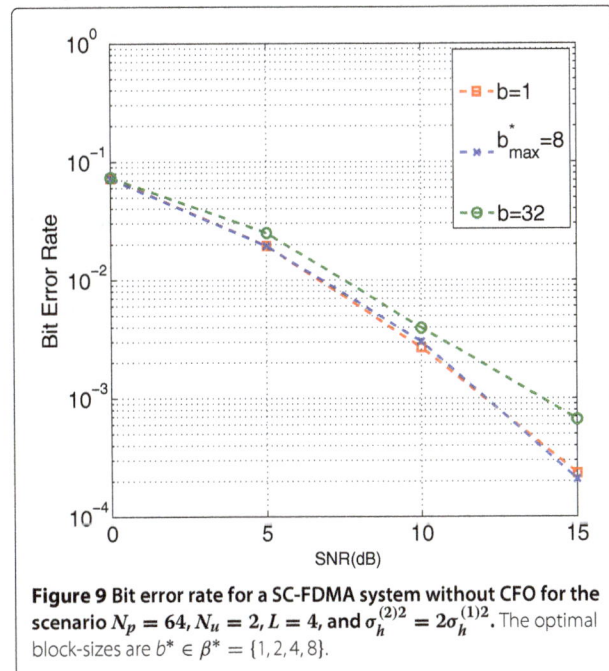

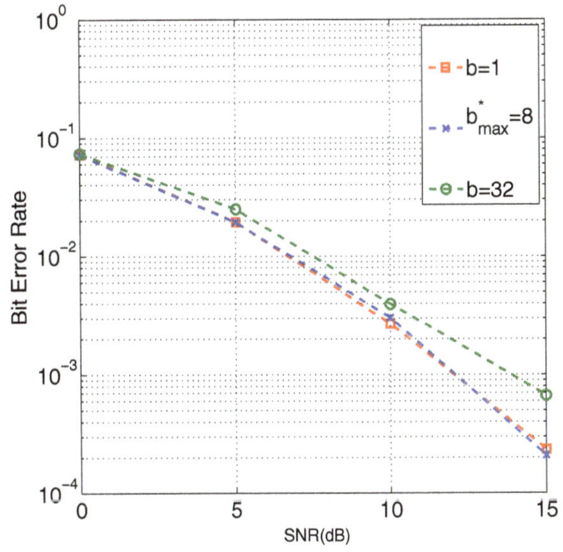

Figure 9 **Bit error rate for a SC-FDMA system without CFO for the scenario $N_p = 64$, $N_u = 2$, $L = 4$, and $\sigma_h^{(2)2} = 2\sigma_h^{(1)2}$.** The optimal block-sizes are $b^* \in \beta^* = \{1, 2, 4, 8\}$.

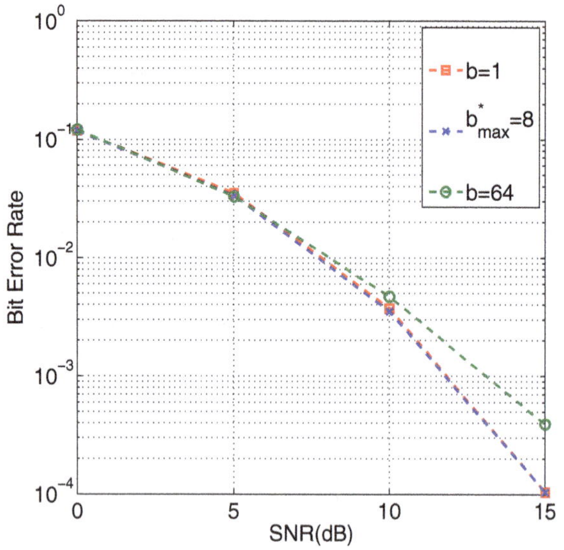

Figure 10 Bit error rate for a SC-FDMA system without CFO for the scenario $N_p = 128, N_u = 2, L = 8$, and exponential power delay profile. The optimal block-sizes are $b^* \in \beta^* = \{1, 2, 4, 8\}$.

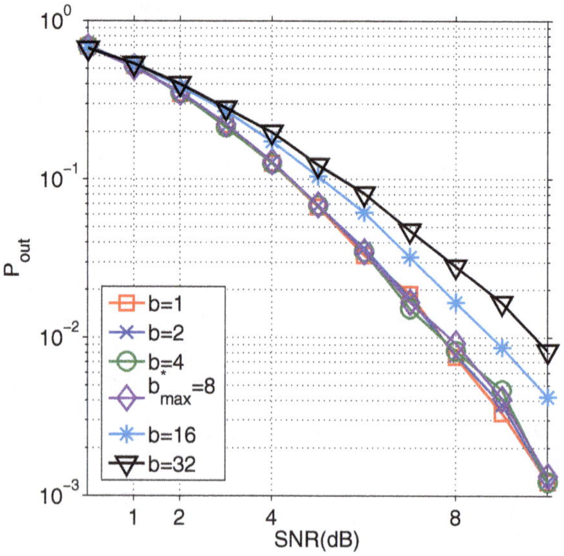

Figure 12 Outage probability for a SC-FDMA system without CFO with $N_p = 64, N_u = 2$, and $L = 4$. The optimal block-sizes are $b^* \in \beta^* = \{1, 2, 4, 8\}$.

decreasing rate as a function of the SNR. We see that the curves with $b^* \in \beta^* = \{1, 2, 4\}$ (in this case $b^*_{\text{max}} = \frac{64}{4\times4} = 4$) are overlapped and they represent the lower outage probability. These block-sizes are the same that minimize the sum-correlation function (see Figure 7).

In Figure 12, we plot the outage probability for the SC-FDMA system without CFO for the scenario $N_p = 64$, $N_u = 2, L = 4$, and $R = 1$ bits/s/Hz so that $b^*_{\text{max}} = \frac{64}{2\times4} = 8$. In this case in which the subcarriers are correlated, we observe that the curves with $b^* \in \beta^* = \{1, 2, 4, 8\}$ have a higher diversity.

Many others simulations, changing the values of the parameters (in particular, N_p and L), have been performed, and similar observations were made. Moreover, we have done simulations choosing the following as a performance metric:

$$\tilde{P}_{\text{out},b} = 1 - \prod_{u=1}^{N_u} \left(1 - P^{(u)}_{\text{out},b}\right). \tag{24}$$

The same observation can be made with this outage metric.

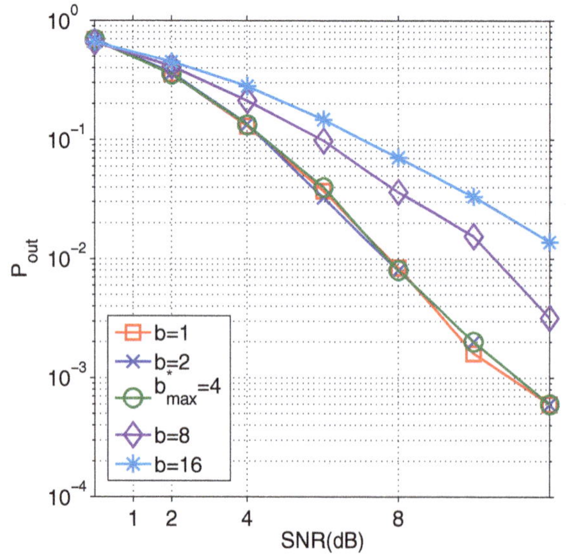

Figure 11 Outage probability for a SC-FDMA system without CFO with $N_p = 64, N_u = 4$, and $L = 4$. The optimal block-sizes are $b^* \in \beta^* = \{1, 2, 4\}$.

4 Robustness to CFO

In this section, we analyze the case of SC-FDMA systems with CFO and the effect of CFO on the optimal block-size. We define the sum-correlation function and we show its robustness to CFO.

We start by describing in details the system model.

4.1 System model

If the system undergoes CFOs, the signal at the input of the receiver DFT is given in (25), and it was introduced in [10],

$$
\begin{pmatrix} y_{N_p-1} \\ \vdots \\ y_0 \\ \vdots \\ y_{-L} \end{pmatrix} = \sum_{u=1}^{N_u} \begin{pmatrix} h_0^{(u)} & \cdots & h_{L-1}^{(u)} & & 0 \\ & \ddots & & \ddots & \\ 0 & & h_0^{(u)} & \cdots & h_{L-1}^{(u)} \end{pmatrix}
\times \begin{pmatrix} \delta_{(N_p+L-1)}^{(u)} & & 0 \\ & \ddots & \\ 0 & & \delta_0^{(u)} \end{pmatrix} \begin{pmatrix} a_{N_p-1}^{(u)} \\ \vdots \\ a_0^{(u)} \\ \vdots \\ a_{-L}^{(u)} \end{pmatrix} + \begin{pmatrix} n_{N_p-1} \\ \vdots \\ n_0 \\ \vdots \\ n_{-L} \end{pmatrix} \quad (25)
$$

The diagonal elements $\delta_k^{(u)}$ are the frequency shift coefficients given by $\delta_k^{(u)} = e^{\frac{j2\pi k \delta f_c^{(u)} T}{N_p}}$, $k \in \{0, \ldots, N_p + L - 1\}$, where $\frac{\delta f_c^{(u)}}{N_p}$ is the normalized CFO of user u.

By discarding the cyclic prefix symbols and rearranging the terms in (25), we have

$$
\begin{pmatrix} y_{N_p-1} \\ \vdots \\ y_0 \end{pmatrix} = \sum_{u=1}^{N_u} \mathbf{h}_{\text{circ}}^{(u)} \underbrace{\begin{pmatrix} \delta_{(N_p+L-1)}^{(u)} & & 0 \\ & \ddots & \\ 0 & & \delta_L^{(u)} \end{pmatrix}}_{\delta^{(u)}} \quad (26)
$$
$$
\times \begin{pmatrix} a_{N_p-1}^{(u)} \\ \vdots \\ a_0^{(u)} \end{pmatrix} + \begin{pmatrix} n_{N_p-1} \\ \vdots \\ n_0 \end{pmatrix}.
$$

The received signal at the BS after the DFT is

$$
\mathbf{r}_{\text{CFO}} = \sum_{u=1}^{N_u} \mathbf{H}^{(u)} \mathbf{\Delta}^{(u)} \mathbf{\Pi}_b^{(u)} \mathbf{x}^{(u)} + \hat{\mathbf{n}}, \quad (27)
$$

where the $N_p \times N_p$ matrix $\mathbf{\Delta}^{(u)} = \mathbf{F}\delta^{(u)}\mathbf{F}^{-1}$ represents the effect of CFO on the interference among subcarriers. In particular, we have the (l, k) element of $\mathbf{\Delta}^{(u)}$:

$$
\begin{aligned}
\Delta_{\ell,k}^{(u)} &= \frac{1}{N_p} \sum_{i=0}^{N_p-1} e^{j2\pi i \delta f / N_p} e^{-j2\pi i(\ell-k)/N_p} \\
&= \frac{1}{N_p} \frac{\sin\left(\pi(\delta f + k - \ell)\right)}{\sin\left(\frac{\pi}{N_p}(\delta f + k - \ell)\right)} e^{\pi j \left(1-\frac{1}{N_p}\right)(\delta f + k - \ell)}. \quad (28)
\end{aligned}
$$

In the sequel, we denote $\tilde{\mathbf{H}}^{(u)} \triangleq \mathbf{H}^{(u)}\mathbf{\Delta}^{(u)}$, which is no longer a diagonal matrix.

4.2 Diversity versus CFO in subcarrier allocation

We consider the following inter-carrier correlation function:

$$
\begin{aligned}
\Gamma_{u,m}^{\text{CFO}}(b, \delta f) &\triangleq \sum_{c_u \in \mathcal{C}_u} \mathbb{E}\left[H_m^{(u)} \Delta_{m,m} [1, \ldots, 1] \tilde{\mathbf{H}}_{c_u}^{(u)\dagger}\right] \\
&= \sum_{c_u \in \mathcal{C}_u} \sum_{k \in \mathcal{C}_u} \mathbb{E}\left[H_m^{(u)} H_{c_u}^{(u)*}\right] \Delta_{m,m} \Delta_{c_u,k}^* \\
&= \sum_{k \in \mathcal{C}_u} \mathbb{E}\left[\left|H_m^{(u)}\right|^2\right] \Delta_{m,m} \Delta_{m,k}^* \\
&\quad + \sum_{\substack{c_u \in \mathcal{C}_u \\ c_u \neq m}} \sum_{k \in \mathcal{C}_u} \mathbb{E}\left[H_m^{(u)} H_{c_u}^{(u)*}\right] \Delta_{m,m} \Delta_{c_u,k}^* \\
&= \mathbb{E}\left[\left|H_m^{(u)}\right|^2\right] \left|\Delta_{m,m}\right|^2 + \sum_{\substack{k \in \mathcal{C}_u \\ k \neq m}} \mathbb{E}\left[\left|H_m^{(u)}\right|^2\right] \Delta_{m,m} \Delta_{m,k}^* + \\
&\quad + \sum_{\substack{c_u \in \mathcal{C}_u \\ c_u \neq m}} \mathbb{E}\left[H_m^{(u)} H_{c_u}^{(u)*}\right] \Delta_{m,m} \Delta_{c_u,c_u}^* \\
&\quad + \sum_{\substack{c_u \in \mathcal{C}_u \\ c_u \neq m}} \sum_{\substack{k \in \mathcal{C}_u \\ k \neq c_u}} \mathbb{E}\left[H_m^{(u)} H_{c_u}^{(u)*}\right] \Delta_{m,m} \Delta_{c_u,k}^* \\
&= \sigma_h^{(u)2} \frac{L_h}{N_p} \left[\frac{1}{N_p^2} \frac{\sin^2(\pi \delta f)}{\sin^2(\frac{\pi}{N_p}\delta f)}\right. \\
&\quad + \sum_{\substack{k \in \mathcal{C}_u \\ k \neq m}} \frac{1}{N_p^2} e^{-\pi j \left(1-\frac{1}{N_p}\right)(k-m)} \frac{\sin(\pi \delta f)}{\sin(\frac{\pi}{N_p}\delta f)} \\
&\quad \left. \times \frac{\sin(\pi(\delta f + k - m))}{\sin(\frac{\pi}{N_p}(\delta f + k - m))}\right] + \\
&\quad + \frac{\sigma_h^{(u)2}}{N_p} \sum_{\substack{c_u \in \mathcal{C}_u \\ c_u \neq m}} e^{-\pi j \frac{(L_h-1)}{N_p}(m-c_u)} \frac{\sin\left[\pi \frac{L_h}{N_p}(m - c_u)\right]}{\sin\left[\frac{\pi}{N_p}(m - c_u)\right]} \times \\
&\quad \times \left[\frac{1}{N_p^2} \frac{\sin^2(\pi \delta f)}{\sin^2(\frac{\pi}{N_p}\delta f)} + \sum_{\substack{k \in \mathcal{C}_u \\ k \neq c_u}} \frac{1}{N_p^2} e^{-\pi j \left(1-\frac{1}{N_p}\right)(k-c_u)}\right. \\
&\quad \left. \times \frac{\sin(\pi \delta f)}{\sin(\frac{\pi}{N_p}\delta f)} \frac{\sin(\pi(\delta f + k - c_u))}{\sin(\frac{\pi}{N_p}(\delta f + k - c_u))}\right]
\end{aligned} \quad (29)
$$

where $m \in \mathcal{C}_u$ is the reference subcarrier, δf represents the CFO of user u, and $\tilde{\mathbf{H}}_{c_u}^{(u)} = \left(H_{c_u}^{(u)} \Delta_{c_u,1}, \ldots, H_{c_u}^{(u)} \Delta_{c_u,N_p}\right)$ represents the c_u-th row of the matrix $\tilde{\mathbf{H}}^{(u)}$.

We define the sum-correlation metric as follows:

$$
\Gamma^{CFO}(b, \delta f) = \sum_{u=1}^{N_u} \sum_{m \in \mathcal{C}_u} \Gamma_{u,m}^{\text{CFO}}(b, \delta f). \quad (30)
$$

In the following, we provide an approximation of the correlation function $\Gamma_{m,u}^{\text{CFO}}(b, \delta f)$ in which the dependance on the CFO values δf is taken into account. In particular, we consider the second order Taylor approximation of $\Gamma^{\text{CFO}}_{m,u}(b,\delta f)$ when $\delta f \to 0$

$$\Gamma^{\text{CFO}}_{m,u}(b,\delta f) \approx \Gamma^{\text{CFO}}_{m,u}(b,0) + \frac{d\Gamma^{\text{CFO}}_{m,u}}{d\delta f}(b,0)\delta f + \frac{1}{2}\frac{d^2\Gamma^{\text{CFO}}_{m,u}}{d(\delta f)^2}(b,0)(\delta f)^2. \tag{31}$$

This first term $\Gamma^{\text{CFO}}_{m,u}(b,0)$ is given by

$$\begin{aligned}\Gamma^{CFO}_{u,m}(b,0) &= \sigma_h^{(u)2}\frac{L_h}{N_p}\left[\frac{1}{N_p^2}N_p^2 + \frac{1}{N_p^2}\sum_{\substack{k\in C_u\\k\neq m}} e^{-\pi j\left(1-\frac{1}{N_p}\right)(k-m)}N_p \times 0\right] + \\ &+ \frac{\sigma_h^{(u)2}}{N_p}\sum_{\substack{c_u\in C_u\\c_u\neq m}} e^{-\pi j\frac{(L_h-1)}{N_p}(m-c_u)}\frac{\sin\left[\pi\frac{L_h}{N_p}(m-c_u)\right]}{\sin\left[\frac{\pi}{N_p}(m-c_u)\right]}\times \\ &\times\left[\frac{1}{N_p^2}N_p^2 + \frac{1}{N_p^2}\sum_{\substack{k\in C_u\\k\neq c_u}} e^{-\pi j\left(1-\frac{1}{N_p}\right)(k-c_u)}N_p \times 0\right] \\ &= \sigma_h^{(u)2}\frac{L_h}{N_p} + \frac{\sigma_h^{(u)2}}{N_p}\sum_{\substack{c_u\in C_u\\c_u\neq m}} e^{-\pi j\frac{(L_h-1)}{N_p}(m-c_u)} \\ &\quad \frac{\sin\left[\pi\frac{L_h}{N_p}(m-c_u)\right]}{\sin\left[\frac{\pi}{N_p}(m-c_u)\right]}.\end{aligned} \tag{32}$$

Indeed, this expression corresponds exactly to $\Gamma_{m,u}(b)$, i.e., the sum-correlation function in the case without CFO in (6).

The first derivative of $\Gamma^{\text{CFO}}_{m,u}(b,\delta f)$ with respect to δf computed in $(b,0)$ is

$$\begin{aligned}\frac{d\Gamma^{\text{CFO}}_{m,u}}{d\delta f}(b,0) &= \sigma_h^{(u)2}\frac{L_h}{N_p}\left[\sum_{\substack{k\in C_u\\k\neq m}}\frac{1}{N_p^2}e^{-\pi j\left(1-\frac{1}{N_p}\right)(k-m)}\right. \\ &\left.\times\frac{\pi N_p\cos(\pi(k-m))}{\sin(\frac{\pi}{N_p}(k-m))}\right] + \\ &+ \frac{\sigma_h^{(u)2}}{N_p}\sum_{\substack{c_u\in C_u\\c_u\neq m}} e^{-\pi j\frac{(L_h-1)}{N_p}(m-c_u)}\frac{\sin\left[\pi\frac{L_h}{N_p}(m-c_u)\right]}{\sin\left[\frac{\pi}{N_p}(m-c_u)\right]}\times \\ &\times\left[\sum_{\substack{k\in C_u\\k\neq c_u}}\frac{1}{N_p^2}e^{-\pi j\left(1-\frac{1}{N_p}\right)(k-c_u)}\frac{\pi N_p\cos(\pi(k-c_u))}{\sin(\frac{\pi}{N_p}(k-c_u))}\right]\end{aligned} \tag{33}$$

The second derivative of $\Gamma^{\text{CFO}}_{m,u}(b,\delta f)$ computed in $(b,0)$ is

$$\begin{aligned}\frac{1}{2}\frac{d^2\Gamma^{\text{CFO}}_{m,u}}{d(\delta f)^2}(b,0) &= \frac{1}{2}\sigma_h^{(u)2}\frac{L_h}{N_p}\left[\frac{2\pi^2(1-N_p^2)}{N_p^2} - \frac{1}{N_p^2}\sum_{\substack{k\in C_u\\k\neq m}} e^{-\pi j\left(1-\frac{1}{N_p}\right)(k-m)}\right. \\ &\left.\times\frac{2\pi^2\cos\left(\frac{\pi}{N_p}(k-m)\right)}{\sin^2\left(\frac{\pi}{N_p}(k-m)\right)}\right] + \\ &+ \frac{1}{2}\frac{\sigma_h^{(u)2}}{N_p}\sum_{\substack{c_u\in C_u\\c_u\neq m}} e^{-\pi j\frac{(L_h-1)}{N_p}(m-c_u)} \\ &\times\frac{\sin\left[\pi\frac{L_h}{N_p}(m-c_u)\right]}{\sin\left[\frac{\pi}{N_p}(m-c_u)\right]}\times \\ &\times\left[\frac{2\pi^2(1-N_p^2)}{N_p^2} - \frac{1}{N_p^2}\sum_{\substack{k\in C_u\\k\neq c_u}} e^{-\pi j\left(1-\frac{1}{N_p}\right)(k-c_u)}\right. \\ &\left.\times\frac{2\pi^2\cos\left(\frac{\pi}{N_p}(k-c_u)\right)}{\sin^2\left(\frac{\pi}{N_p}(k-c_u)\right)}\right]\end{aligned} \tag{34}$$

Therefore, we have

$$\begin{aligned}\Gamma^{\text{CFO}}_{m,u}(b,\delta f) &\approx \frac{\sigma_h^{(u)2}}{N_p}L_h + \frac{\sigma_h^{(u)2}}{N_p}\sum_{c_u\in C_u}\frac{\sin\left[\frac{\pi L_h}{N_p}(m-c_u)\right]}{\sin\left[\frac{\pi}{N_p}(m-c_u)\right]}\times e^{-\pi j\frac{(L_h-1)}{N_p}(m-c_u)} + \\ &+\left\{\sigma_h^{(u)2}\frac{L_h}{N_p}\left[\sum_{\substack{k\in C_u\\k\neq m}}\frac{1}{N_p}e^{-\pi j\left(1-\frac{1}{N_p}\right)(k-m)}\frac{\cos(\pi(k-m))}{\sin(\frac{\pi}{N_p}(k-m))}\right]+\right. \\ &+\frac{\sigma_h^{(u)2}}{N_p}\sum_{\substack{c_u\in C_u\\c_u\neq m}} e^{-\pi j\frac{(L_h-1)}{N_p}(m-c_u)}\frac{\sin\left[\pi\frac{L_h}{N_p}(m-c_u)\right]}{\sin\left[\frac{\pi}{N_p}(m-c_u)\right]}\times \\ &\left.\times\left[\sum_{\substack{k\in C_u\\k\neq c_u}}\frac{1}{N_p}e^{-\pi j\left(1-\frac{1}{N_p}\right)(k-c_u)}\frac{\cos(\pi(k-c_u))}{\sin(\frac{\pi}{N_p}(k-c_u))}\right]\right\}\pi\delta f \\ &+\left\{\sigma_h^{(u)2}\frac{L_h}{N_p}\left[\frac{(1-N_p^2)}{N_p^2} - \frac{1}{N_p^2}\sum_{\substack{k\in C_u\\k\neq m}} e^{-\pi j\left(1-\frac{1}{N_p}\right)(k-m)}\right.\right. \\ &\left.\times\frac{\cos\left(\frac{\pi}{N_p}(k-m)\right)}{\sin^2\left(\frac{\pi}{N_p}(k-m)\right)}\right] + \\ &+\frac{\sigma_h^{(u)2}}{N_p}\sum_{\substack{c_u\in C_u\\c_u\neq m}} e^{-\pi j\frac{(L_h-1)}{N_p}(m-c_u)}\frac{\sin\left[\pi\frac{L_h}{N_p}(m-c_u)\right]}{\sin\left[\frac{\pi}{N_p}(m-c_u)\right]}\times \\ &\times\left[\frac{(1-N_p^2)}{N_p^2} - \frac{1}{N_p^2}\sum_{\substack{k\in C_u\\k\neq c_u}} e^{-\pi j\left(1-\frac{1}{N_p}\right)(k-c_u)}\right. \\ &\left.\left.\times\frac{\cos\left(\frac{\pi}{N_p}(k-c_u)\right)}{\sin^2\left(\frac{\pi}{N_p}(k-c_u)\right)}\right]\right\}\pi^2(\delta f)^2.\end{aligned} \tag{35}$$

We observe in the above approximation the presence of the predominant term represented by the sum-correlation $\Gamma_{m,u}(b)$ without CFO and the first and the second derivatives of $\Gamma^{\mathrm{CFO}}_{m,u}(b,\delta f)$ which are multiplied by the CFO value δf and δf^2, respectively. This last terms may result in a different solution for the optimal block-size or block-sizes that optimize the sum-correlation function than the solution for the case with no CFO. We observe that the first and the second derivatives represent a complex function that implicitly depends on b. Finding the optimal block-size or block-sizes in an analytical manner, as done in the case with no CFO, seems very difficult if at all possible and is left for future investigation.

When the system undergoes CFO, the carrier correlation and CFO affect the system performance simultaneously. We have proposed in [12] the largest optimal block-size $b^*_{\max}$ as the unique optimal block-size: *Since CFO yields a diversity loss, in the presence of moderate values of CFO, the optimal block-size allocation is* b^*_{max} . We have found that the optimal block-sizes that achieve maximum diversity are the ones that minimize the correlation between subcarriers. Moreover, larger block-sizes are preferable to combat the effect of ICI. Also, since $b^*_{\max}$ is the largest block-size between the ones minimizing the correlation, it is also the one that minimizes the negative effects caused by the presence of CFO. Therefore, $b^*_{\max}$ represents a good tradeoff between diversity and CFO. Moreover, the observation is validated also by numerical simulations illustrated in the next subsection.

4.3 Numerical results: CFO impact

In Figure 13, we plot the correlation $\Gamma^{\mathrm{CFO}}(b,\delta f)$ for the scenario: $N_p = 64$, $N_u = 2$, $L_h = 8$, and $\sigma_h^{(1)2} = 0.25$, $\sigma_h^{(2)2} = 0.5$. The considered CFO values are $\delta f \in \{0.1, 0.2, 0.3, 0.4\}$. We observe that $b^*_{\max} = 4$ is the block-size that achieves the minimum value of the correlation function $\Gamma^{\mathrm{CFO}}(b,\delta f)$, validating our conjectured optimal block-size.

In the next two simulations, we use BPSK modulation and $Np = 128$, $Nu = 2$, and $L = 8$. Figure 14 illustrates the bit error rate (BER) curves for a SC-FDMA system with CFO independently and uniformly generated for each user in $[0, 0.03]$. We observe that for these low CFOs we have the optimal block-sizes given by $\beta^* = \{1,2,4,8\}$. Figure 15 illustrates the BER curves for a SC-FDMA system with CFO independently and uniformly generated for each user in $[0, 0.1]$. We observe that, in this case, we have a unique optimal block-size given by $\beta^*_{\max} = 8$. This validates our observations, i.e., when the CFO's values are increasing, the best tradeoff between diversity and CFO is represented by the largest block-size of our proposed set $b^*_{\max}$.

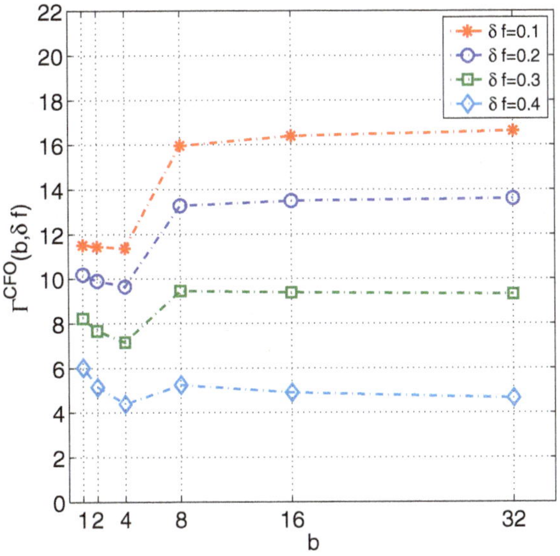

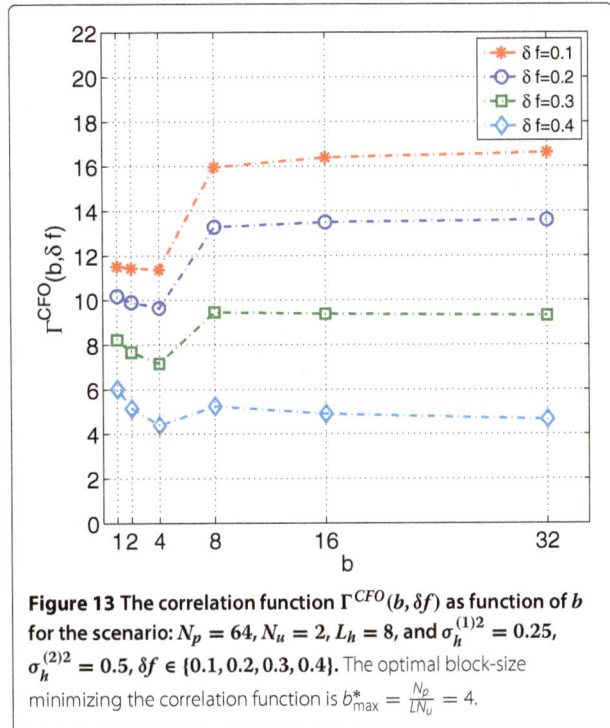

Figure 13 The correlation function $\Gamma^{CFO}(b,\delta f)$ as function of b for the scenario: $N_p = 64, N_u = 2, L_h = 8$, and $\sigma_h^{(1)2} = 0.25$, $\sigma_h^{(2)2} = 0.5$, $\delta f \in \{0.1, 0.2, 0.3, 0.4\}$. The optimal block-size minimizing the correlation function is $b^*_{\max} = \frac{N_p}{LN_u} = 4$.

In Figure 16, we use BPSK modulation in the following scenario: $N_p = 64$, $N_u = 2$, and $L = 4$. We consider the same model proposed in [11] where one symbol is spread over all subcarriers. The CFO is independently and uniformly generated for each user in $[0, 0.1]$. We observe that

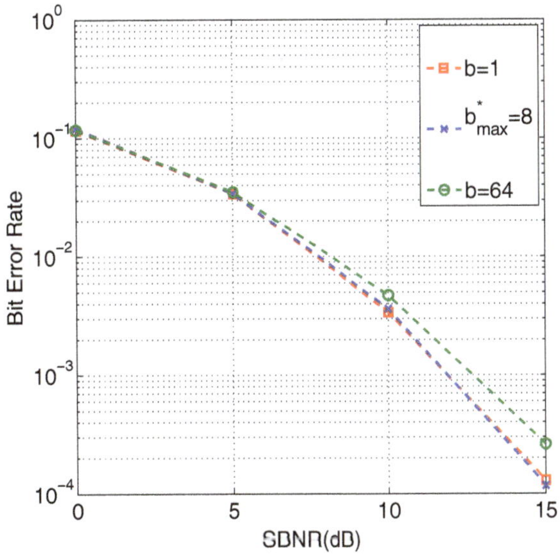

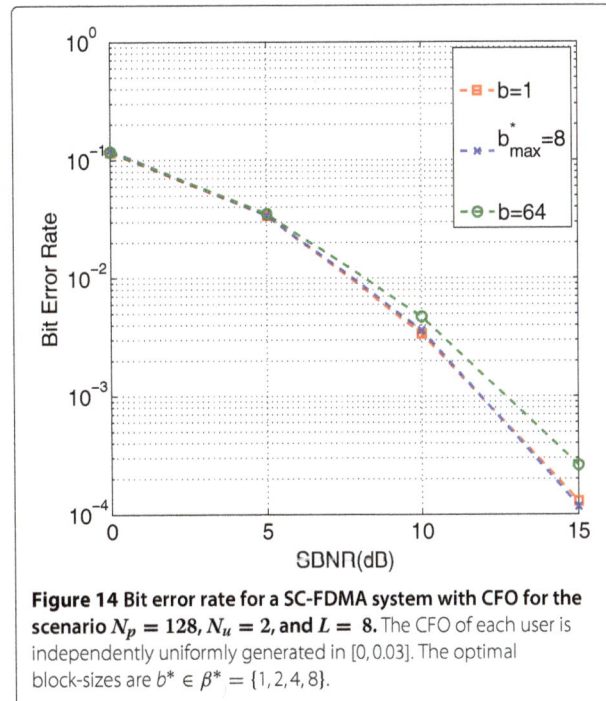

Figure 14 Bit error rate for a SC-FDMA system with CFO for the scenario $N_p = 128, N_u = 2$, and $L = 8$. The CFO of each user is independently uniformly generated in $[0, 0.03]$. The optimal block-sizes are $b^* \in \beta^* = \{1, 2, 4, 8\}$.

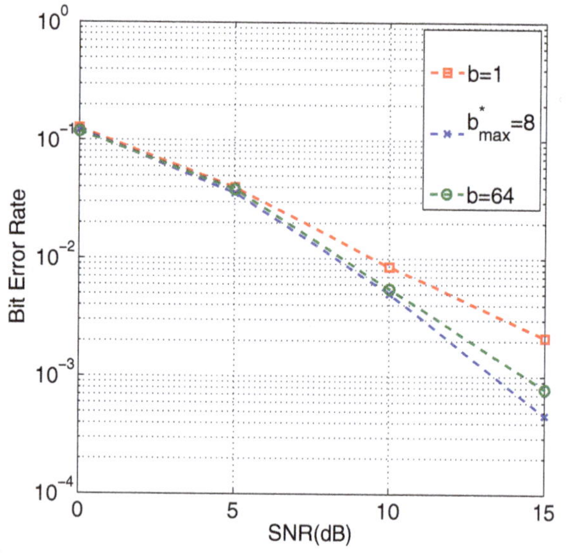

Figure 15 Bit error rate for a SC-FDMA system with CFO for the scenario $N_p = 128, N_u = 2,$ and $L = 8$. The CFO of each user is independently uniformly generated in $[0, 0.1]$. The optimal block-sizes is $b^*_{max} = 8$.

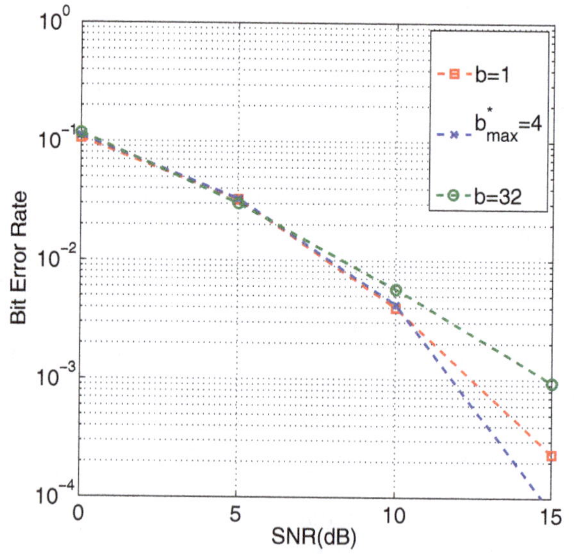

Figure 17 Bit error rate for a SC-FDMA system with CFO for the scenario $N_p = 64, N_u = 2,$ and $L = 8$. The CFO is independently and uniformly generated for each user in $[0, 0.05]$. The optimal block-sizes are $b^* \in \beta^* = \{1, 2, 4\}$.

the optimal block-sizes are in $\beta^* = \{1, 2, 4, 8\}$ for the BER which confirms that our analysis is valid for the model proposed in [11].

In Figure 17, we use BPSK, $N_p = 64$, $N_u = 2$, and $L = 8$ and an exponential power delay profile. The CFO is independently and uniformly generated for each user

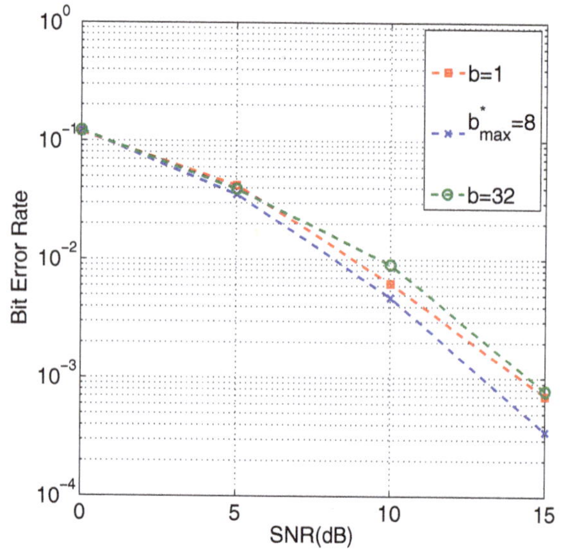

Figure 16 Bit error rate for a SC-FDMA system with CFO for the scenario $N_p = 64, N_u = 2,$ and $L = 4$. The CFO is independently and uniformly generated for each user in $[0, 0.1]$. The optimal block-sizes are $b^* \in \beta^* = \{1, 2, 4, 8\}$.

in $[0, 0.05]$. The set of optimal block-sizes given by $\beta^* = \{1, 2, 4\}$ as shown in the figure.

For different and larger CFO values, as considered in [18], we notice that an error floor is obtained due to the effect of CFO interference. Thus, in such cases, optimizing the block-size is not very relevant as all possibilities obtain such poor results in terms of BER.

In the next simulation, we consider the following scenario: $N_p = 64$, $N_u = 4$, and $L = 8$. In the Figure 18, we plot the outage probability of an SC-FDMA system with CFO (marker lines) against the outage probability of the SC-FDMA system without CFO (dashed lines). The CFO for each user is independently uniformly generated in $\delta f \in [0, 0.01]$, and the rate R is taken equal to 1bits/s/Hz. We can see that the curves in the CFO case fit very well the outage probability curves without CFO. In particular, they appear in a decreasing order of block-size. This validates our analytical analysis on the approximation of the CFO sum-correlation function to the case without CFO when the CFO goes to zero. Moreover, we observe that in the two cases we have the same optimal block-sizes set, given by $\beta^* = \{1, 2\}$.

5 Conclusions

In this work, we have provided the analytical expression of the set of optimal sizes of subcarrier blocks for SC-FDMA uplink systems without CFO and without channel state information. These optimal block-sizes allow us to minimize the sum-correlation between subcarriers and to achieve maximum diversity gain. We have also provided

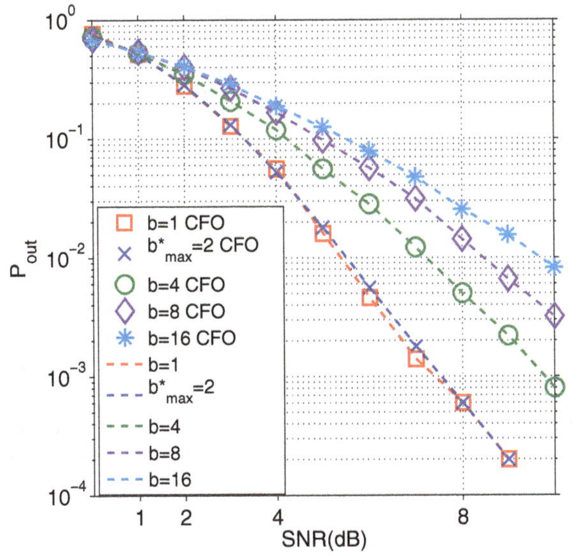

Figure 18 Outage probability for a SC-FDMA system with and without CFO. Outage probability for a SC-FDMA system without CFO (appearing in decreasing order of block-size from up to down) and with CFO (marker lines) for the scenario $\boldsymbol{N_p = 64, N_u = 4}$, and $\boldsymbol{L = 8}$. The CFO of each user is independently uniformly generated in $[0, 0.01]$. The optimal block-sizes are $b^* \in \beta^* = \{1, 2\}$.

the analytical expression of the sum-correlation between subcarriers induced by SC-FDMA/OFDMA. Moreover, we have found an explicit expression of the largest optimal block-size which minimizes the sum-correlation function depending on the system's parameters: number of subcarriers, number of users, and the cyclic prefix length. Interesting properties of this novel sum-correlation function are also presented.

It turns out that the minimal sum-correlation value depends only on the number of subcarriers, number of users, and the variance of the channel impulse response. We validate via numerical simulations that the set of optimal block-sizes achieving maximum diversity minimizes the outage probability in the case without CFO.

Also, in the case where the system undergoes CFO, we consider a sum-correlation function which is robust to CFO. Robustness is induced by the fact that when the CFO goes to zero, the CFO sum-correlation can be well approximated by the sum-correlation function defined in the case without CFO. Therefore, we propose $b^*_{\max} = \frac{N_p}{L_h N_u}$ a good tradeoff between diversity and CFO since it represents the unique optimal block-size that achieves maximum diversity. All these results and observations have been validated via extensive Monte Carlo simulations.

Endnotes

[a] If we do not take this assumption into account, we would have to use $\widetilde{N_p} = MN_u$ instead of N_p to denote the actual allocated number of carriers and $\lfloor \frac{N_p}{N_u} \rfloor$ instead of $\frac{N_p}{N_u}$ as the number of carriers per user.

[b] Here we use the word "distance" as synonym of difference and not for Euclidean distance.

Appendix

Proof of Theorem 1

Proof: From the definition of the set $\mathcal{D}_b$, we can write the function $\Gamma(b)$ as follows:

$$\begin{aligned}
\Gamma(b) &= \sum_{u=1}^{N_u} \frac{N_p}{N_u} \sigma_h^{(u)2} N_p \sum_{d \in \mathcal{D}_b} f(d) \\
&= \sum_{u=1}^{N_u} \frac{\sigma_h^{(u)2}}{N_u} \sum_{i=0}^{b-1} \sum_{k=0}^{\frac{N_p}{bN_u}-1} f(kbN_u + i) \\
&= \sum_{u=1}^{N_u} \frac{\sigma_h^{(u)2}}{N_u} \left(\sum_{k=0}^{\frac{N_p}{bN_u}-1} f(kbN_u) + \sum_{i=1}^{b-1} \sum_{k=0}^{\frac{N_p}{bN_u}-1} f(kbN_u + i) \right) \\
&= \sum_{u=1}^{N_u} \frac{\sigma_h^{(u)2}}{N_u} \left(L_h + \sum_{k=1}^{\frac{N_p}{bN_u}-1} f(kbN_u) + \sum_{i=1}^{b-1} \sum_{k=0}^{\frac{N_p}{bN_u}-1} f(kbN_u + i) \right).
\end{aligned} \tag{36}$$

First of all, we analyze the term: $\sum_{i=1}^{b-1} \sum_{k=0}^{\frac{N_p}{bN_u}-1} f(kbN_u+i)$ and, in particular, its ith term

$$\begin{aligned}
&\sum_{k=0}^{\frac{N_p}{bN_u}-1} f(kbN_u + i) = \\
&\qquad = \sum_{k=0}^{\frac{N_p}{bN_u}-1} \frac{\sin\left(\pi \frac{L_h}{N_p}(kbN_u + i)\right)}{\sin\left(\frac{\pi}{N_p}(kbN_u + i)\right)} e^{-j\pi \frac{(L_h-1)}{N_p}(kbN_u+i)} \\
&\qquad = \sum_{k=0}^{\frac{N_p}{bN_u}-1} \frac{1 - e^{-j2\pi \frac{L_h}{N_p} kbN_u} e^{-j2\pi \frac{L_h}{N_p} i}}{1 - e^{-j2\pi \frac{1}{N_p} kbN_u} e^{-j2\pi \frac{1}{N_p} i}} \\
&\qquad = \sum_{k=0}^{\frac{N_p}{bN_u}-1} \frac{1 - \left(\alpha_i z^k\right)^{L_h}}{1 - \alpha_i z^k}
\end{aligned} \tag{37}$$

with $\alpha_i := e^{-j2\pi \frac{1}{N_p} i}$ and $z := e^{-j2\pi \frac{1}{N_p} bN_u}$. Using the decomposition of a geometric series of radius $\alpha_i z^k$, we further obtain

$$\sum_{k=0}^{\frac{N_p}{bN_u}-1} f(kbN_u + i) = \sum_{k=0}^{\frac{N_p}{bN_u}-1} \left[\sum_{\ell=0}^{L_h-1} \left(\alpha_i z^k\right)^{\ell} \right]$$

and, inverting the two sums, we have

$$\sum_{k=0}^{\frac{N_p}{bN_u}-1} f(kbN_u+i) = \sum_{\ell=0}^{L_h-1} \alpha_i^\ell \sum_{k=0}^{\frac{N_p}{bN_u}-1} \left(z^\ell\right)^k = \frac{N_p}{bN_u} + \sum_{\ell=1}^{L_h-1} \sum_{k=0}^{\frac{N_p}{bN_u}-1} \alpha_i^\ell (z^\ell)^k. \tag{38}$$

Now, we look at the term $\sum_{k=1}^{\frac{N_p}{bN_u}-1} f(kbN_u)$ in Equation (36) and observe that, by using the same reasoning, we can write

$$\sum_{k=1}^{\frac{N_p}{bN_u}-1} f(kbN_u) = \frac{N_p}{bN_u} - 1 + \sum_{\ell=1}^{L_h-1} \sum_{k=1}^{\frac{N_p}{bN_u}-1} (z^\ell)^k. \tag{39}$$

In what follows, we consider two different cases:
(a) The case in which $z^\ell \neq 1$. In this case, we observe that $\forall \ell \in \{1, 2, \ldots, L-1\}$

$$\sum_{k=0}^{\frac{N_p}{bN_u}-1} \alpha_i^\ell (z^\ell)^k = \alpha_i^\ell \frac{1 - z^{\frac{N_p}{bN_u}}}{1 - z^\ell} = \alpha_i^\ell \frac{1 - e^{-j2\pi\ell}}{1 - e^{-j2\pi \frac{1}{N_p} bN_u \ell}} = 0 \quad \text{if } z^\ell \neq 1. \tag{40}$$

and

$$\sum_{k=1}^{\frac{N_p}{bN_u}-1} (z^\ell)^k = \sum_{k=0}^{\frac{N_p}{bN_u}-1} (z^\ell)^k - 1 = -1. \tag{41}$$

Therefore, using Equations (38), (39), (40), and (41), Equation (36) becomes

$$\Gamma(b) = \sum_{u=1}^{N_u} \frac{\sigma_h^{(u)2}}{N_u} \left[L_h + \frac{N_p}{bN_u} - 1 - (L_h - 1) + (b-1)\frac{N_p}{bN_u} \right] = \sum_{u=1}^{N_u} \frac{\sigma_h^{(u)2}}{N_u} \left[b \frac{N_p}{bN_u} \right] = \sum_{u=1}^{N_u} \sigma_h^{(u)2} \frac{N_p}{N_u^2}. \tag{42}$$

(b) The case in which $z^\ell = 1$. We observe that if there exists an integer ℓ in $\{1, 2, \ldots, L_h - 1\}$ such that $z^\ell = 1$, then we have

$$\sum_{k=0}^{\frac{N_p}{bN_u}-1} \alpha^\ell (z^\ell)^k = \sum_{k=0}^{\frac{N_p}{bN_u}-1} \alpha^\ell (1)^k = \alpha^\ell \frac{N_p}{bN_u} \neq 0. \tag{43}$$

and

$$\sum_{k=1}^{\frac{N_p}{bN_u}-1} (z^\ell)^k = \frac{N_p}{bN_u} - 1 \neq -1. \tag{44}$$

Therefore, from the definition of $z = e^{-j2\pi \frac{1}{N_p} bN_u}$, we have that $z^\ell = 1$ when $\frac{nu}{N_p} \in \mathbb{Z}^+$. Without loss of generality, we look at the smallest integer in $\mathbb{Z}^+$, and we see that

$$\frac{bN_u}{N_p}\ell = 1 \quad \Leftrightarrow \quad \ell = \frac{N_p}{bN_u}. \tag{45}$$

Hence, since $\ell \in \{1, 2, \ldots, L_h - 1\}$ we have that

$$\frac{N_p}{bN_u} < L_h \tag{46}$$

which is equivalent to $b > \frac{N_p}{LN_u}$. Therefore, when $b > \frac{N_p}{L_h N_u}$, we can have at least one sum of the form

$$\sum_{k=0}^{\frac{N_p}{bN_u}-1} \alpha^\ell (z^\ell)^k > 0$$

(and $\sum_{k=1}^{\frac{N_p}{bN_u}-1} (z^\ell)^k > -1$). From Equations (38), (39), and (42), we can conclude that

$$\Gamma(b) > \sum_{u=1}^{N_u} \sigma_h^{(u)2} \frac{N_p}{N_u^2}, \quad \forall b > \frac{N_p}{L_h N_u}. \tag{47}$$

To conclude our proof, from the analysis of cases (a) and (b), we can state the following result:

$$\Gamma_m(1) = \cdots = \Gamma_m\left(\frac{N_p}{L_h N_u}\right) = \sum_{u=1}^{N_u} \sigma_h^{(u)2} \frac{N_p}{N_u^2} \tag{48}$$

and

$$\Gamma_m(b) > \Gamma_m\left(\frac{N_p}{L_h N_u}\right) \tag{49}$$

for all $b > \frac{N_p}{L_h N_u}$. ∎

Competing interests

The authors declare that they have no competing interests.

Acknowledgements

The work of the first author has been done while she was with ETIS/ENSEA - University of Cergy Pontoise - CNRS Laboratory, Cergy-Pontoise, France.

References

1. HG Myung, J Lim, DJ Goodman, Single carrier FDMA for uplink wireless transmission. IEEE Vehicular Technol. **1**(3), 30–38 (2006)
2. AF Molisch, A Mammela, Taylor D P, *Wideband Wireless Digital Communication*. (Prentice Hall PTR, Upper Saddle River, NJ, USA, 2001)
3. H Ekstrom, Technical solutions for the 3G Long-Term Evolution. IEEE Commun. Mag. **44**(3), 38–45 (2006)
4. HG Myung, in *Proceedings of the 15th European Signal Processing Conference*. Introduction to Single Carrier FDMA (Poznan, Poland, 2007), pp. 2144–2148
5. PH Moose, A technique for orthogonal frequency division multiplexing frequency offset correction. IEEE Trans. Commun. **42**(10), 2908–2914 (1994)
6. H Sari, G Karam, I Jeanclaude, in *Proceedings of the 6th Tirrenia International Workshop on Digital Communications*. Channel equalization and carrier synchronization in OFDM systems (Tirrenia, Italy, 1993), pp. 191–202
7. Y Zuh, B Letaief, in *Proceedings of Global Telecommunication Conference*. CFO estimation and compensation in single carrier interleaved FDMA systems (Honolulu, Hawaii, USA, 2009), pp. 1–5
8. A Sohl, A Klein, in *Proceedings of the 15th European Signal Processing Conference*. Comparison of localized, interleaved, and block-interleaved FDMA in terms of pilot multiplexing and channel estimation (Poznan, Poland, 2007)
9. L Koffman, V Roman, Broadband wireless access solutions based on OFDM access in IEEE 802.16. IEEE Commun. Mag. **40**(4), 96–103 (2002)
10. B Aziz, I Fijalkow, M Ariaudo, in *Proceedings of Global Telecommunications Conference (GLOBECOM 2011)*. Tradeoff between frequency diversity and robustness to carrier frequency offset in uplink OFDMA system (Houston, Texas, 2011), pp. 1–5
11. SH Song, GL Chen, KB Letaief, Localized or interleaved? A tradeoff between diversity and CFO interference in multipath channels. IEEE Trans. Wireless Commun. **10**(9), 2829–2834 (2011)
12. AM Masucci, I Fijalkow, EV Belmega, in *IEEE International Symposium on Personal, Indoor and Mobile Radio Communications (PIMRC)*. Subcarrier allocation in coded OFDMA uplink systems: Diversity versus CFO (London, United Kingdom, 2013)
13. E Telatar, Capacity of multi-antenna gaussian channels. Eur. Trans. Telecommun. **10**, 585–595 (1999)
14. D Tse, P Viswanath, *Fundamentals of Wireless Communications*. (Cambridge University Press, New York, NY, USA, 2004)
15. W Yu, W Rhee, S Boyd, JM Cioffi, Iterative water-filling for gaussian vector multiple-access channels. Inform. Theory, IEEE Trans. **50**(1), 145–152 (2004)
16. L Zheng, DNC Tse, Diversity and multiplexing: a fundamental tradeoff in multiple-antenna channels. IEEE Trans. Inform. Theory. **49**(5), 1073–1096 (2003)
17. M Godavarti, A Hero, in *Proceedings of IEEE International Conference on Acoustic, Speech and Signal Processing*. Diversity and degrees of freedom in wireless communications (Orlando, FL, USA, 2002), pp. 2861–2864
18. M-O Pun, M Morelli, CCJ Kuo, Maximum-likelihood synchronization and channel estimation for OFDMA uplink transmissions. IEEE Trans. Commun. **54**(4), 726–736 (2006)

2

Texture classification using rotation invariant models on integrated local binary pattern and Zernike moments

Yu Wang[1*], Yongsheng Zhao[2] and Yi Chen[1]

Abstract

More and more attention has been paid to the invariant texture analysis, because the training and testing samples generally have not identical or similar orientations, or are not acquired from the same viewpoint in many practical applications, which often has negative influences on texture analysis. Local binary pattern (LBP) has been widely applied to texture classification due to its simplicity, efficiency, and rotation invariant property. In this paper, an integrated local binary pattern (ILBP) scheme including original rotation invariant LBP, improved contrast rotation invariant LBP, and direction rotation invariant LBP is proposed which can effectively overcome the deficiency of original LBP that is ignoring contrast and direction information. In addition, for surmounting another major drawback of LBP such as locality which can result in the lack of shape and space expression of the holistic texture image, Zernike moment features are fused into the improved LBP texture features in the proposed method because they comprise orthogonal and rotation invariant property and can be easily and rapidly calculated to an arbitrary high order. Experimental results show that the proposed method can be remarkably superior to the other state-of-the-art methods when rotation invariant texture features are extracted and classified.

Keywords: Local binary pattern; Texture classification; Zernike moments; Rotation invariance

1 Introduction

Texture analysis is an attractive topic in image processing and pattern recognition. It plays a vital role in many important applications such as object tracking or recognition, remote sensing, image retrieval based on similarity, and so on [1-4]. Guo et al. [5] summarized four primary problems about texture analysis which are respectively image classification based on texture content, image segmentation of homogeneous texture regions, texture synthesis for graphics applications, and shape information acquisition from texture cue.

It is a very difficult problem to analyze existing texture in the real world mainly because of some uncertain factors such as inhomogeneity, illumination changes, and variability of texture appearance, etc. In the early stage, researchers focus on using statistical features to classify texture images. Haralick et al. [6] firstly proposed to use cooccurrence statistics to describe texture features. In the nineties, the Gabor filtering method of Manjunath and Ma [7] is credited as the current excellent technique in texture analysis. Although these methods obtained good performance, generally they need be made an explicit or implicit assumption that the training and testing samples have identical or similar orientations or are acquired from the same viewpoint [8]. In many practical applications, however, this assumption often cannot be guaranteed. Based on the practical experience, this phenomenon can be found that no matter how to rotate the texture images, these texture images always can be exactly classified from human vision point of view. Therefore, invariant texture analysis is highly demanded in both theoretical research and practical application.

More and more attention has been paid to the invariant texture analysis. An excellent review is summarized by Zhang and Tan [8]. Among these methods, Kashyap and Khotanzad [9] firstly researched rotation invariant texture classification by using a circular autoregressive model whose parameters are invariant to image rotation.

* Correspondence: wangyu@btbu.edu.cn
[1]Department of Computer and Information engineering, Beijing Technology and Business University, Beijing, China
Full list of author information is available at the end of the article

Choe and Kashyap [10] proposed an autoregressive fractional difference model to possess rotation invariant parameters. Hidden Markov model [11] also was used to explore rotation invariant texture classification. In addition, wavelet analysis is an excellent tool to obtain rotation invariant texture feature. For example, Jafari-Khouzani and Soltanian-Zadeh [12] proposed to extract wavelet energy features containing the texture orientation information to classify the texture images. In addition, a polar, analytic form of a two-dimensional Gabor wavelet [13] was used to deduce rotation invariant texture feature. Recently, some methods based on statistical learning was proposed by Varma and Zisserman [14,15], in which a rotation invariant texton library is first built using a training set, and then a testing texture image is classified according to its texton distribution. Crosier and Griffin [16] use basic image features (BIF) for texture classification and obtain excellent results. Furthermore, some pioneering work on scale and affine invariant texture classification has been done by using fractal analysis [17] and affine adaptation [18].

Local binary pattern (LBP) has been being reputable due to its effectiveness, speed, and rotation invariant property since it was mentioned by Harwood et al. [19]. Later it was introduced to the public by Ojala et al. [20]. Many researchers developed LBP methods based on Ojala's idea. For example, Zhao et al. [21], Maani et al. [22], and Ahonen et al. [23] respectively improved the LBP method using frequency domain analysis methods. Mäenpää [24] pointed out that texture can be regarded as a two-dimensional phenomenon characterized by two orthogonal properties: patterns and the strength of the patterns, and these two measures are supplementary to each other in a very useful way. However, it is 'the strength of the pattern' that the original LBP ignores besides direction information. Guo et al. [5] proposed an adaptive LBP method including the directional statistical information of texture for rotation invariant texture classification. Motivated by their work, original rotation invariant LBP, improved contrast rotation invariant LBP, and direction rotation invariant LBP are combined, called integrated LBP (ILBP) shown using the dashed line and box in Figure 1, to represent the texture information of the image, which can effectively overcome the inherent deficiency of original LBP that is ignoring contrast and direction information.

Although an LBP descriptor can get an excellent performance, it only describes the difference of local gray level and lacks the shape and space expression of the holistic texture image. Furthermore, compared to homogeneous textures such as bricks or sands which have the uniform statistical features, inhomogeneous textures like clouds or flowers generally cannot be extracted robust texture features using conventional algorithms focusing on homogeneous textures [25]. In effectively making up the missed shape and space information of the holistic texture image when LBP texture features are extracted, Zernike moment is a desirable choice.

Moments and functions of moments have been successfully utilized as pattern features in many applications such as image recognition [26] and image retrieval [25] which can capture global information of the image. Zernike moments are deduced based on the theory of orthogonal polynomials. Khotanzad and Hong [26] have suggested that orthogonal moments like Zernike moments are better than other types of moments in terms of information redundancy and image representation. Compared to other orthogonal moments, Zernike moments are possessed of rotation invariant property and can be easily and rapidly calculated to an arbitrary high order.

Therefore, a promising rotation invariant texture classification method is proposed which combines ILBP features with Zernike moment rotation invariant features. These two features respectively describe local and holistic information of texture images. Using the fusion strategy effectively, excellent performances are obtained by means of elaborate experiments and comprehensive texture databases including the Columbia-Utrecht Reflection and Texture (CUReT) database [27], the Outex database [28], and the KTH-TIPS database [29]. The framework of the proposed method is shown using a solid line and box in Figure 1.

The rest of the paper is organized as follows. Section 2 explains the original LBP. Section 3 presents the proposed method in which contrast and direction information of LBP are considered, and shape and space information of the holistic image obtained by Zernike moments are fused during the course of feature extraction. The experimental results of the proposed method and the other compared methods are shown in Section 4. Finally, a conclusion is drawn.

2 Original LBP texture model

2.1 Basic LBP model

Ojala et al. [20] used LBP as a texture descriptor of the image as shown in Figure 2, which is composed by central pixel and neighborhoods. Considering the central pixel as the threshold of texton, LBP code can be described using the following equation:

$$LBP(x_c, y_c) = \sum_{p=0}^{P-1} s\left(g_p - g_c\right)2^P \quad (1)$$

where $s(x)$ is a signal function, and $s(x) = \begin{cases} 1 & x\geq 0 \\ 0 & x < 0 \end{cases}$. (x_c, y_c) is the allowable position as the central pixel. g_c is

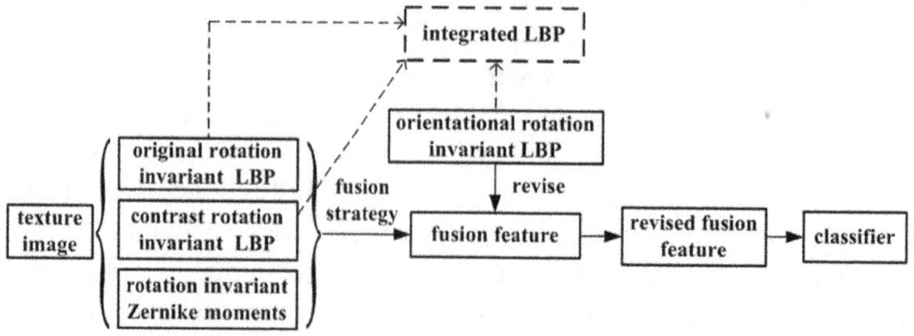

Figure 1 The framework of the proposed method. (The integrated LBP is shown using the dashed line and box).

the central pixel, g_p is the pixel value of neighborhood, P is the number of the neighbors.

By making statistics about the frequencies of the occurred LBP codes at all allowable positions in the image, the texture spectrum histogram $S[h]$ ($h = 0, 1, \ldots, 2^P$) can be obtained using the following equation:

$$S[h] = \frac{\sum_{x_c=0}^{u-1}\sum_{y_c=0}^{v-1} f(x_c, y_c)}{u \times v} \quad (2)$$

where $f(x_c, y_c) = \begin{cases} 1 & LBP(x_c, y_c) = h \\ 0 & \text{otherwise} \end{cases}$, $u \times v$ is the size of image.

Subsequently, Ojala et al. [1] improved the square LBP to be a circular form with discretionary radius R and neighborhoods P. Supposing that the coordinate of central pixel g_c is (x_c, y_c), then the coordinate of the neighbor g_p is $(x_c + R\cos(2\pi i/P),\ y_c - R\sin(2\pi i/P))$. The pixel values of the neighbors which are not in the image grids can be calculated using an interpolation method. The relative position of central pixel and neighbors is shown in the Figure 3.

2.2 Uniform and rotation invariant LBP

A hidden trouble exists in the abovementioned LBP. As the number of neighbors increases, the dimension of the histogram grows rapidly. For example, if P is 16, then the dimension of the histogram is $2^{16} = 65{,}536$. Therefore, the texture spectrum is so long that it is inconvenient to be applied in practice.

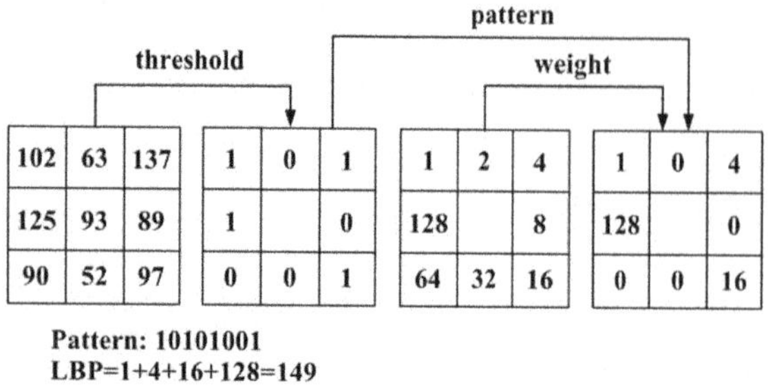

Figure 2 An example of the pattern and LBP.

In the LBP code, the number of spatial transitions (bitwise 0/1 changes) can be described as:

$$U(\mathrm{LBP}_{P,R}) = |s(g_P - g_c) - s(g_1 - g_c)| + \sum_{i=2}^{P} |s(g_i - g_c) - s(g_{i-1} - g_c)| \quad (3)$$

When $U(\mathrm{LBP}_{P,R}) \le 2$, the LBP pattern is defined as uniform patterns $\mathrm{LBP}_{P,R}^{u2}$ which has $P(P-1)+2$ discriminative patterns [1]. Although the histogram spectrum feature can be simplified using the uniform pattern, this processing way is feasible. By experiments and observation, uniform LBPs are fundamental properties of texture, providing the vast majority of patterns, sometimes over 90%. Detailed experimental results are listed in Section 4.

Furthermore, by observing, it is not difficult to find that no matter how to rotate the LBP, its structure is identical, which means that the original LBP and the rotated LBP have the same order and bitwise 0/1 changes as shown in Figure 4. For obtaining the rotation

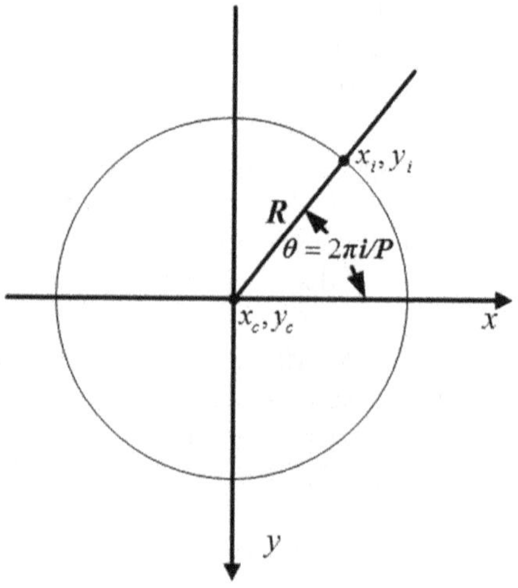

Figure 3 The relative position of central pixel and neighbors.

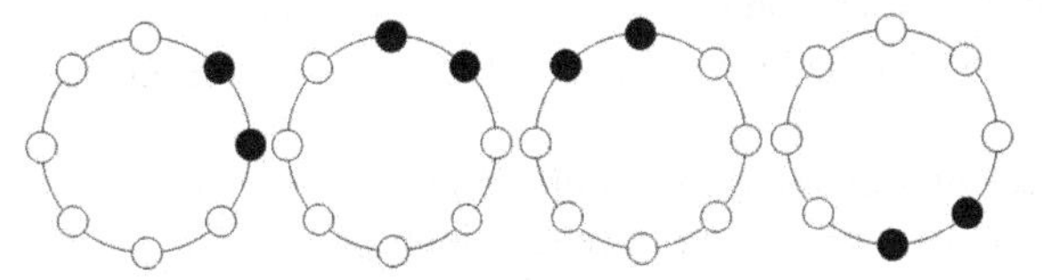

Figure 4 Some LBPs belonging to the same family.

invariant texture description, Ojala et al. [1] gave the following definition:

$$\mathrm{LBP}^{ri}_{P,R} = \min\{\mathrm{ROR}(\mathrm{LBP}_{P,R}, p)\}, (p = 0, ..., P\text{-}1) \quad (4)$$

where ri means the rotation invariance, ROR(x, p) represents that the LBP code x is rotated p times around the center pixel. That is to say, using the LBP with the minimal decimal value stands for other LBPs belonging to the same family. Figure 4 shows some LBPs pertaining to the same family. The rotation invariant uniform LBP $\mathrm{LBP}^{\mathrm{riu2}}_{P,R}$ can be calculated using the following equation:

$$\mathrm{LBP}^{\mathrm{riu2}}_{P,R} = \begin{cases} \sum_{p=0}^{P-1} s(g_p - g_c) & if \quad U(\mathrm{LBP}_{P,R}) \leq 2 \\ P + 1 & \text{otherwise} \end{cases} \quad (5)$$

where riu2 means rotation invariant uniform pattern which has $P + 2$ discriminative patterns. Thus the dimension of texture spectrum histogram is greatly simplified. By making statistics about the frequencies of the occurred $\mathrm{LBP}^{\mathrm{riu2}}_{P,R}$ at all allowable pixel positions in the image, the texture spectrum histogram S_{original} can be obtained.

3 Integrated LBP and Zernike moments model

As mentioned above, attention is paid to the detailed information when texture features are extracted by LBP. But the major drawback of LBP texture analysis is its locality. Zernike moment features are just opposite. That is to say, Zernike moments emphasize holistic and shape information of images but lack specific information. Therefore, LBP and Zernike moments complement each other in terms of information description of images. What is more, these two measure ways can be both described as a histogram spectrum, so it is very convenient to fuse them.

3.1 Integrated rotation invariant LBP model

Other two kinds of rotation invariant LBPs are proposed besides the original rotation invariant pattern $\mathrm{LBP}^{\mathrm{riu2}}_{P,R}$. They are respectively contrast rotation invariant LBPs represented by $C_{\mathrm{L}}\mathrm{BP}^{\mathrm{riu2}}_{P,R}$ and direction rotation invariant LBP represented by $O_{\mathrm{L}}\mathrm{BP}^{\mathrm{ri}}_{P,R}$. These three kinds of rotation invariant LBPs are collectively referred to as an ILBP model.

3.1.1 Contrast rotation invariant LBP

Although rotation invariant pattern $\mathrm{LBP}^{\mathrm{riu2}}_{P,R}$ can obtain an excellent performance, this kind of LBP texture representation only describes the change between the central pixel and neighbors. As to how much change occurs between them on earth, $\mathrm{LBP}^{\mathrm{riu2}}_{P,R}$ cannot give an explicit description. For example, both of the central pixels are 50 in two local textons whose neighbors are respectively {82,90,30,75,124, 69,39,104} and {79,68,24,82,136,73,45,233}. Although their LBP codes are both {1,1,0,1,1,1,0,1}, the absolute values of their contrast change between the central pixel and neighbors are different which are respectively {32,40,20,25,74, 19,11,54} and {29,18,26,32,86,23,5,183}. For supplementing these missed information, contrast rotation invariant LBP is added to describe the texture images besides the original $\mathrm{LBP}^{\mathrm{riu2}}_{P,R}$. Using C_p represents the absolute value of contrast change between the central pixel and neighbors in every texton, i.e., $C_p = |g_p - g_c|$; LBP of C_p can be obtained by the following equation:

$$C_{\text{-LBP}_{P,R}}(x_c, y_c) = \sum_{p=0}^{P-1} s(C_p - \mu_C)2^P, \quad (6)$$

where μ_C is the mean of the absolute value C_p of contrast change between the central pixel and neighbors in every texton, and $\mu_C = \frac{1}{P}\sum_{p=0}^{P-1} C_p$.. If the similar processing method such as (5) is applied to $C\_\mathrm{LBP}_{P,R}$, the contrast rotation invariant $C_{\mathrm{L}}\mathrm{BP}^{\mathrm{riu2}}_{P,R}$ can be obtained. By making statistics about the frequencies of the occurring $C_{\mathrm{L}}\mathrm{BP}^{\mathrm{riu2}}_{P,R}$ at all allowable pixel positions in the image, the texture spectrum histogram S_C can be obtained.

3.1.2 Direction rotation invariant LBP

For the stochastic texture images as shown in Figure 5a, the direction information is not apparent. But for the periodic or partly periodic texture images as shown in Figure 5b, the direction information is obvious. In the real world, most of the texture images contain the directional cue, so supplementing direction information in the discriminative features is worth trying.

The mean μ_{Op} and variance σ_{Op} of C_p in whole texture image are used to describe the direction information along the orientation $2\pi p/P$. The specific equations are shown below.

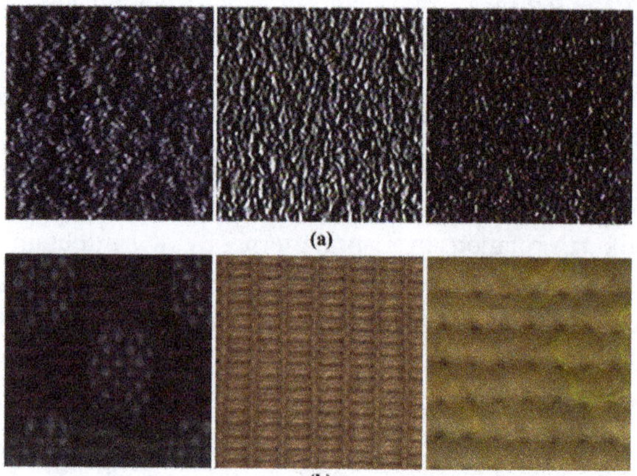

Figure 5 Some examples of (a) stochastic texture images and (b) periodic or partly periodic texture images.

$$\mu_{Op} = \frac{1}{u \times v}\sum_{i=1}^{u}\sum_{j=1}^{v} C_p, (p = 1, ..., P) \quad (7)$$

$$\sigma_{Op} = \sqrt{\frac{1}{u \times v}\sum_{i=1}^{u}\sum_{j=1}^{v}\left(C_p - \mu_{Op}\right)^2}, (p = 1, ..., P) \quad (8)$$

Therefore, two vectors $\boldsymbol{\mu}_O = [\mu_{O1}, \mu_{O2}, ..., \mu_{OP}]$ and $\boldsymbol{\sigma}_O = [\sigma_{O1}, \sigma_{O2}, ..., \sigma_{OP}]$ representing direction information can be obtained. Figure 6 shows an example of directional information $\boldsymbol{\mu}_O$ and $\boldsymbol{\sigma}_O$ about one texture image and corresponding rotated image with a 90° angle, respectively. By the observation, it can be found that $\boldsymbol{\mu}_O$ and $\boldsymbol{\sigma}_O$ contain strong directional information and can be used to revise the histogram spectrum feature of the images so that more similarities between the image and its rotated images are mined. $\boldsymbol{\mu}_O$ and $\boldsymbol{\sigma}_O$ can be respectively converted into rotation invariant LBP using the means of $\boldsymbol{\mu}_O$ and $\boldsymbol{\sigma}_O$ as the thresholds. Direction rotation invariant information $O_{\mu_L}\mathrm{BP}^{ri}_{P,R}$ and $O_{\sigma L}\mathrm{BP}^{ri}_{P,R}$ of the holistic texture image can be obtained using the following equations:

$$O_{\mu}\_\mathrm{LBP}^{ri}_{P,R} = \min\left\{\mathrm{ROR}\left(\sum_{p=0}^{P-1} s\left(\mu_{Op} - \bar{\mu}_{Op}\right)2^P, p\right)\right\} \quad (9)$$

$$O_{\sigma}\_\mathrm{LBP}^{ri}_{P,R} = \min\left\{\mathrm{ROR}\left(\sum_{p=0}^{P-1} s(\sigma_{Op} - \bar{\sigma}_{Op})2^P, p\right)\right\} \quad (10)$$

where $\bar{\mu}_{Op} = \frac{1}{P}\sum_{p=0}^{P-1}\mu_{Op}, \bar{\sigma}_{Op} = \frac{1}{P}\sum_{p=0}^{P-1}\sigma_{Op}$ $O_{\mu}\_\mathrm{LBP}^{ri}_{P,R}$ and $O_{\sigma}\_\mathrm{LBP}^{ri}_{P,R}$ are used to together represent direction rotation invariant $O_L\mathrm{BP}^{ri}_{P,R}$ of the whole texture image. As to how to revise the histogram spectrum feature of the image using direction rotation invariant $O_L\mathrm{BP}^{ri}_{P,R}$, the processing method will be detailedly introduced in the following section.

3.2 Rotation invariant Zernike moments model

Although LBP is an excellent method in both performance and efficiency, it ignores the shape and space information of the holistic texture image. For supplementing the missed information, Zernike moment rotation invariant features are used and fused. Because the basis set of ordinary moments is not orthogonal, Zernike [30] introduced a set of complex polynomials which makes a complete orthogonal set denoted by $\{V_{nm}(x, y)\}$ over the interior of the unit circle, i.e., $x^2 + y^2 = 1$. The form of these polynomials is described as:

$$V_{nm}(x, y) = V_{nm}(\rho, \theta) = R_{nm}(\rho)\exp(jm\theta) \quad (11)$$

where n is positive integer or zero, m is positive and negative integers subject to constraints that $n - |m|$ is even, and $|m| \leq n$. ρ is the length of vector from origin to (x, y) pixel, and θ is the angle between vector ρ and x axis in counterclockwise direction, and $R_{nm}(\rho)$ is radial polynomial shown as the following equation:

$$R_{nm}(\rho) = \sum_{s=0}^{n-|m|/2} (-1)^s \cdot \frac{(n-s)!}{s!\left(\frac{n+|m|}{2}-s\right)!\left(\frac{n-|m|}{2}-s\right)!}\rho^{n-2s} \quad (12)$$

And $R_{n,-m}(\rho) = R_{nm}(\rho)$. At the same time, these polynomials are orthogonal and satisfy:

$$\iint_{x^2+y^2\leq 1}[V_{nm}(x,y)]V_{pq}(x,y)dxdy = \frac{\pi}{n+1}\delta_{np}\delta_{mq} \quad (13)$$

where $\delta_{ab} = \begin{cases}1 & a = b\\ 0 & \text{otherwise}\end{cases}$ Zernike moments are the projection of the image function onto these orthogonal basis functions. So Zernike moment of nth order with the repetition m for a texture image $f(x, y)$ is:

$$A_{nm} = \frac{n+1}{\pi}\iint_{x^2+y^2\leq 1} V_{nm}(\rho, \theta)f(x, y)dxdy \quad (14)$$

For a digital image, the above equation can be changed into the following form:

$$A_{nm} = \frac{n+1}{\pi}\sum_x\sum_y V_{nm}(\rho, \theta)f(x, y), x^2 + y^2 \leq 1 \quad (15)$$

When calculating the Zernike moments of a given image, the center of the image is taken as the origin and pixel coordinates are mapped into the unit circle. The pixels falling outside the circle are not used, and $A_{nm} = A_{n,-m}$. By theoretical testifying, Zernike moments have the rotation invariant property, that is to say, if the Zernike moments of an image and its rotated image with

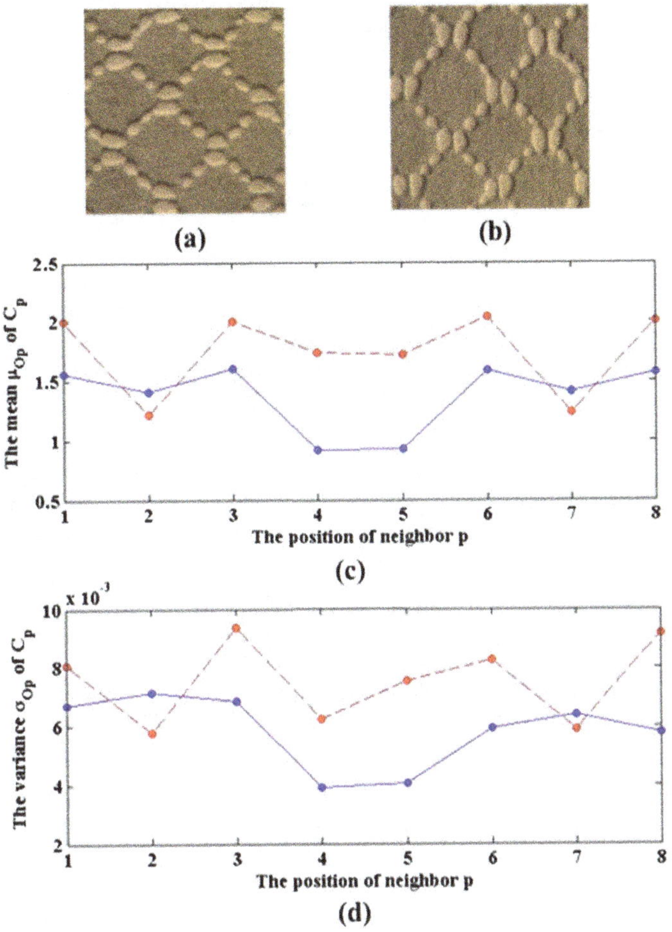

Figure 6 Texture image and directional information. An example of **(a)** 0° texture image, **(b)** 90° rotated image, and **(c)** corresponding mean μ_{Op} and **(d)** variance σ_{Op} of C_p. (Solid line and dashed line respectively denote 0° and 90° image. Here, $P = 8$ and $R = 1$).

an angle θ are respectively denoted using A_{nm} and A'_{nm}, they have the following relation:

$$A'_{nm} = A_{nm} \exp(-jm\theta) \tag{16}$$

If the image is preprocessed using some simple methods [26], Zernike moments are also invariant to translation and scale besides rotation. Using (15), the Zernike moments of different orders can be obtained such as A_{00}, A_{11}, A_{20}, A_{22}, and so on. The vector $\boldsymbol{S_Z}$ composed of Zernike moments of different orders is used as the histogram spectrum feature to describe the image information, and the specific form is:

$$S_Z = [A_{00}, A_{11}, A_{20}, A_{22}, \ldots., A_{nm}] \tag{17}$$

3.3 Construction and revise of fusion feature

After the ILBP and Zernike moment features of the image are respectively obtained through the above description, the fusion feature is constructed and revised, then a final classification decision is made.

3.3.1 Construction of fusion feature

Because the features of LBP and Zernike moments are both histogram spectrum form, it is very convenient to fuse them. In fact, a lot of experiments are made including serial, parallel, and jointly methods. However, the serial method can obtain more stable and excellent performance. The serial method is very simple and can be described as:

$$\boldsymbol{F} = \left[\boldsymbol{S}_{\text{original}} \; \boldsymbol{S}_C \; \boldsymbol{S}_Z\right] \tag{18}$$

where $\boldsymbol{F}$ denotes the fused histogram spectrum feature. Actually, the histogram spectrum $\boldsymbol{S}_{\text{original}}$ of original rotation invariant $\text{LBP}_{P,R}^{\text{riu2}}$ and the histogram spectrum $\boldsymbol{S}_C$ of contrast rotation invariant $C_L\text{BP}_{P,R}^{\text{riu2}}$ can also be serially fused. The related experimental results will be given in Section 4.

3.3.2 Revise of fusion feature

In the preceding section, a method for acquiring directional information of the image is proposed. Here the revise method of fused histogram spectrum feature using the direction rotation invariant $O_L BP^{ri}_{P,R}$ including $O_{\mu L} BP^{ri}_{P,R}$ and $O_{\sigma L} BP^{ri}_{P,R}$ will be elaborated. The equation is described as:

$$F' = F \cdot \left(1 + c_1 \cdot \exp\left(-c_2 \cdot \left(O_\mu\_LBP^{ri}_{P,R} - \mu(O_\mu)\right)/\sigma(O_\mu)\right)\right) \cdot \left(1 + c_1 \cdot \exp\left(-c_2 \cdot \left(O_\sigma\_LBP^{ri}_{P,R} - \mu(O_\sigma)\right)/\sigma(O_\sigma)\right)\right) \quad (19)$$

where F' is the revised fusion histogram spectrum feature. $\mu(O_\mu)$ and $\sigma(O_\mu)$ are respectively the mean and variance of the direction rotation invariant $O_{\mu L} BP^{ri}_{P,R}$ training images; $\mu(O_\sigma)$ and $\sigma(O_\sigma)$ are respectively the mean and standard of the direction rotation invariant $O_{\sigma L} BP^{ri}_{P,R}$ of all training images. c_1 and c_2 are positive parameters. In fact, besides fusion histogram spectrum feature F, $O_L BP^{ri}_{P,R}$ can also revise other histogram spectrum features such as $S_{original}$ generated by original rotation invariant $LBP^{riu2}_{P,R}$, S_C generated by contrast rotation invariant $C_L BP^{riu2}_{P,R}$, even S_Z calculated by rotation invariant Zernike moments.

3.4 Classifier and multiscale fusion idea

Nearest neighbor is a kind of effective and simple classification criterion. There are many good measures to estimate the difference and similarity between two histograms such as log-likelihood ratio and chi-square statistic [1]. The chi-square distance function in the experiments is chosen due to its excellent performance in terms of both speed and good recognition rates which is described as:

$$d(F'_{train}, F'_{test}) = \sum_{i=1}^{N} \left(F'_{train,i} - F'_{test,i}\right)^2 / \left(F'_{train,i} + F'_{test,i}\right) \quad (20)$$

where d is the chi-square distance between the revised fusion histogram F'_{train} of the training image and the revised fusion histogram F'_{test} of the testing image. Subscript i is the corresponding bin, and N is the number of bins.

In fact, multiscale fusion idea could be used to improve the classification accuracy in the proposed method, i.e., multiple descriptors of various (P, R) are used simultaneously. Because different scale operators support different structure space of the image, multiple scale descriptors can capture richer and completer texture information.

4 Experimental results

Many experiments have been elaborately designed and executed with the aim of demonstrating the effectiveness of the proposed method.

4.1 The database

Two large and comprehensive texture databases in the study are chosen which are respectively the CUReT database [27], the Outex database [28], and the KTH-TIPS database [29]. The CUReT database includes 61 classes of real-world textures, and each corresponds to different combinations of illumination and viewing angle. The same as the literature proposed by Guo [5], 92 sufficiently large images in each class with a viewing angle less than 60° are selected in the experiments. Among them, the first 23 images in each class were used as training images. Therefore, there are 1,403 (61 × 23 = 1,403) training models and 4,209 (61 × 69 = 4,209) testing samples. This design may be regarded as an analog about the situation with a small number of and less comprehensive training images.

In the Outex database, each texture is captured using six spatial resolutions (100, 120, 300, 360, 500, and 600 dpi), nine rotation angles (0°, 5°, 10°, 15°, 30°, 45°, 60°, 75°, and 90°), and three different simulated illuminants ('horizon', 'inca', and 'TL84'). The experimental images include canvas (46 classes), cardboard (1 classes), carpet (12 classes), chips (12 classes), and wallpaper (17 classes), i.e., 99 classes texture images all together in the Outex database. Each class texture images contains 27 images (3 illuminants, 9 angles, and spatial resolution of 600 dpi). The first 9 images ('horizon' illuminant, 9 angles, and spatial resolution of 600 dpi) in each class are chosen as training images. Therefore, there are 891 (99 × 9 = 891) training models and 1,782 (99 × 18 = 1,782) testing samples.

The KTH-TIPS database contains 10 texture classes such as crumpled aluminum foil, sponge, brown bread, etc. Each texture is captured under 9 scales, 3 different illumination directions, and 3 different poses. Therefore, there are 81 images per material. The first 21 images in each class are chosen as training images. Therefore, there are 210 (10 × 21 = 210) training models and 600 (10 × 60 = 600) testing samples.

The proposed method are compared with the state of the art LBP methods including $LBP^{riu2}_{P,R}$ [1], variance method ($VAR_{P,R}$) [1], $LBP^{riu2}_{P,R}/VAR_{P,R}$ [1], adaptive LBP method $\left(ALBPF^{riu2}_{P,R}\right)$ [5] and LBP histogram Fourier (LBPHF) method [21] (concatenating sign LBP histogram Fourier and magnitude LBP histogram Fourier). Because $VAR_{P,R}$ and $LBP^{riu2}_{P,R}/VAR_{P,R}$ were set as 128 and 16 bins. All the images are converted to grey scale. For removing the effect of global intensity and contrast, each texture image was normalized to have an average intensity 128 and a standard deviation 20 [1].

4.2 The feasibility of uniform LBP

For showing the effectiveness on dimensionality reduction using $LBP^{u2}_{P,R}$, the proportions of frequencies of $LBP^{u2}_{P,R}$ are calculated. Some statistic results are shown in Table 1, and the images are selected from the Outex database.

As can be seen from the Table 1, the uniform LBP occupies the vast majority of a local binary pattern, sometimes over 90%. Therefore, it is feasible to use the uniform LBP to reduce the dimensionality of histogram spectrum.

4.3 Experimental results on CUReT database

In the experiments, different combination on three kinds of rotation invariant LBP operators and rotation invariant Zernike moments are compared. '/O' denotes revising the histogram spectrum by direction rotation invariant LBP. 'C' represents capturing the histogram spectrum features by contrast rotation invariant LBP. 'Z' is Zernike moments method. And '_' denotes connecting two or three kinds of histogram spectrum features in series. For example, $LBP^{riu2}_{P,R}C_Z$ represents serially connecting original rotation invariant $LBP^{riu2}_{P,R}$, contrast rotation invariant $C_L BP^{riu2}_{P,R}$ and Zernike moments rotation invariant A_{nm}. $LBP^{riu2}_{P,R}C_Z/O$ represents revising the fusion feature $LBP^{riu2}_{P,R}C_Z$ by direction rotation invariant $O_L BP^{ri}_{P,R}$. The number 5, 8, or 10 denotes the order of Zernike moments. VZ_MR4 and VZ_MR8 respectively denote MR4 and MR8 of MR filter banks method. Table 2 lists experimental results on CUReT database using different methods.

As can be seen from the Table 2, firstly, the recognition rate of contrast rotation invariant $C_L BP^{riu2}_{P,R}$ (represented by 'C' in the Table 2) alone is worse than that of original rotation invariant $LBP^{riu2}_{P,R}$. For example, the recognition rates of $LBP^{riu2}_{P,R}$ can respectively reach at 62.25%, 64.93%, and 68.33% when P and R are respectively (8,1), (16,2), and (24,3). Whereas the results of $C_L BP^{riu2}_{P,R}$ are respectively 52.58%, 51.41%, and 50.18% in the same case. It shows that the information which is contained by $LBP^{riu2}_{P,R}$ is richer than that contained by $C_L BP^{riu2}_{P,R}$.

Table 1 The proportions of frequencies of $LBP^{u2}_{P,R}$ ($P = 8$, $R = 1$)

Images	$LBP^{u2}_{P,R}$ (%)
Canvas 006	87.06
Cardboard 001	81.32
Carpet 005	83.05
Chips 007	87.90
Wallpaper 008	90.52

Secondly, the role of contrast information, not only $VAR_{P,R}$ but also $C_L BP^{riu2}_{P,R}$ decreases as the number of neighbors and the size of texton increase. It states that the reliability of difference value between the central pixel and the neighbors reduces as the size of texton augments. But the recognition rate of original rotation invariant $LBP^{riu2}_{P,R}$ grows as the number of the neighbors and the size of texton increase.

Thirdly, among the compared methods with respect to LBP, LBPHF and adaptive LBP method obtain better results. And for non-LBP method, the results of MR8 method are better than ones of MR4 because of the richer feature representation.

Fourthly, for Zernike moment features, the recognition rate grows as the order increases. The reason for this phenomenon is that the higher the order is, the richer the detailed information contained by the Zernike moment histogram spectrum is. Fourthly, directional information can improve the recognition results of different features including LBP, Zernike moments, and fusion histogram spectrum.

Finally, fusion modes can effectively boost the recognition results. For example, when $P = 8$ and $R = 1$, the recognition rates are respectively 62.25%, 52.58%, and 36.07% obtained alone by $LBP^{riu2}_{P,R}$, $C_L BP^{riu2}_{P,R}$ and Zernike moments (10 order). However, when fusion features $LBP^{riu2}_{P,R}C$ and $LBP^{riu2}_{P,R}C_Z$ are used, the recognition rates can reach at 67.31% and 76.36%, respectively.

By applying the multiscale idea mentioned above in Section 3, better results can be obtained. For example, recognition rates respectively reach at 77.33% and 81.94% when different radius and different neighbors fusion features $LBP^{riu2}_{P,R}C_{8,1+16,2+24,3}$ and $LBP^{riu2}_{P,R}C_{Z_{8,1+16,2+24,3}}$ are used. And recognition rates respectively reach at 81.84% and 78.33% when different radius and same neighbors fusion features $LBP^{riu2}_{P,R}C_{16,1+16,2+16,3}$ and same radius and different neighbors fusion features $LBP^{riu2}_{P,R}C_{Z_{8,2+16,2+24,2}}$ are used. Here, Zernike moment features are gotten using 10 order moments, and different scale fusion features are obtained by simply connecting the histogram features of different scales. Better performance can be expected if more ingenious fusion strategies are used [31]. Because the results on LBPHF method are more stable among these compared methods, we also calculated the recognition rate of different radius and different neighbors fusion features $LBPHF_{8,1+16,2+24,3}$ which reaches at 71.77%.

4.4 Experimental results on Outex database

In this section, all the experiments are done using the same methods, and the results are listed in Table 3. Because the images in the Outex database are larger than those in the

Table 2 Recognition rates of different methods

Methods		$P=8, R=1$			$P=16, R=2$			$P=24, R=3$		
		Recognition rate (%)	Bins	c_1/c_2	Recognition rate (%)	Bins	c_1/c_2	Recognition rate (%)	Bins	c_1/c_2
$VAR_{P,R}$		45.17	128	-	41.15	128	-	38.92	128	-
$LBP_{P,R}^{riu2}/VAR_{P,R}$		66.48	10/16	-	70.56	10/16	-	71.04	10/16	-
$ALBPF_{P,R}^{riu2}$		69.73	10	-	73.49	18	-	73.63	26	-
LBPHF		68.40	76	-	73.34	276	-	73.91	604	-
VZ_MR4		67.55	1,220	-	67.55	1,220	-	67.55	1,220	-
VZ_MR8		71.25	1,220	-	71.25	1,220	-	71.25	1,220	-
$LBP_{P,R}^{riu2}$		62.25	10	-	64.93	18	-	68.33	26	-
$LBP_{P,R}^{riu2}/O$		62.77	10	0.1/0.15	65.17	18	0.1/0.15	68.64	26	0.1/0.15
C		52.58	10	-	51.41	18	-	50.18	26	-
C/O		53.27	10	0.1/0.15	53.15	18	0.1/0.15	52.27	26	0.1/0.15
$LBP_{P,R}^{riu2}C$		67.31	20	-	68.76	36	-	71.20	52	-
$LBP_{P,R}^{riu2}C/O$		67.36	20	0.01/0.015	68.76	36	0.01/0.15	71.35	52	0.1/0.15
Z	5	27.54	12	-	27.54	12	-	27.54	12	-
	8	34.33	25		34.33	25		34.33	25	
	10	36.07	36		36.07	36		36.07	36	
Z/O	5	30.39	12	0.1/0.15	32.62	12	0.1/1.5	33.43	12	0.1/1.5
	8	37.06	25		37.80	25		37.99	25	
	10	38.54	36		39.96	36		39.23	36	
$LBP_{P,R}^{riu2}C_Z$	5	74.79	32	-	76.41	48	-	77.19	64	-
	8	75.53	45		76.88	61		77.60	77	
	10	76.36	56		77.22	72		77.86	88	
$LBP_{P,R}^{riu2}C_Z/O$	5	73.27	32	0.01/0.015	76.41	48	0.01/0.015	77.19	64	0.01/0.015
	8	74.48	45		76.93	61		77.62	77	
	10	76.38	56		77.22	72		77.86	88	

CURet database, the results of many methods show 'out of memory' besides those of $VAR_{P,R}$ and $LBP_{P,R}^{riu2}/VAR_{P,R}$, when the number of neighbors P is 24 and radius R is 3. Therefore, the results on the scale of $P=24$ and $R=3$ are not listed in Table 3.

As can be seen from the Table 3, firstly, the results of original rotation invariant $LBP_{P,R}^{riu2}$ are better than those of contrast rotation invariant $C_LBP_{P,R}^{riu2}$. Secondly, the results of $LBP_{P,R}^{riu2}$ improve as the number of neighbors and the size of texton increase; however, the results of $C_LBP_{P,R}^{riu2}$ are the opposite.

Thirdly, for the Zernike moment method, the recognition rate grows as the order increases. The change trend is the same as that of the CUReT database. In addition, it can be found that the results of Zernike moments are very excellent mainly because of two factors. On the one hand, angle changes are highly emphasized for the images in the Outex database. On the other hand, Zernike moment features are possessed of a rotation invariant property and can well describe shape and space information of the image, so they are very suitable to be used to recognize the images with different rotation angles. It is the direction information of the image that has been fully mined by Zernike moments; therefore, the proposed direction rotation invariant LBP can hardly affect the original feature histogram.

Finally, the fusion method can remarkably improve the results. For example, when P and R are respectively 16 and 2, $LBP_{P,R}^{riu2}$ and $C_LBP_{P,R}^{riu2}$ respectively obtain the recognition rate of 31.03% and 15.38%, but fusion features $LBP_{P,R}^{riu2}C$ and $LBP_{P,R}^{riu2}C_Z$ can respectively reach at 32.72% and 71.16%. Here, Zernike moments are calculated using a 10-order parameter. However, it can be found that the fusion results are worse than the results of Zernike moments. It is not difficult to explain this phenomenon from the signal processing point of view. When the quality difference between two signal sources is too big, then the fusion result would be bad because the relatively worse signal may disturb the relatively better signal resembling the noise. Therefore, the recognition rates of fusion feature $LBP_{P,R}^{riu2}C_Z$ are worse than those of

Table 3 Recognition rates of different methods

Methods		$P = 8, R = 1$			$P = 16, R = 2$		
		Recognition rate (%)	Bins	c_1/c_2	Recognition rate (%)	Bins	c_1/c_2
$VAR_{P,R}$		34.68	128	-	43.77	128	-
$LBP^{riu2}_{P,R}/VAR_{P,R}$		38.95	10/16	-	52.86	10/16	-
$ALBPF^{riu2}_{P,R}$		17.00	10	-	30.02	18	-
LBPHF		38.83	76	-	56.29	276	-
VZ_MR4		32.38	1,980	-	32.38	1,980	-
VZ_MR8		35.97	1,980	-	35.97	1,980	-
$LBP^{riu2}_{P,R}$		20.71	10	-	31.03	18	-
$LBP^{riu2}_{P,R}/O$		21.10	10	0.1/0.15	31.43	18	0.1/0.15
C		16.55	10	-	15.38	18	-
C/O		17.12	10	0.1/0.15	16.33	18	0.01/0.015
$LBP^{riu2}_{P,R}C$		23.34	20	-	32.72	36	-
$LBP^{riu2}_{P,R}C/O$		24.97	20	0.1/0.15	33.56	36	0.1/0.15
Z	5	86.20	12	-	86.20	12	-
	8	92.93	25		92.93	25	
	10	94.33	36		94.33	36	
Z/O	5	86.20	12	0.1/0.015	86.14	12	0.1/0.015
	8	92.93	25		92.93	25	
	10	94.39	36		94.33	36	
$LBP^{riu2}_{P,R}C_Z$	5	58.02	32	-	61.73	48	-
	8	66.95	45		67.79	61	
	10	69.58	56		71.16	72	
$LBP^{riu2}_{P,R}C_Z/O$	5	58.24	32	0.1/0.15	61.73	48	0.01/0.015
	8	67.06	45	0.1/0.015	67.85	61	
	10	69.58	56		71.16	72	

alone Zernike moments but greatly better than those of alone texture feature such as $LBP^{riu2}_{P,R}$ or contrast LBP and even the fusion feature $LBP^{riu2}_{P,R}C$.

Multiscale method in the Outex database is also tried, and an excellent performance is obtained. For example, the recognition rates can respectively reach at 72.17%, 68.86%, and 74.41% when different radius and different neighbors fusion feature $LBP^{riu2}_{P,R}Cz_{8,1+16,2}$, different radius and same neighbors fusion features $LBP^{riu2}_{P,R}Cz_{16,1+16,2}$ and same radius and different neighbors fusion features LB $P^{riu2}_{P,R}Cz_{8,2+16,2}$ are used. Here, Zernike moments are calculated using a 10-order parameter. Furthermore, we also calculated the recognition rate of the LBPHF method with different radius and different neighbors fusion features $LBPHF_{8,1+16,2}$ which reaches at 56.73%.

4.5 Experimental results on KTH-TIPS database

In this section, all the experiments are done using the same methods, and the results are listed in Table 4. Because the trends of most of the results are similar to those of the CURet and Outex databases, here, only some different phenomena are given. Firstly, the recognition rates of $ALBPF^{riu2}_{P,R}$ and LBPHF methods decrease as the number of the neighbors and the size of texton increase. Secondly, compared with the results on the CURet and Outex databases, the role of contrast information is very obvious, sometimes even better than the ones of $LBP^{riu2}_{P,R}$. The reason may be that the images in the KTH-TIPS database contain sharp scale changes.

In addition, the multiscale method can further improve the results. For example, the recognition rates can respectively reach at 64.50%, 62.33%, and 63.83% when different radius and different neighbors fusion features $LBP^{riu2}_{P,R}\_C_{8,1+16,2+24,3}$, different radius and same neighbors fusion features $LBP^{riu2}_{P,R}\_C_{16,1+16,2+16,3}$, and same radius and different neighbors fusion features $LBP^{riu2}_{P,R}\_C\_Z_{8,2+16,2+24,2}$ are used. Here, Zernike moments are calculated using a 10-order parameter. Furthermore, we also calculated the recognition rate of the LBPHF method with different radius and different neighbors fusion features $LBPHF_{8,1+16,2+24,3}$ which reaches at 55.83%.

Table 4 Recognition rates of different methods

Methods		$P=8, R=1$			$P=16, R=2$			$P=24, R=3$		
		Recognition rate (%)	Bins	c_1/c_2	Recognition rate (%)	Bins	c_1/c_2	Recognition rate (%)	Bins	c_1/c_2
$VAR_{P,R}$		34.50	128	-	30.83	128	-	38.50	128	-
$LBP^{riu2}_{P,R}/VAR_{P,R}$		41.17	10/16	-	42.32	10/16	-	47.17	10/16	-
$ALBPF^{riu2}_{P,R}$		53.33	10	-	45.00	18	-	44.67	26	-
LBPHF		60.67	76	-	53.50	276	-	51.83	604	-
VZ_MR4		45.50	200	-	45.50	200	-	45.50	200	-
VZ_MR8		49.00	200	-	49.00	200	-	49.00	200	-
$LBP^{riu2}_{P,R}$		48.50	10	-	42.50	18	-	44.83	26	-
$LBP^{riu2}_{P,R}/O$		49.33	10	0.01/0.015	43.17	18	0.1/0.015	45.50	26	0.1/0.15
C		49.83	10	-	41.83	18	-	44.17	26	-
C/O		50.17	10	0.1/0.15	42.17	18	0.01/0.015	47.67	26	0.1/0.15
$LBP^{riu2}_{P,R}C$		57.67	20	-	51.5	36	-	50.27	52	-
$LBP^{riu2}_{P,R}C/O$		58.17	20	0.01/0.015	51.83	36	0.01/0.15	50.85	52	0.1/0.15
Z	5	19.00	12	-	19.00	12	-	19.00	12	-
	8	17.33	25		17.33	25		17.33	25	
	10	19.50	36		19.50	36		19.50	36	
Z/O	5	20.67	12	0.1/0.15	20.67	12	0.1/0.15	20.67	12	0.1/0.15
	8	23.33	25		23.33	25		23.33	25	
	10	25.33	36		25.33	36		25.33	36	
$LBP^{riu2}_{P,R}C_Z$	5	62.00	32	-	55.33	48	-	52.83	64	-
	8	62.17	45		55.17	61		53.17	77	
	10	62.50	56		55.50	72		53.17	88	
$LBP^{riu2}_{P,R}C_Z/O$	5	62.17	32	0.01/0.015	55.33	48	0.01/0.015	53.00	64	0.01/0.015
	8	62.33	45		56.17	61		53.50	77	
	10	62.50	56		56.00	72		53.67	88	

In a word, the proposed method in this paper obtained more exact, stable, and robust results compared with other methods including $LBP^{riu2}_{P,R}$, $VAR_{P,R}$, $LBP^{riu2}_{P,R}/VAR_{P,R}$, $ALBPF^{riu2}_{P,R}$, LBPHF and MR methods. Although the results of alone Zernike moment features in the Outex database are very outstanding, they are not stable compared with the proposed method because the results in the CUReT and KTH-TIPS databases are very bad. In addition, multiscale idea can further notably improve the recognition results.

5 Conclusions

LBP is an excellent tool for texture classification because of its simplicity, efficiency, and rotation invariant property. However, two mainly adverse factors weaken its performance, which are respectively ignoring contrast and direction information and lacking the shape and space expression of the holistic texture image. To effectively make up for the missed information, the rotation invariant contrast and direction information are added to the original rotation invariant LBP texture feature, which is called ILBP. In addition, Zernike moments are fused into the improved LBP texture features when representing images because they can effectively describe shape and space information of the holistic image, are possessed of orthogonal and rotation invariant properties, and can be easily and rapidly calculated to an arbitrary high order. Experimental results show that the proposed method can obtain a superior performance in terms of the large and comprehensive CUReT, Outex, and KTH-TIPS texture databases compared with other classic LBP and non-LBP methods, and multiscale idea can further remarkably improve the recognition results.

Competing interests

Pattern recognition, image processing and computer vision.

Acknowledgements

The authors sincerely thank Postdoctor Zhenhua Guo from Tinghua University and Professor Guoying Zhao from University of Oulu for sharing the source codes on adaptive LBP method and LBP histogram Fourier method. This work was supported by the national natural science foundation of China (NSFC) under Grant No. 61171068.

Author details
[1]Department of Computer and Information engineering, Beijing Technology and Business University, Beijing, China. [2]Department of Mechanical Engineering, Yanshan University, Qinhuangdao City, Hebei Province, China.

References

1. T. Ojala, M. Pietikainen, T. Mäenpää, Multiresolution gray-scale and rotation invariant texture classification with local binary patterns. IEEE Trans. Pattern Anal. Machine Intell. **24**, 971–987 (2002)
2. L. Zhang, B. Zou, J. Zhang, Y. Zhang, Classification of Polarimetric SAR Image based on Support Vector Machine using Multiple-Component Scattering Model and Texture Features. EURASIP J. Adv. Signal Proc. **3**, 1–10 (2010)
3. L. Sajn, I. Kononenko, Multiresolution image parameterization for improving texture classification. EURASIP J. Adv. Signal Proc. **2**, 1–13 (2008)
4. Y. Wang, D.J. He, C.C. Yu, T.Q. Jiang, Z.W. Liu, Multimodal biometrics approach using face and ear recognition to overcome adverse effects of pose changes. J. Electron. Imaging **21**, 043026-1–043026-11 (2012)
5. Z.H. Guo, L. Zhang, D. Zhang, S. Zhang, Rotation invariant texture classification using adaptive LBP with directional statistical features, in *Proceedings of the 7th IEEE International Conference on Image Processing* (IEEE, Hong Kong, China, 2010), pp. 285–288
6. R.M. Haralick, K. Shanmngam, I. Dinstein, Texture feature for image classification. IEEE Trans. Syst. Man Cy. **3**, 610–621 (1973)
7. B. Manjunath, W. Ma, Texture features for browsing and retrieval of image data. IEEE Trans. Pattern Anal. Machine Intell. **18**, 837–842 (1996)
8. J.G. Zhang, T.N. Tan, Brief review of invariant texture analysis methods. Pattern Recogn. **35**, 735–747 (2002)
9. R.L. Kashyap, A. Khotanzed, A model-based method for rotation invariant texture classification. IEEE Trans. Pattern Anal. Machine Intell. **8**, 472–481 (1986)
10. Y. Choe, R.L. Kashyap, 3-D shape from a shaded and textural surface image. IEEE Trans. Pattern Anal. Machine Intell. **13**, 907–918 (1991)
11. W.R. Wu, S.C. Wei, I.E.E.E. Trans, Rotation and gray-scale transform invariant texture classification using spiral resampling, subband decomposition, and hidden Markov model. Image Process. **5**, 1423–1434 (1996)
12. K. Jafari-Khouzani, H. Soltanian-Zadeh, Radon transform orientation estimation for rotation invariant texture analysis. IEEE Trans. Pattern Anal. Machine Intell. **27**, 1004–1008 (2005)
13. G.M. Haley, B.S. Manjunath, Rotation-invariant texture classification using a complete space-frequency model. IEEE Trans. Image Process. **8**, 255–269 (1999)
14. M. Varma, A. Zisserman, Texture classification: Are filter banks necessary? in *Proceedings of IEEE Computer Society Conference on Computer Vision and Pattern Recognition* (IEEE, Madison, USA, 2003), pp. 691–698
15. M. Varma, A. Zisserman, A statistical approach to texture classification from single images. Int. J. Comput. Vision **62**, 61–81 (2005)
16. M. Crosier, L.D. Griffin, Using Basic Image Features for Texture Classification. Int. J. Comput. Vision **88**, 447–460 (2010)
17. Y. Xu, H. Ji, C. Fermuller, Viewpoint invariant texture description using fractal analysis. Int. J. Comput. Vision **38**, 85–100 (2005)
18. S. Lazebnik, C. Schmid, J. Ponce, A sparse texture representation using local affine regions. IEEE Trans. Pattern Anal. Machine Intell. **27**, 1265–1278 (2005)
19. D. Harwood, T. Ojala, M. Pietikäinen, S. Kelman, L. Davis, Texture classification by center-symmetric auto-correlation, using Kullback discrimination of distributions. Pattern Recogn. Lett. **16**, 1–10 (1995)
20. T. Ojala, M. Pietikäinen, D. Harwood, A comparative study of texture measures with classification based on featured distributions. Pattern Recogn. **29**, 51–59 (1996)
21. G.Y. Zhao, T. Ahonen, J. Matas, M. Pietikäinen, Rotation-invariant image and video description with local binary pattern features. IEEE Trans. Image Process. **21**, 1465–1477 (2012)
22. R. Maani, S. Kalra, Y.H. Yang, Rotation invariant local frequency descriptors for texture classification. IEEE Trans. Image Process. **22**, 2409–2419 (2013)
23. T. Ahonen, J. Matas, C. He, M. Pietikäinen, Rotation invariant image description with local binary pattern histogram fourier features, in *Image Analysis* (Springer, Berlin Heidelberg, 2009), pp. 61–70
24. T. Mäenpää, *The local binary pattern approach to texture analysis-extensions and applications. Ph.D. dissertation* (Dept. Elect. Inf. Eng., University of Oulu, Oulu, Finland, 2003)
25. C.Y. Kim, O.J. Kwon, S. Choi, A practical system for detecting obscene videos. IEEE Trans. Consum. Electr. **57**, 646–650 (2011)
26. A. Khotanzad, Y.H. Hong, Invariant image recognition by Zernike moments. IEEE Trans. Pattern Anal. Machine Intell. **12**, 489–497 (1990)
27. K.J. Dana, B. van Ginneken, S.K. Nayar, J.J. Koenderink, Reflectance and texture of real world surfaces. ACM Trans. Graphic. **18**, 1–34 (1999)
28. T. Ojala, T. Mäenpää, M. Pietikäinen, J. Viertola, J. Kyllönen, S. Huovinen, Outex-new framework for empirical evaluation of texture analysis algorithm, in *Proceedings of the International Conference on Pattern Recognition* (IEEE, Quebec, Canada, 2002), pp. 701–706
29. B. Caputo, E. Hayman, M. Fritz, J.O. Eklundh, Classifying materials in the real world. Image Vis. Comput. **28**(1), 150–163 (2010)
30. F. Zernike, Diffraction theory of the cut procedure and its improved form, the phase contrast method. Physica **1**, 689–704 (1934)
31. K. Woods, W.P. Kegelmeyer Jr., K. Bowyer, Combination of multiple classifiers using local accuracy estimates. IEEE Trans. Pattern Anal. Machine Intell. **19**, 405–410 (1997)

3

Linking speech enhancement and error concealment based on recursive MMSE estimation

Balázs Fodor*, Florian Pflug and Tim Fingscheidt

Abstract

Speech enhancement and error concealment have seen a considerable progress over the past decades. Although both fields deal with distorted speech signals, there has rarely been an attempt to relate respective approaches to each other. In this paper, for the first time, a clear synopsis of recursive minimum mean square error (MMSE) estimation in both fields is provided. Our work intentionally does not propose a certain algorithm furthering the state of the art, nor does it provide simulation results of algorithms. Instead, our aim is threefold: First we revisit the basics of Bayes estimation in a recursive manner, covering both kinds of distortion acoustic noise as well as transmission channel noise. Second, we present recursive MMSE estimation applied to speech enhancement (in the frequency domain, as typical) and applied to error concealment (in the time domain, as typical) in strictly coherent notations and provide respective overview diagrams. Finally, we discuss commonalities and differences between both approaches, identify a particular strength of error concealment in general, and provide possible research directions for speech enhancement. A particularly interesting observation is that noise introduced by error concealment is far from being Gaussian and that additive acoustic noise can be expressed in terms of bit errors in DFT coefficients providing a potential interface to error concealment approaches.

Keywords: MMSE estimation; Speech enhancement; Error concealment

1 Introduction

Minimum mean square error (MMSE) estimation is omnipresent in a wide range of research fields and applications. It belongs to the family of Bayesian estimators which are based on the following model: The unobservable quantity to be estimated is considered to be the outcome of a random process such as a speech sample [1,2]. These outcomes can be measured through a channel introducing distortions, resulting in so-called observations. Besides the observations, Bayes estimators use *a priori* knowledge about the aforementioned random process and the channel, resulting in improved estimation results [2].

Based on this model, MMSE estimators minimize the estimation error variance conditioned on the observations. As an example, the widely used Wiener filter is optimal with respect to the MMSE error criterion [2]. In speech enhancement, it is widely assumed that the observations are statistically independent of each other, therefore, MMSE estimation of speech is carried out sequentially by means of the current observation only [3]. Signal history, such as the last speech estimate, is merely used for smoothing purposes in a practical system (cf. [3], Section V). Assuming, however, a dynamic signal model in the form of an autoregressive (AR) speech process, e. g., in conjunction with the source-filter model [4,5], the optimal MMSE estimator is able to exploit signal redundancy by employing both the current and previous observations [6]. In this case, under certain assumptions, the estimation can be carried out *recursively* and can be split into two steps typically decreasing computational complexity and relaxing memory requirements [7]. The first step exploits signal history in the form of previous observations providing an *a priori* estimate which is subsequently corrected in a second step taking into account

*Correspondence: b.fodor@tu-bs.de
Technische Universität Braunschweig, Institute for Communications Technology, Schleinitzstr. 22, 38106, Braunschweig, Germany

also the current observation, resulting in an *a posteriori* estimate.

In speech enhancement, the following model is widely employed for MMSE estimation: The unobservable speech is distorted by the unobservable acoustic noise, resulting in the observations in the form of noisy speech. The aim of speech enhancement is to estimate the speech by means of some *a priori* knowledge and the observations. State-of-the-art speech enhancement is often carried out in the short-time Fourier transform (STFT) domain and it is assumed that the discrete Fourier transform (DFT) coefficients of the speech and the acoustic noise are statistically independent. Thus, the *a priori* knowledge employed for MMSE estimation is a specific distribution of the speech (Gaussian [3,8,9], super-Gaussian [10-12]) and the acoustic noise (typically Gaussian). Classical MMSE estimators for speech enhancement under a Gaussian speech assumption are, e. g., the Wiener filter (e. g., [13]), the MMSE short-time speech amplitude (STSA) estimator [3], or the MMSE log-spectral amplitude (LSA) estimator [9]. These estimators mainly differ with respect to the estimation domain, i. e., instead of the complex-valued DFT coefficients, an arbitrary function of it is estimated (typically its amplitude or the logarithm of its amplitude).

In speech enhancement, the widely used Kalman filter [6] can be employed as a *recursive* MMSE estimator. In Kalman filtering, a Gaussian assumption is made for the acoustic noise and the error of the *a priori* speech estimate. In [14], a Kalman filter was proposed for time-domain speech enhancement. The proposed approach models the speech as an AR process based on a predictor, therefore, the estimation employs the current and the previous observations and, according to Kalman theory, the estimation of the speech is carried out recursively in two steps. A DFT-domain Kalman filter for speech enhancement was introduced, e. g., in [15-17]. In [15], the Kalman filter operates on complex-valued DFT coefficients and the speech was modeled as an AR process, while the noise process was assumed to be memoryless. In [16], different approaches to calculate the prediction coefficients and different estimators for calculating the *a posteriori* speech were investigated. Additionally, the memoryless assumption for the noise was replaced by an AR noise model in [17].

In error concealment, the following model is widely utilized: On the transmitter side, speech samples or source-coded parameters are quantized and mapped to corresponding bit combinations. These are transmitted over digital error-prone transmission channels and are received as bit-wise log-likelihood ratios (LLRs) comprising channel reliability information by the decoder. The aim of error concealment is the disguise of transmission errors which would otherwise lead to an unacceptable degradation of speech quality on the receiver side. This is achieved by MMSE estimation of unobservable speech samples or source-coded parameters, requiring *a posteriori* probabilities. These are proportional to both a likelihood term resulting from the received LLRs and a prior term comprising the available inherent signal redundancy as *a priori* knowledge at decoding time. This approach has been employed for robust source decoding of speech signals [18-23], source-coded audio signals [24,25], and uncompressed audio [26] that exploit signal redundancy in sample values or various source codec parameters (e. g., scaling factors, line spectral frequencies (LSFs) vectors, vector-quantized gains, adaptive codebook indices). Thereby, the signal redundancy is exploited by a time-variant modeling of the prior either using Markov chains [18-24] or employing approaches based on linear prediction in [25,26]. Typical applications are speech and audio transmission systems such as mobile phones or digital wireless microphones.

The aim of this paper is to reveal links between the fields of speech enhancement and error concealment focusing on recursive MMSE estimation approaches. We will show that the main structure of recursive MMSE estimation is the same in both disciplines which allows for drawing interesting links between tackling acoustic noise and transmission channel noise. In speech enhancement, the well-known Kalman filter can be employed as optimal recursive MMSE estimator which assumes that the acoustic noise is Gaussian distributed. Transmission channel noise as found in distorted speech signals, however, is far from a Gaussian distribution and — motivated by the nature of digital transmission — is rather modeled on *bit level* in error concealment which allows for exploiting powerful bit reliability information. This extra *a priori* knowledge can be identified as a definite strength of error concealment compared to speech enhancement. Based on this finding, this paper sketches as outlook new research directions for speech enhancement exploiting bit likelihoods.

This paper is structured as follows: Section 2 gives an introduction to recursive MMSE estimation. Section 3 shows how recursive MMSE estimation is commonly used for *speech enhancement* in the STFT domain. Section 4 gives an example for employing recursive MMSE estimation in *error concealment* in the time domain, presented in strong analogy to Section 3. Section 5 discusses links between speech enhancement and error concealment based on the recursive MMSE approaches from the previous Sections 3 and 4. In addition, some new research recommendations for speech enhancement motivated by error concealment are sketched. Finally, Section 6 closes the paper with conclusions.

2 On recursive MMSE estimation

As pointed out in Section 1, the aim of recursive MMSE estimation is to estimate the unobservable speech samples $s(n)$, with n being the discrete time index, which are transmitted through either an acoustic or a transmission channel. This channel distorts the speech signal by superimposing unobservable noise samples $d(n)$ being modeled as statistically independent, resulting in the observed noisy speech signal (cf. Figure 1)[a]

$$y(n) = s(n) + d(n). \tag{1}$$

The estimation of the clean speech $s(n)$ is carried out by means of all previous and the current observations $\mathbf{y}_0^n = [y(0), y(1), \ldots, y(n-1), y(n)]^T$, with $(\cdot)^T$ denoting the transpose operation, as well as by some *a priori* knowledge about the speech signal and the channel, resulting in the clean speech estimate $\hat{s}(n)$ (cf. Figure 1).

The *a priori* knowledge about the speech includes the following autoregressive model [14]: The current speech sample $s(n)$ is assumed to be a sum of the predicted speech $s^+(n)$ and the prediction error $e(n)$ (cf. Figure 1), the latter being a zero-mean (random) signal which is statistically independent of $s^+(n)$. The predicted speech is generated by a predictor of the order N_p as $s^+(n) = \mathbf{a}^T \cdot \mathbf{s}_{n-N_p}^{n-1}$ with $\mathbf{s}_{n-N_p}^{n-1} = [s(n-N_p), s(n-N_p+1), \ldots, s(n-1)]^T$ and $\mathbf{a} = [a_{N_p}, a_{N_p-1}, \ldots, a_1]^T$ being the so-called prediction coefficients. Please note that these coefficients are time-variant in practice, therefore, they need to be estimated. However, assuming a slow time variability, $\mathbf{a}$ is treated as a constant for the moment.

According to the speech signal model, the current speech sample $s(n)$ is statistically dependent not only on the current observation $y(n)$ but also on the previous ones $\mathbf{y}_0^{n-1}$, thus, these are also included in the estimation process [6]. This paper deals with *recursive* estimation, therefore, the estimation process is typically split into two parts, namely the *propagation* step and the *update* step. The propagation step exploits the previous observations to provide an *a priori* estimate of the current speech sample $s(n)$. Since this estimate can yield a relatively high error variance, the update step improves it incorporating the current observation $y(n)$, resulting in the *a posteriori* speech estimate $\hat{s}(n)$. Hence, the information carried by the previous observations becomes successively part of the *a priori* knowledge during the estimation process.

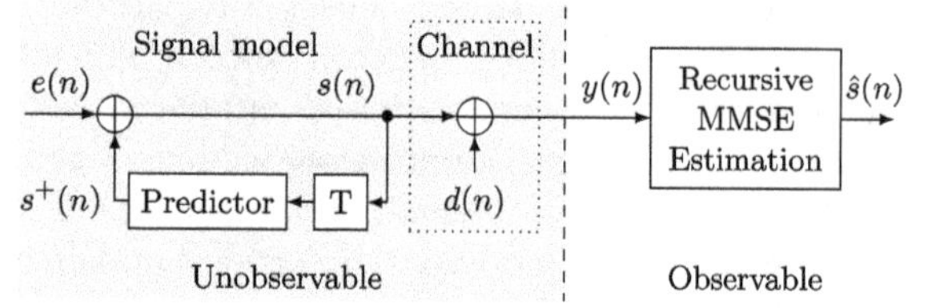

Figure 1 Signal model, channel, and recursive MMSE estimation.

2.1 The estimator

The recursive MMSE estimation with the underlying signal model as in Figure 1 yields

$$\hat{s}(n) = E\left\{s\,\middle|\,y(n), \mathbf{y}_0^{n-1}\right\} = \int_{\mathbb{R}} s \cdot p\left(s\,\middle|\,y(n), \mathbf{y}_0^{n-1}\right) ds \tag{2}$$

with $E\{\cdot\}$ being the expectation operator, $p(\cdot)$ being a probability density function (pdf), and $p(s(n)|y(n), \mathbf{y}_0^{n-1}) = p\left(s(n)|\mathbf{y}_0^n\right)$ being the so-called posterior. Usually, the posterior is computed by means of Bayes' rule, therefore, (2) can be rewritten as

$$\hat{s}(n) = \frac{\int_{\mathbb{R}} s \cdot p\left(y(n)\,\middle|\,s, \mathbf{y}_0^{n-1}\right) \cdot p\left(s\,\middle|\,\mathbf{y}_0^{n-1}\right) ds}{p\left(y(n)\,\middle|\,\mathbf{y}_0^{n-1}\right)} \tag{3}$$

with $p\left(y(n)\,\middle|\,s(n), \mathbf{y}_0^{n-1}\right)$, $p\left(s(n)\,\middle|\,\mathbf{y}_0^{n-1}\right)$, and $p\left(y(n)\,\middle|\,\mathbf{y}_0^{n-1}\right) = \int_{\mathbb{R}} p\left(y(n)\,\middle|\,s, \mathbf{y}_0^{n-1}\right) \cdot p\left(s\,\middle|\,\mathbf{y}_0^{n-1}\right) ds$ being the so-called likelihood, the prior, and the evidence, respectively. Please note that the evidence is typically calculated by marginalizing the pdf product of the numerator in (3).

2.2 The prior

As can be seen above, the prior is a function of the previous observations $\mathbf{y}_0^{n-1}$. However, the prior can also be determined recursively by marginalization using the signal model in Figure 1 as

$$p\left(s(n)\,\middle|\,\mathbf{y}_0^{n-1}\right) = \int \cdots \int_{\mathbb{R}^{N_p}} p\left(s(n)\,\middle|\,\mathbf{s}_{n-N_p}^{n-1}\right) \cdot p\left(\mathbf{s}_{n-N_p}^{n-1}\,\middle|\,\mathbf{y}_0^{n-1}\right) d\mathbf{s}_{n-N_p}^{n-1}. \tag{4}$$

The first pdf in the integral is a predictor pdf, the second one is the joint pdf of the last N_p posteriors. Therefore, the current prior is obviously dependent on the (distribution of the) last N_p estimates. Moreover, the mean of the prior using the signal model in Figure 1 turns out to be

$$\begin{aligned}
E\left\{s(n)\,\middle|\,\mathbf{y}_0^{n-1}\right\} &= E\left\{e(n)\,\middle|\,\mathbf{y}_0^{n-1}\right\} + E\left\{s^+(n)\,\middle|\,\mathbf{y}_0^{n-1}\right\} \\
&= 0 + \mathbf{a}^T \cdot \begin{pmatrix} E\left\{s(n-N_p)\,\middle|\,\mathbf{y}_0^{n-N_p}\right\} \\ E\left\{s(n-N_p+1)\,\middle|\,\mathbf{y}_0^{n-N_p+1}\right\} \\ \vdots \\ E\left\{s(n-2)\,\middle|\,\mathbf{y}_0^{n-2}\right\} \\ E\left\{s(n-1)\,\middle|\,\mathbf{y}_0^{n-1}\right\} \end{pmatrix} \\
&= \mathbf{a}^T \cdot \hat{\mathbf{s}}_{n-N_p}^{n-1} = \hat{s}^+(n)
\end{aligned} \tag{5}$$

which is the result of the propagation step. Please note that the prediction error $e(n) = s(n) - s^+(n)$ (also called innovation, e. g., [7]) is orthogonal to the previous speech samples and, therefore, also to the previous observations (e. g., [7,27]). Thus, we obtain $E\{e(n)|\mathbf{y}_0^{n-1}\} = E\{e(n)\} = 0$.

Assuming that this *a priori* speech estimate $\hat{s}^+(n) = f\left(\mathbf{y}_0^{n-1}\right)$ is a sufficient statistic for $s(n)$, the prior turns out to be

$$p(s(n)|\mathbf{y}_0^{n-1}) = p(s(n)|\hat{s}^+(n)). \quad (6)$$

This pdf describes the clean speech given the *a priori* speech estimate, in other words, it is the pdf of the propagation error $\bar{e}(n) = s(n) - \hat{s}^+(n)$ which can be written as

$$p\left(s(n)\left|\hat{s}^+(n)\right.\right) = p_{\bar{e}}\left(\bar{e}(n) = s(n) - \hat{s}^+(n)\right). \quad (7)$$

Using the signal model in Figure 1, the variance of the propagation error turns out to be [28]

$$\begin{aligned} E\{(s(n) - \hat{s}^+(n))^2\} &= E\{e^2(n)\} + E\{(s^+(n) - \hat{s}^+(n))^2\} \\ &= \sigma_e^2(n) + \sigma_{e^+}^2(n) = \sigma_{\bar{e}}^2(n) \end{aligned} \quad (8)$$

with $\sigma_e^2(n) = E\left\{e^2(n)\right\}$ being the prediction error variance and $\sigma_{e^+}^2(n) = E\left\{\left(s^+(n) - \hat{s}^+(n)\right)^2\right\}$. Please note that (8) holds *independently* of the type of the propagation error pdf.

2.3 The likelihood

The *a priori* knowledge about the (acoustic or transmission) channel is contained in the likelihood. This means that the additive noise $d(n)$, which distorts the clean speech $s(n)$ (cf. Figure 1), is modeled by the likelihood $p\left(y(n)\left|s, \mathbf{y}_0^{n-1}\right.\right)$. Assuming a memoryless (acoustic or transmission) channel, meaning that the noise samples $d(n)$, $d(n-1)$, ... are statistically independent of each other, the likelihood is [29]

$$p(y(n)|s(n), \mathbf{y}_0^{n-1}) = p(y(n)|s(n)) \quad (9)$$

being the pdf of $y(n)$ given a specific speech sample $s(n)$. Therefore, this is the pdf of the additive noise $d(n) = y(n) - s(n)$ as [3]

$$p\left(y(n)\left|s(n)\right.\right) = p_d\left(d(n) = y(n) - s(n)\right). \quad (10)$$

Since the variance of the noise $\sigma_d^2(n) = E\left\{d^2(n)\right\}$ cannot be measured directly in practice, it is usually estimated by a noise power estimator or a channel quality estimator.

3 Application to speech enhancement

Assuming that the speech and the noise processes are at least quasi-stationary along the time frames ℓ in each frequency bin k and statistically independent of each other, the signal model in Figure 1 is also valid in the STFT domain. Accordingly, the aim of recursive MMSE estimation in speech enhancement is to estimate the clean speech DFT coefficients $S_\ell(k)$ which are generated by an autoregressive process as sketched in Section 2: The clean speech $S_\ell(k)$ is modeled by the sum of the predicted speech $S_\ell^+(k)$ and a statistically independent, zero-mean prediction error $E_\ell(k)$, i. e., $S_\ell(k) = S_\ell^+(k) + E_\ell(k)$. $S_\ell^+(k)$ is calculated by a predictor of the order L_p as $S_\ell^+(k) = \mathbf{A}^H(k) \cdot \mathbf{S}_{\ell-L_p}^{\ell-1}(k)$ with $\mathbf{A}(k) = \left[A_{L_p}(k), A_{L_p-1}(k), \ldots, A_1(k)\right]^T$ being the complex-valued prediction coefficients and $\mathbf{S}_{\ell-L_p}^{\ell-1}(k) = \left[S_{\ell-L_p}(k), S_{\ell-L_p+1}(k), \ldots, S_{\ell-1}(k)\right]^T$. It is assumed that the prediction coefficients change very slowly along the frames ℓ and are, therefore, modeled as constants for the moment. The acoustic channel distorts the speech spectrum by superimposing some statistically independent acoustic noise $D_\ell(k)$, so that $Y_\ell(k) = S_\ell(k) + D_\ell(k)$ with $Y_\ell(k)$ being the noisy speech DFT coefficients. The estimated speech DFT coefficients $\widehat{S}_\ell(k)$ are calculated by the recursive MMSE estimator employing the underlying observations $\mathbf{Y}_0^\ell(k) = [Y_0(k), Y_1(k), \ldots, Y_\ell(k)]^T$ as (cf. (2))

$$\widehat{S}_\ell(k) = E\left\{S_\ell(k)\left|Y_\ell(k), \mathbf{Y}_0^{\ell-1}(k)\right.\right\}. \quad (11)$$

A block diagram of recursive MMSE estimation for speech enhancement assuming a memoryless acoustic channel is given in Figure 2. Please note that the upper signal path is related to the prior computation, the block in the center is the MMSE estimator (11), and the lower signal path refers to the likelihood computation. Starting in the lower left-hand corner, windowed segments of the noisy speech signal $y(n) = s(n) + d(n)$ are transformed into the DFT domain, followed by the likelihood computation.

3.1 The likelihood

Since a memoryless acoustic channel is assumed, the likelihood turns out to be (cf. (9) and (10))

$$\begin{aligned} p\left(Y_\ell(k)\left|S_\ell(k), \mathbf{Y}_0^{\ell-1}(k)\right.\right) &= p\left(Y_\ell(k)\left|S_\ell(k)\right.\right) \\ &= p_D\left(D_\ell(k) = Y_\ell(k) - S_\ell(k)\right). \end{aligned} \quad (12)$$

As can be seen, the likelihood is a function of the noisy speech (cf. Figure 2). Moreover, after computing the likelihood by means of the current observation $Y_\ell(k)$, the likelihood remains a function $g(S)$ of the unknown speech DFT coefficient S (cf. output of 'Likelihood Computation' in Figure 2). Furthermore, as we will see later, S will be the integration variable of the MMSE estimator (cf. (11)).

Assuming that the (complex-valued) additive noise $D_\ell(k)$ is a zero-mean Gaussian process with independent and identically distributed (i. i. d.) real and imaginary parts, the likelihood turns out to be [3]

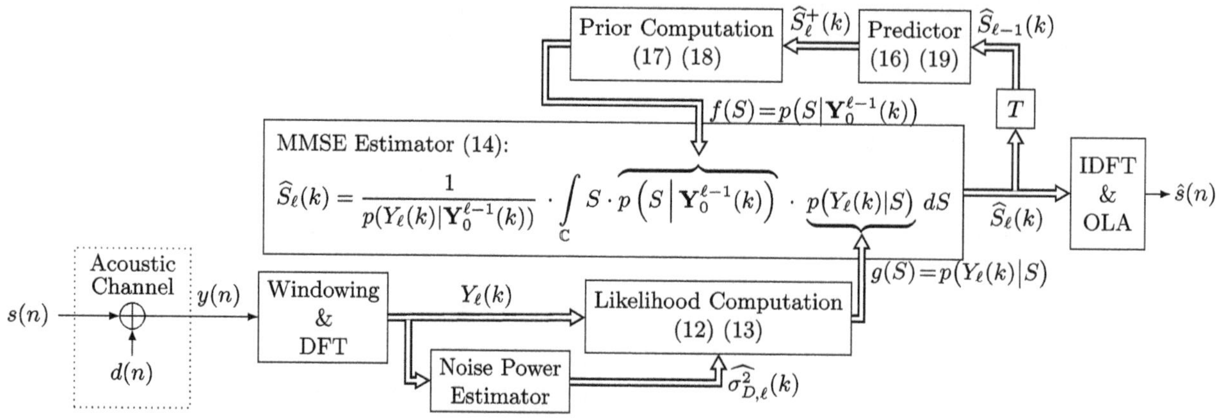

Figure 2 STFT domain recursive MMSE estimation for speech enhancement assuming a memoryless acoustic channel.

$$p(Y_\ell(k)|S_\ell(k)) = \frac{1}{\pi\sigma_{D,\ell}^2(k)} \cdot \exp\left(-\frac{|Y_\ell(k) - S_\ell(k)|^2}{\sigma_{D,\ell}^2(k)}\right) \quad (13)$$

with $\sigma_{D,\ell}^2(k)$ being the variance of the quasi-stationary noise process D. Please note that also for non-Gaussian noise pdfs, the likelihood is always a function of the noise power $\sigma_{D,\ell}^2(k)$ (cf. connection between 'Noise Power Estimator' and 'Likelihood Computation' in Figure 2). Therefore, in practice, its estimate $\widehat{\sigma_{D,\ell}^2}(k)$ is calculated by a noise power estimator using the noisy speech DFT coefficients [30-32].

3.2 The estimator

Using the likelihood (12), the clean speech DFT coefficients $\widehat{S}_\ell(k)$ are estimated as (cf. (2), (3), and (11))

$$\widehat{S}_\ell(k) = \frac{\int_{\mathbb{C}} S \cdot p(S|\mathbf{Y}_0^{\ell-1}(k)) \cdot p(Y_\ell(k)|S)\, dS}{p(Y_\ell(k)|\mathbf{Y}_0^{\ell-1}(k))} \quad (14)$$

with the evidence

$$p(Y_\ell(k)|\mathbf{Y}_0^{\ell-1}(k)) = \int_{\mathbb{C}} p(S|\mathbf{Y}_0^{\ell-1}(k)) \cdot p(Y_\ell(k)|S)\, dS \quad (15)$$

and the prior $p\left(S\middle|\mathbf{Y}_0^{\ell-1}(k)\right)$. Please note that both the prior and the likelihood are a function of the integration variable S, namely $f(S)$ (cf. upper signal path in Figure 2) and $g(S)$ (cf. lower signal path in Figure 2), respectively. The estimated clean speech signal $\hat{s}(n)$ is obtained by taking the inverse DFT (IDFT) of $\widehat{S}_\ell(k)$ from (14) and performing, e. g., an overlap-add (OLA) step.

3.3 The prior

As discussed in Section 2, the *a priori* speech estimate is calculated as (cf. (5))

$$\widehat{S}_\ell^+(k) = \mathbf{A}^H(k) \cdot \widehat{\mathbf{S}}_{\ell-L_p}^{\ell-1}(k). \quad (16)$$

This step is denoted as 'Predictor' in the upper signal path in Figure 2. Please note that the predictor employs previous speech estimates which is reflected by the delay unit denoted by 'T' in Figure 2. Assuming that the *a priori* speech estimate $\widehat{S}_\ell^+(k)$ is a sufficient statistic for the speech $S_\ell(k)$, the prior turns out to be the pdf of the propagation error $\bar{E}_\ell(k) = S_\ell(k) - \widehat{S}_\ell^+(k)$ (cf. (6) and (7))

$$p(S_\ell(k)|\mathbf{Y}_0^{\ell-1}(k)) = p_{\bar{E}}(\bar{E}_\ell(k) = S_\ell(k) - \widehat{S}_\ell^+(k)). \quad (17)$$

Employing a specific *a priori* speech estimate $\widehat{S}_\ell^+(k)$, the prior remains a function of the speech $f(S)$ and can be fed into the MMSE estimator (14) (cf. connection between 'Prior Computation' and 'MMSE Estimator' in Figure 2).

Assuming that the (complex-valued) propagation error is a zero-mean Gaussian process with i. i. d. real and imaginary parts, the prior is calculated [15,16]

$$p_{\bar{E}}\left(\bar{E}_\ell(k)\right) = \frac{1}{\pi\sigma_{\bar{E},\ell}^2(k)} \cdot \exp\left(-\frac{|\bar{E}_\ell(k)|^2}{\sigma_{\bar{E},\ell}^2(k)}\right) \quad (18)$$

with $\sigma_{\bar{E},\ell}^2(k)$ being the propagation error variance which cannot be measured in practice, thus, it has to be estimated [15,16].

Since the prediction coefficients $\mathbf{A}(k)$ in (16) are not accessible in practice, they need to be estimated as well, e. g., by the widely used normalized least-mean-squares (NLMS) algorithm [7]. Introducing again time variability, the prediction coefficients for the next frame are calculated recursively as

$$\widehat{\mathbf{A}}_{\ell+1}(k) = \widehat{\mathbf{A}}_\ell(k) + \mu \cdot \frac{\widehat{E}_\ell^*(k)}{||\widehat{\mathbf{S}}_{\ell-L_p}^{\ell-1}(k)||^2 + \Delta} \cdot \widehat{\mathbf{S}}_{\ell-L_p}^{\ell-1}(k) \quad (19)$$

with $\widehat{E}_\ell(k) = \widehat{S}_\ell(k) - \widehat{S}_\ell^+(k)$, as well as with μ, $(\cdot)^*$, Δ, and $||\cdot||$ denoting the step size constant, the complex conjugate, the regularization parameter, and the Euclidean norm, respectively.

3.4 The Kalman filter

The estimator (14) requires calculating two integrals over the whole complex plane for each time-frequency unit (ℓ, k). Fortunately, this estimator can be obtained in closed form by solving these integrals, reducing the computational complexity for practical implementations. Assuming a Gaussian distribution for both the propagation error (18) and the acoustic noise (13), the MMSE estimator (14) turns out to be a sum of the *a priori* estimate and the update (the derivation can be found in the Appendix) in the form of the Kalman filter equations (cf. [15], Equation 12)

$$\widehat{S}_\ell(k) = \widehat{S}_\ell^+(k) + \widehat{E}_\ell(k) \qquad (20)$$

with

$$\widehat{E}_\ell(k) = K_\ell(k) \cdot R_\ell(k), \qquad (21)$$

$$K_\ell(k) = \frac{\zeta_\ell(k)}{1+\zeta_\ell(k)}, \qquad (22)$$

$$R_\ell(k) = Y_\ell(k) - \widehat{S}_\ell^+(k) \qquad (23)$$

where $K_\ell(k)$ is the so-called Kalman gain and $\zeta_\ell(k) = \sigma^2_{E,\ell}(k)/\sigma^2_{D,\ell}(k)$ (cf. the *a priori* SNR in [3]). The latter can be estimated by [16]

$$\hat{\zeta}_\ell(k) = \beta \frac{|\widehat{E}_{\ell-1}(k)|^2}{\widehat{\sigma^2_{D,\ell-1}}(k)} + (1-\beta)\max\left\{\frac{|R_\ell(k)|^2}{\widehat{\sigma^2_{D,\ell-1}}(k)} - 1, 0\right\} \qquad (24)$$

with β being a smoothing factor, typically chosen close to one.

Please note that the recursive nature of MMSE estimation is well reflected by (20): The *a priori* estimate $\widehat{S}_\ell^+(k)$ utilizing the previous observations is corrected by the term $\widehat{E}_\ell(k)$ employing the current observation $Y_\ell(k)$, resulting in the *a posteriori* speech estimate $\widehat{S}_\ell(k)$.

4 Application to error concealment

The aim of error concealment is to estimate transmitter-sided (speech) samples, e. g., by a recursive MMSE estimator, in order to conceal distortions due to residual bit errors after demodulation or channel decoding. Bit error concealment of PCM audio or speech could theoretically be carried out by the equations in Section 2 using the hard-decoded receiver-sided samples. However, in order to exploit more information for improved estimation results, it is often advantageous to employ error concealment using reliability information on a bit level as in [19,26]. A block diagram of such a soft-decision decoding scheme based on recursive MMSE estimation is given in Figure 3. Similar to Figure 2, the likelihood computation is related to the lower signal path, the estimator can be found in the center, and the prior computation is performed in the upper signal path.

4.1 The likelihood

In error concealment, it is assumed that each transmitter-sided sample $s(n)$, being processed as introduced in Section 2, is quantized with M bit and, therefore, can bijectively be mapped to a natural-binary bit combination $\mathbf{x}(n) = [x_0(n), x_1(n), \ldots, x_m(n), \ldots, x_{M-1}(n)]$ (see 'Quantization and Bit Mapping' in Figure 3). Assuming further binary phase-shift keying (BPSK) modulation, each transmitted bit (BPSK symbol) $x_m(n) \in \{-1, 1\}$ is more or less distorted by the channel, modeled by the real-valued channel noise $d_m(n)$ (cf. 'Transmission Channel' in Figure 3). For the demodulation of the received real-valued noisy symbols $y_m(n)$, the so-called energy per bit to noise power spectral density ratio E_b/N_0 is needed which is calculated by the channel estimator (cf. connection between 'Channel Estimator' and 'Demodulator' in Figure 3). The demodulator then calculates the LLR which is defined as

$$L(\hat{x}_m(n)) = \ln \frac{P(\hat{x}_m(n)|x_m(n) = +1)}{P(\hat{x}_m(n)|x_m(n) = -1)} \qquad (25)$$

with the hard-decided receiver-sided bit $\hat{x}_m(n) = \mathrm{sign}(y_m(n))$ representing the receiver-sided observation. In practice, the LLR is calculated by means of E_b/N_0 and $y_m(n)$ (cf. the two inputs of the 'Demodulator' in Figure 3)[b] which reflects the likelihood of a possibly transmitted bit $x_m(n)$. The bit-error probabilities $\mathrm{BER}_m(n)$ describe the probability that a transmitted bit was distorted through the channel and can be calculated by means of the LLRs as [33]

$$\mathrm{BER}_m(n) = \frac{1}{1 + e^{|L(\hat{x}_m(n))|}}. \qquad (26)$$

Once each transmitter-sided sample $s(n)$ is quantized, it assumes a discrete value $s^{(i)}$ with $i \in \{0, 1, \ldots, 2^M - 1\}$. Moreover, each $s^{(i)}$ can be mapped to one corresponding bit combination $\mathbf{x}^{(i)}$. The bit-wise transition probabilities define the *bit* likelihood given bit $x_m^{(i)}$ of the ith quantization table entry [19]

$$P\left(\hat{x}_m(n)\,\middle|\,x_m^{(i)}\right) = \begin{cases} \mathrm{BER}_m(n), & \text{if } \hat{x}_m(n) \neq x_m^{(i)}, \\ 1 - \mathrm{BER}_m(n), & \text{else.} \end{cases} \qquad (27)$$

Assuming that the transmission channel is memoryless and that the bit distortions $d_m(n)$ are statistically independent of each other along the bit indices m, the *sample* likelihood is computed as (cf. 'Likelihood Computation' in Figure 3 and, e. g., [19])

$$P\left(\hat{\mathbf{x}}(n)\,\middle|\,\mathbf{x}^{(i)}\right) = \prod_{m=0}^{M-1} P\left(\hat{x}_m(n)\,\middle|\,x_m^{(i)}\right). \qquad (28)$$

Please note that until this step, the recursive MMSE estimator operates on bit level. After computing the *sample*

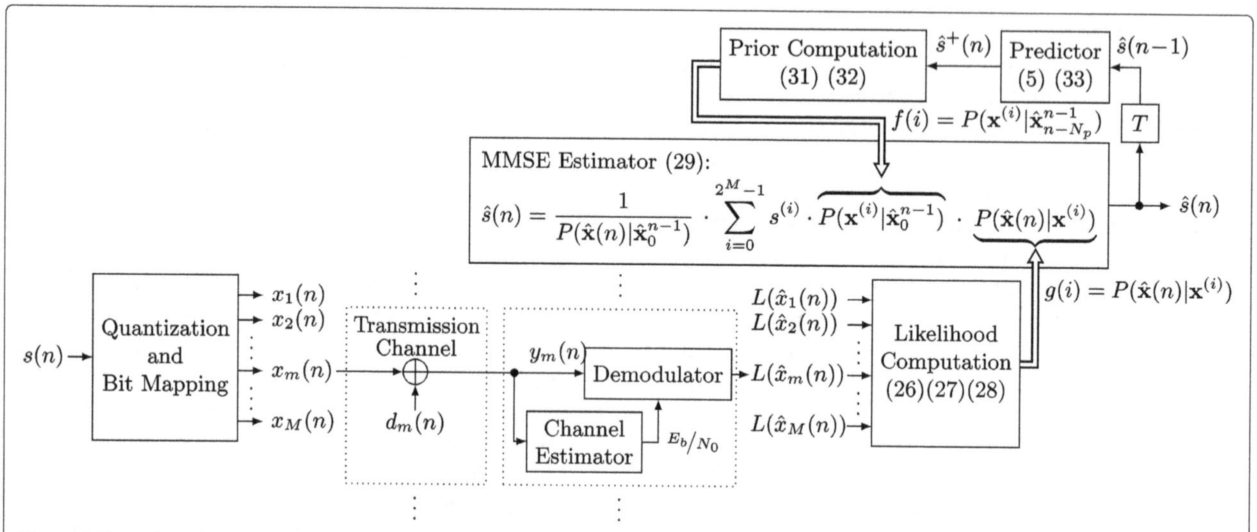

Figure 3 **Time domain recursive MMSE estimation for error concealment assuming a memoryless transmission channel.**

likelihood (28) and for any further processing, however, the estimator deals with samples again.

4.2 The estimator

The recursive MMSE estimator (3) turns out to be a sum

$$\hat{s}(n) = \frac{\sum_{i=0}^{2^M-1} s^{(i)} \cdot P(\mathbf{x}^{(i)}|\hat{\mathbf{x}}_0^{n-1}) \cdot P(\hat{\mathbf{x}}(n)|\mathbf{x}^{(i)})}{P(\hat{\mathbf{x}}(n)|\hat{\mathbf{x}}_0^{n-1})} \tag{29}$$

with the evidence

$$P\left(\hat{\mathbf{x}}(n)\left|\hat{\mathbf{x}}_0^{n-1}\right.\right) = \sum_{i=0}^{2^M-1} P\left(\mathbf{x}^{(i)}\left|\hat{\mathbf{x}}_0^{n-1}\right.\right) \cdot P\left(\hat{\mathbf{x}}(n)\left|\mathbf{x}^{(i)}\right.\right), \tag{30}$$

the *sample* likelihood (28), and the prior $P\left(\mathbf{x}^{(i)}\left|\hat{\mathbf{x}}_0^{n-1}\right.\right)$. Please note that both the prior and the *sample* likelihood are a function of the summation index i in (29), namely $f(i)$ (cf. upper signal path in Figure 3) and $g(i)$ (cf. lower signal path in Figure 3), respectively. The result of the summation is the speech estimate $\hat{s}(n)$ (cf. right-hand side of the MMSE estimator in Figure 3).

4.3 The prior

As discussed in Section 2, the *a priori* speech estimate is calculated as $\hat{s}^+(n) = \mathbf{a}^T \cdot \hat{\mathbf{s}}_{n-N_p}^{n-1} = f\left(\hat{\mathbf{x}}_0^{n-1}\right)$ (cf. (5)). This step is carried out in the block 'Predictor' in Figure 3. Just as in Section 3, the predictor incorporates the previously estimated speech samples, reflected by the delay unit 'T' in Figure 3. Assuming that the *a priori* speech estimate $\hat{s}^+(n)$ is a sufficient statistic for $s^{(i)}$ and using that each bit combination $\mathbf{x}^{(i)}$ can bijectively be mapped to one corresponding $s^{(i)}$, the prior turns out to be

$$P\left(\mathbf{x}^{(i)}\left|\hat{\mathbf{x}}_0^{n-1}\right.\right) = P\left(s^{(i)}\left|\hat{s}^+(n)\right.\right). \tag{31}$$

However, since $s^{(i)}$ is a quantized quantity, the probability of its ith value can be calculated as [34]

$$P(s^{(i)}|\hat{s}^+(n)) = \int_{I_i} p_{\tilde{e}}(s - \hat{s}^+(n))\, ds \tag{32}$$

with $p_{\tilde{e}}(\cdot)$ being the propagation error pdf and with the PCM sample quantization intervals I_i, $i = 0, 1, \ldots, 2^M-1$. Employing a specific *a priori* speech estimate $\hat{s}^+(n)$, the prior remains a function of summation index i as $f(i)$ and can be fed into the MMSE estimator (29) (cf. connection between 'Prior Computation' and 'MMSE Estimator' in Figure 3).

Please note that in [26], the quantity $E\left\{\left(s^+(n) - \hat{s}^+(n)\right)^2\right\}$ in (8) was assumed to be zero. Furthermore, assuming that the prediction error $e(n) = s(n) - s^+(n)$ is stationary, the propagation error pdf can be determined by a histogram measurement in a training process. Moreover, online integration of $p_{\tilde{e}}(\cdot)$ is not necessary if the integrations over I_i intervals are performed beforehand and $\hat{s}^+(n)$ is quantized with M bits, leading to discrete probabilities $P_{\tilde{e}}(\cdot)$ [26]. Thus, employing a lookup table containing the precomputed $P_{\tilde{e}}(\cdot)$ values in the block 'Prior Computation' in Figure 3, the table entries can be indexed by the quantized *a priori* speech estimate $\hat{s}^{+(i)}(n)$. Hence, the resulting prior $P(s^{(i)}\,|\hat{s}^{+(i)}(n))$ being a function of the summation index i can be obtained in a computationally efficient way.

Introducing again time variability, the prediction coefficients $\mathbf{a}$ in (5) have to be estimated which can be done recursively, e. g., by means of the NLMS algorithm [7]

$$\hat{\mathbf{a}}(n+1) = \hat{\mathbf{a}}(n) + \mu \cdot \frac{\hat{e}(n)}{||\hat{\mathbf{s}}_{n-N_p}^{n-1}||^2 + \Delta} \cdot \hat{\mathbf{s}}_{n-N_p}^{n-1} \quad (33)$$

with $\hat{e}(n) = \hat{s}(n) - \hat{s}^{+}(n)$ as well as with μ and Δ being the step size constant and the regularization parameter, respectively. Please note that the prediction coefficients can alternatively be obtained by a slightly modified NLMS algorithm [26,35].

5 Links between speech enhancement and error concealment

So far, we have introduced an application example of recursive MMSE estimation for both speech enhancement and error concealment. In this section, we aim at showing links between the presented estimators.

As can be seen above, while the estimator (14) used for speech enhancement (cf. Figure 2) utilizes continuous distributions, the estimator (29) employed for error concealment deals with discrete ones (cf. Figure 3). In error concealment, the samples of the signal to be transmitted are *quantized* typically by 16 bit or 24 bit and thus are from a finite set of elements, whereas in speech enhancement the digital signals are assumed to be quantized fine enough (even though quantization with 64, 32, or 16 bit may take place), therefore, the codomain of the samples is assumed to be *continuous*.

5.1 The likelihood

This also influences the channel model. While in speech enhancement, the noise is modeled on a *sample* (or *coefficient*) level (1), in digital transmission and error concealment the noise occurs on a modulation symbol level or, for BPSK equivalently, on bit level (26), (27): The transmitted binary source bits (BPSK: symbols) $x_m(n)$ are distorted by the real-valued channel noise $d_m(n)$. Using the received value (symbol) $y_m(n)$, the demodulator calculates a corresponding LLR $L(\hat{x}_m(n))$ by means of E_b/N_0, which is a normalized SNR measure being a function of the channel noise power. Therefore, the likelihood (28) being computed by means of the LLRs (cf. (26), (27), and (28)) is a function of the channel noise power as it is the case in speech enhancement (cf. (13)). Thus, in both speech enhancement and error concealment, the current observation ($Y_\ell(k)$ or $y_m(n)$) and the channel noise power (density) ($\sigma^2_{D,\ell}(k)$ or N_0) are needed for likelihood computation (cf. Figures 2 and 3).

However, there remains a distinct difference between speech enhancement and error concealment concerning the estimation of the noise power. In speech enhancement, the noise power is estimated by means of the noisy speech often assuming that noise is more stationary than speech. Typical noise power estimators are, e. g., approaches based on minimum statistics (MS) [30], (improved) minima-controlled recursive averaging ((I)MCRA) [31,36], or approaches based on speech presence probability [32]. In digital transmission or error concealment, however, the amount of the noise is dependent on the distance between the received symbol and all possibly transmitted symbols in the constellation diagram, the latter having fixed positions depending on the modulation scheme. Thus, implicitly, in error concealment one has more information about possible channel inputs which is a clear advantage over speech enhancement.

The likelihood in speech enhancement (13) is usually modeled by a common pdf, typically by a Gaussian which is more or less justified by the central limit theorem [3]. In error concealment, however, the noise pdf does not follow any typical distribution and depends on the employed bit mapping. For the further discussion, we define the transmission channel noise in error concealment similar to the noise in speech enhancement: The noise $d(n)$ in error concealment will be the difference between the hard-decided speech samples and the transmitted (quantized) one. The histogram of the transmission channel noise turns out to be spiky as can be seen on the right-hand side of Figure 4 for 16 bit uniform PCM quantization, natural-binary bit mapping, and BPSK transmission. The bottom and top spikes in the histograms and the higher amplitudes in the waveforms are evoked by bit errors at bit positions close to the most significant bit (MSB).

In the case of acoustic noise, increasing the noise power (decreasing the SNR) naturally results in higher noise signal levels and an increasing width of the time-domain noise histogram (cf. car noise example on the left hand-side of Figure 4). In the case of the transmission channel noise, with increasing noise power (decreasing SNR), more bit errors occur, resulting in higher peaks belonging to $d \neq 0$ in the histogram on the right-hand side of Figure 4. However, this increase of noise power scales the height of all high spikes in the histogram (referring to single bit errors) belonging to $d \neq 0$ approximately equally (cf. right-hand side of Figure 4). Thus, while in the case of acoustic noise the noise power is typically associated with the *width* of the noise pdf, in error concealment, the noise power can be related to the *height* of spikes in the noise histogram belonging to $d \neq 0$ (here: for uniform PCM quantization, natural-binary bit mapping, and BPSK modulation).

5.2 The estimator

Although the structure of the estimators in Sections 3 and 4 is very similar (cf. Figures 2 and 3), their implementation may considerably differ. In speech enhancement, the clean speech is estimated by an integral over the whole complex plane (14). The online numerical computation of this integral is typically hard to manage in practice due to the two-dimensional pdfs, however, employing a common

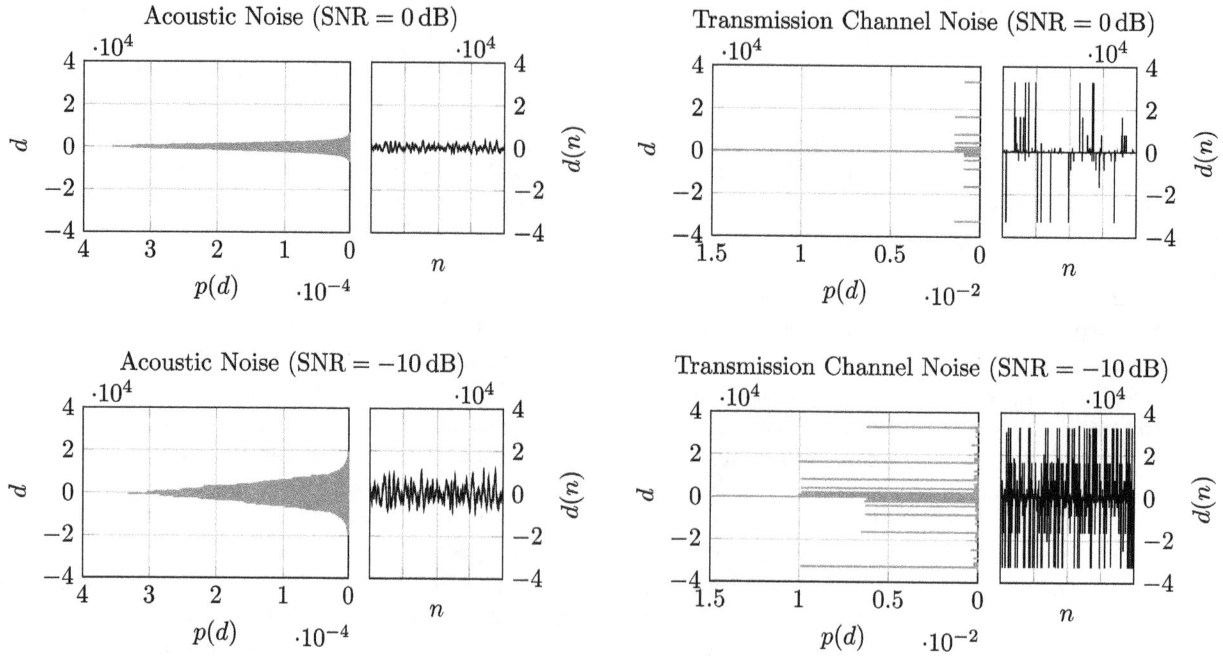

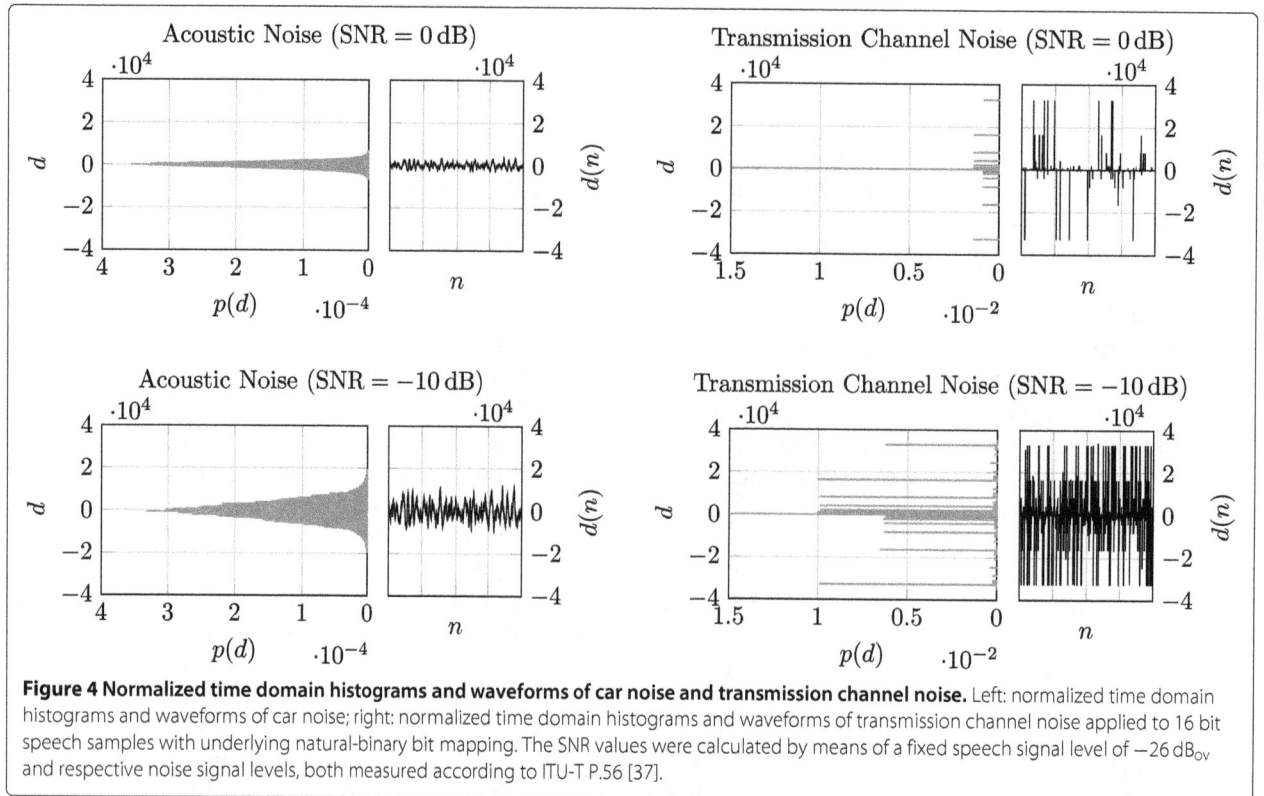

Figure 4 Normalized time domain histograms and waveforms of car noise and transmission channel noise. Left: normalized time domain histograms and waveforms of car noise; right: normalized time domain histograms and waveforms of transmission channel noise applied to 16 bit speech samples with underlying natural-binary bit mapping. The SNR values were calculated by means of a fixed speech signal level of −26 $\mathrm{dB_{ov}}$ and respective noise signal levels, both measured according to ITU-T P.56 [37].

distribution for the prior and the likelihood allows for a closed-form solution. Accordingly, a Gaussian assumption for both the prior and the likelihood results in the Kalman filter Eqs. (20), (21), (22), (23) (cf. Section 3).

In error concealment, the clean speech estimator (29) is a sum due to quantization and the respective finite number of transmittable bit combinations. This estimator is computationally complex but manageable in practice due to the one-dimensional pdfs (cf. discrete terms (28), (31)). Thus, the sum (29) can explicitly be computed at runtime [26]. Interestingly, a closed-form solution of the sum cannot be achieved due to the likelihood which cannot be approximated by a common pdf, as outlined in Section 5.1.

5.3 The prior

As can be seen in Section 2, the prior is the pdf of the propagation error (7). In Section 3, it was assumed that the propagation error is Gaussian distributed, therefore, the prior is modeled by a bivariate Gaussian (18) with a complex-valued argument. In [15], the Gaussian assumption is justified with the tradeoff between mathematical manageability and pdf model mismatch. It was shown in [16] that the histogram of the propagation error DFT coefficients differ from a Gaussian (cf. left-hand side of Figure 5). Therefore, in [16] the histogram of the propagation error was measured and a parametric pdf was trained and then employed as prior. Furthermore, since the speech estimates strongly depend on the channel, the propagation error also depends on the channel. Accordingly, in [38] the SNR dependency of the propagation error histogram was reported and an SNR-dependent estimator was proposed.

Although in [26] the variance $E\left\{\left(s^+ - \hat{s}^+\right)^2\right\}$ is assumed to be zero, the non-Gaussianity of the propagation error pdf $p_{\tilde{e}}\left(\tilde{e} = s - \hat{s}^+\right)$ was also observed in the context of error concealment. As can be seen on the right-hand side of Figure 5, $p_{\tilde{e}}$ in (32), being measured in a training process, turns out to be rather super-Gaussian. Furthermore, in [26] the dependency of $p_{\tilde{e}}\left(\tilde{e} = s - \hat{s}^+\right)$ on $\hat{s}^+$ was investigated revealing different shapes and variances dependent on the amplitude of $\hat{s}^+$, while $\hat{s}^+ = s^+$ was assumed there.

The propagation error in error concealment is quasi-continuous, while the prior is discrete. Thus, an integration step (32) is needed in order to discretize the propagation error pdf to obtain the prior. Fortunately, this discretization step can be done during a training process, resulting in a lookup table which nicely reduces the computational complexity [26].

Please note that usually the NLMS algorithm is employed for calculating the prediction coefficients both in speech enhancement and in error concealment due to its robustness and low computational complexity. However, there are other algorithms which can also be utilized,

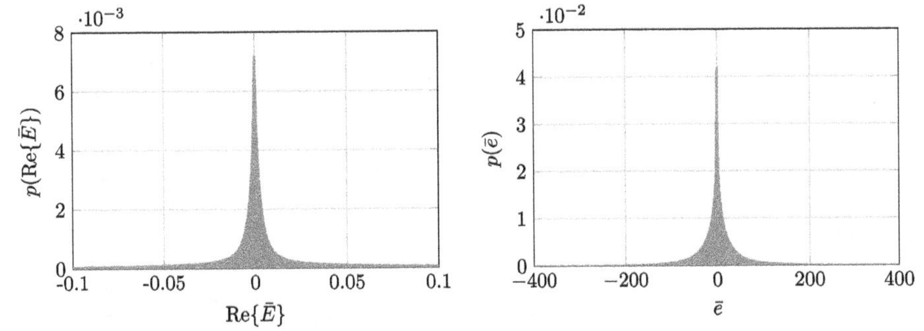

Figure 5 Propagation error histograms measured in a speech enhancement system and in an error concealment system. Left: normalized frequency domain histogram of the propagation error $\bar{E}_\ell(k) = S_\ell(k) - \widehat{S}_\ell^+(k)$ measured in a speech enhancement system from Section 3; right: normalized time domain histogram of the propagation error $\bar{e}(n) = s(n) - \hat{s}^+(n)$ measured in an error concealment system from Section 4 assuming that $E\{(s^+ - \hat{s}^+)^2\}$ is zero [26].

as reported in [16]. Please note that the propagation error is also dependent on the algorithm for determining the prediction coefficients, therefore, the change of the algorithm involves a new propagation error pdf training in both disciplines.

5.4 Outlook

In this section, we briefly sketch further possible research directions for speech enhancement inspired by error concealment. Since one of the key success factors in error concealment is that bit reliability information is exploited, the speech enhancement approach from Section 3 could benefit from using bit likelihoods. This means that instead of (13) the DFT coefficient likelihood (12) is calculated by means of bit likelihoods as in error concealment (cf. (27) and (28)).

In the following, we will estimate the real and imaginary parts of the complex-valued speech DFT coefficients separately, as in, e. g., [10]. Assuming that the resulting real-valued real and imaginary parts of STFT-domain quantities are quantized, their natural-binary representation is possible. In the following, we will introduce the processing steps for the real part only, denoted by superscript 're'. Of course, the imaginary part can be treated in the same way employing the same processing steps.

In order to be able to employ a bit-level model for the acoustic channel, we assume that the real part of the speech and the noisy speech DFT coefficients are quantized by M bit. Therefore, those quantities can bijectively be mapped into the bit combinations $\mathbf{S}_\ell^{\mathrm{re}}(k) = \left[S_{0,\ell}^{\mathrm{re}}(k), S_{1,\ell}^{\mathrm{re}}(k), \ldots, S_{m,\ell}^{\mathrm{re}}(k), \ldots, S_{M-1,\ell}^{\mathrm{re}}(k)\right]$ and $\mathbf{Y}_\ell^{\mathrm{re}}(k) = \left[Y_{0,\ell}^{\mathrm{re}}(k), Y_{1,\ell}^{\mathrm{re}}(k), \ldots, Y_{m,\ell}^{\mathrm{re}}(k), \ldots, Y_{M-1,\ell}^{\mathrm{re}}(k)\right]$, respectively. Due to the fact that speech is distorted by acoustic noise while passing through the acoustic channel, the observed bit combinations at the acoustic channel output $\mathbf{Y}_\ell^{\mathrm{re}}(k)$ may differ from those at the channel input $\mathbf{S}_\ell^{\mathrm{re}}(k)$. Accordingly, a bit error at bit position $m \in \{0, 1, \ldots, M-1\}$ occurs if the received bit $Y_{m,\ell}^{\mathrm{re}}(k)$ is not equal to the transmitted one $S_{m,\ell}^{\mathrm{re}}(k)$. The bit error rate $\mathrm{BER}_m^{\mathrm{re}}(k)$ can be measured within a training process by comparing $S_{m,\ell}^{\mathrm{re}}(k)$ to $Y_{m,\ell}^{\mathrm{re}}(k)$ for all bit positions m, individually for each frequency bin k. Please note that the bit errors also depend on the local SNR in the current time-frequency unit (ℓ, k), therefore, the SNR has to be taken into account during the training process. Accordingly, the training steps can be summarized as follows: Using speech and car noise data we generated noisy speech signals at different signal SNR levels. Then, we calculated the short-time spectra of the clean speech, the noise, and the noisy speech signals resulting in $S_\ell(k)$, $D_\ell(k)$, and $Y_\ell(k) = S_\ell(k) + D_\ell(k)$, respectively. Using the resulting DFT coefficients, we calculated the true speech power $\sigma_{S,\ell}^2(k) = E\left\{|S_\ell(k)|^2\right\}$ and the true noise power $\sigma_{D,\ell}^2(k) = E\left\{|D_\ell(k)|^2\right\}$. Using those two power spectra, we obtained the true *a priori* SNR as $\xi_\ell(k) = \sigma_{S,\ell}^2(k)/\sigma_{D,\ell}^2(k)$. Then, we quantized the real part of the clean speech and noisy speech DFT coefficients by 16 bit resulting in the bit combinations $\mathbf{S}_\ell^{\mathrm{re}}(k)$ and $\mathbf{Y}_\ell^{\mathrm{re}}(k)$, respectively. By this means, the whole training data was processed and $S_{m,\ell}^{\mathrm{re}}(k) \in \mathbf{S}_\ell^{\mathrm{re}}(k)$ and $Y_{m,\ell}^{\mathrm{re}}(k) \in \mathbf{Y}_\ell^{\mathrm{re}}(k)$ were compared to each other. Bit errors $S_{m,\ell}^{\mathrm{re}}(k) \neq Y_{m,\ell}^{\mathrm{re}}(k)$ were counted at bit position m, frequency bin k, and in dependence of the *a priori* SNR. For the latter, the ideal *a priori* SNR $\xi_\ell(k)$ was quantized resulting in discrete *a priori* SNR values ξ_q.

The resulting bit error rates $\mathrm{BER}_m^{\mathrm{re}}(k, \xi_q)$ were stored in a lookup table; examples can be seen in Figure 6: As expected, at higher SNRs less bits are in error. Typical for car noise, at higher frequencies less bits are in error as compared to lower frequencies.

A speech enhancement approach as described in Section 3 can be modified to include bit likelihoods which can be calculated by the bit error rates from the previous training step. The resulting frequency-dependent

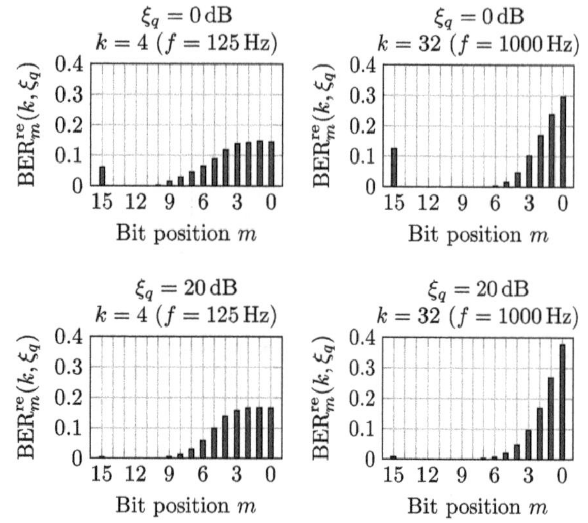

Figure 6 Bit error probabilities $\text{BER}_m^{\text{re}}(k, \xi_q)$ for different *a priori* SNRs ξ_q and frequencies k. Natural-binary bit representation with $M = 16$ bit quantization and the MSB (bit position $M-1 = 15$) being the sign bit is used.

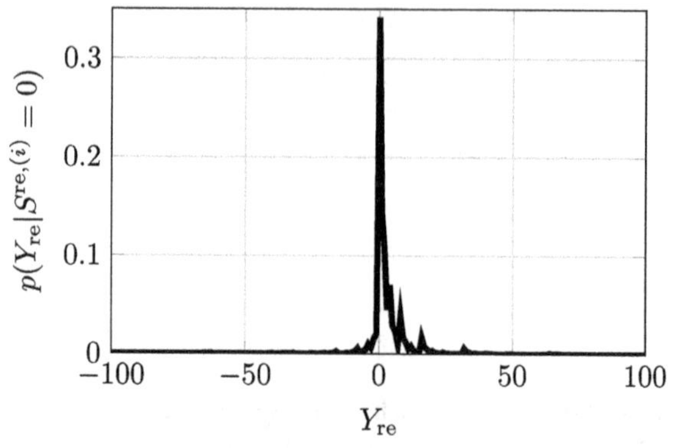

Figure 7 Coefficient likelihood example. Coefficient likelihood $P(\mathbf{Y}_\ell^{\text{re}}(k)|S^{\text{re},(i)})$ for the specific channel input $S^{\text{re},(i)} = 0$ calculated by (34) and (35) using trained bit error probabilities $\text{BER}_m^{\text{re}}(k, \xi_q)$.

lookup tables are then addressed by a quantized *a priori* SNR estimate $\hat{\xi}_{q,\ell}(k)$, computed, e. g., by the well-known decision-directed *a priori* SNR estimator [3] and the same quantization intervals as for the training process, resulting in $\text{BER}_m^{\text{re}}(k, \hat{\xi}_{q,\ell}(k))$ values. Since each transmitter-sided DFT coefficient $S_\ell(k)$ is quantized, it assumes a discrete value $S^{\text{re},(i)}$ with $i \in \{0, 1, \ldots, 2^M - 1\}$. Furthermore, each $S^{\text{re},(i)}$ can be mapped to one corresponding bit combination $\mathbf{S}^{\text{re},(i)}$. Then, using the resulting bit error rate $\text{BER}_m^{\text{re}}\left(k, \hat{\xi}_{q,\ell}(k)\right)$, the bit likelihood with the given bit $S_m^{\text{re},(i)} \in \mathbf{S}^{\text{re},(i)}$ of the ith quantization table entry can be obtained similar to (27) as

$$P\left(Y_{m,\ell}^{\text{re}}(k) \middle| S_m^{\text{re},(i)}\right) = \begin{cases} \text{BER}_{m,\ell}^{\text{re}}\left(k, \hat{\xi}_{q,\ell}(k)\right), & \text{if } Y_{m,\ell}^{\text{re}}(k) \neq S_m^{\text{re},(i)}, \\ 1 - \text{BER}_{m,\ell}^{\text{re}}\left(k, \hat{\xi}_{q,\ell}(k)\right), & \text{else.} \end{cases} \tag{34}$$

The bit likelihood describes the probability of an observed bit $Y_{m,\ell}^{\text{re}}(k)$ given a possible channel input $S_m^{\text{re},(i)}$. The coefficient likelihood can be obtained using the bit likelihoods similar to (28) as

$$P\left(\mathbf{Y}_\ell^{\text{re}}(k) \,\middle|\, \mathbf{S}^{\text{re},(i)}\right) = \prod_{m=0}^{M-1} P\left(Y_{m,\ell}^{\text{re}}(k) \,\middle|\, S_m^{\text{re},(i)}\right). \tag{35}$$

By this means, the simple Gaussian assumption for the noise DFT coefficients (13) can be replaced by such a new approach which allows for an environment-specific processing (cf. [39]). This is also illustrated in Figure 7 for an *a priori* SNR of 20 dB and the frequency bin $k = 32$ (corresponding to $f = 1\,\text{kHz}$): As can be seen, the resulting likelihood is a sharp pdf unlike a Gaussian pdf. The Gaussian assumption for noise is typically justified with the central limit theorem assuming that the span of correlation of the noise samples is sufficiently short compared to the frame length [40]. Although this assumption is better fulfilled by a wide range of noise types than by speech signals, it can generally be said that it is not fulfilled perfectly in practice by noise signals. Accordingly, the likelihood turns out to be a sharper pdf than a Gaussian such as in Figure 7. Please note that the non-Gaussianity of noise DFT coefficients was also reported in [41]. Furthermore, there are publications dealing with speech estimators based on a non-Gaussian assumption for the noise, e. g., [10,12]. Moreover, it was shown in these papers that a more realistic likelihood function (just as the proposed one obtained by training) offers a more precise (acoustic) channel model which can improve estimation results.

The speech prior $P\left(\mathbf{S}^{\text{re},(i)} \mid S_\ell^{+,\text{re}}(k)\right)$ (cf. (17)) can be obtained by integrating (18) according to the PCM quantization intervals (i) (cf. (32)). Here, $S_\ell^{+,\text{re}}(k)$ is the real part of the *a priori* speech estimate $S_\ell^{+}(k)$ calculated by, e. g., (16). Using the coefficient likelihood (35) and a speech prior, the recursive MMSE estimation formula turns out to be (cf. (14) and (29))

$$\widehat{S}_\ell^{\text{re}}(k) = \frac{\sum\limits_{i=0}^{2^M-1} S^{\text{re},(i)} \cdot P\left(\mathbf{Y}_\ell^{\text{re}}(k) \mid \mathbf{S}^{\text{re},(i)}\right) \cdot P\left(\mathbf{S}^{\text{re},(i)} \mid S_\ell^{+,\text{re}}(k)\right)}{\sum\limits_{i=0}^{2^M-1} P\left(\mathbf{Y}_\ell^{\text{re}}(k) \mid \mathbf{S}^{\text{re},(i)}\right) \cdot P\left(\mathbf{S}^{\text{re},(i)} \mid S_\ell^{+,\text{re}}(k)\right)}. \tag{36}$$

The imaginary part of the speech estimate $\widehat{S}_\ell^{\text{im}}(k)$ can be obtained in a similar way as $\widehat{S}_\ell^{\text{re}}(k)$ and the final (complex-valued) speech estimate is calculated by $\widehat{S}_\ell(k) = \widehat{S}_\ell^{\text{re}}(k) + j\widehat{S}_\ell^{\text{im}}(k)$. The *a priori* speech estimate $S_\ell^{+}(k)$ is gained by (16) which is then used for the update step.

6 Conclusions

This paper provides new insights into links between speech enhancement and error concealment based on recursive MMSE estimation. It turns out that recent approaches to bit error concealment based on a predictor are well comparable to iterative approaches in speech enhancement, such as the Kalman filter approach. The main difference between both disciplines are the channel model and the noise pdf. In error concealment, powerful bit reliability information can be exploited in order to obtain robust estimation results, while in speech enhancement the channel is modeled on a sample level without reliable reference. On the other hand, the autoregressive model of speech and the prior computation are well comparable in both disciplines. Finally, some new research directions are identified for speech enhancement, inspired by error concealment.

Endnotes

[a]Please note that the presented signal model and the equations are valid in analogy also in the STFT domain.

[b]Please note that for binary phase-shift keying (BPSK) modulation, the LLR is obtained by $L(\hat{x}_m(n)) = 4 \cdot E_b/N_0 \cdot y_m(n)$ [33].

Appendix

For ease of readability, we will omit the indices ℓ and k in this section. In this appendix, we aim at showing that assuming a Gaussian distribution for the propagation error and the acoustic noise, the recursive MMSE estimator (14) turns out to be the Kalman filter as in (20), (21), (22), and (23). Employing (17) and (12), (14) turns out to be

$$\widehat{S} = \frac{\int_{\mathbb{C}} S \cdot p_{\bar{E}}(S - \widehat{S}^+) \cdot p_D(Y - S)\, dS}{\int_{\mathbb{C}} p_{\bar{E}}(S - \widehat{S}^+) \cdot p_D(Y - S)\, dS}. \tag{37}$$

Introducing a new integration variable $\bar{E} = S - \widehat{S}^+$, (37) can be rewritten as

$$\begin{aligned}\widehat{S} &= \widehat{S}^+ + \frac{\int_{\mathbb{C}} \bar{E} \cdot p_{\bar{E}}\left(\bar{E}\right) \cdot p_D\left(Y - \bar{E} - \widehat{S}^+\right) d\bar{E}}{\int_{\mathbb{C}} p_{\bar{E}}\left(\bar{E}\right) \cdot p_D\left(Y - \bar{E} - \widehat{S}^+\right) d\bar{E}} \\ &= \widehat{S}^+ + \widehat{E}.\end{aligned} \tag{38}$$

As can be seen, the recursive MMSE estimator turns out to be the sum of the *a priori* speech estimate $\widehat{S}^+$ and a fraction with each an integral in the numerator and denominator $\widehat{E} = f\left(Y - \widehat{S}^+\right)$. Please note that this fraction is a classical (non-recursive) MMSE estimator, however, with an extra term '$-\widehat{S}^+$' in the pdf $p_D(\cdot)$. Assuming a Gaussian distribution for $p_{\bar{E}}(\cdot)$ and $p_D(\cdot)$, this classical MMSE estimator turns out to be the Wiener filter as we will see later.

Employing (13) for $p_D(\cdot)$ and (18) for $p_{\bar{E}}(\cdot)$ as well as canceling the constant factors to the exponential functions, the fraction in (38) turns out to be

$$\widehat{E} = \frac{\int_{\mathbb{C}} \bar{E} \cdot e^{-\frac{|\bar{E}|^2}{\sigma_{\bar{E}}^2}} \cdot e^{-\frac{|Y - \bar{E} - \widehat{S}^+|^2}{\sigma_D^2}}\, d\bar{E}}{\int_{\mathbb{C}} e^{-\frac{|\bar{E}|^2}{\sigma_{\bar{E}}^2}} \cdot e^{-\frac{|Y - \bar{E} - \widehat{S}^+|^2}{\sigma_D^2}}\, d\bar{E}}. \tag{39}$$

Employing polar integration with $\bar{E} = |\bar{E}| e^{j\epsilon}$ and $d\bar{E} = |\bar{E}|\, d|\bar{E}|\, d\epsilon$, as well as employing $R = Y - \widehat{S}^+ = |R| e^{j\rho}$, we obtain

$$\widehat{E} = \frac{\int_0^\infty \int_0^{2\pi} |\bar{E}|^2 e^{j\epsilon} \cdot e^{-\frac{|\bar{E}|^2}{\sigma_{\bar{E}}^2}} \cdot e^{-\frac{|\bar{E}|^2 + |R|^2 - 2|\bar{E}||R|\cos(\epsilon - \rho)}{\sigma_D^2}}\, d\epsilon\, d|\bar{E}|}{\int_0^\infty \int_0^{2\pi} |\bar{E}| \cdot e^{-\frac{|\bar{E}|^2}{\sigma_{\bar{E}}^2}} \cdot e^{-\frac{|\bar{E}|^2 + |R|^2 - 2|\bar{E}||R|\cos(\epsilon - \rho)}{\sigma_D^2}}\, d\epsilon\, d|\bar{E}|}. \tag{40}$$

Integrating with respect to ϵ using [42], (40) turns out to be

$$\widehat{E} = \frac{e^{j\rho} \int_0^\infty |\bar{E}|^2 \cdot e^{-|\bar{E}|^2 \left[\frac{1}{\sigma_{\bar{E}}^2} + \frac{1}{\sigma_D^2}\right]} \cdot I_1\left(\frac{2|\bar{E}||R|}{\sigma_D^2}\right) d|\bar{E}|}{\int_0^\infty |\bar{E}| \cdot e^{-|\bar{E}|^2 \left[\frac{1}{\sigma_{\bar{E}}^2} + \frac{1}{\sigma_D^2}\right]} \cdot I_0\left(\frac{2|\bar{E}||R|}{\sigma_D^2}\right) d|\bar{E}|} \tag{41}$$

with $I_0(\cdot)$ and $I_1(\cdot)$ being the modified Bessel function of zeroth and first order, respectively. Integrating with respect to $|\bar{E}|$ using [42], (41) turns out to be

$$\widehat{E} = \frac{\sigma_{\bar{E}}^2}{\sigma_{\bar{E}}^2 + \sigma_D^2} \cdot |R| e^{j\rho} = \frac{\sigma_{\bar{E}}^2 \sigma_D^2}{\sigma_{\bar{E}}^2 \sigma_D^2 + 1} \cdot \left(Y - \widehat{S}^+\right). \tag{42}$$

Thus, substituting (42) for $\widehat{E}$ in (38) and defining $\zeta = \sigma_{\bar{E}}^2 / \sigma_D^2$ results in (cf. (20)-(23))

$$\widehat{S} = \widehat{S}^+ + \frac{\zeta}{1 + \zeta} \cdot \left(Y - \widehat{S}^+\right). \tag{43}$$

Competing interests
The authors declare that they have no competing interests.

Acknowledgements
The authors would like to thank Gerald Enzner for inspiring initial discussions toward this paper.

References
1. Trees Van HL, *Detection, Estimation and Modulation Theory. Vol 1. Detection, Estimation and Linear Modulation Theory*. (Wiley, Hoboken, NJ, USA, 1968)
2. S Kay, *Fundamentals of Statistical Signal Processing. Vol 1. Estimation Theory*. (Prentice Hall, Upper Saddle River, NJ, USA, 1993)
3. Y Ephraim, D Malah, Speech enhancement using a minimum-mean square error short-time spectral amplitude estimator. IEEE Trans. Acoustics Speech Signal Process. **32**(6), 1109–1121 (1984)

4. L Rabiner, R Schafer, *Digital Processing of Speech Signals*. (Prentice Hall, Upper Saddle River, NJ, USA, 1978)
5. P Vary, R Martin, *Digital Speech Transmission: Enhancement, Coding and Error Concealment*. (Wiley, Hoboken, NJ, USA, 2006)
6. R Kalman, A new approach to linear filtering and prediction problems. Trans. ASME J. Basic Eng. **82**, 35–45 (1960)
7. S Haykin, *Adaptive Filter Theory*. (Prentice Hall, Upper Saddle River, NJ, USA, 2002)
8. R McAulay, M Malpass, Speech enhancement using a soft-decision noise suppression filter. IEEE Trans. Acoustics Speech Signal Process. **28**(2), 137–145 (1980)
9. Y Ephraim, D Malah, Speech enhancement using a minimum mean-square error log-spectral amplitude estimator. IEEE Trans. Acoustics Speech Signal Process. **33**(2), 443–445 (1985)
10. R Martin, in *Proc. of IEEE International Conference on Acoustics, Speech, and Signal Processing (ICASSP)*. Speech enhancement using MMSE short time spectral estimation with gamma distributed speech priors, vol. 1 (Orlando, FL, USA, 2002), pp. 253–256
11. I Andrianakis, PR White, in *Proc. of IEEE International Conference on Acoustics, Speech, and Signal Processing (ICASSP)*. MMSE speech spectral amplitude estimators with chi and gamma speech priors, vol. 3 (Toulouse, France, 2006), pp. 1068–1071
12. JS Erkelens, RC Hendriks, R Heusdens, J Jensen, Minimum mean-square error estimation of discrete fourier coefficients with generalized gamma priors. IEEE Trans Audio Speech Lang. Process. **15**(6), 1741–1752 (2007)
13. P Scalart, JV Filho, in *Proc. of IEEE International Conference on Acoustics, Speech, and Signal Processing (ICASSP)*. Speech enhancement based on a priori signal to noise estimation, vol. 2 (Atlanta, GA, USA, 1996), pp. 629–632
14. KK Paliwal, A Basu, in *Proc. of IEEE International Conference on Acoustics, Speech, and Signal Processing (ICASSP)*. A speech enhancement method based on Kalman filtering (Dallas, TX, USA, 1987), pp. 177–180
15. E Zavarehei, S Vaseghi, in *Proc. of ISCA INTERSPEECH 2005*. Speech enhancement in temporal DFT trajectories using Kalman filters (Lisbon, Portugal, 2005), pp. 2077–2080
16. T Esch, P Vary, in *Proc. of IEEE International Conference on Acoustics, Speech, and Signal Processing (ICASSP)*. Speech enhancement using a modified Kalman filter based on complex linear prediction and supergaussian priors (Las Vegas, NV, USA, 2008), pp. 4877–4880
17. T Esch, P Vary, in *Proc. of International Workshop on Acoustic Echo and Noise Control (IWAENC)*. Modified Kalman filter exploiting interframe correlation of speech and noise magnitudes (Seattle, WA, USA, 2008), pp. 1–4
18. N Görtz, in *Proc. of IEEE International Symposium on Information Theory*. Joint source channel decoding using bit-reliability information and source statistics (Cambridge, MA, USA, 1998), p. 9
19. T Fingscheidt, P Vary, Softbit speech decoding: a new approach to error concealment. IEEE Trans. Speech Audio Process. **9**(3), 240–251 (2001)
20. F Lahouti, AK Khandani, Soft reconstruction of speech in the presence of noise and packet loss. IEEE Trans. Audio Speech Lang. Process. **15**(1), 44–56 (2007)
21. AM Pourmir, F Lahouti, in *Proc. of 16th International Conference on Software, Telecommunications, and Computer Networks (SoftCOM)*. Joint source channel speech decoding using long-term residual redundancy (Split, Croatia, 2008), pp. 329–333
22. AJ Jameel, H Adnan, Y Xiaohu, A Hussain, in *Proc. of Developments in eSystems Engineering (DeSE)*. Error concealment of EVRC speech decoder using residual redundancy (Abu Dhabi, United Arab Emirates, 2009), pp. 84–88
23. S Han, F Pflug, T Fingscheidt, in *Proc. of 21th European Signal Processing Conference (EUSIPCO)*. Improved AMR wideband error concealment for mobile communications (Marrakech, Morocco, 2013), pp. 1–5
24. M Adrat, J Spittka, S Heinen, P Vary, in *Proc. of IEEE Workshop on Speech Coding (SCW)*. Error concealment by near optimum MMSE-estimation of source codec parameters (Delavan, WI, USA, 2000), pp. 84–86
25. F Pflug, T Fingscheidt, in *Proc. of 134th International Audio Engineering Society (AES) Convention*. Delayless robust DPCM audio transmission for digital wireless microphones (Rome, Italy, 2013), pp. 1–8
26. F Pflug, T Fingscheidt, Robust ultra-low latency soft-decision decoding of linear PCM audio. IEEE Trans. Audio Speech Lang. Process. **21**(11), 2324–2336 (2013)
27. A Papoulis, U Pillai, *Probability, Random Variables and Stochastic Processes*, 4th edn. (McGraw-Hill, New York, NY, USA, 2002)
28. T Esch, Model-based speech enhancement exploiting temporal and spectral dependencies. PhD thesis, Rheinisch-Westfälische Technische Hochschule Aachen, Aachen, Germany (2012). http://darwin.bth.rwth-aachen.de/opus3/volltexte/2012/4035/pdf/4035.pdf
29. R Martin, PU Heute, Eds Antweiler C, *Advances in Digital Speech Transmission*. (Wiley, Hoboken, NJ, USA, 2008)
30. R Martin, Noise power spectral density estimation based on optimal smoothing and minimum statistics. IEEE Trans. Speech Audio Process. **9**(5), 504–512 (2001)
31. I Cohen, Noise spectrum estimation in adverse environments: Improved minima controlled recursive averaging. IEEE Trans. Speech Audio Process. **11**(5), 466–475 (2003)
32. T Gerkmann, RC Hendriks, Unbiased MMSE-based noise power estimation with low complexity and low tracking delay. IEEE Trans. Audio Speech Lang. Process. **20**(4), 1383–1393 (2012)
33. J Hagenauer, Source-controlled channel decoding. IEEE Trans. Commun. **43**(9), 2449–2457 (1995)
34. T Fingscheidt, P Vary, in *Proc. IEEE International Conference on Acoustics, Speech, and Signal Processing (ICASSP)*. Robust speech decoding: A universal approach to bit error concealment, vol. 3 (Munich, Germany, 1997), pp. 1667–1670
35. GDT Schuller, B Yu, D Huang, B Edler, Perceptual audio coding using adaptive pre- and post-filters and lossless compression. IEEE Trans. Speech Audio Process. **10**(6), 379–390 (2002)
36. I Cohen, B Berdugo, Noise estimation by minima controlled recursive averaging for robust speech enhancement. IEEE Signal Process. Lett. **9**(1), 12–15 (2002)
37. Telecommunication Standardization Sector of the International Telecommunication Union (ITU-T). Recommendation ITU-T P.56, Objective Measurement of Active Speech Level (1993). http://www.itu.int/rec/T-REC-P/
38. T Esch, P Vary, in *Proc. of IEEE International Conference on Acoustics, Speech, and Signal Processing (ICASSP)*. Model-based speech enhancement using SNR dependent MMSE estimation (Prague, Czech Republic, 2011), pp. 4073–4076
39. T Fingscheidt, S Suhadi, S Stan, Environment-optimized speech enhancement. IEEE Trans. Audio Speech Lang. Process. **16**(4), 825–834 (2008)
40. T Lotter, P Vary, Speech enhancement by MAP spectral amplitude estimation using a super-gaussian speech model. EURASIP J. Appl. Signal Process. **7**, 1110–1126 (2005)
41. B Fodor, T Fingscheidt, in *Proc. of European Signal Processing Conference (EUSIPCO)*. MMSE Speech Spectral Amplitude Estimation Assuming Non-Gaussian Noise (Barcelona, Spain, 2011), pp. 2314–2318
42. IS Gradshteyn, IM Ryzhik, *Table of Integral, Series, and Products*, 4th edn. (Academic Press, New York, NY, USA, 1965)

4

Systematic analysis of uncertainty principles of the local polynomial Fourier transform

Xiumei Li[1*] and Guoan Bi[2]

Abstract

In this paper, we show that there are a number of uncertainty principles for the local polynomial Fourier transform and local polynomial periodogram. Systematic analysis of uncertainty principles is given, explicit expressions of the uncertainty relations are derived, and an example using the chirp signal and the Gaussian window function is given to verify the expressions.

1 Introduction

Time-frequency representations are of significant importance to better describe time-varying signals, i.e, signals with time-varying frequencies. Among the representations, the short-time Fourier transform (STFT) is the simplest and easiest one to implement. However, because the STFT assumes that the frequencies within a signal segment are not changing with time, the resolution in the time and/or frequency domain is often limited for practical applications. To overcome this drawback, the local polynomial Fourier transform (LPFT), as a generalized form of the STFT, was proposed [1]. The kernel of the LPFT uses extra parameters to approximate the signal's phase into a polynomial form. Therefore, the LPFT can describe the time-varying signals with better accuracy, and the resolution representing the signal components in the time-frequency domain can be significantly improved compared to that achieved by the STFT.

The uncertainty principle plays an important role in signal processing [2]. In general, the more concentrated the signal is, the wider band its Fourier transform occupies. It is impossible to arbitrarily concentrate both a time-domain signal and its Fourier transform. This trade-off can be formalized in the form of the uncertainty principle. Similarly, the time-frequency concentration of the transforms belonging to the Cohen class is restricted and related to an uncertainty principle [3]. The STFT is also limited by the uncertainty principle [4], and it is understood that a shorter window used to capture the signal segment leads to a poor resolution to represent the signal in the frequency domain, and vice verse. It is not possible to arbitrarily increase the resolution in both domains simultaneously. The standard formulation of the uncertainty principle, known as the global uncertainty principle, is in terms of global standard deviations to involve the entire time range and the entire frequency range of the signal. With regard to the local behavior of the signal, the local uncertainty principle is invoked to present the uncertainty limits on local signal, by defining local quantities as conditional standard deviations [5]. The conditional standard deviations can be considered as the local widths or measures of the local spread in the time and/or frequency domain. The local uncertainty product of the spectrogram and a large class of bilinear time-frequency distributions were considered [5]. It shows that the local uncertainty product of the spectrogram has a lower bound due to the windowing approach and cannot be arbitrarily small, while for a large class of bilinear time-frequency distributions, the local uncertainty product is always less than or equal to the global uncertainty product and can be arbitrarily small.

It has been observed that the resolution of the LPFT in the time-frequency domain is influenced by the window length which controls the trade-off of bias and variance [1,6]. A comprehensive study on the uncertainty principle for the LPFT has been reported in [7]. It was shown that when the Gaussian window is used to segment the signal, the uncertainty product of the LPFT is time-independent if the polynomial coefficients are estimated correctly. However, the work reported in [7] was mainly

*Correspondence: lixiumei@pmail.ntu.edu.sg
[1]School of Information Science and Engineering, 58 Haishu Road, Hangzhou Normal University, 311121 Hangzhou, China
Full list of author information is available at the end of the article

focused on the uncertainty product obtained by multiplying the duration and bandwidth of the local signal. Other kinds of uncertainty products of the LPFT, such as the global uncertainty products, have not been reported in the literature.

In this paper, systematic analysis regarding the uncertainty principles of the LPFT is given. Several kinds of uncertainty principles are discussed, including the global uncertainty principle, the uncertainty principle of local duration and conditional standard deviation, the uncertainty principle of local bandwidth and conditional standard deviation, and the uncertainty principle of the conditional standard deviations in time and frequency.

The rest of the paper is organized as follows. After the review on the uncertainty principles of the STFT in Section 2, the characteristic functions of the second-order local polynomial periodogram and the uncertainty principles of the second-order LPFT are discussed in Section 3. Section 4 presents an example of the uncertainty principles of the second-order LPFT by using the chirp signals. Section 5 discusses the uncertainty principles of the Mth-order LPFT. Finally, conclusions are given in Section 6.

2 Review on the uncertainty principles of STFT

The uncertainty principles of the STFT were derived in [8], and the definitions and equations will be reviewed in this section.

Let $h(t)$ represent a window function that segments an input signal $s(t)$. By multiplying the input signals with the window function that is peaked around time t, the local signal is defined as

$$s_t(\tau) = s(\tau)h(\tau - t). \tag{1}$$

The normalized local signal at the time instant t is

$$\eta_t(\tau) = \frac{s(\tau)h(\tau - t)}{\sqrt{P(t)}}, \tag{2}$$

where $P(t) = \int |s(\tau)h(\tau - t)|^2 d\tau$. For simplicity in the rest of the paper, the integral without limits implies that the integration is from $-\infty$ to ∞.

The STFT is the Fourier transform of the local signal

$$\begin{aligned} S_t(\omega) &= \frac{1}{\sqrt{2\pi}} \int s_t(\tau)e^{-j\omega(\tau-t)} d\tau \\ &= \frac{1}{\sqrt{2\pi}} \int s(\tau)h(\tau - t)e^{-j\omega(\tau-t)} d\tau. \end{aligned} \tag{3}$$

Similarly, the local spectrum is defined as

$$F_\omega(w) = S(w)H(\omega - w), \tag{4}$$

and the normalized local spectrum is

$$\mu_\omega(w) = \frac{S(w)H(\omega - w)}{\sqrt{P(\omega)}}, \tag{5}$$

where $P(\omega) = \int |S(w)H(\omega - w)|^2 dw$.

The short-frequency time transform is

$$s_\omega(t) = \frac{1}{\sqrt{2\pi}} \int S(w)H(\omega - w)e^{jwt} dw. \tag{6}$$

The Fourier transform pairs of the signal and window are normalized as follows:

$$\begin{aligned} S(\omega) &= \frac{1}{\sqrt{2\pi}} \int s(t)e^{-j\omega t} dt, \\ H(\omega) &= \frac{1}{\sqrt{2\pi}} \int h(t)e^{-j\omega t} dt. \end{aligned} \tag{7}$$

The spectrum of the two-dimensional time-frequency distribution, the spectrogram $P_{SP}(t, \omega)$, is defined as

$$P_{SP}(t, \omega) = |S_t(\omega)|^2 = |s_\omega(t)|^2, \tag{8}$$

with its marginals expressed as

$$\begin{aligned} P_{SP}(t) &= \int P_{SP}(t, \omega) d\omega, \\ P_{SP}(\omega) &= \int P_{SP}(t, \omega) dt. \end{aligned} \tag{9}$$

2.1 Properties of the local signal and spectrum

The mean time and duration of the normalized local signal in (2) are

$$\begin{aligned} \langle \tau \rangle_t &= \int \tau |\eta_t(\tau)|^2 d\tau, \\ T_t^2 &= \int (\tau - \langle \tau \rangle_t)^2 |\eta_t(\tau)|^2 d\tau. \end{aligned} \tag{10}$$

Similarly, from the windowed spectrum, the local frequency and local bandwidth can be defined as follows:

$$\begin{aligned} \langle w \rangle_\omega &= \int w |\mu_\omega(w)|^2 dw, \\ B_\omega^2 &= \int (w - \langle w \rangle_\omega)^2 |\mu_\omega(w)|^2 dw. \end{aligned} \tag{11}$$

These quantities pertain to the local signal and local spectrum. They should not be confused with the local properties of the spectrogram which is to be considered in the next subsection.

2.2 Local and global quantities for the spectrogram

2.2.1 *Local quantities*

The mean averages and conditional standard deviations for the spectrogram are [8]

$$
\begin{aligned}
\langle\omega\rangle_t &= \frac{1}{P(t)}\int \omega P_{SP}(t,\omega)d\omega \\
&= \int \eta_t^*(\tau)\frac{1}{j}\frac{d}{d\tau}\eta_t(\tau)d\tau \\
\langle t\rangle_\omega &= \frac{1}{P(\omega)}\int tP_{SP}(t,\omega)dt \\
&= \int \mu_\omega^*(w)\frac{1}{j}\frac{d}{dw}\mu_\omega(w)dw \\
\sigma_{\omega|t}^2 &= \frac{1}{P(t)}\int(\omega-\langle\omega\rangle)^2P_{SP}(t,\omega)d\omega \\
&= \int \eta_t^*(\tau)\left(\frac{1}{j}\frac{d}{d\tau}-\langle\omega\rangle_t\right)^2\eta_t(\tau)d\tau \\
&= \int\left|\left(\frac{1}{j}\frac{d}{d\tau}-\langle\omega\rangle_t\right)\eta_t(\tau)\right|^2d\tau \\
\sigma_{t|\omega}^2 &= \frac{1}{P(\omega)}\int(t-\langle t\rangle)^2P_{SP}(t,\omega)dt \\
&= \int \mu_\omega^*(w)\left(\frac{1}{j}\frac{d}{dw}-\langle t\rangle_\omega\right)^2\mu_\omega(w)dw \\
&= \int\left|\left(\frac{1}{j}\frac{d}{dw}-\langle t\rangle_\omega\right)\mu_\omega(w)\right|^2dw. \qquad (12)
\end{aligned}
$$

2.2.2 *Global quantities*

The spectrogram is constructed from the signal under observation and the window function. Their mean times and bandwidth of the signal and the window function are [8]

$$
\begin{aligned}
\langle t\rangle^s &= \int t|s(t)|^2dt \\
\langle t\rangle^h &= \int t|h(t)|^2dt \\
\langle\omega\rangle^s &= \int \omega|S(\omega)|^2d\omega \\
\langle\omega\rangle^h &= \int \omega|H(\omega)|^2d\omega \\
T_s^2 &= \int\left(t-\langle t\rangle^s\right)^2|s(t)|^2dt \\
T_h^2 &= \int\left(t-\langle t\rangle^h\right)^2|h(t)|^2dt \\
B_s^2 &= \int\left(\omega-\langle\omega\rangle^s\right)^2|S(\omega)|^2d\omega \\
B_h^2 &= \int\left(\omega-\langle\omega\rangle^h\right)^2|H(\omega)|^2d\omega. \qquad (13)
\end{aligned}
$$

As a two-dimensional density, the spectrogram also has mean durations and mean standard deviations which are defined as

$$
\begin{aligned}
\langle t\rangle^{SP} &= \int tP_{SP}(t,\omega)dtd\omega \\
\langle\omega\rangle^{SP} &= \int \omega P_{SP}(t,\omega)dtd\omega \\
T_{SP}^2 &= \int\left(t-\langle t\rangle^{SP}\right)^2P_{SP}(t,\omega)dtd\omega \\
B_{SP}^2 &= \int\left(\omega-\langle\omega\rangle^{SP}\right)^2P_{SP}(t,\omega)dtd\omega. \qquad (14)
\end{aligned}
$$

It has been shown that these quantities can be expressed using the corresponding quantities of the signal and window as [4]

$$
\begin{aligned}
\langle t\rangle^{SP} &= \langle t\rangle^{(s)}-\langle t\rangle^h \\
\langle\omega\rangle^{SP} &= \langle\omega\rangle^{(s)}+\langle\omega\rangle^h \\
T_{SP}^2 &= T_s^2+T_h^2 \\
B_{SP}^2 &= B_s^2+B_h^2. \qquad (15)
\end{aligned}
$$

3 Uncertainty principles of the second-order LPFT

The Mth-order LPFT of the signal $s(t)$ is defined as [1]:

$$
\begin{aligned}
&\mathrm{LPFT}_s(t,\omega) \qquad (16)\\
&= \frac{1}{\sqrt{2\pi}}\int s(t+\tau)h(\tau) \\
&\quad\cdot\exp\left\{-j\omega\tau-j\sum_{m=2}^{M}\omega_{m-1}\tau^m/m!\right\}d\tau, \\
&= \frac{1}{\sqrt{2\pi}}\int s(\tau)h(\tau-t) \\
&\quad\cdot\exp\left\{-j\omega(\tau-t)-j\sum_{m=2}^{M}\omega_{m-1}(\tau-t)^m/m!\right\}d\tau,
\end{aligned}
$$

where $h(t)$ is the window function to segment the signal, M is the order of the polynomial function, $\omega_1, \omega_2, ...,$ and ω_{M-1} are the first-order derivative and other higher-order derivative of the instantaneous frequency of the analyzed signal. These parameters can be estimated by using the polynomial time frequency transform [9] or the Lv's distribution [10]. The energy distribution of the LPFT is called the local polynomial periodogram (LPP) which is defined as $|\mathrm{LPFT}_s(t,\omega)|^2$. When $M = 2$, we can get the second-order LPFT, which has achieved improved performance in many applications such as radar imaging [11,12], nonstationary interference excision in DSSS communications [13,14], chirp signal detection [15], and source localization and tracking in nonstationary environment [16]. A review on the developments and applications of the LPFT can be referred to [17,18]. In the following, we will focus on the uncertainty principles of the second-order LPFT and then generalize to those of the Mth-order LPFT.

The characteristic function is a powerful tool for the study and construction of densities [4]. The two-dimensional characteristic function of the spectrogram $M_{SP}(\theta,\tau)$ is defined as

$$M_{SP}(\theta,\tau) = \int\int e^{j\theta t + j\tau\omega}|S_t(\omega)|^2 dt d\omega$$
$$= A_s(\theta,\tau)A_h(-\theta,\tau), \tag{17}$$

where

$$A_s(\theta,\tau) = \int s\left(t+\frac{\tau}{2}\right)s^*\left(t-\frac{\tau}{2}\right)e^{j\theta t}dt$$

is the ambiguity function of the signal, and A_h is the ambiguity function of the window function.

The distribution function may be obtained from $M_{SP}(\theta,\tau)$ by the Fourier inversion,

$$P_{SP}(t,\omega) = |S_t(\omega)|^2 = \int\int M_{SP}(\theta,\tau)e^{-j(\theta t+\tau\omega)}d\theta d\tau. \tag{18}$$

Proposition 1. The distribution function of the LPP $P_{\text{LPP}}(t,\omega)$ can be presented as

$$P_{\text{LPP}}(t,\omega) = |\text{LPFT}_s(t,\omega)|^2 \tag{19}$$
$$= \int\int M_{\text{LPP}}(\theta,\tau)e^{-j(\theta t+\tau\omega)}d\theta d\tau,$$

where $M_{\text{LPP}}(\theta,\tau)$ is the characteristic function of the second-order LPP as

$$M_{\text{LPP}}(\theta,\tau) = \int\int e^{j\theta t+j\tau\omega}|\text{LPFT}_s(t,\omega)|^2 dt d\omega$$
$$= A_s(\theta,\tau)A_h(\omega_1\tau+\theta,\tau). \tag{20}$$

We will prove Proposition 1 as follows.

Proof. Let us expand the right hand side of (19).

$$\int\int\int\int h\left(x+\frac{\tau}{2}\right)h^*\left(x-\frac{\tau}{2}\right)e^{j(\omega_1\tau+\theta)x}$$
$$s\left(y+\frac{\tau}{2}\right)s^*\left(y-\frac{\tau}{2}\right)e^{j\theta y}e^{-j\theta t-j\tau\omega}dxdyd\tau d\theta$$
$$= \int\int\int \frac{1}{2\pi}h\left(x+\frac{\tau}{2}\right)h^*\left(x-\frac{\tau}{2}\right)e^{j(\omega_1\tau)x}$$
$$s\left(y+\frac{\tau}{2}\right)s^*\left(y-\frac{\tau}{2}\right)e^{-j\tau\omega}\delta(t-x-y)\,dxdyd\tau$$
$$= \int\int h\left(x+\frac{\tau}{2}\right)h^*\left(x-\frac{\tau}{2}\right)e^{j(\omega_1\tau)x}$$
$$s\left(t-x+\frac{\tau}{2}\right)s^*\left(t-x-\frac{\tau}{2}\right)e^{-j\tau\omega}dxd\tau. \tag{21}$$

□

Let $a = x+\frac{\tau}{2}$ and $b = x-\frac{\tau}{2}$, then

$$x = \frac{a+b}{2}, \quad \tau = a-b.$$

With $d\tau dx = |J|dadb$, where the Jacobian determinant is

$$J = \begin{vmatrix} \frac{\partial x}{\partial a} & \frac{\partial x}{\partial b} \\ \frac{\partial \tau}{\partial a} & \frac{\partial \tau}{\partial b} \end{vmatrix} = -1,$$

(21) becomes

$$\frac{1}{\sqrt{2\pi}}\int h(a)s^*(t-a)\exp(-ja\omega)\exp\left(ja^2\frac{\omega_1}{2}\right)da$$
$$\cdot\frac{1}{\sqrt{2\pi}}\int h^*(b)s(t-b)\exp(jb\omega)\exp\left(-jb^2\frac{\omega_1}{2}\right)db$$
$$= |\text{LPFT}_s(t,\omega)|^2.$$

From Proposition 1, we can easily get the following Corollary 1 including (22) to (24).

Corollary 1. The total energy can also be given by the characteristic function evaluated at zero.

$$\int\int P_{\text{LPP}}(t,\omega)dt d\omega = M_{\text{LPP}}(0,0)$$
$$= A_s(0,0)A_h(0,0)$$
$$= \int|s(t)|^2dt\cdot\int|h(t)|^2dt. \tag{22}$$

The time marginal is obtained by integrating over frequency,

$$P_{\text{LPP}}(t) = \int P_{\text{LPP}}(t,\omega)d\omega$$
$$= \int|L(\tau)|^2|h(\tau-t)|^2d\tau, \tag{23}$$

where $L(\tau) = s(\tau)e^{j\omega t}e^{-j\frac{\omega_1}{2}(\tau-t)^2}$.

Similarly, the frequency marginal is

$$P_{\text{LPP}}(\omega) = \int P_{\text{LPP}}(t,\omega)dt$$
$$= \int|S'(w)|^2|H(\omega-w)|^2dw, \tag{24}$$

where $S'(w)$ is the Fourier transform of signal $L(\tau)$.

Since the second-order LPFT can be considered as the STFT with modulated window function, it is expected that uncertainty limits for LPP can be derived from corresponding SP limits, that is

$$P_{\text{LPP}:s}(t,\omega) = P_{SP:L}(t,\omega). \tag{25}$$

Therefore, we can use the uncertainty principles of the STFT for signal $L(\tau)$ to discuss the uncertainty principles of the LPFT for signal $s(\tau)$.

It should be noted that the subscript $_{\text{LPP}:s}$ has the same meaning as the subscript $_{\text{LPP}}$, which means the LPP for the signal $s(\tau)$. The subscript $_{SP}$ has the same meaning as the subscript $_{SP:s}$, which means the SP for the signal $s(\tau)$. The subscript $_{\text{LPP}:s}$ is used together with subscript $_{SP:L}$ with the same meaning, to indicate the LPP of signal $s(\tau)$ and the spectrogram of signal $L(\tau)$, respectively.

We apply the results and equations in [6] to the case in the LPFT/LPP and get the uncertainty principles of the LPFT/LPP with the Gaussian window as shown in Table 1. Details of the uncertainty principles are discussed as follows.

The first type of uncertainty principle: the global uncertainty principle

Considering signal $L(\tau)$ and the window $h(t)$ as two separate functions, each satisfying the uncertainty principle [4]

$$T_L^2 B_L^2 \geq \frac{1}{4},$$
$$T_h^2 B_h^2 \geq \frac{1}{4}. \qquad (26)$$

Now, we consider the uncertainty principle for the global duration and spread of the spectrogram of signal $L(\tau)$, which is the LPP of the signal $s(\tau)$. We can relate the global uncertainty product of the LPP to the uncertainties of the signal and window. Therefore, we have

$$\begin{aligned} &T_{\mathrm{LPP}}^2 B_{\mathrm{LPP}}^2 \\ &= \left(T_L^2 + T_h^2\right)\left(B_L^2 + B_h^2\right) \\ &= T_L^2 B_L^2 + T_h^2 B_h^2 + T_L^2 B_h^2 + T_h^2 B_L^2. \end{aligned} \qquad (27)$$

Using (26) and noting that the last two terms in (27) are manifestly positive, we always have that

$$T_{\mathrm{LPP}}^2 B_{\mathrm{LPP}}^2 \geq \frac{1}{2}. \qquad (28)$$

However, the last two terms in (27) can never be zero, and hence we could obtain a stronger inequality. Assume that we have a Gaussian signal and window. In this case, $T_L^2 B_L^2 = \frac{1}{4}$ and $T_h^2 B_h^2 = \frac{1}{4}$ and therefore,

$$T_{\mathrm{LPP}}^2 B_{\mathrm{LPP}}^2 = \frac{1}{2} + \frac{1}{4}\left(\frac{T_L^2}{T_h^2} + \frac{T_h^2}{T_L^2}\right).$$

Because the minimum is achieved when $T_L^2 = T_h^2$, the global uncertainty principle for the LPP of signal $s(t)$ becomes

$$T_{\mathrm{LPP}}^2 B_{\mathrm{LPP}}^2 \geq 1. \qquad (29)$$

The second type of uncertainty principle: local duration-conditional standard deviation

For the second-order LPFT, the normalized local signal,

$$\eta_t(\tau) = \frac{s(\tau)h(\tau - t)\exp\left\{j\omega t - j\frac{\omega_1}{2}(\tau - t)^2\right\}}{\sqrt{P(t)}},$$

which has T_t as its duration, and $P(t) = \int |s(\tau)h(\tau - t)|^2 d\tau$.

The Fourier transform of the local signal is

$$F_t(\omega) = \frac{1}{\sqrt{2\pi}} \int \eta_t(\tau) \exp\left\{-j\omega\tau\right\} d\tau, \qquad (30)$$

whose bandwidth is

$$\begin{aligned} &\int (\omega - \langle\omega\rangle_t)^2 |F_t(\omega)|^2 d\omega \\ &= \frac{1}{P(t)} \int (\omega - \langle\omega\rangle_t)^2 |S_t(\omega)|^2 d\omega \\ &= \sigma_{\omega|t}^2. \end{aligned} \qquad (31)$$

That is, the bandwidth of the local signal is the conditional standard deviation of the LPP. Hence, the second type of uncertainty principle is expressed as

$$\sigma_{\omega|t}^2 T_t^2 \geq \frac{1}{4}. \qquad (32)$$

The third type of uncertainty principle: local bandwidth-conditional standard deviation

The local spectrum is defined as

$$F_\omega(w) = S'(w)H(\omega - w). \qquad (33)$$

The normalized local spectrum is

$$\mu_\omega(w) = \frac{S'(w)H(\omega - w)}{\sqrt{P(\omega)}}, \qquad (34)$$

where $P(\omega) = \int |S'(w)H(\omega - w)|^2 dw$. The local spectrum has the following signal as its Fourier transform

$$f_\omega(t) = \frac{1}{\sqrt{2\pi}} \frac{1}{\sqrt{P(\omega)}} \int S'(w)H(\omega - w)e^{jwt} dw. \qquad (35)$$

The bandwidth of this signal is B_ω as given in (11). Its duration is

$$\begin{aligned} &\int (t - \langle t\rangle_\omega)^2 |f_\omega(t)|^2 dt \\ &= \frac{1}{P(\omega)} \int (t - \langle t\rangle_\omega)^2 |s_\omega(t)|^2 dt \\ &= \sigma_{t|\omega}^2, \end{aligned} \qquad (36)$$

which is the conditional standard deviation of time for a given frequency of the LPP. Therefore, we have

$$\sigma_{t|\omega}^2 B_\omega^2 \geq \frac{1}{4}. \qquad (37)$$

The fourth type of uncertainty principle: conditional standard deviations in time and frequency

We now try to obtain the uncertainty relation which directly relates the two conditional standard deviations of the LPP, $\sigma_{t|\omega}^2$ and $\sigma_{\omega|t}^2$. Following the procedure in [8], we get

$$\sigma_{t|\omega}^2 \sigma_{\omega|t}^2 \geq \int \left(\frac{d}{d\tau}|\eta_t(\tau)|\right)^2 d\tau \int \left(\frac{d}{dw}|\mu_\omega(w)|\right)^2 dw, \qquad (38)$$

Table 1 Expressions of the uncertainty principles for the second-order LPFT

Type of uncertainty principle	Uncertainty product
1. Global uncertainty principle	$T_{\mathrm{LPP}}^2 B_{\mathrm{LPP}}^2 \geq 1$
2. Local duration-conditional standard deviation	$\sigma_{\omega\vert t}^2 T_t^2 \geq \frac{1}{4}$
3. Local bandwidth-conditional standard deviation	$\sigma_{t\vert\omega}^2 B_\omega^2 \geq \frac{1}{4}$
4. Conditional standard deviations in time and frequency	$\sigma_{t\vert\omega}^2 \sigma_{\omega\vert t}^2 > \frac{1}{4}$

which is called the local uncertainty product in [5]. Following the procedure in [5], the local uncertainty product of the LPP for a Gaussian window is achieved as

$$\sigma^2_{t|\omega}\sigma^2_{\omega|t} > \frac{1}{4}. \tag{39}$$

It is shown in [5] that for the general case as well as a large subset of bilinear distributions such as Wigner-Ville distribution, the local uncertainty product is upper bounded by the global uncertainty product, and it can be arbitrarily small even though the product of the global variance cannot. However, for the STFT/spectrogram, the local uncertainty product cannot be arbitrarily small, and there is a lower bound on the local uncertainty product of the spectrogram due to the windowing operation. This limitation is an inherent property of the spectrogram and is not a property of the signal or a fundamental limit. Since the LPFT/LPP also uses the windowing operation, its local uncertainty product has a lower bound as the spectrogram. In Section 4, examples are given to show that the local uncertainty product of the LPFT/LPP has a lower bound and cannot be arbitrarily small.

4 Example

Let us consider an example to verify the expressions in Section 3 by using the chirp signal

$$s(t) = \left(\frac{\alpha}{\pi}\right)^{1/4} \exp\left\{\frac{-\alpha t^2}{2} + ja_0 t + \frac{ja_1 t^2}{2}\right\}, \tag{40}$$

as the input signal of the second-order LPFT. A Gaussian window is used to obtain the segments of the signal $s(t)$

$$h(t) = \left(\frac{a}{\pi}\right)^{1/4} \exp\left\{-\frac{at^2}{2}\right\}, \tag{41}$$

where $a > 0$ is the parameter to control the window width.

Therefore, we have

$$\begin{aligned} L(\tau) &= s(\tau)\exp\{j\omega t\}\exp\left\{-j\frac{\omega_1}{2}(\tau - t)^2\right\} \\ &= (\frac{\alpha}{\pi})^{1/4}\exp\left\{\frac{-\alpha\tau^2}{2} + ja_0\tau + \frac{ja_1\tau^2}{2}\right\} \\ &\quad \cdot \exp\left\{j\omega t - j\frac{\omega_1}{2}(\tau - t)^2\right\}. \end{aligned} \tag{42}$$

According to the definitions in (23) and (24), we can calculate that

$$\begin{aligned} P_{\mathrm{LPP}}(t) &= \int |L(\tau)h(\tau - t)|^2 d\tau \\ &= \left(\frac{a\alpha}{\pi(a+\alpha)}\right)^{1/2} \exp\left[-\frac{a\alpha}{a+\alpha}t^2\right], \\ P_{\mathrm{LPP}}(\omega) &= \int |S'(w)H(\omega - w)|^2 dw \\ &= \left(\frac{a'\alpha'}{\pi(a'+\alpha')}\right)^{1/2} \exp\left[-\frac{a'\alpha'}{a'+\alpha'}(\omega - a_0 - \omega_1 t)^2\right]. \end{aligned}$$

with

$$\begin{aligned} \alpha' &= \frac{\alpha}{\alpha^2 + (a_1 - \omega_1)^2}, \\ \beta' &= \frac{a_1 - \omega_1}{\alpha^2 + (a_1 - \omega_1)^2}, \\ a' &= \frac{1}{a}. \end{aligned}$$

Since

$$|\eta_t(\tau)|^2 = \left(\frac{\alpha + a}{\pi}\right)^{1/2} \exp\left\{-(\alpha + a)\left(\tau - \frac{at}{\alpha + a}\right)^2\right\},$$

the spectra of the signal and the window are

$$\begin{aligned} S'(\omega) &= \exp\left\{j\omega t - j\frac{\omega_1}{2}t^2\right\} \\ &\quad \sqrt{\frac{\sqrt{\alpha/\pi}}{\alpha - ja_1}} \exp\left\{-\frac{(\omega - a_0 - \omega_1 t)}{2(\alpha - ja_1)}\right\}, \\ H(\omega) &= \sqrt[4]{1/a\pi}\exp\left\{-\frac{\omega^2}{2a}\right\}. \end{aligned}$$

Therefore,

$$\begin{aligned} |S'(\omega)| &= \left(\alpha'/\pi\right)^{1/4}\exp\left\{-\frac{1}{2}\alpha'(\omega - a_0 - \omega_1 t)^2\right\}, \\ |H(\omega)| &= \left(a'/\pi\right)^{1/4}\exp\left\{-\frac{1}{2}a'\omega^2\right\}. \end{aligned}$$

By using these relations and definitions from (10) to (12), and using $P_{\mathrm{LPP}}(t)$ and $P_{\mathrm{LPP}}(\omega)$ for calculation, we obtain

$$\begin{aligned} \langle\tau\rangle_t &= \frac{a}{a+\alpha}t \\ T_t^2 &= \frac{1}{2(a+\alpha)} \\ \langle\omega\rangle_t &= \frac{a(a_1 - \omega_1)}{a+\alpha}t + a_0 + \omega_1 t \\ \sigma^2_{\omega|t} &= \frac{1}{2}(a+\alpha) + \frac{1}{2}\frac{(a_1 - \omega_1)^2}{a+\alpha} \\ \langle w\rangle_\omega &= \frac{a'\omega + \alpha'(a_0 + \omega_1 t)}{a' + \alpha'} \\ B_\omega^2 &= \frac{1}{2(a' + \alpha')} \\ \langle t\rangle_\omega &= \frac{a'\beta'}{a' + \alpha'}(\omega - a_0 - \omega_1 t) \\ \sigma^2_{t|\omega} &= \frac{1}{2}\left(a' + \alpha'\right) + \frac{1}{2}\frac{\beta'^2}{a' + \alpha'}. \end{aligned}$$

Global uncertainty principle

Using (13), we can calculate

$$\begin{aligned} T_L^2 &= \frac{1}{2\alpha} \\ T_h^2 &= \frac{1}{2a} \\ B_L^2 &= \frac{1}{2\alpha'} \\ B_h^2 &= \frac{1}{2a'}. \end{aligned} \tag{43}$$

Therefore, we have

$$\begin{aligned} T_{\mathrm{LPP}}^2 &= \frac{1}{2}\left(\frac{1}{a}+\frac{1}{\alpha}\right) \\ B_{\mathrm{LPP}}^2 &= \frac{1}{2}\left(\frac{1}{a'}+\frac{1}{\alpha'}\right)=\frac{1}{2}\left(a+\alpha+\frac{(a_1-\omega_1)^2}{\alpha}\right). \end{aligned} \tag{44}$$

Then

$$T_{\mathrm{LPP}}^2 B_{\mathrm{LPP}}^2 = 1+\frac{(\alpha-a)^2}{4a\alpha}+\frac{1}{4}\left(\frac{1}{a}+\frac{1}{\alpha}\right)\frac{(a_1-\omega_1)^2}{\alpha} \geq 1, \tag{45}$$

which is consistent with (29). The equality can be achieved when $\alpha = a$, and the parameter ω_1 is estimated correctly, that is $\omega_1 = a_1$.

The second type of uncertainty principle

The duration and conditional standard deviations are given by

$$\begin{aligned} T_t^2 &= \frac{1}{2(a+\alpha)} \\ \sigma_{\omega|t}^2 &= \frac{1}{2}(a+\alpha)+\frac{1}{2}\frac{(a_1-\omega_1)^2}{a+\alpha}. \end{aligned} \tag{46}$$

Therefore,

$$\sigma_{\omega|t} T_t = \frac{1}{2}\sqrt{1+\frac{(a_1-\omega_1)^2}{(a+\alpha)^2}} \geq \frac{1}{2}, \tag{47}$$

which is consistent with the uncertainty principle given in (32).

The third type of uncertainty principle

We also have the bandwidth and conditional standard deviations as

$$\begin{aligned} B_\omega^2 &= \frac{1}{2(a'+\alpha')} \\ \sigma_{t|\omega}^2 &= \frac{1}{2}(a'+\alpha')+\frac{1}{2}\frac{\beta'^2}{a'+\alpha'}. \end{aligned}$$

Therefore,

$$\sigma_{t|\omega} B_\omega = \frac{1}{2}\sqrt{1+\frac{\beta'^2}{(a'+\alpha')^2}} \geq \frac{1}{2} \tag{48}$$

which is consistent with the uncertainty principle given in (37).

The fourth type of uncertainty principle

The fourth type of uncertainty principle deals directly with both conditional standard deviations. It states that

$$\sigma_{t|\omega}^2 \sigma_{\omega|t}^2 \geq \int \left(\frac{d}{d\tau}|\eta_t(\tau)|\right)^2 d\tau \int \left(\frac{d}{dw}|\mu_\omega(w)|\right)^2 dw. \tag{49}$$

By direct calculation, we have

$$\begin{aligned} \int \left(\frac{d}{d\tau}|\eta_t(\tau)|\right)^2 d\tau &= \frac{1}{2}(a+\alpha) \\ \int \left(\frac{d}{dw}|\mu_\omega(w)|\right)^2 dw &= \frac{1}{2}(a'+\alpha'). \end{aligned}$$

Therefore,

$$\sigma_{t|\omega}^2 \sigma_{\omega|t}^2 \geq \frac{1}{4}(a+\alpha)(a'+\alpha'). \tag{50}$$

The exact calculations of $\sigma_{t|\omega}^2$ for this example are

$$\begin{aligned} \sigma_{t|\omega}^2 \sigma_{\omega|t}^2 = \frac{1}{4}&\left[\frac{(a+\alpha)^2+(a_1-\omega_1)^2}{a+\alpha}\right] \\ \times &\left[\frac{(a'+\alpha')^2+\beta'^2}{a'+\alpha'}\right], \end{aligned} \tag{51}$$

which implies that (50) is satisfied.

As α goes to zero, the signal in (40) becomes a chirp signal with constant amplitude, and the local uncertainty product of the LPP becomes

$$\sigma_{t|\omega}^2 \sigma_{\omega|t}^2 > \frac{1}{4} \quad \text{as} \quad \alpha \to 0. \tag{52}$$

It can be easily shown that for arbitrary values of the signal parameters, the minimum of the local uncertainty product for the LPP in (51) can be achieved when the Gaussian window width parameter satisfies $a = \sqrt{\alpha^2+(a_1-\omega_1)^2}$.

With different width parameters of the Gaussian window, Figure 1 shows the time-frequency representations achieved by using the second-order LPFT to process the signal consisting of chirp components. Following the definition in (40), the signal components are Gaussian-modulated chirp components. Expressions of the first two chirp components and the second two chirp components are as follows:

$$\begin{aligned} s_1(t) = &\left(\frac{0.0001}{\pi}\right)^{1/4} \exp\left(\frac{-0.0001(t-260)^2}{2}\right) \\ &\times \exp\left(j2\pi\left(0.5t-0.000481t^2\right)\right) \\ &+\left(\frac{0.0001}{\pi}\right)^{1/4} \exp\left(\frac{-0.0001(t-260)^2}{2}\right) \\ &\times \exp\left(j2\pi\left(0.4t-0.000481t^2\right)\right), \end{aligned} \tag{53}$$

$$\begin{aligned} s_2(t) = &\left(\frac{0.0001}{\pi}\right)^{1/4} \exp\left(\frac{-0.0001(t-260)^2}{2}\right) \\ &\times \exp\left(j2\pi\left(0.35t-0.000481t^2\right)\right) \\ &+\left(\frac{0.0001}{\pi}\right)^{1/4} \exp\left(\frac{-0.0001(t-260)^2}{2}\right) \\ &\times \exp\left(j2\pi\left(0.25t-0.000481t^2\right)\right), \end{aligned} \tag{54}$$

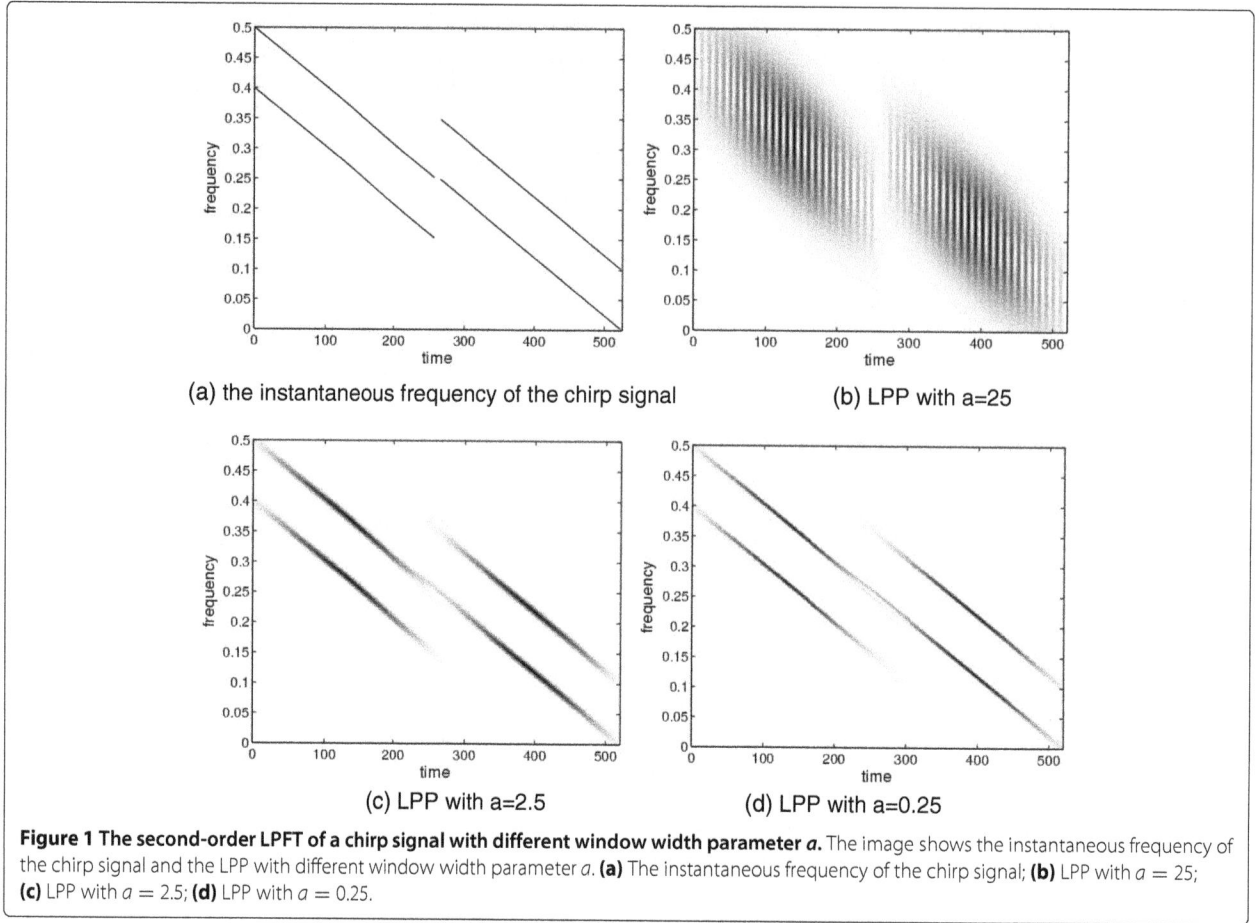

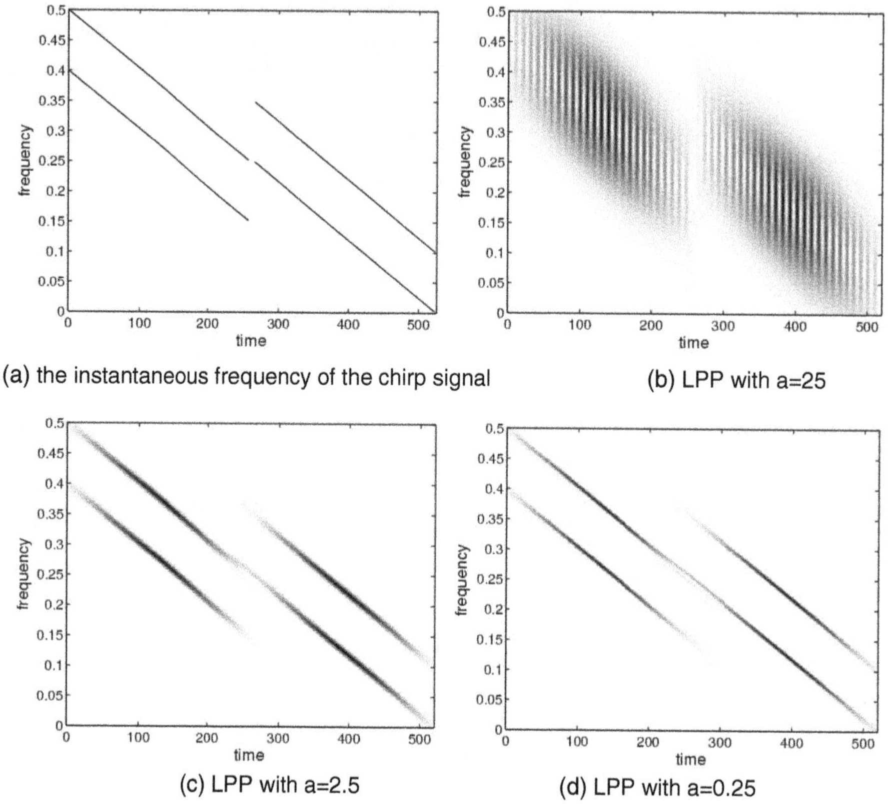

Figure 1 The second-order LPFT of a chirp signal with different window width parameter *a*. The image shows the instantaneous frequency of the chirp signal and the LPP with different window width parameter a. **(a)** The instantaneous frequency of the chirp signal; **(b)** LPP with $a = 25$; **(c)** LPP with $a = 2.5$; **(d)** LPP with $a = 0.25$.

and there are five-point spaces between $s_1(t)$ and $s_2(t)$. Without loss of generality, the sampling interval is assumed to be unit. In Figure 1, truncated Gaussian window is used, and the window length is 1/4 of the signal length. It is seen that as a decreases or the window width increases, the chirp components become more concentrated in the frequency direction, or equivalently, the resolution of the signal representation in the frequency direction is increased. As for the resolution in the time direction, the signal components in Figure 1b can be clearly separated in the time instant 260. As the parameter a decreases, such separation in time direction disappears. Therefore, from Figure 1b,d, we can observe that the resolution in the time direction decreases as the parameter a decreases. This observation is consistent with the derivation in (45). For example, decreasing the window parameter a leads to the increasing of the signal duration T_t^2, or equivalently, the decreasing of signal time resolution, as shown in Figure 2a,b,c by sampling the LPP at a particular frequency instant $f = 0.25$. At the same time, the conditional standard deviation $\sigma^2_{\omega|t}$ decreases as a decreases so that the signal frequency resolution is increased, as shown in Figure 2d,e,f by sampling the LPP at a particular time instant $t = 400$.

5 Uncertainty principle of the *M*th-order LPFTs

The uncertainty product can be similarly derived for higher-order LPFTs whose input signals are the same order polynomial phase signals, as shown in Table 2. For simplicity, only the major steps of derivation are presented in this section.

Let us consider an Mth-order polynomial phase signal defined as

$$s(t) = \left(\frac{\alpha}{\pi}\right)^{1/4} e^{-\alpha t^2/2} \exp\left\{j\sum_{m=1}^{M} \frac{a_{m-1}t^m}{m!}\right\}, \quad (55)$$

where the phase of the signal is

$$\Phi(t) = \sum_{m=1}^{M} \frac{a_{m-1}t^m}{m!}. \quad (56)$$

Because the local signal segment is

$$L(\tau) = s(\tau)e^{j\omega t}e^{-j\sum_{m=2}^{M}\frac{\omega_{m-1}}{m!}(\tau-t)^m}, \quad (57)$$

we have

$$|S'(\omega)| = \left(\frac{1}{\pi\alpha}\right)^{\frac{1}{4}} e^{-\frac{\left(\omega - j\sum_{m=2}^{M}\frac{\omega_{m-1}}{(m-1)!}t^{m-1}\right)^2}{2\alpha}}. \quad (58)$$

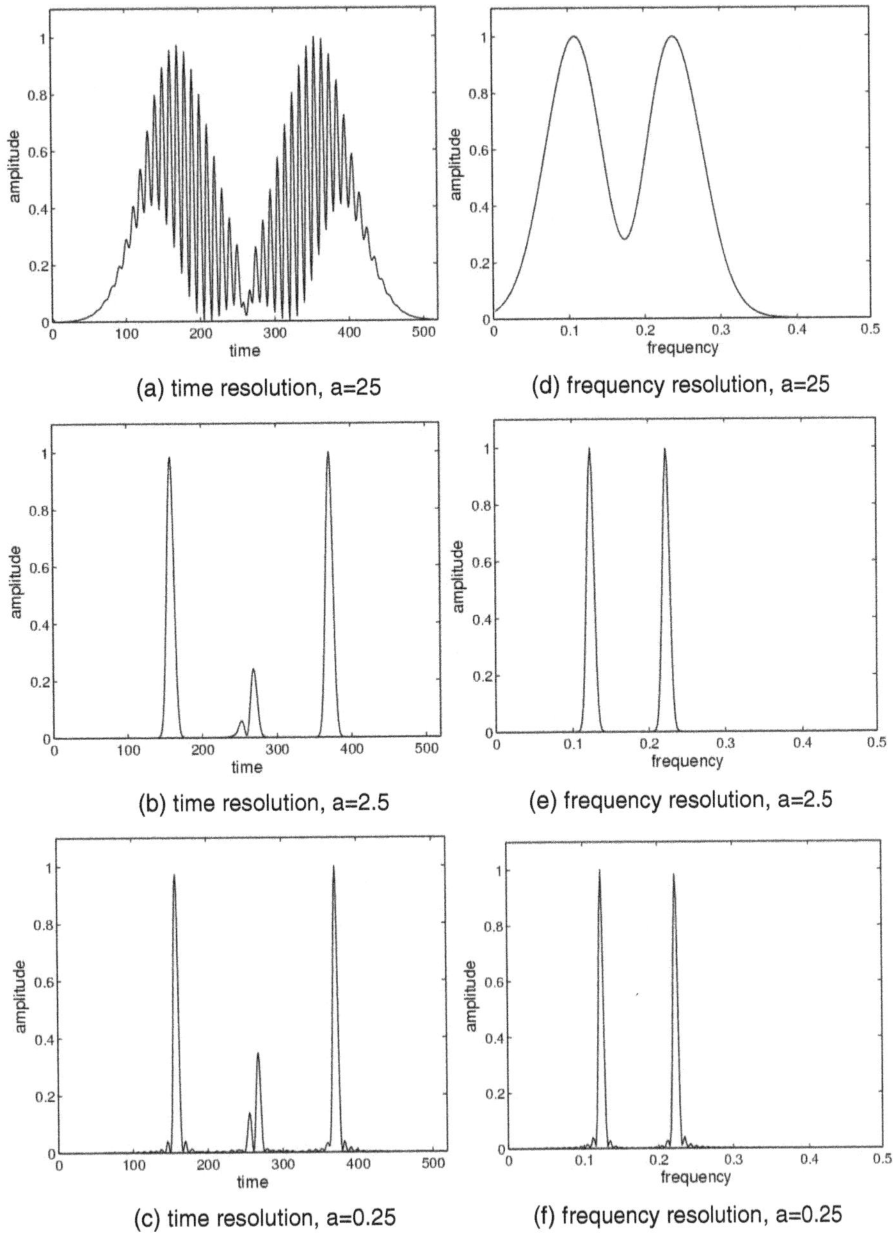

Figure 2 Signal resolution for a chirp signal with different window width parameter *a*. (a-c) The signal time resolution for $f = 0.25$. **(d-f)** The signal frequency resolution for $t = 400$.

The normalized local signal segment is

$$\begin{aligned}
\eta_t(\tau) &= \frac{s(\tau)h(\tau-t)\exp\left\{j\omega t - j\sum_{m=2}^{M}\frac{\omega_{m-1}}{m!}(\tau-t)^m\right\}}{\sqrt{\int |s(\tau)h(\tau-t)|^2 d\tau}} \\
&= \left(\frac{\alpha+a}{\pi}\right)^{1/4}\exp\left\{-\frac{a^2t^2}{2(\alpha+a)}\right\} \\
&\exp\left\{-\frac{(\alpha+a)\tau^2}{2} + a\tau t + j\sum_{m=1}^{M}\frac{a_{m-1}t^m}{m!}\right\} \\
&\exp\left\{j\omega t - j\sum_{m=2}^{M}\frac{\omega_{m-1}(\tau-t)^m}{m!}\right\},
\end{aligned}$$

and

$$|\eta_t(\tau)|^2 = \left(\frac{\alpha+a}{\pi}\right)^{1/2}\exp\left\{-(\alpha+a)\left(\tau - \frac{at}{\alpha+a}\right)^2\right\}.$$

The mean time $\langle\tau\rangle_t$ and duration T_t^2 of the Mth-order LPFT for the Mth-order polynomial phase signal are the same as given in Section 4, that is,

$$\langle\tau\rangle_t = \frac{at}{a+\alpha},$$

Table 2 Expressions of the uncertainty principles for the *M*th-order LPFT

Type of uncertainty principle	Uncertainty product
1. Global uncertainty principle	$T^2_{\mathrm{LPP}}B^2_{\mathrm{LPP}} \geq 1$
2. Local duration-conditional standard deviation	$\sigma^2_{\omega\|t}T^2_t \geq \frac{1}{4}$
3. Local bandwidth-conditional standard deviation	$\sigma^2_{t\|\omega}B^2_\omega \geq \frac{1}{4}$
4. Conditional standard deviations in time and frequency	$\sigma^2_{t\|\omega}\sigma^2_{\omega\|t} > \frac{1}{4}$

and

$$T_t^2 = \frac{1}{2(a+\alpha)}.$$

When the parameters of the LPFT such as $\omega_1, \omega_2, \ldots, \omega_{M-1}$ are estimated correctly, that is, $\omega_{M-1} = \Phi^{(M)} = a_{M-1}, \ldots, \omega_m = \Phi^{(m+1)}, \ldots, \omega_1 = \Phi^{(2)} = \sum_{m=1}^{M} \frac{a_{m-1}t^{m-2}}{(m-2)!}$, where the values of superscripts in $\Phi^{(m)}$ are the derivative orders of Φ, we have

$$\eta^*(\tau)\frac{d}{d\tau}\eta_t(\tau) = \left(\frac{\alpha+a}{\pi}\right)\exp\left\{-\frac{a^2t^2}{a+\alpha}\right\}\exp\left\{-(a+\alpha)\tau^2+2at\tau\right\} \cdot \left\{-(a+\alpha)\tau + at + j\sum_{m=1}^{M}\frac{a_{m-1}t^{m-1}}{(m-1)!}\right\}.$$

Therefore,

$$\begin{aligned}\langle\omega\rangle_t &= \left(\frac{\alpha+a}{\pi}\right)\exp\left\{-\frac{a^2t^2}{a+\alpha}\right\} \cdot \int \exp\left\{-(a+\alpha)\tau^2+2at\tau\right\} \\ &\quad \cdot \left\{j(a+\alpha)\tau - jat + \sum_{m=1}^{M}\frac{a_{m-1}t^{m-1}}{(m-1)!}\right\}d\tau \\ &= \sum_{m=1}^{M}\frac{a_{m-1}t^{m-1}}{(m-1)!} \\ &= \Phi^{(1)},\end{aligned} \tag{59}$$

which is the instantaneous frequency of the Mth-order polynomial phase signals.

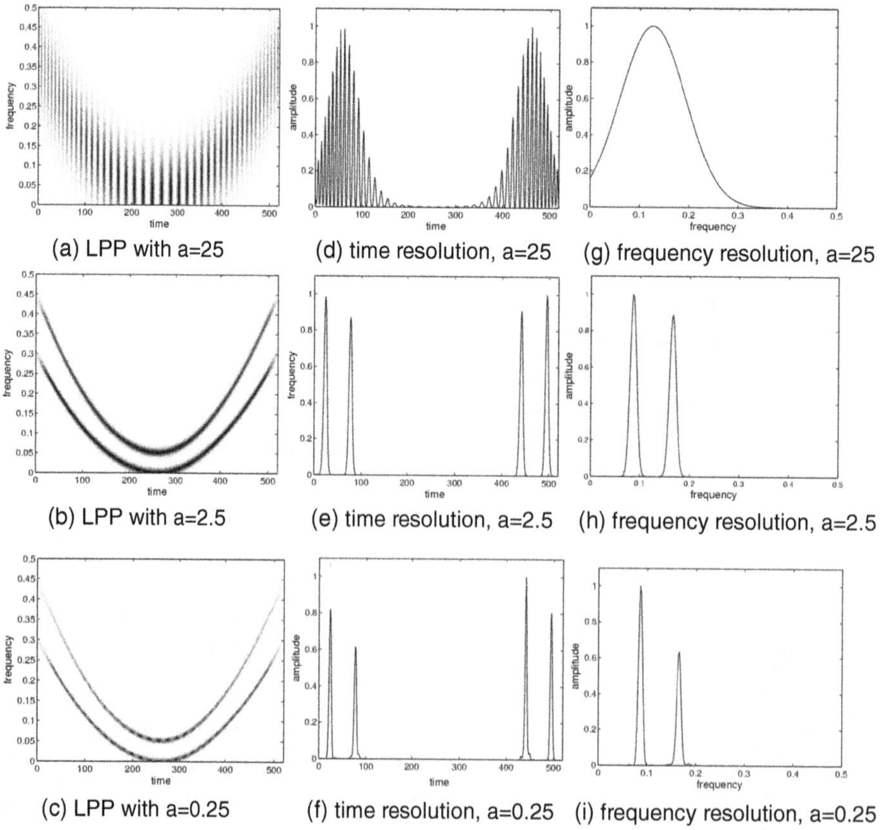

Figure 3 Signal resolution for a third order polynomial phase signal with different window width parameter *a*. (a)-(c): the third order LPP. **(d)-(f)**: the signal time resolution for $f = 0.25$. **(g)-(i)**: the signal frequency resolution for $t = 400$.

Similar with $\omega_{M-1} = \Phi^{(M)} = a_{M-1}, \ldots, \omega_m = \Phi^{(m+1)}$, $\ldots, \omega_1 = \Phi^{(2)} = \sum_{m=1}^{M} \frac{a_{m-1}t^{m-2}}{(m-2)!}$, we have

$$\begin{aligned}\langle \omega^2 \rangle &= -\int \eta^*(\tau) \frac{d^2 \eta_t(\tau)}{d\tau^2} d\tau \\ &= \int \left\{ a + \alpha + \left[(a+\alpha)\tau - at - j \sum_{m=1}^{M} \frac{a_{m-1}t^{m-1}}{(m-1)!} \right]^2 \right\} \\ &\quad \cdot \exp\left\{ -(a+\alpha)\tau^2 + 2at\tau \right\} d\tau \\ &= \frac{a+\alpha}{2} + \left[\sum_{m=1}^{M} \frac{a_{m-1}t^{m-1}}{(m-1)!} \right]^2 . \end{aligned} \tag{60}$$

Based on the above derivation, we can conclude that if the coefficients of the polynomial phase are accurately estimated, the bandwidth obtained by the LPFTs of any order $M > 1$ is

$$\begin{aligned}\sigma^2_{\omega|t} = B_t^2 &= \langle \omega^2 \rangle - \langle \omega \rangle_t^2 \\ &= \frac{a+\alpha}{2}.\end{aligned}$$

Similarly, we can get that

$$\begin{aligned} B_\omega^2 &= \frac{1}{2\left(\frac{1}{a} + \frac{1}{\alpha}\right)}, \\ \sigma^2_{t|\omega} &= \frac{1}{2}\left(\frac{1}{a} + \frac{1}{\alpha}\right). \end{aligned} \tag{61}$$

Therefore, we have the global uncertainty principle as

$$T^2_{\mathrm{LPP}} B^2_{\mathrm{LPP}} = \frac{1}{4}\left(\frac{1}{a} + \frac{1}{\alpha}\right)(a+\alpha) \geq 1, \tag{62}$$

and

$$\begin{aligned} \sigma^2_{\omega|t} T_t^2 &= \frac{1}{4} \\ \sigma^2_{t|\omega} B_\omega^2 &= \frac{1}{4} \\ \sigma^2_{\omega|t}\sigma^2_{t|\omega} &= \frac{1}{4}(a+\alpha)\left(\frac{1}{a} + \frac{1}{\alpha}\right) > \frac{1}{4} \quad \text{as} \quad \alpha \to 0. \end{aligned} \tag{63}$$

It means that for the higher-order polynomial phase signals segmented by a Gaussian window function, the uncertainty products of the Mth-order LPFT are also consistent with the theoretical analysis in Section 3 if the phase parameters are accurately estimated.

An example using the third-order LPFT to process the third-order polynomial phase signal is given in Figure 3a,b,c. Figure 3d,e,f shows the signal time resolution by sampling the LPP at a particular frequency instant $f = 0.25$, and Figure 3g,h,i shows the signal frequency resolution by sampling the LPP at a particular time instant $t = 400$. From Figure 3, we have similar observations to those from Figures 1 and 2 for the second-order LPFT processing the chirp signal, that is, as the parameter a decreases, the time resolution is decreased while the frequency resolution is increased.

6 Conclusions

In this paper, systematic analysis of the uncertainty principles of the LPFT are demonstrated to show that as a windowed transform, the LPFT is limited by the uncertainty principles, including the global uncertainty principle, the uncertainty principle of local duration and conditional standard deviation, the uncertainty principle of local bandwidth and conditional standard deviation, and the uncertainty principle of the conditional standard deviations in time and frequency. Explicit expressions of the uncertainty principles for the second-order LPFT are derived, and an example is given in which the uncertainty relations for a chirp signal is obtained to show that the relations match with the derived expressions. The uncertainty principles of Mth-order LPFT are also discussed to show that they are consistent with the theoretical analysis.

Abbreviations

LPFT: local polynomial Fourier transform; LPP: local polynomial periodogram; STFT: short-time Fourier transform.

Competing interests

The authors declare that they have no competing interests.

Acknowledgement

This research work is supported by the National Natural Science Foundation of China (No. 61102164), the Open Project of Zhejiang Key Laboratory for Signal Processing (ZJKL_4_SP – OP2013 – 02), and the Scientific Research Foundation for the Returned Overseas Chinese Scholars, State Education Ministry.

Author details

[1]School of Information Science and Engineering, 58 Haishu Road, Hangzhou Normal University, 311121 Hangzhou, China. [2]School of Electronic and Electrical Engineering, 50 Nanyang Ave., Nanyang Technological University, Singapore 639798, Singapore.

References

1. V Katkovnik, A new form of Fourier transform for time-frequency estimation. Signal Process. **47**(2), 187–200 (1995)
2. D Gabor, textbf93, (26), Theory of communication. J. Inst. Electrical Eng, 429–457 (1946)
3. P Korn, Some uncertainty principles for time-frequency transforms of the Cohen class. IEEE Trans. Signal Process. **53**(2), 523–527 (2005)
4. L Cohen, *Time-Frequency Analysis*. (Prentice Hall, NJ, 1995)
5. P Loughlin, L Cohen, The uncertainty principle: global, local, or both?. IEEE Trans. Signal Process. **52**(5), 1218–1227 (2004)
6. V Katkovnik, Discrete-time local polynomial approximation of the instantaneous frequency. IEEE Trans. Signal Process. **46**(10), 2626–2637 (1998)
7. X Li, G Bi, On uncertainty principle of the local polynomial Fourier transform. EURASIP J. Adv. Signal Process. 120 (2012)
8. L Cohen, The uncertainty principles of the short-time Fourier transform. Proc. SPIE. **2563**, 80–90 (1995)
9. X Xia, Discrete chirp-Fourier transform and its applications to chirp rate estimation. IEEE Trans. Signal Process. **48**(11), 3122–3133 (2000)
10. X Lv, G Bi, C Wan, M Xing, Lv's distribution: principle, implementation, properties and performance. IEEE Trans. Signal Process. **59**(8), 3576–3591 (2011)

11. X Li, G Bi, Y Ju, Quantitative SNR analysis for ISAR imaging using local polynomial Fourier transform. IEEE Trans. Aerospace Electronic Syst. **45**(3), 1241–1248 (2009)
12. I Djurovic, T Thayaparan, L Stankovic, Adaptive local polynomial Fourier transform in ISAR. EURASIP J. Appl. Signal Process. **2006**, 1–15 (2006)
13. S Djukanovic, M Dakovic, L Stankovic, Local polynomial Fourier transform receiver for nonstationary interference excision in DSSS communications. IEEE Trans. Signal Process. **54**(4), 1627–1636 (2008)
14. L Stankovic, S Djukanovic, Order adaptive local polynomial FT based interference rejection in spread spectrum communication systems. IEEE Trans. Instrum. Meas. **54**(6), 2156–2162 (2005)
15. G Bi, X Li, C See, LFM signal detection using LPP-Hough transform. Signal Process. **91**(6), 1432–1443 (2011)
16. V Katkovnic, A Gershman, A local polynomial approximation based beamforming for source localization and tracking in nonstationary environments. IEEE Signal Process. Lett. **7**(1), 3–5 (2000)
17. X Li, G Bi, S Stankovic, A Zoubir, Local polynomial Fourier transform: a review on recent developments and applications. Signal Process. **91**(6), 1370–1393 (2011)
18. Y Wei, G Bi, Efficient analysis of time-varying muliticomponent signals with modified LPTFT. EURASIP J. Appl. Signal Process. **2005**(1), 1261–1268 (2005)

5

Analysis of narrowband power line communication channels for advanced metering infrastructure

José Antonio Cortés*, Alfredo Sanz, Pedro Estopiñán and José Ignacio García

Abstract

This paper analyzes the characteristics of narrowband power line communication (NB-PLC) channels and assesses their performance when used for advanced metering infrastructure (AMI) communications. This medium has been traditionally considered too hostile. However, the research activities carried out in the last decade have shown that it is a suitable technology for a large number of applications. This work provides a statistical characterization of NB-PLC channels in the CENELEC-A band. The presented results have been obtained from a set of 106 links measured in urban, suburban, and rural scenarios. The study covers the input impedance of the power line network, the channel response and the noise. The analysis of the channel response examines the delay spread, the coherence bandwidth, and the attenuation, while the assessment of the noise considers both its spectral and temporal characteristics. Since low voltage (LV) distribution networks consists of several conductors, they can be simultaneously used to set up multiple-input multiple-output (MIMO) communication links. This paper investigates the correlation between the MIMO streams. The bit rates that can be attained both in the single-input single-output (SISO) and in the MIMO cases are estimated and discussed.

Keywords: Power line communications; Narrowband; Smart Grid; Advanced metering infrastructure

1 Introduction

The conventional paradigm of electricity networks 'generate what is consumed' is shifting towards the new 'consume what is produced' [1,2]. This change is motivated by facts like the increased use of renewable sources, which have a much more decentralized structure than conventional ones and whose generating capacity is subject to unpredictable factors, and by new consumption patterns like electric vehicles charging, which complicate the demand forecasting.

The traditional electricity network must evolve into the so-called smart grid to support this change. Advanced metering infrastructure (AMI) is considered a constituent part of the Smart Grid. It enables applications such as automatic meter reading (AMR), demand side response, and distribution automation [1,3]. AMI requires bidirectional communication links between the medium voltage to low voltage (MV/LV) transformer stations and the costumers, usually known as the *last mile*. Recent studies suggest that power line communications (PLC) is the most cost-effective technology for ARM [4]. In addition, it easily enables power quality measurements and distribution automation functions; it gives utilities full control of the communication network and seems to be the most appropriate technology for the communication between the on-board charging system of electric vehicles and the grid [5].

PLC technology can be classified in terms of the employed bandwidth into narrowband (NB) and broadband (BB) [5,6]. Data rates estimated for last-mile AMI applications suggest that they can be delivered by NB-PLC in a more inexpensive way than with BB-PLC [7,8]. Examples of suitable systems for this end are the ones defined in the ITU-T Recommendations G.9902 (known as G.hnem), G.9903 and G.9904, and the IEEE P1901.2 [9-12]. ITU-T G.9903 and G.9904 are based on the industry specifications G3-PLC and Powerline Intelligent Metering Evolution (PRIME), respectively.

*Correspondence: jose_antonio.cortes@atmel.com
Atmel Spain, Torre C2, Polígono Puerta Norte, A-23 Zaragoza, Spain

Nowadays, the CENELEC-A band (3 to 95 kHz) is the most widespread one in NB-PLC [13]. However, the quantitative knowledge of the channel in this frequency range is still imprecise [14]. This is clearly reflected by the significant differences among the physical layer parameters of the latest NB-PLC systems [5]. Recent noise measurements and models incorporated into the IEEE P1901.2 have provided much insight into the noise features in the frequency band above 100 kHz [15]. Nevertheless, their suitability for the CENELEC-A band has not been assessed. The same uncertainty applies to the 'fading modeling method' stated in the IEEE P1901.2 for the channel response. It results from a particularization of the well-known model proposed in [16], but the appropriateness of the selected parameters to generate responses in the CENELEC-A band has not been evaluated. Similarly, measurements performed in selected scenarios have provided much information about the qualitative features of the channel response [15,17,18], but the absence of a statistical knowledge leads to a large uncertainty in the expected performance of NB-PLC.

LV distribution networks consist of several conductors which can be simultaneously used to set up multiple-input multiple-output (MIMO) communication links. This strategy is being successfully employed in wireless communications and in indoor BB-PLC [19], where the phase, neutral, and protective earth conductors are used for the MIMO. However, there are almost no available works on NB-PLC MIMO, and they are limited to explore indoor NB-PLC channels [20] or to plain tests using existing single phase PRIME devices [21].

In this context we make three main contributions:

- We provide a statistical analysis of the channel characteristics in the CENELEC-A band. These results are compared to the ones obtained with the more recent channel model for this band, proposed in the IEEE P1901.2 standard.
- We analyze the correlation between the channels of the 3 × 3 MIMO links that could be established by injecting and receiving the communication signal between each of the phases and the neutral conductor.
- We estimate the data rates that could be achieved both in single-input single-output (SISO) and MIMO communications. These results will be useful to clarify whether there is a need for using BB-PLC in the last mile of AMIs.

The rest of the paper is organized as follows. Section 2 provides a brief description of the employed measurement setup and signal processing algorithms. Sections 3 and 4 are devoted to the characterization of the input impedance of the power line network and its channel response, respectively. Section 5 is the noise counterpart of Section 4. Based on the presented characterization, Section 6 provides an estimate of the achievable performance. Finally, Section 7 summarizes the main conclusions.

2 Measurement methodology

2.1 Measurement setup

The measurement setup consists of two equal signal generation and data acquisition systems, as shown in Figure 1. They are controlled by a laptop, which also stores the acquired signals. Measurements are differentially performed from the three phases to the neutral conductor. The signal generation system (SGS) comprises a signal generation board (SGB), a high power amplifier (implemented using a parallel structure), and a coupling circuit. The SGS has very low output impedance, its real part is about 220 mΩ, and is able to deliver 116 dBμV over 2 Ω load. The data acquisition system (DAS) has a coupling circuit, a band pass filter (BPF), and a 16-bits analog to digital converter (ADC) with a sampling frequency of $f_s = 1$ Msamples/s. The use of a high-resolution ADC and a large oversampling factor avoids the need for an amplifier. The reason is that the power spectral density (PSD) of the quantization noise is -33.14 dBμV/$\sqrt{\text{Hz}}$, which is between 42 dB (at low frequencies) and 28 dB (at high frequencies) lower than the PSD of the weakest measured noise, as it will be shown in Section 5. The input impedance of the DAS is about 1.2 kΩ. Hence, the loading effect is negligible, since the largest input impedance values of the power line network are on the order of tenths of Ω.

Estimates of the channel response are computed from the input and output signals $v_T(t)$ and $v_R(t)$ shown in Figure 1. Therefore, the attenuation due to both coupling circuits (about 1 dB in the passband) and the coupling loss between the SGS and the power line network are measured as part of the channel. The latter effect could be compensated, since the input impedance of the channels has also been measured. In fact, this must be done when the results are to be used in a channel emulator which separately models both magnitudes, channel response and input impedance [17]. However, the characterization accomplished in this work is intended for assessing the performance of actual communication systems. Hence, all the effects that are present in a real situation have to be taken into account, including impedance mismatch.

The characterization of the 3 × 3 MIMO links is done using the sequential measurements of the nine SISO channels accomplished with the aforementioned setup. Since the input phases not involved in the SISO channel that is being measured are left open circuit, this does not exactly model the actual situation in which these phases would be loaded with the impedance of the MIMO transmitter.

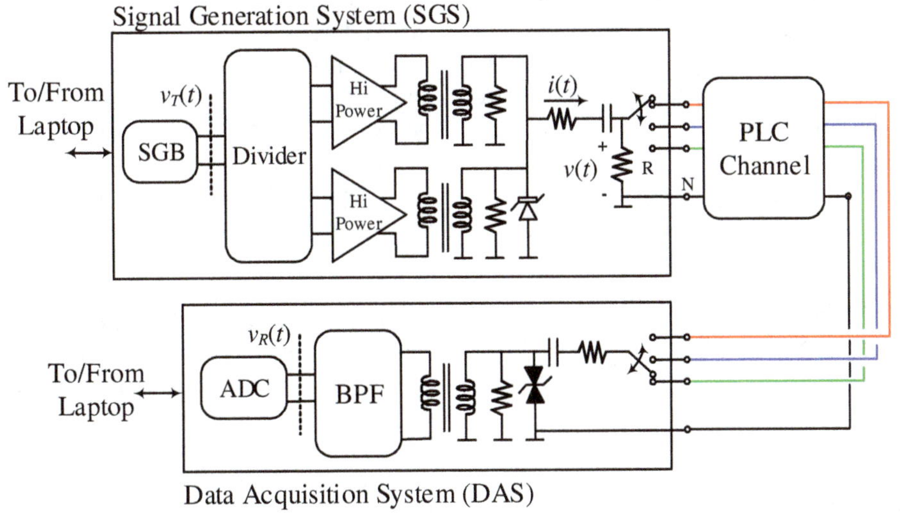

Figure 1 Simplified scheme of the measurement setup.

However, this does not limit the validity of the results since the influence of the impedance connected to the unused phases has proven to be negligible. Simulations using multiconductor transmission line (MTL) theory have shown that the response of the MIMO streams and their spatial correlation is essentially determined by the cable characteristics, by the topology of the underlying LV line, and by the loads connected to the same phase.

Measurements have been carried out in nine networks in the Center and North West of Spain. Table 1 summarizes the information of each site. A total of 106 links have been measured in three environments: rural (27), suburban (29), and urban (50). Each of them is characterized as a 3 × 3 MIMO. Hence, the analysis involves 954 channel responses. Measurements are always performed in the LV part of the network, generally between the LV busbar of the MV/LV transformer and the meter position. However, measurements are performed at intermediate locations in cases like pole-mounted transformers. As shown in Table 1, cables are generally deployed underground (UG) in the urban scenario, overhead (OH) in the rural one, and a mix of both in the semiurban scenario. In rural areas, all customers are connected to a single LV line, while in urban ones, the number of lines deployed from the transformer station is much larger. Table 1 also indicates the linear distance between the transformer station and the closest and the farthest meters.

2.2 Measurements processing

The noise signal in the jth port, $n^j(t)$, is registered during $C = 26$ European mains cycles. Since NB-PLC noise has cyclostationary nature, magnitudes employed for its characterization must be computed synchronously with the mains. To this end, the jitter of the mains signal in the cth cycle, $\tau^j(c)$, must be taken into account. Hence, each mains cycle is divided into L intervals with $N_L = \frac{f_s}{f_0 L}$

Table 1 Description of the measured scenarios

Site	Scenario	Cabling	Number of LV lines	Number of customers	Min. dist. (m)	Max. dist. (m)	Number of MIMO links
1	Rural	OH	2	126	6	427	10
2	Rural	OH	1	105	14	359	6
3	Rural	OH	1	66	10	710	5
4	Rural	OH	1	67	12	1063	6
5	Semiurban	OH	2	124	16	615	10
6	Semiurban	UG	4	132	5	163	6
7	Semiurban	OH	2	183	34	423	13
8	Urban	OH & UG	24	737	4	340	36
9	Urban	UG	12	329	19	65	14

samples, where f_o denotes the frequency of the mains signal. The nominal and actual lengths of the cth mains cycle are $N_C = f_s/f_o$ and $N_C + \tau^j(c)$ samples, respectively. The noise captured in the jth port during the ℓth interval of the cth cycle can then be written as:

$$n^j_{c,\ell}(n) = n^j(n + cN_c + \sum_{i=0}^{c} \tau^j(i) + \ell N_L), \tag{1}$$

with $0 \leq n \leq N_L - 1$.

This framework is employed to assess the spectral and temporal characteristics of the noise. In this paper, the former is accomplished in terms of the PSD, which is estimated by means of the periodogram:

$$P^j_c(\ell, k) = \frac{1}{UN_L} \left| \sum_{n=0}^{N_L-1} w(n) n^j_{c,\ell}(n) e^{-j\frac{2\pi}{N_L}kn} \right|^2, \tag{2}$$

where $w(n)$ is a Hanning window of N_L samples and U is the normalization factor that removes the estimation bias [22]. The cyclostationary and periodic components of the noise have a frequency multiple of the mains. Hence, an estimate of the frequency sampled version of the PSD can be obtained by performing an averaging of the periodograms in (2) with $L = 2$:

$$\widehat{S}_{N^j}(k) = \widehat{S}_{N^j}(f)\big|_{f=k\frac{f_s}{N_L}} = \frac{1}{CL} \sum_{c=0}^{C-1} \sum_{\ell=1}^{L} P^j_c(\ell, k). \tag{3}$$

The employed value of L yields a frequency resolution of 100 Hz.

In order to characterize the noise variation, each mains cycle has now been divided into $L = 80$, and the average value of the energy (along multiple mains cycles) in these intervals is estimated as:

$$E^j(\ell) = \frac{1}{C} \sum_{c=1}^{C} \sum_{n=0}^{N_L-1} \left| n^j_{c,\ell}(n) \right|^2. \tag{4}$$

The selected value of L leads to a time resolution of 250 µs.

Channel response measurements are obtained by transmitting an orthogonal frequency division multiplexing (OFDM)-like sounding signal generated using a 2048-point DFT. Its lengths also equals $C = 26$ European mains cycles. An estimate of the frequency response is then obtained by averaging the least squares (LS) estimations obtained from each symbol. The moderate length of the acquired signal has obliged to accomplish an asynchronous averaging (with respect to the mains) of the LS estimates in order to achieve a reasonable signal-to-noise ratio (SNR). This provides an estimate of the average channel response in the frequency range 40 to 91 kHz, veiling possible periodic variations in the channel response [23].

The input impedance of the power line channel is estimated during the transmission of the OFDM signal used for channel sounding. The current and voltage signals $i(t)$ and $v(t)$ shown in Figure 1 have been employed for this purpose, following a similar approach to the one in [24]. To this end, the resistance R is fixed to a much larger value than the one of the PLC grid. For the sake of clarity, the circuits used for conditioning and digitizing $i(t)$ and $v(t)$ are not shown.

3 Impedance characterization

The input impedance of the power line network in the considered band is frequency selective. Its magnitude generally increases with frequency, reaching maximum values of tenths of Ω. Illustrative shapes can be found in [18,24].

Figure 2 depicts the cumulative distribution function (CDF) of the frequency-averaged magnitude of the impedance values measured in each scenario. As seen, the median values (50% probability) are always below 10 Ω. Moreover, in the semiurban and urban scenarios, they are even below 5 Ω. These scenarios exhibit lower impedance values, with respect to the rural one, mainly because of the loading effect caused by the customers located in the same metering room.

The magnitudes shown in Figure 2 pose a twofold problem in the design of NB-PLC systems. The first one is the difficulty of injecting signal levels of up to 5 V, as the ones allowed by the EN 50065-1 [13], into such low impedance values. The second one is the minimization of the coupling loss from the transmitter to the PLC network. To this end, the most desirable situation is to make the output impedance of the transmitter negligible with respect to the input impedance of the PLC grid. However, this obliges to make an output impedance on the order of a few mΩ, which is in the range of the resistance of some printed circuit board traces or the transformers wiring. On the other hand, conjugate impedance matching is also technologically difficult because of the aforementioned frequency-selective behavior of the impedance and its large variation between locations.

4 Channel response characterization

NB-PLC channel responses are frequency selective. Figure 3 depicts the channel response of six actual channels, two per measured scenario. One of the channels can be assumed to have an 'average' response, and the other is among the worst channels measured in the corresponding scenario. The frequency selectivity is caused by reflections at discontinuities, which lead to a multipath propagation phenomenon and by the loading effect of the in-home network. However, in Europe, the latter seems to be more important in the CENELEC-A band. The reason is that links are usually too short for the conductors to behave as transmission lines. As a rule of thumb, this occurs when

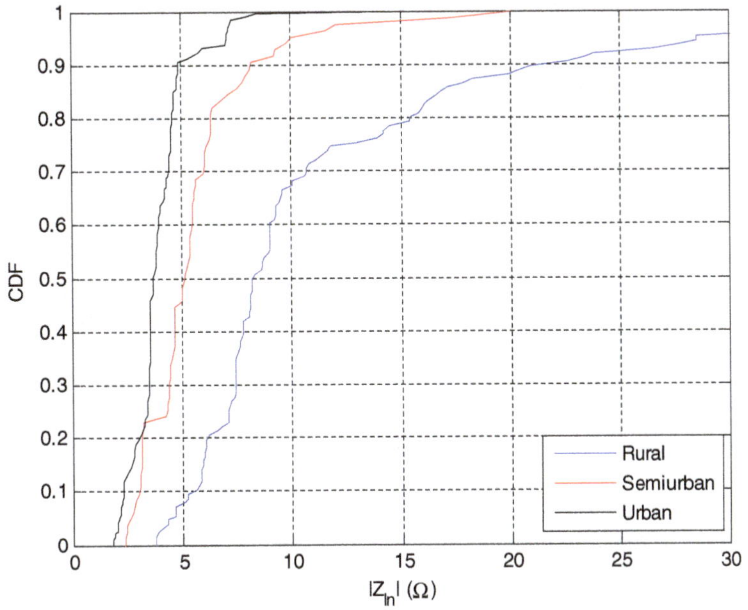

Figure 2 CDF of the frequency-averaged magnitude of the input impedance.

the wavelength of the transmitted signal is, at least, ten times larger than the involved distances. In the employed frequency band (40 to 91 kHz), this translates into 5.25 and 2.33 km, respectively, which are much longer than the distances usually involved in European grids [25].

4.1 Statistical analysis

The coherence bandwidth and the delay spread are the most widespread parameters used to characterize the frequency selectivity of a channel response and its time-domain counterpart, the time dispersion. In this paper, the former is computed as the frequency separation for which the spaced-frequency correlation function falls down to 0.9 [26]. Both magnitudes have a plain relation to the parameters of OFDM communication systems, like the ones currently used in NB-PLC. In particular, distortion in the OFDM signal is avoided if the cyclic prefix is larger than the length of the channel impulse response.

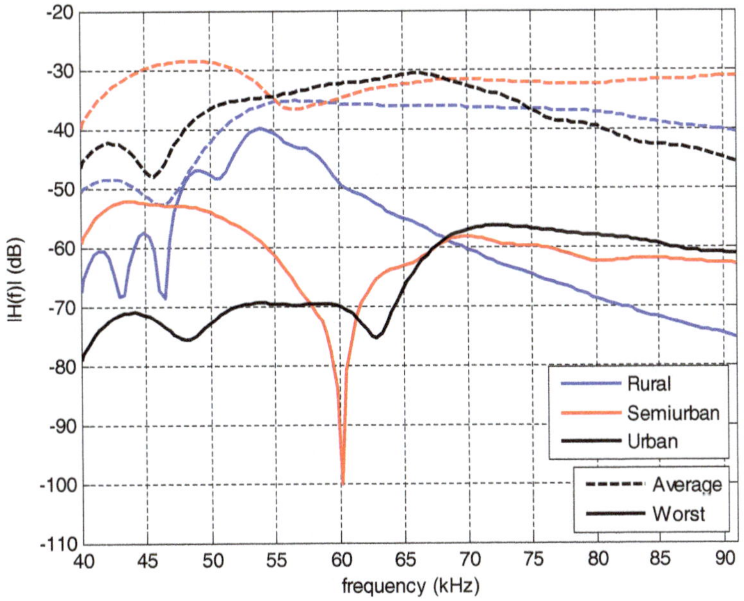

Figure 3 Attenuation profiles of example channels.

Table 2 System parameters of NB-PLC systems standardized by the ITU-T

Parameter	G.9902 (G.hnem)	G.9903 (G3-PLC)	G.9904 (PRIME)
Cyclic prefix length (μs)	60/120	55	120
Intercarrier spacing (Hz)	1562.5	1562.5	480
FFT window length (ms)	0.640	0.640	2048
Max bit rate (kbit/s)	101.3	55.5	64.3

Nevertheless, since increasing the cyclic prefix decreases the symbol rate, its optimum value is usually shorter than the channel impulse response length [27]. The delay spread is a root mean squared (rms) measure of the latter. Hence, the cyclic prefix length should generally be several times larger than the delay spread.

When an insufficient cyclic prefix is used, the magnitude of the distortion increases with the ratio of the carrier bandwidth to the coherence bandwidth of the channel. Hence, the coherence bandwidth can be also used to design the intercarrier spacing of an OFDM system. In order to assess the suitability of the parameters employed in the NB-PLC systems standardized by the ITU-T, their cyclic prefix, intercarrier spacing, FFT window length (i.e., symbol length excluding the cyclic prefix), and maximum bit rate are shown in Table 2. The bit rate indicated for PRIME corresponds to convolutionally coded transmissions. Larger bit rates can be achieved with uncoded transmissions.

Figure 4 depicts the CDF of the delay spread values corresponding to the channels measured in each scenario. In order to assess the suitability of the 'fading modeling method' proposed in the IEEE P1901.2 standard, the CDF computed from 300 channels generated according to this model is also drawn. As seen, rural channels have larger delay spread values than urban and semiurban. This is likely due to the attenuation of the cables, which has a low-pass behavior. It can be also noticed that measured values are larger than modeled ones. In fact, the median value of the semiurban channels is about 45% larger than the median of the modeled ones. This figure rises up to 82% in the rural scenario. Conversely, measured channels have lower coherence bandwidth than modeled ones, as shown in Figure 5, where the CDF of the coherence bandwidth and the average attenuation is depicted. The comparison of the delay spread and the coherence bandwidth with the parameters shown in Table 2 reveals that all the systems have intercarrier spacings lower than the measured channel coherence bandwidth, although the value in the G.hnem/G3-PLC systems is about three times larger than in PRIME[a]. Regarding the cyclic prefix, it seems that the value used in PRIME might be too large for the majority of channels, while the G3-PLC one seems to be too small for the worst channels. In terms of performance, G.hnem seems to provide the most appropriate solution.

Regarding the average attenuation of the measured and modeled channels, Figure 5 shows that semiurban channels experience lower attenuation than rural and urban ones. This might be due to the fact that semiurban scenarios use to involve shorter distances than rural ones and less derivations than urban environments. As seen, the spread of the attenuation in the modeled channels (around

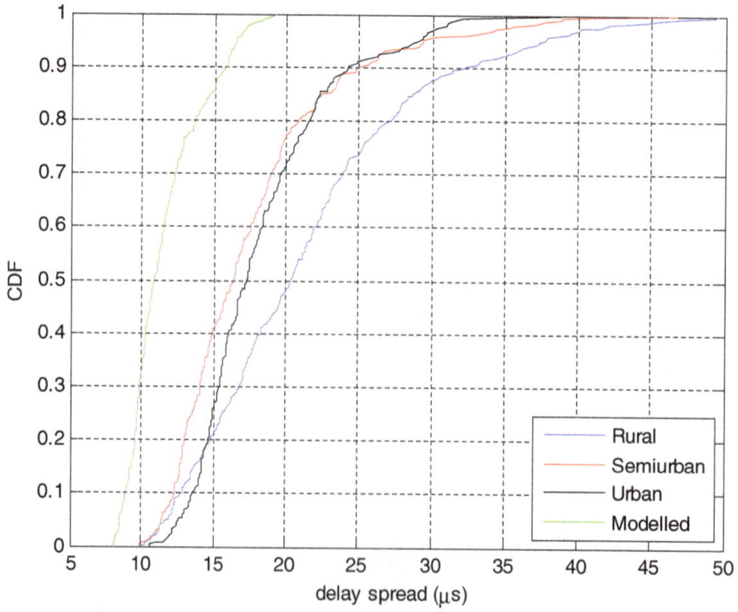

Figure 4 CDF of the delay spread of the measured and modeled channels.

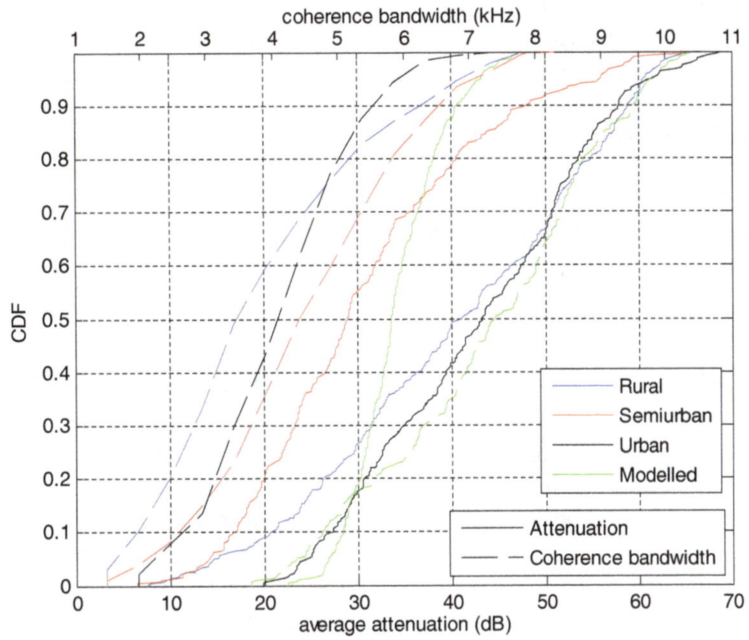

Figure 5 CDF of the average attenuation and the coherence bandwidth of the measured and modeled channels.

25 dB) is much smaller than in the measured ones. This can also be observed in Figure 6, where a scatter plot of the attenuation vs. the delay spread is drawn. As shown, the dispersion of the values in the measured channels is much larger than that in the modeled ones. It is also interesting to highlight that there is no clear correlation between attenuation and delay spread, as it happens in indoor BB-PLC channels. This might be due to the reduced influence that the multipath propagation phenomenon has in NB-PLC, where derivations cause a much flatter power loss than in BB-PLC.

The presented results indicate that the parameters of the 'fading modeling method' are unsuitable for the CENELEC-A band, although they might be appropriate

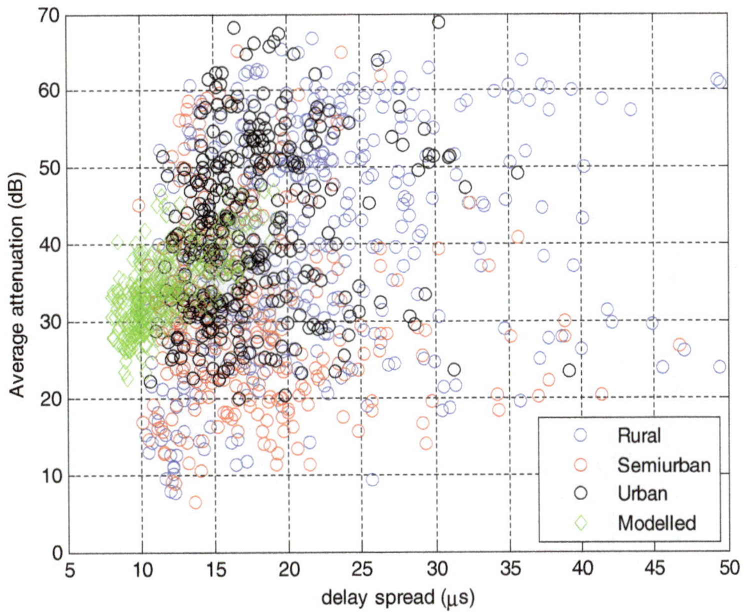

Figure 6 Relation between the average attenuation and the delay spread of the measured and modeled channels.

for the 154.69 to 487.5 kHz range, where the channel response is known to have better transmission characteristics. This band is defined in the IEEE P1901.2 for transmissions in the Federal Communications Commission (FCC) and the Association of Radio Industries and Businesses (ARIB) bands.

The characterization of the 3×3 MIMO links is accomplished by using the channel matrix at each frequency point, **H**, where the frequency index is omitted for the sake of clarity. It is defined as:

$$\mathbf{H} = \begin{pmatrix} H^{11} & H^{12} & H^{13} \\ H^{21} & H^{22} & H^{23} \\ H^{31} & H^{32} & H^{33} \end{pmatrix}, \qquad (5)$$

where H^{ij} denotes the frequency response of the channel between input port i and output port j at the corresponding frequency. The singular value decomposition (SVD) of **H** can be expressed as:

$$\mathbf{H} = \mathbf{U}\mathbf{D}\mathbf{V}^H, \qquad (6)$$

where **U** and **V** are unitary matrices and **D** is a diagonal matrix whose values are the singular values, σ_i, and $(\cdot)^H$ denotes the Hermitian operator. The singular values are related to the eigenvalues of $\mathbf{H}\mathbf{H}^H$, λ_i, as $\sigma_i = \sqrt{\lambda_i}$.

Hence, the SVD decomposes the MIMO channel into a set of orthogonal SISO channels, or streams, with amplitude σ_i. The ratio of the singular values can be used as a measure of the correlation between the constituent channels of the MIMO. If fact, the ratio of the maximum to the minimum singular value is the condition number of the matrix **H**, denoted by $\kappa = \max(\sigma_i)/\min(\sigma_i)$. When the constituent channels of the MIMO are perfectly correlated, all singular values except one will be zero and the ratio is infinite. On the other side, when channels are absolutely uncorrelated, all the singular values have the same value and its ratio equals one.

Figure 7 depicts the values of κ (dB) for the measured MIMO channels at each frequency point, computed as $20\log_{10}(\kappa)$. As seen, it has a quite homogeneous behavior along the considered band in most channels. In addition, it can be observed that most values are larger than 10 dB, which indicates that at least one of the MIMO stream is highly correlated with the others. Figure 8 corroborates this end. It shows the relevant part of the CDF of the singular values corresponding to the three MIMO streams and to the SISO channels. As seen, the singular values of the third stream are about 15 dB below the ones of the first stream. The median values of the first stream are between 4 and 6 dB larger than those of the SISO. Similarly, the median values of the second stream are between 1 and 4 dB lower than the SISO ones. The lowest singular values of the second and third MIMO streams, with respect to the SISO ones, occur in the rural environment. Hence, MIMO gains are expected to be lower here than in the other scenarios.

5 Noise characterization

Noise in NB-PLC is caused by the electrical devices connected to the power grid and to external signals coupled via radiation or via conduction. It is composed of three main components: background noise, narrowband interference (NBI), and impulsive noise. The former is

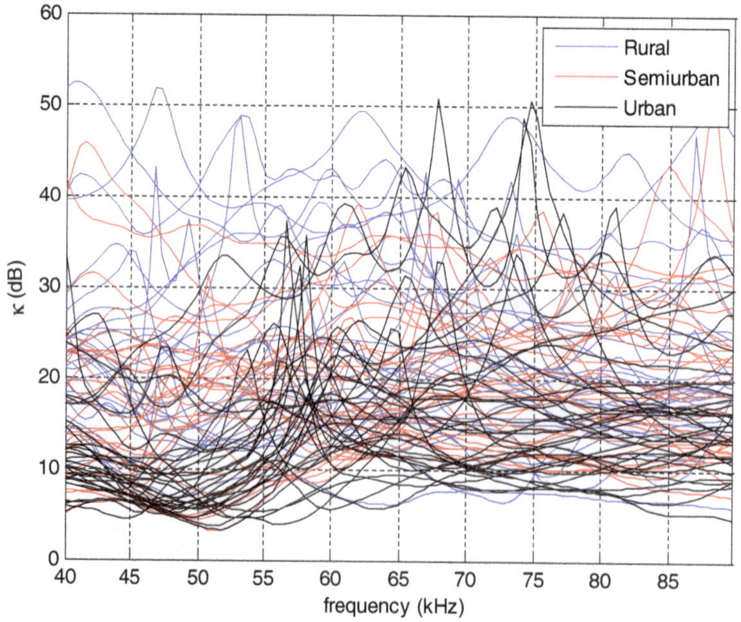

Figure 7 Ratio of the maximum to the minimum singular values.

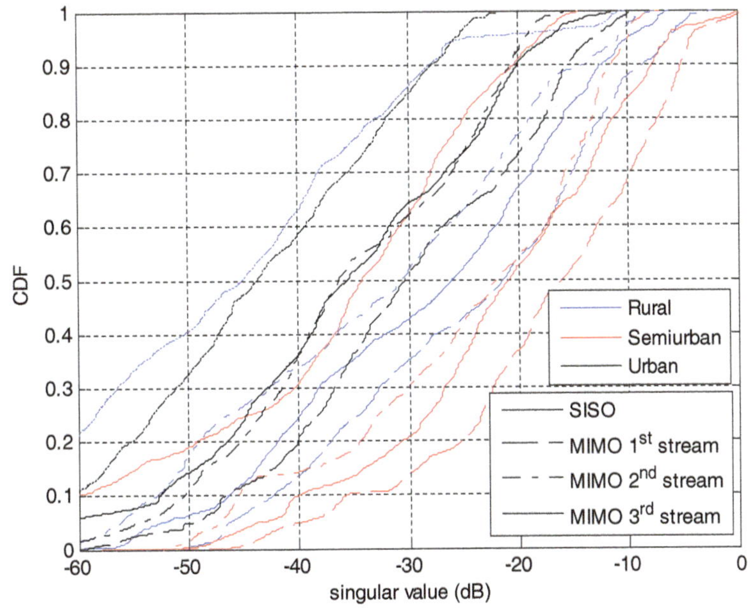

Figure 8 CDF of the singular values.

generally cyclostationary with a frequency multiple of the mains cycle. The latter consists of sporadic (aperiodic) impulses as well as components that are synchronous to the mains [15]. Figure 9 depicts the spectrogram and the waveform of a measured noise register. The periodic impulsive term is clearly observable in both plots. In addition, it shows three high-level NBI around 59, 62, and 73 kHz and a sporadic impulse at around 35 ms.

5.1 Statistical analysis

The first magnitude used to characterize the noise is the PSD, synchronously computed as indicated in Section 2.

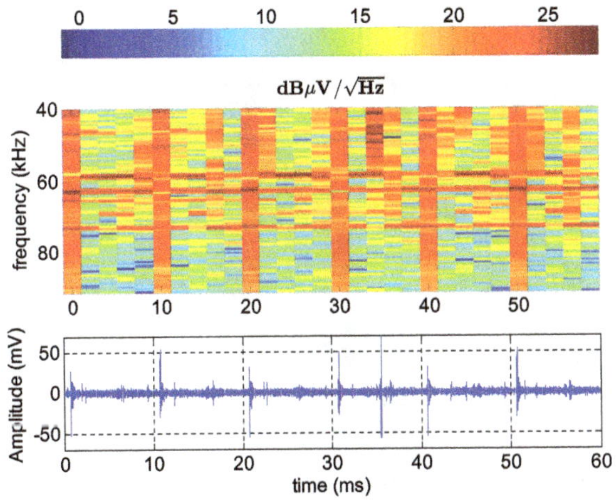

Figure 9 Example of a measured noise waveform and its corresponding spectrogram.

Figure 10 depicts the PSD of the heaviest and the weakest noise register measured in each scenario. The average PSDs have also been included along with their fitting lines obtained using a robust regression,

$$\mathrm{PSD}(f) = \mathrm{PSD}_0 - \Delta\mathrm{PSD} \cdot f(\mathrm{kHz}) \quad (\mathrm{dB}\mu V/\sqrt{Hz}), \quad (7)$$

where PSD_0 and $\Delta\mathrm{PSD}$ are given in Table 3. As seen, they have the well-known $1/f$ decay [15].

Figure 10 also shows the PSD of the 12 noise patterns described in the IEEE P1901.2 model. In this case, the synchronous averaging used for the estimation of the PSD has taken into account that they correspond to a mains frequency of 60 Hz. Their background level is higher than that in the measured ones, especially at high frequencies, where differences can be up to 40 dB. In addition, the modeled noise shows no trace of the NBI. This is somehow surprising because these patterns do clearly reflect NBI in the frequency range above 100 kHz (not shown in the figure).

In order to characterize the time variation of the noise level, Figure 11 depicts the CDF of the peak and rms variation of the energy along the mains cycle, $E^j(\ell)$, computed as described Section 2. Values corresponding to the noise patterns defined in the IEEE P1901.2 have been marked with circles. In this case, the window length has been scaled to take into account that they correspond to a mains frequency of 60 Hz. It can be observed that 50% of the measured noise registers exhibit peak and rms values higher than 13 and 3.4 dB, respectively. The suitability of the noise model given in the IEEE P1901.2 to reflect the range of the noise level variation can be assessed by its

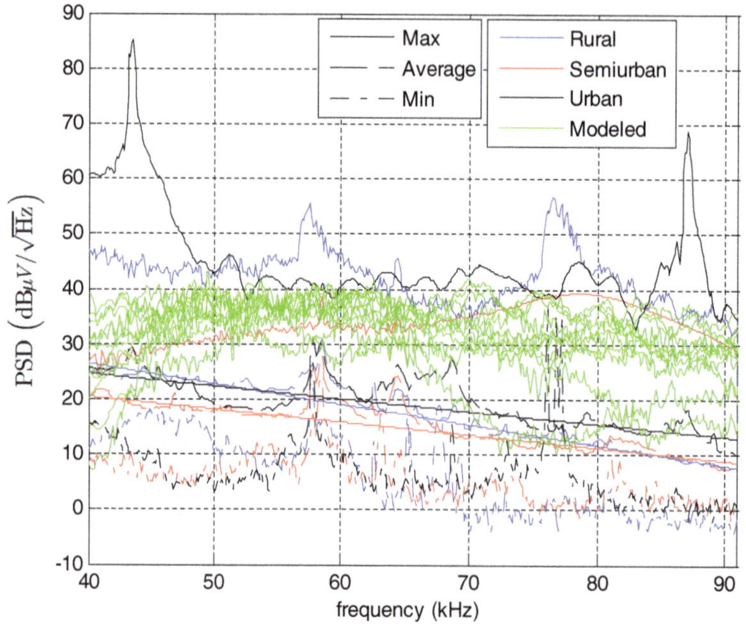

Figure 10 **PSDs of the measured and modeled noise.**

sampling of the measured curves, which is quite good in almost all the quartiles.

Figure 11 has shown the amplitude of the noise level variation along the mains cycle. However, from a communication system perspective, the rate of this variation is even more interesting. It provides an indication of the data rate gain that could be achieved by using an adaptive modulation strategy [28]. This issue has already been explored in the context of NB-PLC systems in [29]. To this end, the duration of the region in which energy variations are lower than 3 dB (which would imply a change in the number of bits/symbol when using QAM modulations) has been computed. The CDFs of these values are shown in Figure 12 for the three scenarios, along with the ones of the IEEE P1901.2 noise model. As seen, the rate at which the energy varies in the modeled noise patterns is larger than that in the measured ones. Thus, only one pattern covers 50% of the measured values. Hence, while the model reflects the magnitude of the measured noise level variations, it underestimates their rate of change. The comparison of the values in Figures 11 and 12 with the FFT window lengths shown in Table 2 confirms that the use of an adaptive modulation strategy can provide fair data rate gains, as already reported in [29].

Table 3 **Values of the noise PSD fitting curves**

Scenario	PSD_0	ΔPSD
Rural	41.412	$371.863 \cdot 10^{-3}$
Semiurban	29.879	$233.782 \cdot 10^{-3}$
Urban	33.939	$230.935 \cdot 10^{-3}$

6 Performance assessment

The most straightforward measure of the performance that can be achieved in a given channel is the capacity. However, it requires precise knowledge of the noise statistics, which are still unknown in NB-PLC. Moreover, simple closed-form expressions are known only for certain distributions, like the Gaussian one. An alternative

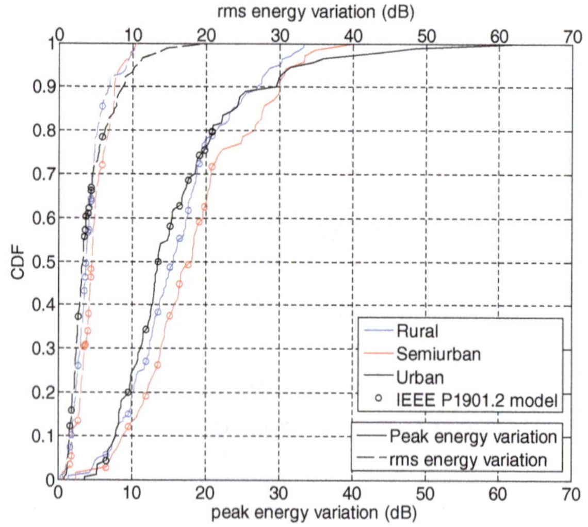

Figure 11 **CDF of the peak and rms energy variation along the mains cycle for the measured and modeled noise.**

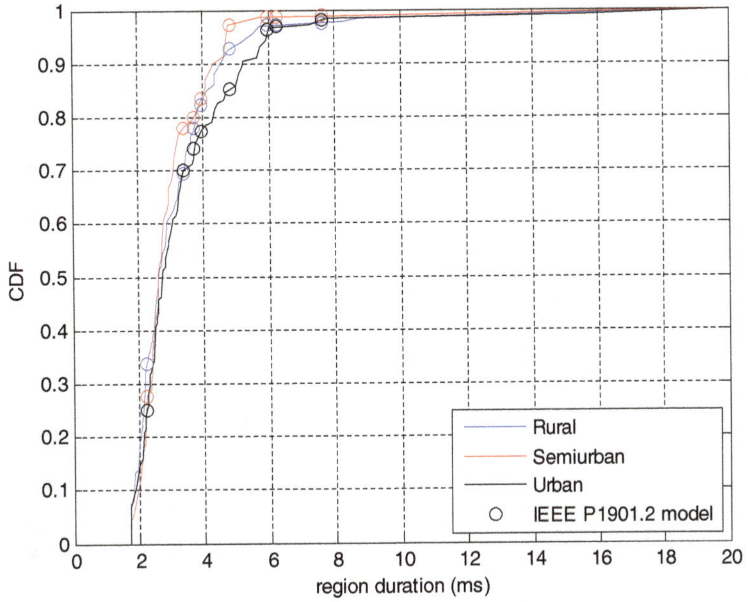

Figure 12 CDF of the duration of the regions in which energy variations are lower than 3 dB.

approach could be to assess the performance that can be attained by standardized systems. Results obtained in this way do not really inform of the potentiality of the channel, which might be under underutilized. Moreover, significant performance differences can be found depending on the algorithms employed at the receiver.

When the noise distribution is unknown, a lower bound on the performance can be obtained by assuming that it is Gaussian-distributed and using the well-known expression of the capacity of a set of parallel flat fading channels with bandwidth Δf and stationary Gaussian noise [30]:

$$R_{\mathrm{SISO}} = \sum_{k\in K} \Delta f \log_2\left(1 + \frac{P_k|H_k|^2}{N_k}\right), \quad (8)$$

where P_k, H_k, and N_k are the injected power, the frequency response, and the noise power in the kth channel, respectively. K denotes the set of employed frequency indexes. In the case of the MIMO channel, capacity is computed as [30]:

$$R_{\mathrm{MIMO}} = \sum_{j=1}^{3}\sum_{k\in K} \Delta f \log_2\left(1 + \frac{P_k^j\lambda_k^j}{N_k^j}\right), \quad (9)$$

where j denotes the index of the MIMO stream and λ_k^j, N_k^j, and P_k^j are the eigenvalue, the noise, and the input power at the kth frequency index of the jth stream, respectively.

This is the approach used in this paper. For the sake of simplicity, the values of N_k and N_k^j have been taken from the fitting PSDs whose parameters are shown in Table 3. Noise at the MIMO ports is assumed to be uncorrelated. However, to provide values as close to the state-of-the-art technology as possible, the following practical constraints have been taken into account: a SNR gap of 5 dB has been included to model the SNR loss caused by the use of practical constellations; the back-off of the power amplifier at the transmitter has been assumed to be 8 dB, and the maximum number of bits per constellation symbol has been fixed to 6. In order to explore the performance limit of the channel, bit rate values achieved without the latter constraint have also been computed.

Limits for the transmitted level are defined in the EN 50065-1 [13]. It fixes both the signal level (134 dBμV) and a PSD mask (120 dBμV/200 Hz). When using MIMO communications, the maximum level that can be injected on any phase is 6 dB lower than that in the SISO case. The PSD constraint is the most restrictive one because transmitting at 120 dBμV/200 Hz in the 40 to 90 kHz band results in a signal level that exceeds 134 dBμV. Hence, in this paper, a flat PSD of 110 dBμV/200 Hz has been employed.

Figure 13 depicts the relevant part of the CDF of the bit rates attained in the measured SISO and MIMO channels. Curves denoted as 'unconstrained' have been obtained without restricting the number of bits per constellation symbol. Values corresponding to the modeled channels have been computed by combining each of the 300 generated channels whose characteristics have been presented in Section 4 with the 12 noise patterns shown in Section 5. As seen, performance estimated with the models are unrealistically high because of the underestimation of the channel attenuation (see Figure 5), which is larger than

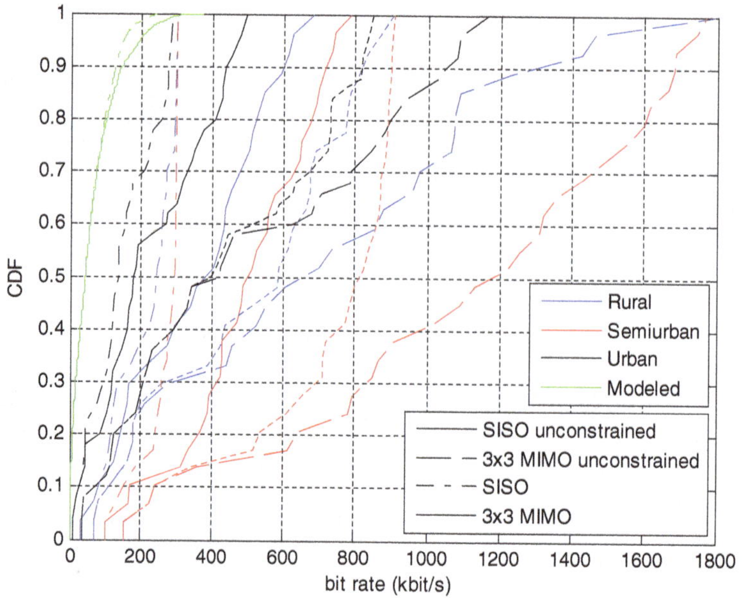

Figure 13 CDF of the bit rate achieved by the MIMO and SISO systems.

the overestimation of the noise levels (see Figure 10). Regarding the measured values, the lower bit rates are obtained in the urban and rural channels. These results are in accordance with the attenuation and noise levels shown in Figures 5 and 10, respectively. They show that the noise level is similar in all scenarios, but rural and urban ones experience higher attenuation. The median bit rates of SISO links in the urban, semiurban, and rural environments are 137, 295, and 246 kbit/s, respectively. It these scenarios, 90% of the SISO links can deliver more than 28, 153, and 80 kbit/s, respectively.

The comparison of the bit rates in Figure 13 and the ones achieved by the actual systems given in Table 2 reveals that their performance is largely limited by complexity (cost) constraints. In fact, they use constellations with at most 4 bits/symbol. In PRIME, all carriers must employ the same constellation. The same approach has been adopted in G3-PLC, except for the possibility of avoiding transmitting in groups of adjacent carriers with very low SNR. Since the noise in NB-PLC is strongly colored and the channel response is frequency selective, using the same constellation in all employed carriers causes a severe performance degradation. Additionally, to reduce the memory size, the efficiency of the physical layer frames (data symbols length/frame length) has been penalized. In G3-PLC, for instance, the impossibility of fitting more than two Reed-Solomon blocks per frame leads to an efficiency of 78% when D8PSK is employed.

Figure 13 also shows the bit rate that could be attained with a 3 × 3 MIMO system in which the maximum power level allowed by the EN 50065-1 is injected in all phases. As seen, MIMO can offer significant bit rate improvements when the complexity of the system is constrained. The minimum bit rate attained in 90% of the channels in the urban, semiurban, and rural environments improves to 73 kbit/s (2.61×), 241 kbit/s (1.57×), and 108 kbit/s (1.45×), respectively. Performance gains in 50% of the urban, semiurban, and rural channels are 2.89×, 2.75×, and 2.28×, respectively. It should be taken into account that despite the weak contribution of the third stream, part of the MIMO gain comes from the larger singular values of the first stream with respect to the SISO one. As expected from Figure 8, the lowest gains are obtained in the rural scenario. Removing the constraint on the number of bits per constellation symbol reduces the median values of the MIMO gain in the urban, semiurban, and rural channels to 2.34×, 2.35×, and 1.72×, respectively. As expected, it does affect the gains achieved in links with bad transmission characteristics.

7 Conclusions

This paper has presented a statistical analysis of the characteristics of NB-PLC channels in the CENELEC-A band and has assessed their performance when used for advanced metering infrastructure (AMI) communications in rural, semiurban, and urban scenarios. The accomplished study has included the noise, the channel response, and the input impedance of the power line network. The analysis of the former has examined both its spectral and temporal characteristics. The channel response has been studied in terms of the delay spread, the coherence bandwidth, and the attenuation. An estimation of the data rates that could be achieved has been accomplished. The obtained results indicate that the highest performance is achieved in semiurban scenarios and the lowest in urban ones. In the former, 90% of the channels can deliver more than 153 kbit/s, while in the latter, this figure goes down to 28 kbit/s. Some of the reasons that prevent current NB-PLC systems from achieving these performance have been highlighted. In addition, the use of the three phase conductors for MIMO communications has been explored. It has been shown that a practical 3×3 MIMO system can give performance improvements larger than $2.61\times$, $1.57\times$, and $1.45\times$ in 90% of the urban, semiurban, and rural channels, respectively.

Endnote

[a]The low intercarrier spacing in PRIME is motivated by the use of a differential-in-frequency modulation.

Competing interests

The authors declare that they have no competing interests.

Acknowledgments

The authors thank Atmel corporation for the support provided to perform all the tests referred in the paper.

The authors would also like to thank the anonymous reviewers for their valuable comments and suggestions.

References

1. H Farhangi, The path of the Smart Grid. IEEE Power Energy Mag. **8**(1), 18–28 (2010)
2. Kema Limited. GB Demand Response. Report 1: The Opportunities for Demand Response. Commissioned by the Energy Association. (Version 1.0) (Kema Limited, Arnhem, 2011)
3. G Strbac, C Kim Gan, M Aunedi, V Stanojevic, P Djapic, J Dejvises, P Mancarella, A Hawkes, D Pudjianto, S Le Vine, J Polak, D Openshaw, S Burns, P West, D Brogden, A Creighton, A Claxton. Benefits of Advanced Smart Metering for Demand Response based Control of Distribution Networks. ENA/SEDG/Imperial College report on Benefits of Advanced Smart Metering (Version 2.0) (Energy Networks Association, London, 2010)
4. H Edelmann, T Kästner. Cost-Benefit Analysis for the Comprehensive Use of Smart Metering. Technical report, Ernst & Young On Behalf of the Federal Ministry of Economics and Technology (Ernst & Young GmbH, Germany, 2013)
5. S Galli, A Scaglione, Z Wang, For the grid and through the grid: the role of power line communications in the Smart Grid. Proc. IEEE. **99**(6), 998–1027 (2011)
6. The OPEN meter Consortium, Description of current state-of-the-art of technology and protocols description of state-of-the-art of PLC-based access technology. European Union Project Deliverable FP7-ICT-2226369, d 2.1 Part 2, Version 2.3, March 2009. http://www.openmeter.com/files/deliverables/OPEN-Meter%20WP2%20D2.1%20part2%20v2.3.pdf
7. A Mengi, M Waechter, M Koch, in *Proceedings of the IEEE International Symposium on Power Line Communications and Its Applications (ISPLC)*. 500 kHz G3-PLC access technology for the roll-outs in Germany (Glasgow, 2014)
8. Engage Consulting Limited. High-level smart meter data traffic analysis. Technical report for Energy Networks Association (Energy Networks Association, London, 2010)
9. ITU.T Recommendation G.9902. Narrowband orthogonal frequency division multiplexing power line communication transceivers for ITU-T G.hnem networks (ITU, Geneva, 2012)
10. ITU-T Recommendation G.9903. Narrowband orthogonal frequency division multiplexing power line communication transceivers for G3-PLC networks (ITU, Geneva, 2013)
11. ITU-T Recommendation G.9904. Narrowband orthogonal frequency division multiplexing power line communication transceivers for PRIME networks (ITU, Geneva, 2012)
12. IEEE P1901.2. IEEE Standard for Low-Frequency (less than 500 kHz) Narrowband Power Line Communications for Smart Grid Applications (IEEE, New York, 2013)
13. European Committee for Electrotechnical Standardization (CENELEC). Signalling on low-voltage electrical installations in the frequency range 3 kHz to 148,5 kHz - part 1: general requirements, frequency bands and electromagnetic disturbances. European Standard EN 50065-1 (CENELEC, Brussels, 2001)
14. S Galli, Recent developments on the international standardization of narrowband PLC for smart grid applications. Keynote given at the IEEE Internationaly Symposium on Power Line Communications and its Applications (ISPLC) (2012). http://www.ieee-isplc.org/2012/program_44_1704874382.pdf
15. M Nassar, J Lin, Y Mortazavi, A Dabak, IH Kim, BL Evans, Local utiliy power line communications in the 3-500 khz band: channel impairments, noise, and standards. IEEE Signal Process. Mag. **29**(5), 116–127 (2012)
16. M Zimmermann, K Dostert, A multipath model for the powerline channel. IEEE Trans. Commun. **50**(4), 553–559 (2002)
17. W Liu, M Sigle, K Dostert, Channel characterization and system verification for narrowband power line communication in Smart Grid applications. IEEE Commun. Mag. **49**(12), 28–35 (2011)
18. G Chu, J Li, W Liu, in *Proceedings of the, International Symposium on Power Line Communications and Its Applications (ISPLC)*. Narrow band power line channel characteristics for low voltage access network in China (Johannesburg, 2013)
19. LT Berger, A Schwager, P Pagani, Scheneider DM (eds.), *MIMO Power Line Communications: Narrow and Broadband Standards, EMC and Avanced Processing*, 1st edn. (CRC Press, Boca Raton, FL, 2014)
20. Y Sugiura, T Yamazato, M Katayama, in *Proceedings of the IEEE International Symposium on Power Line Communications and Its Applications*. Measurement of narrowband channel characteristics in single-phase three-wire indoor power-line channels (Jeju, 2008), pp. 18–23
21. A Sendin, A Llano, A Arzuaga, I Berganza, in *Proceedings of the IEEE International Symposium on Power Line Communications and Its Applications*. Strategies for PLC signal injection in electricity distribution grid transformers (Udine, 2011), pp. 346–351
22. AV Oppenheim, RW Schafer, J Buck, *Discrete-time Signal Processing*. (Prentice Hall, New Jersey, 1999)
23. FJ Cañete, JA Cortés, L Díez, JT Entrambasaguas, Analysis of the cyclic short-term variation of indoor power line channels. IEEE J. Selected Areas Commun. **24**(7), 1327–1338 (2006)
24. M Sigle, W Liu, K Dostert, in *Proceedings of the, IEEE International Symposium on Power Line Communications and Its Applications*. On the impedance of the low-voltage distribution grid at frequencies up to 500 kHz (Beijing, 2012), pp. 30–34
25. A Sendin, I Peña, P Angueira, Strategies for power line communications smart metering network deployment. Energies. **2014**(7), 2377–2420 (2014)
26. JG Proakis, *Digital Communications*. (McGraw-Hill, Massachusetts, 1995)
27. JA Cortés, L Díez, FJ Cañete, JJ Sánchez-Martínez, JT Entrambasaguas, Performance analysis of OFDM modulation on indoor broadband PLC channels. EURASIP J. Adv. Signal Process. **2011**(78), 1–12 (2011)
28. A Goldsmith, S Chua, Variable-rate variable-power MQAM for fading channels. IEEE Trans. Commun. **45**(10), 1218–1230 (1997)
29. KF Nieman, J Lin, M Nassar, K Waheed, BL Evans, in *Proceedings of the International Symposium on Power Line Communications and Its Applications (ISPLC)*. Cyclic spectral analysis of power line noise in the 3-200 kHz band (Johannesburg, 2013)
30. A Goldsmith, *Wireless Communications*. (Cambridge University Press, New York, 2005)

An efficient method to include equality constraints in branch current distribution system state estimation

Carlo Muscas*, Marco Pau, Paolo Attilio Pegoraro and Sara Sulis

Abstract

Distribution system state estimation is a fundamental tool for the management and control functions envisaged for future distribution grids. The design of accurate and efficient algorithms is essential to provide estimates compliant with the needed accuracy requirements and to allow the real-time operation of the different applications. To achieve such requirements, peculiarities of the distribution systems have to be duly taken into account. Branch current-based estimators are an efficient solution for performing state estimation in radial or weakly meshed networks. In this paper, a simple technique, which exploits the particular formulation of the branch current estimators, is proposed to deal with zero injection and mesh constraints. Tests performed on an unbalanced IEEE 123-bus network show the capability of the proposed method to further improve efficiency performance of branch current estimators.

Keywords: Distribution systems; State estimation; Branch current estimator; Weighted least squares; Equality constraints

Introduction

In the smart grid (SG) scenario, where control and management activities of the electric distribution network are expected to play a relevant and increasing role, distribution system state estimation (DSSE) is conceived as a fundamental monitoring tool. In fact, control systems, such as distribution management systems (DMSs), must rely on a possibly complete and accurate knowledge of the state of the network given by DSSE [1].

In the current evolving scenario, the increasing presence of distributed generation (DG), storage devices and flexible loads to be controlled leads to unforeseen dynamics, which require suitable measurement responsiveness. In this context, DSSE techniques able to work at high reporting rates, while safeguarding the estimation accuracies required by specific applications, are needed. DSSE has to be able to include all the different measurement types, provided by traditional and modern measurement devices, which can be available with different frequencies and accuracies. In particular, the phasor measurement units (PMUs) [2,3], which give phasor measurements of both voltages and currents synchronized with respect to a common time reference (the so-called synchrophasors) are becoming increasingly widespread in transmission systems [4-6] and are expected to be widely used also in distribution systems, along with new-generation power meters.

For these reasons, innovative and dedicated solutions able to estimate the operating point of the future distribution network are increasingly needed.

DSSE techniques are based on measurement models that link measurements with the state variables to be estimated. They are intended to elaborate the measurements acquired from the field, by means of distributed measurement systems, along with all the *a priori* information that can be collected about load and generator activity.

In fact, it is impractical and economically infeasible to have a fully monitored distribution network, where each node is equipped with a measurement device connected to the monitoring infrastructure. The DSSE thus reaches observability by relying on the so-called pseudo-measurements [7], which include historical or forecast data on generator production and load consumption.

*Correspondence: carlo@diee.unica.it
Department of Electrical and Electronic Engineering, University of Cagliari, Piazza d'Armi 1, 09123 Cagliari, Italy

DSSE approaches proposed in the literature are mainly based on weighted least squares (WLS) algorithms [7-12] and mainly differ between each other in the chosen state variables and in the way the different measurement types are included. In particular, two main classes of DSSE exist, based on two different choices of the underlying state vector: node voltage state estimators, NV-DSSEs (as in [7,13]) and branch current estimators, BC-DSSEs (as [8-12]). BC-DSSE is suitably designed to better keep into account the peculiar characteristics of distribution systems, as the radial or weakly meshed topology and the high r/x ratios, and it is usually faster with respect to those based on voltage state (see [14]).

An important topic, which can influence both accuracy and speed of DSSE based on WLS methods, concerns the constraints given by *a priori* knowledge on network operation and topology. For instance, *a priori* information exploitable by DSSE includes also the identification of the so-called 'zero injections', that is of the nodes that are surely known to have no power consumption or generation. Zero injections are frequent in a distribution grid, in particular, because in a three-phase unbalanced context, some nodes may have no loads or generators connected to some of the phases. Besides, the inactivity of a load or generator, if it represents an absolutely sure information, could be also translated in a zero injection constraint. Additional constraints can be also present. As an example, in a branch current formulation, possible meshes have to be duly considered.

Such constraints can be treated in different ways. The simplest method is to consider them as virtual pseudo-measurements and to use a large weight to enforce them. This choice can lead to numerical conditioning problems [15], and thus other options have been advanced in the literature. In [16], for instance, it is proposed to include them using constraints expressed through Lagrange multipliers, while in [17], they are treated as normal measurements with a low weight and the constraints are re-imposed between the subsequent iterations of the WLS.

In this paper, a simple way to deal with the equality constraints, well-suited to BC-DSSE (and in particular to the efficient formulation presented in [12]) and based on state vector reduction, is proposed. This approach is compared with other traditional and commonly used techniques to underline the advantages by means of simulation results obtained on a IEEE 123-buses three-phase test network.

Branch current state estimation

State estimation techniques are based on mathematical relations between system state variables and measurements collected from the distributed measurement system. The measurements in a distribution grid can be the traditional ones, as voltage and current magnitudes, real and reactive power flows, and power injections at buses, or the current and voltage synchrophasors provided by PMUs. Usually, distribution networks are only partially monitored. As a consequence, prior information on the loads (the so-called pseudo-measurements) are necessary to perform state estimation. Thus, the forecasts of power injections usually constitute the majority of the measurements available for DSSE.

As aforementioned, different state variables can be considered for the estimation algorithm and in particular node voltages or branch currents. Such variables can be represented either in polar or rectangular coordinates. In this paper, the enhanced branch current-based estimator proposed in [12] is adopted, as it was shown to be as accurate as those traditionally based on node voltages and more efficient in the practical application of distribution networks. This algorithm will be referred to in the following as BC-DSSE.

The general measurement model adopted for state estimation is:

$$\mathbf{z} = \mathbf{h}(\mathbf{x}) + \mathbf{e} \tag{1}$$

where $\mathbf{z} = [z_1 \dots z_M]^{\mathrm{T}}$ is the vector of the measurements obtained from real instrumentation in the network and of the chosen pseudo-measurements, $\mathbf{h} = [h_1 \dots h_M]^{\mathrm{T}}$ is the vector of measurement functions, $\mathbf{x} = [x_1 \dots x_N]^{\mathrm{T}}$ is the vector of the chosen state variables and $\mathbf{e}$ is the measurement error vector, which is a zero-mean random vector with covariance matrix $\mathbf{\Sigma_z}$. Measurement functions in $\mathbf{h}$ can be nonlinear, depending on the type of considered measurements, and are strictly influenced by the topology and the parameters (impedances) of the network.

The state vector $\mathbf{x}$, in the BC-DSSE, is given by the branch currents of all the N_{br} network branches, in rectangular coordinates, and the voltage v_s at a reference node, for instance the slack bus. In a three-phase framework, $\mathbf{x}$ is $\left[\mathbf{x}_A^{\mathrm{T}}, \mathbf{x}_B^{\mathrm{T}}, \mathbf{x}_C^{\mathrm{T}}\right]^{\mathrm{T}}$, with $\mathbf{x}_\phi$ ($\phi = A, B, C$) equal to $\left[v_{s\phi}^r, v_{s\phi}^x, i_{1\phi}^r \dots i_{N_{\mathrm{br}\phi}}^r, i_{1\phi}^x \dots i_{N_{\mathrm{br}\phi}}^x\right]^{\mathrm{T}}$, under the hypothesis that synchronized measurements are present. This formulation exploits the absolute phase angles provided by PMU measurements, which use the common time reference of the coordinated universal time (UTC). Such time synchronization can be obtained by means of Global Positioning system (GPS) or other synchronization sources (see for instance [18,19]).

Pseudo-measurements, in the model (1), are handled as measurements that are assigned with a higher standard deviation σ to highlight the lower accuracy due to the fact they are not based on real measurements but rather on historical and forecast data.

In BC-DSSE, the estimation of the state $\mathbf{x}$ is obtained by an iterative algorithm. Each iteration consists of:

- definition/update of measurements and residuals;
- branch current estimation applying a WLS method;
- network voltage state computation through a forward sweep calculation.

For each iteration k, in the first step of BC-DSSE, power measurements are translated into equivalent phasor current measurements using the node voltages estimated in the previous iteration. This approach allows including power measurements (and above all pseudo-measurements) easily in the estimator, since equivalent current injections are linearly linked to the branch current variables. Using the updated vector $\mathbf{z}_k$ of the measurements and the previously estimated state $\hat{\mathbf{x}}_{k-1}$, the measurement residuals $\mathbf{r}_k = \mathbf{z}_k - \mathbf{h}(\hat{\mathbf{x}}_{k-1})$ are computed. In the first iteration, when estimates are not still available, an initialization of the state variables is needed.

The WLS step is then used, at each iteration, to find the state variable variation $\Delta\mathbf{x}_k = \hat{\mathbf{x}}_k - \hat{\mathbf{x}}_{k-1}$ that minimizes the weighted sum of the squares of the residuals, by solving the following normal equations:

$$\mathbf{H}_k^{\mathrm{T}}\mathbf{W}\mathbf{H}_k\Delta\mathbf{x}_k = \mathbf{H}_k^{\mathrm{T}}\mathbf{W}\mathbf{r}_{k-1} \tag{2}$$

where $\mathbf{H}_k = \mathbf{H}(\hat{\mathbf{x}}_{k-1})$ is the Jacobian of the measurement functions at iteration k and $\mathbf{W}$ is the weighting matrix, equal to the inverse of $\boldsymbol{\Sigma}_{\mathbf{z}}$.

Matrix $\mathbf{G} = \mathbf{H}^{\mathrm{T}}\mathbf{W}\mathbf{H}$ (subscript k will be dropped in the following, for the sake of simplicity) represents the so-called gain matrix, which has to be inverted or factorized to find the solution of (2). Such matrix and its characteristics play thus a key role in the estimation process.

In the last step of the algorithm, network voltages for each node are computed, by a simple evaluation of voltage drops along the lines, starting from the estimated voltage in the reference bus. With a matrix expression, the node voltage phasors $\mathbf{v}_k$, at iteration k, are obtained as follows:

$$\mathbf{v}_k = \mathbf{Z}_{\text{paths}}\mathbf{x}_k \tag{3}$$

where $\mathbf{Z}_{\text{paths}}$ is the matrix that contains, for each row i, the branch impedances z_j that belong to the path that links v_i to the reference bus v_s.

It is worth noting that, in the case of meshed networks, a radial tree of the network can be considered in order to identify the paths linking the slack bus to each node of the grid and to make the forward sweep step possible. The chosen tree can be whatever, since the inclusion of the mesh constraints in the preceding WLS step ensures the final achievement of the same voltage results independently from the particular choice of the path.

The procedure is repeated until a given threshold in estimated state variation is reached.

Formulation of the equality constraints

Zero injections are the most common case of equality constraints that can be found in a distribution system. They can be generally represented by:

$$\mathbf{c}(\mathbf{x}) = \mathbf{0} \tag{4}$$

where $\mathbf{c}(\cdot)$ is a N_c size vector that represents the constraints to be kept into account in state estimation. It is worth noting that these constraints can be nonlinear, depending on the chosen state variables of the system, but they are linear in the case of rectangular BC-DSSE. In a similar way, possible presence of meshes also leads to equality constraints that, in the branch current formulation, have to be suitably considered. The way to handle such constraints in the DSSE can affect the performance of the estimator, in terms of both accuracy and speed. Different methods have been used in the literature. In the following, virtual measurements and Lagrange multipliers are first described, in order to present the equality constraints issue in a self-contained discussion. Then, a new method, based on state vector reduction, is proposed.

Virtual measurement method

Zero injections can be treated as virtual measurements given by (4), that is as measurements to be included into the measurement model (1) with a very low measurement uncertainty (represented by the virtual standard deviation σ_{zi}). Such approach leads to a WLS where a high weight is attributed to the power measurement of zero injection nodes. It is important to remark that the resulting weight is not representative of a real uncertainty (which should be equal to zero), but it is adopted only to point out the higher reliability of the information associated to the virtual measurements with respect to the other available measurements. In the DSSE, the measurement vector becomes $\mathbf{z} = \left[\mathbf{z}_m^{\mathrm{T}}, \mathbf{z}_{zi}^{\mathrm{T}}\right]^{\mathrm{T}}$, where $\mathbf{z}_m$ and $\mathbf{z}_{zi}$ are the proper measurement vector and the virtual measurement vector, respectively. Indicating with $\mathbf{C}$ the Jacobian of the zero injection constraints, and using the previous notation for the weighting matrix and the residuals, they can also be divided as follows:

$$\mathbf{H} = \begin{bmatrix} \mathbf{H}_m \\ \mathbf{C} \end{bmatrix} \tag{5}$$

$$\mathbf{W} = \begin{bmatrix} \mathbf{W}_m & \mathbf{0} \\ \mathbf{0} & \mathbf{W}_{zi} \end{bmatrix} \tag{6}$$

$$\mathbf{r} = \begin{bmatrix} \mathbf{r}_m \\ \mathbf{r}_{zi} \end{bmatrix} \tag{7}$$

The zero injection weighting matrix is $\mathbf{W}_{zi} = \sigma_{zi}^{-2}\mathbf{I}_{N_c}$, where $\mathbf{I}_{N_c}$ is an identity matrix, whose size is equal to the number of virtual measurements. The residual vectors are $\mathbf{r}_m = \mathbf{z}_m - \mathbf{h}_m(\hat{\mathbf{x}})$ and $\mathbf{r}_{zi} = -\mathbf{c}(\hat{\mathbf{x}})$.

In the BC-DSSE context, each zero injection leads to two virtual measurements, for the real and imaginary parts of the corresponding current injection. The Jacobian rows of the ith zero injection appear as follows:

$$\mathbf{C}^i = \begin{bmatrix} \cdots +1 \cdots -1 \cdots & \cdots \\ \cdots & \cdots +1 \cdots -1 \cdots \end{bmatrix} \tag{8}$$

where the only non-zero elements are those corresponding to the real and imaginary parts of the branch currents to and from the considered node (the sign depends on the direction assumed for the currents).

The WLS estimation step of DSSE is thus performed by means of the following N-dimensional system (where N is the number of unknowns in $\Delta\mathbf{x}$) derived from (2):

$$\left(\mathbf{H}_m^{\mathrm{T}}\mathbf{W}_m\mathbf{H}_m + \mathbf{C}^{\mathrm{T}}\mathbf{W}_{zi}\mathbf{C}\right)\Delta\mathbf{x} = \mathbf{H}_m^{\mathrm{T}}\mathbf{W}_m\mathbf{r}_m - \mathbf{C}^{\mathrm{T}}\mathbf{W}_{zi}\mathbf{c} \tag{9}$$

Possible meshes in the network can be expressed as additional constraints among the branch current state variables, using the Kirchoff voltage law, as $\mathbf{m}(\mathbf{x}) = \mathbf{Mx} = \mathbf{0}$, where each couple of rows j and $j+1$ refers to the Kirchoff voltage law of a mesh, expressed in real and imaginary parts, respectively. Matrix $\mathbf{M}$ thus contains, for each mesh, the resistances or inductances of the branches involved in the mesh itself.

Even in this case, the constraints can be included in the estimator model as virtual measurements, leading to the following:

$$\begin{aligned}&\left(\mathbf{H}_m^{\mathrm{T}}\mathbf{W}_m\mathbf{H}_m + \mathbf{C}^{\mathrm{T}}\mathbf{W}_{zi}\mathbf{C} + \mathbf{M}^{\mathrm{T}}\mathbf{W}_{zm}\mathbf{M}\right)\Delta\mathbf{x} = \\ &\mathbf{H}_m^{\mathrm{T}}\mathbf{W}_m\mathbf{r}_m - \mathbf{C}^{\mathrm{T}}\mathbf{W}_{zi}\mathbf{c} - \mathbf{M}^{\mathrm{T}}\mathbf{W}_{zm}\mathbf{m}\end{aligned} \tag{10}$$

where $\mathbf{M}$ and $\mathbf{W}_{zm}$ are the Jacobian and the weighting sub-matrix (with large weights) of the mesh constraints, respectively.

Lagrangian method

It is known that using large weights can lead to ill-conditioning problems in the system of normal equations to be solved. To avoid the use of very large weights, an alternative is represented by the explicit formulation of the constraints (4) in the problem. Such constrained minimization problem can be faced through the Lagrangian method [16]. According to this approach, the normal equation system (2) is extended associating at each equality constraint a Lagrange multiplier. In the case of zero injections, for a generic iteration, (2) becomes:

$$\begin{bmatrix} \mathbf{G}_m & \mathbf{C}^{\mathrm{T}} \\ \mathbf{C} & 0 \end{bmatrix}\begin{bmatrix} \Delta\mathbf{x} \\ -\lambda_{zi} \end{bmatrix} = \begin{bmatrix} \mathbf{H}^{\mathrm{T}}\mathbf{W}\mathbf{r} \\ -\mathbf{c}(\hat{\mathbf{x}}) \end{bmatrix} \tag{11}$$

where $\mathbf{C}$ is the Jacobian of the zero injection constraints, λ_{zi} is the N_c vector of the associated Lagrange multipliers and, using the same notation as in (5), $\mathbf{G}_m = \mathbf{H}_m^{\mathrm{T}}\mathbf{W}_m\mathbf{H}_m$ represents the gain matrix referred only to the set of measurements and pseudo-measurements. With this approach, the number of the unknowns increases because of the multipliers, but the sparsity of system (11) is higher with respect to the case of virtual measurements.

It is worth underlining that in [15], in order to significantly reduce the numerical conditioning of the system, the use of a normalization coefficient $\gamma = 1/\max(W_{ii})$ for the weighting matrix is suggested. Thus, the equation system (11) has to be modified in the following form:

$$\begin{bmatrix} \gamma\mathbf{G}_m & \mathbf{C}^{\mathrm{T}} \\ \mathbf{C} & 0 \end{bmatrix}\begin{bmatrix} \Delta\mathbf{x} \\ -\lambda_{zi} \end{bmatrix} = \begin{bmatrix} \gamma\mathbf{H}^{\mathrm{T}}\mathbf{W}\mathbf{r} \\ -\mathbf{c}(\hat{\mathbf{x}}) \end{bmatrix} \tag{12}$$

Similarly to the zero injections, even the mesh constraints can be managed by using Lagrangian multipliers. In this case, (12) has to be rewritten as:

$$\begin{bmatrix} \gamma\mathbf{G}_m & \mathbf{C}^{\mathrm{T}} & \mathbf{M}^{\mathrm{T}} \\ \mathbf{C} & 0 & 0 \\ \mathbf{M} & 0 & 0 \end{bmatrix}\begin{bmatrix} \Delta\mathbf{x} \\ -\lambda_{zi} \\ -\lambda_{zm} \end{bmatrix} = \begin{bmatrix} \gamma\mathbf{H}^{\mathrm{T}}\mathbf{W}\mathbf{r} \\ -\mathbf{c}(\hat{\mathbf{x}}) \\ -\mathbf{m}(\hat{\mathbf{x}}) \end{bmatrix} \tag{13}$$

where λ_{zm} is the vector of Lagrange multipliers included to deal with the mesh constraints $\mathbf{m}(\mathbf{x})$.

Proposed method: state vector reduction

An efficient way to include equality constraints, well-suited to BC-DSSE formulation, is proposed here. Each zero injection corresponds, as aforementioned, to an equivalent phasor current injection equal to zero. Since the real and imaginary parts of the branch currents are included in the state vector, it is straightforward to express the constraint on a node i as:

$$\sum_{j\in\Theta_i}\alpha_j i_j^r = 0,\ \sum_{j\in\Theta_i}\alpha_j i_j^x = 0 \tag{14}$$

where Θ_i is the set of branches incident to node i, while α_j is a coefficient equal to $+1$ or -1, depending on the direction assumed for the currents in the state vector. It is easy to understand how a simple variable elimination can be performed from (14), writing one current as a function of the other incident ones. In fact, for each zero injection, a current can be expressed as a function of the other state variables and eliminated from the state. The state $\mathbf{x}$ can be thus written as a function of a new reduced state vector $\tilde{\mathbf{x}}$ (of length $\tilde{N} = N - N_c$), replacing N_c variables by a simple combination of the remaining state variables, as follows:

$$\mathbf{x} - \begin{bmatrix} \tilde{\mathbf{x}} \\ \mathbf{x}_{zi} \end{bmatrix} = \begin{bmatrix} \mathbf{I}_{\tilde{N}} \\ \Gamma_{zi} \end{bmatrix}\tilde{\mathbf{x}} \tag{15}$$

where $\mathbf{I}_{\tilde{N}}$ is a $\tilde{N} \times \tilde{N}$ identity matrix and $\mathbf{\Gamma}_{zi}$ is a $N_c \times \tilde{N}$ sparse matrix of ±1 elements linking the eliminated variables $\mathbf{x}_{zi}$ to the remaining ones.

Referring to the reduced state vector $\tilde{\mathbf{x}}$, the normal equation system (2) can be rewritten as:

$$\tilde{\mathbf{H}}_m^{\mathrm{T}} \mathbf{W}_m \tilde{\mathbf{H}}_m \Delta\tilde{\mathbf{x}} = \tilde{\mathbf{H}}_m^{\mathrm{T}} \mathbf{W}_m \mathbf{r}_m \tag{16}$$

where the new Jacobian $\tilde{\mathbf{H}}_m$ is:

$$\tilde{\mathbf{H}}_m = \mathbf{H}_m \begin{bmatrix} \mathbf{I}_{\tilde{N}} \\ \mathbf{\Gamma}_{zi} \end{bmatrix} \tag{17}$$

and the residual vector is computed as $\mathbf{r}_m = \mathbf{z}_m - \mathbf{h}\left(\left[\tilde{\mathbf{x}}^{\mathrm{T}}, \tilde{\mathbf{x}}^{\mathrm{T}} \mathbf{\Gamma}_{zi}^{\mathrm{T}}\right]^{\mathrm{T}}\right)$.

As for the mesh constraints, even in this case it is possible to express one of the currents involved in the mesh in terms of the remaining ones. In general, the mesh constraint is:

$$\mathbf{m}(\mathbf{x}) = \sum_{j \in \Omega} \mathbf{Z}_j \mathbf{i}_j = 0 \tag{18}$$

where Ω identifies the set of branches in the mesh, and $\mathbf{Z}_j$ and $\mathbf{i}_j$ are the impedance and the current of the jth branch, respectively. It is worth noting that, in a three-phase framework, $\mathbf{Z}$ is a 3×3 impedance matrix, which includes also the mutual impedance terms, while $\mathbf{i}$ is the vector of the phase currents.

From (18), it is easy to find:

$$\mathbf{i}_k = -\mathbf{Z}_k^{-1} \sum_{\substack{j \in \Omega \\ j \neq k}} \mathbf{Z}_j \mathbf{i}_j \tag{19}$$

where k is the index of the branch whose currents will be eliminated from the state variables.

Considering this additional reduction of the state vector, it is possible to rewrite (15) in the following way:

$$\mathbf{x} = \begin{bmatrix} \tilde{\mathbf{x}} \\ \mathbf{x}_{zi} \\ \mathbf{x}_{zm} \end{bmatrix} = \begin{bmatrix} \mathbf{I}_{\tilde{N}} \\ \mathbf{\Gamma}_{zi} \\ \mathbf{\Gamma}_{zm} \end{bmatrix} \tilde{\mathbf{x}} \tag{20}$$

where $\mathbf{\Gamma}_{zm}$ is a $N_m \times \tilde{N}$ matrix linking the N_m removed current variables $\mathbf{x}_{zm}$ to the remaining ones. Thus, in this case, the resulting state vector has a reduced length equal to $\tilde{N} = N - N_c - N_m$. Once the transformation matrix linking the starting state vector to the reduced one is defined, the same considerations involved in (16) and (17) hold for the execution of the estimation algorithm.

The performed transformation leads to a lower sparsity of the system, reflecting the fact that each eliminated variable is expressed in terms of more remaining variables. However, since distribution grids usually have a large number of zero injections, the transformation also allows a significant reduction of the dimensions of the equation system to be solved. It is worth underlining that $\mathbf{\Gamma}_{zi}$ and $\mathbf{\Gamma}_{zm}$ are constant matrices that can be built *a priori* knowing the operation and topology of the network; thus, there is no need to compute them at each run of the DSSE.

It is important highlighting that such approach can be applied in an efficient way only because of the chosen state variables: a similar logic does not apply, with the same simplicity, to traditional estimators based on polar node voltages.

The proposed approach allows an efficient implementation even for the bad data detection and identification functions. In fact, the same techniques traditionally adopted in WLS estimators, based on the computation of the normalized residuals, can be conveniently implemented here, thanks to the reduced sizes of the Jacobian and gain matrices that are involved in the computation of the residual covariance matrix (see [15] for further details). Identified bad measurement data are removed, and the BC-DSSE estimation steps previously described are performed again on the reduced measurement set. It is worth noting that, in order to avoid the computation of the residual covariance matrix when bad data are not present, the bad data detection can be also implemented by using the well-known χ^2 test [15].

Tests and results

Test assumptions and metrics

Several tests have been performed on the unbalanced three-phase IEEE 123-bus system (Figure 1) to assess the performance of the proposed method. Data of the network can be found in [20]. Different measurement configurations have been considered in order to analyze various possible scenarios.

In order to achieve significant results from a statistical standpoint, tests have been carried out following a Monte Carlo approach. For each test, at the beginning, the true reference quantities of the network are computed by means of a load flow calculation. Then, for each trial of the Monte Carlo simulation, measurements are extracted adding random noise to the reference values, according to the uncertainty characteristics assumed for each measurement. At the end of each trial, estimation results are stored to be available for the subsequent analysis. The following assumptions are used in the tests:

- number of Monte Carlo trials: $N_{\mathrm{MC}} = 25,000$;
- pseudo-measurements available for all the loads of the network and characterized by normally distributed uncertainty with maximum deviation $3\sigma = 50\%$ of the nominal value;
- PMU measurements characterized by uncertainty with uniform distribution and variance σ^2 equal to one third of the squared accuracy value; in particular, an accuracy equal to 0.7% and 0.7 crad (i.e. 0.7×10^{-2} rad) is used for magnitude and phase-angle measurements, respectively, in order to simulate the accuracy limits specified in the synchrophasor standard [3].

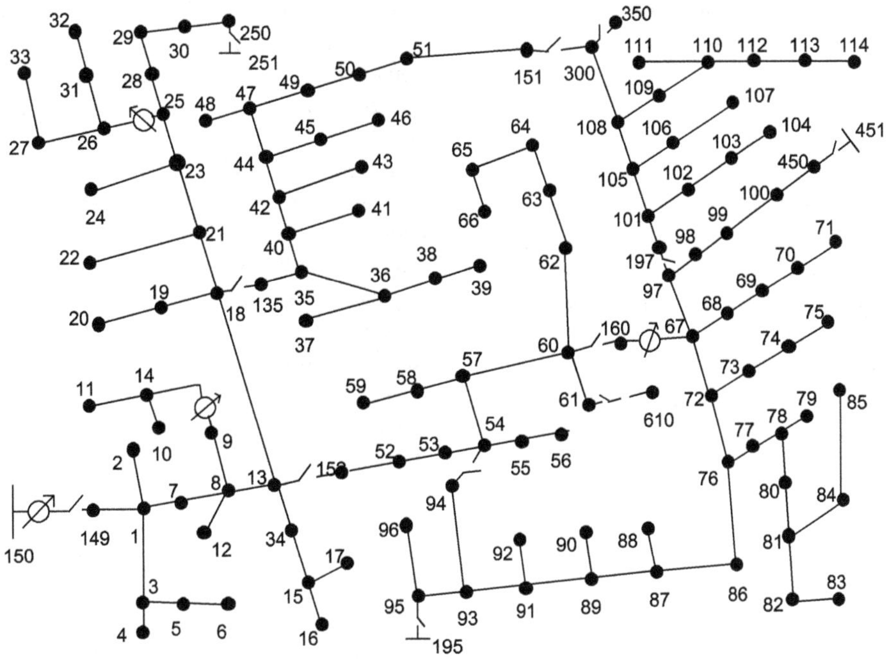

Figure 1 IEEE 123-bus test system.

Results of the tests have been analysed to assess the performance of the proposed approach in comparison to virtual measurements and Lagrange multiplier methods. In particular, accuracy, computational properties and efficiency of the different approaches have been analysed in order to have an overall performance evaluation.

Estimation accuracy

To assess the estimation uncertainty of a given quantity, the root mean square error (RMSE) has been used. Given a generic electrical quantity y, the RMSE is defined as:

$$\mathrm{RMSE}_y = \sqrt{\frac{1}{N_{\mathrm{MC}}}\sum_{i=1}^{N_{\mathrm{MC}}}(\hat{y}_i - y_{\mathrm{true}})^2} \tag{21}$$

where $\hat{y}_i$ is the estimation of y at the ith Monte Carlo trial and y_{true} is the true value of the quantity. In this paper, as overall index for the whole network, the mean RMSE, obtained averaging the RMSEs of all the nodes or branches (depending on the considered electrical quantity), has been used.

Since the focus of the paper is on the processing of the equality constraints, a second parameter is also monitored, that is the sum of the estimated power injections in the zero injection nodes. This index allows to evaluate the capability of the method to fully satisfy the zero injection constraints. In [17], it is shown that a bad handling of the constraints in the estimator could lead to the estimation of significant power injections in the zero injection nodes, consequently affecting the overall accuracy of the estimated quantities. Equation 22 shows the definition of the indexes for both active and reactive power (with B representing the set of the zero injection nodes of the network):

$$P_{0\mathrm{inj}} = \sum_{i \in B} |P_i|, \qquad Q_{0\mathrm{inj}} = \sum_{i \in B} |Q_i| \tag{22}$$

Numerical properties

An important issue for the estimator design is the numerical conditioning of the equation system. In fact, as described in [15], due to ill-conditioning, small errors in the different entries of the equation system may be translated in large errors for the solution vector. As a consequence, the accuracy and the convergence properties of the algorithm can be significantly affected and numerical instabilities can appear.

In all the analysed approaches, the equation systems to be solved at each iteration of the algorithm ((9), (12) and (16)) can be written in the following compact form:

$$\mathbf{G}_{\mathbf{tot}} \boldsymbol{\Delta} \mathbf{x}_{\mathbf{tot}} = \mathbf{u} \tag{23}$$

where $\mathbf{G}_{\mathbf{tot}}$ is a coefficient matrix, $\boldsymbol{\Delta}\mathbf{x}_{\mathbf{tot}}$ is the total vector of the unknowns (note that for the Lagrangian method also the Lagrange multipliers are included in this vector) and $\mathbf{u}$ is a vector resulting from the measurement residuals.

To evaluate the numerical properties of the system, the condition number k of the coefficient matrix $\mathbf{G}_{\mathbf{tot}}$ is considered, that is:

$$k(\mathbf{G}_{\mathbf{tot}}) = \|\mathbf{G}_{\mathbf{tot}}\| \cdot \|\mathbf{G}_{\mathbf{tot}}^{-1}\| \quad (24)$$

where $\|\cdot\|$ is the 2-norm of the matrix.

Other interesting properties from the computational point of view are the density and the size of the coefficient matrix. A low density (defined as the ratio between non-zero terms and total number of elements in the matrix) implies a large number of zero elements in the matrix and thus the possibility to use sparse matrix techniques for the calculations. The size, instead, is obviously associated to the number of unknowns and represents the dimension of the equation system to be solved. In the following, all these parameters will be taken into account to discuss the obtained results.

Computational efficiency

The efficiency of the estimator is a crucial factor for the real-time management and control of future distribution systems. For this reason, in the following discussion, the average execution times of the different methods (obtained averaging among the N_{MC} Monte Carlo trials) will be compared. Moreover, since the execution times strictly depend on the number of iterations needed for the algorithm convergence, also the average number of iterations of the different approaches will be evaluated. Tests have been performed under Matlab environment and run on a 2.4-GHz quad-core processor with 8-GB RAM.

Test results

In this section, the performance of the approaches virtual measurement method (VM), Lagrangian method (LM) and state vector reduction (SVR) will be analysed and discussed.

First, a test has been carried out considering a possible realistic measurement configuration for the network. In particular, three measurement points have been supposed to be available in nodes 150 (primary station), 18 and 67. Each measurement point is composed of a voltage synchrophasor measurement on the node and of current synchrophasor measurements on all the branches converging to that node.

Results of the test show that all the analysed methods provide really similar accuracy performance. The mean RMSE of the branch active power is, as an example, 5.7 kW for all the approaches. A similar behaviour has been found also for the mean RMSEs of voltage and current. As for the estimated power injection in zero injection nodes, simulations prove that a really low total power injection can be found with a suitable setting of the algorithms, despite the large number of zero injections (119 nodes over 227 total nodes for all the three phases). In particular, both LM and SVR always give a null sum of the considered power injections. As for VM, instead, results strictly depend on the choice of the weights used for the constraints. Table 1 shows the values of P_{0inj} and Q_{0inj} achieved using different weights. It is possible to observe that increases of an order of magnitude in the weights cause corresponding decreases on the P_{0inj} index. Similar results have been found also for the reactive powers. In any case, it is worth noting that, since the unbalances are different in sign, the mean RMSEs basically do not change in the different scenarios.

Table 2 shows the efficiency performance of the different methods. In this case, significant differences can be found depending on the approach used to deal with the zero injection constraints. In particular, it is possible to note that SVR provides the best results in terms of execution times, with an enhancement of the computational efficiency with respect to VM and LM larger than 30% and 40%, respectively. Furthermore, due to the correct modeling of the zero injection constraints, a slightly better performance can be observed for SVR even for the convergence properties (in terms of average number of iterations).

The reasons of such enhancement in the algorithm speed can be found looking at the numerical features of each approach. Table 3 reports the density properties and the size of the coefficient matrix involved in the WLS step of the estimation process. It is clear that SVR allows a large reduction of the coefficient matrix due to the elimination of all the zero injection constraints. Such reduction, despite the lower sparsity of the system, allows a faster factorization of the coefficient matrix (in this case, no sparse matrix techniques have been used for SVR) for the solution of (16). On the other hand, in LM, the introduction of the Lagrange multipliers leads to a larger size of the coefficient matrix, and the solution of the equation system, even if manageable with sparse matrix techniques, is slower.

An additional advantage guaranteed by the use of SVR concerns the condition number of the coefficient matrix. As it can be observed in Table 3, SVR gives the lowest conditioning. It is worth recalling that, in general, VM suffers possible ill-conditioning problems because of the use of very large weights to enforce the constrained measurements. As a confirmation of such an impact, the condition number obtained changing the weights to 10^8 and 10^{12}

Table 1 Variation of P_{0inj} [kW] and Q_{0inj} [kvar] in VM

	Virtual measurement weight			
	10^9	10^{10}	10^{11}	10^{12}
P_{0inj}	$1.9 \cdot 10^{-1}$	$1.9 \cdot 10^{-2}$	$1.9 \cdot 10^{-3}$	$1.9 \cdot 10^{-4}$
Q_{0inj}	$3.0 \cdot 10^{-1}$	$3.0 \cdot 10^{-2}$	$3.0 \cdot 10^{-3}$	$3.0 \cdot 10^{-4}$

Table 2 Average iteration numbers and execution times

Method	Iteration number	Execution time [ms]
VM	3.75	18.2
LM	3.75	21.5
SVR	3.64	12.8

has been checked: in both the cases, a consequent variation of the condition number ($3.74 \cdot 10^4$ and $3.02 \cdot 10^8$, respectively) has been found.

Several tests have been performed also to verify the proper operation of the aforementioned bad data detection and identification function when using the proposed approach. For instance, tests have been performed by adding intentional errors (of 5%, 10% and 20%) to the voltage magnitude measurement at node 18 (on the first phase). In such a scenario, all the analysed methods allow the detection of the bad data, through the χ^2 test, and the proper identification of the erroneous voltage, by means of the largest normalized residual technique. It is important to note that, since the presence of the bad data implies the computation of the residual covariances and the need to repeat the estimation process, the aforementioned improvements brought by the SVR method on the execution times further increase. It is also worth noting that, in some cases, depending on measurement configuration, when considering the bad data on other measurements (for example, on PMU currents), the bad data identification function could be unable to properly identify the erroneous measurement. However, as known from the literature, this is a general issue that can occur due to the low redundancy of the measurements, and it does not depend on the particular method used to handle the equality constraints. In fact, several tests (not reported here for the sake of brevity) have been performed changing the corrupted measurement, and as expected, test results prove that all the approaches exhibit exactly the same behaviour, with the same identification results, in all the different scenarios. Since possible identification problems are generally related to the measurement system deployed in the distribution grid, a deeper analysis on this issue is out of the scope of this paper.

Table 3 Numerical properties of the coefficient matrix

Method	Coefficient matrix density (%)	Coefficient matrix size	Condition number
VM[a]	3.26	454 × 454	$3.02 \cdot 10^6$
LM	1.48	692 × 692	$4.63 \cdot 10^4$
SVR	21.17	216 × 216	$1.02 \cdot 10^4$

[a]Weight used for the virtual measurements = 10^{10}.

Impact of the measurement configuration

Further tests have been performed to assess the performance of the methods with different measurement types and configurations.

First of all, the general validity of the previous considerations has been tested using different measurement devices. To this purpose, voltage and current phasor measurements have been replaced by voltage magnitude and active and reactive power measurements (with accuracy equal to 1% and 3%, respectively). It is worth noting that in this case, since synchronized measurements are not available, the estimator model has to be suitably adapted to take into account the absence of an absolute phase angle reference (see [12] for more details).

Tables 4 and 5 show the results about the efficiency performance and the numerical properties of the different methods in this case with traditional measurements. It is possible to observe that all the previously made considerations still hold, with SVR providing the best performance in terms of execution time and numerical conditioning. As for the accuracy performance, even in this case, all the methods provide really similar results.

As for the measurement configuration, the attention has been mainly focused on the impact of additional voltage measurements, since they can significantly affect numerical properties and efficiency of the branch current-based formulation of DSSE. In fact, voltage measurements lead to non-zero terms in the Jacobian $\mathbf{H_m}$ corresponding to all the derivatives with respect to the branch currents included in the path between the bus used as reference in the state vector and the measured node (for details, see [12]). This, in turn, causes a lower sparsity of the coefficient matrix in (23), thus affecting the efficiency of the equation system solution.

As an example, Tables 6 and 7 show the results about the numerical properties and the computational efficiency when two voltage measurements (in nodes 86 and 105) are added with respect to the configuration used in the previous tests.

As expected, the presence of additional voltage measurements affects the coefficient matrix leading to higher densities: this is the main reason for the increased execution times shown in Table 7 (with respect to the results in Table 2). Moreover, it is worth highlighting that the increased density brings different impacts on the different

Table 4 Average iteration numbers and execution times with traditional measurements

Method	Iteration number	Execution time [ms]
VM	5.30	23.3
LM	5.30	28.4
SVR	5.28	17.8

Table 5 Numerical properties of the coefficient matrix with traditional measurements

Method	Coefficient matrix density (%)	Coefficient matrix size	Condition number
VM[a]	3.09	451 × 451	$5.75 \cdot 10^6$
LM	1.40	689 × 689	$2.26 \cdot 10^4$
SVR	20.56	213 × 213	$5.41 \cdot 10^3$

[a]Weight used for the virtual measurements = 10^{10}.

Table 7 Average iteration numbers and execution times with two additional voltage measurements

Method	Iteration number	Execution time [ms]
VM	3.62	22.3
LM	3.62	27.7
SVR	3.54	14.0

methods. In fact, since the solution of the equation system in SVR is managed without using sparse matrix techniques, the impact on this approach is smaller with respect to the other methods. This is confirmed by the enhancements obtained on the execution times that, in this scenario, rise up to almost 37% and 50% with respect to VM and LM, respectively.

To emphasize and check such result, a further test has been carried out adding four supplementary voltage measurements (in nodes 25, 42, 48 and 91). Table 8 shows the results concerning coefficient matrix density and execution times. It is possible to observe that further increases in the density of the coefficient matrix, with consequent further improvement of the computational efficiency of SVR, are achieved. In this case, the time saved through SVR is larger than 45% with respect to VM and 60% with respect to LM.

As for the accuracy and the numerical conditioning of the analysed methods, considerations similar to those made for the first test can be derived also in these cases: all the methods provide very similar accuracy results, and SVR shows the best conditioning properties.

Impact of the size of the network

One of the main issues involved in the handling of distribution systems is the large size of these networks. This aspect is particularly critical from the standpoint of the execution times, since it implies a significant increase of the size of the equation system to be solved. In such a situation, possible drawbacks can arise for the SVR method due to the fill-ins resulting from the elimination of the state variables.

To assess the impact of such issue, additional tests have been performed with networks having a larger number of nodes. In particular, in order to simulate different sizes of the grid, new networks with an increasing number of feeders have been built, where each feeder replicates the topology of the previously considered 123-bus network. For all the tests, the measurement configuration is supposed to be composed of a measurement point in the substation and, for each feeder, of two measurement points in the buses corresponding to the nodes 18 and 67 of the original 123-bus network (see Figure 1). In order to obtain different loading conditions in the feeders, power consumptions of the loads have been modified adding random variations.

Performed tests show an obvious decrease for the density of the coefficient matrices in all the tested methods. For instance, in the SVR approach, the coefficient matrix density drops to 4.42% and to 2.22% when five and ten feeders are considered, respectively. In such a situation, even considering the huge size of the coefficient matrix, the use of sparse matrix techniques is a forced choice also for the SVR approach. Even in this case, the SVR method maintains the already cited advantages in terms of execution times. As an example, Table 9 shows the obtained results, for iteration numbers and execution times, when ten feeders are considered in the network (the resulting three-phase grid is, in this case, composed of more than 2,000 nodes). It is possible to observe that, because of the large size of the equation system to be solved, the execution times are now significantly higher than the previous cases. However, despite of the increased size of the network, the SVR method still allows a significant enhancement of the computational performance, leading to an improvement of about 22% and 37% with respect to the VM and the LM approaches, respectively.

Impact of the mesh constraints

In this section, the performance of the proposed method are tested for the case of weakly meshed networks,

Table 6 Numerical properties of the coefficient matrix with two additional voltage measurements

Method	Coefficient matrix density (%)	Coefficient matrix size
VM	5.95	454 × 454
LM	2.68	692 × 692
SVR	24.52	216 × 216

Table 8 Coefficient matrix density and execution times with six additional voltage measurements

Method	Coefficient matrix density (%)	Execution time [ms]
VM	10.36	28.1
LM	4.64	39.0
SVR	31.49	15.2

Table 9 Average iteration numbers and execution times with ten-feeder network

Method	Iteration number	Execution time [ms]
VM	4.00	166.7
LM	4.00	206.0
SVR	4.00	129.2

Table 11 Average iteration numbers and execution times with weakly meshed network

Method	Iteration number	Execution time [ms]
VM	3.89	33.8
LM	3.89	32.8
SVR	3.82	18.2

referring to the original 123-bus grid. To this purpose, the presence of a branch between nodes 151 and 300 and between nodes 54 and 94 of the benchmark network has been supposed in order to create two meshes. Simulations have been carried out considering the base monitoring configuration composed of the three measurement points in nodes 150, 18 and 67.

Table 10 shows the numerical properties of the coefficient matrix for the different methods.

It is possible to observe that the density of the coefficient matrix in VM and SVR is significantly affected by the presence of the meshes. In fact, as clear from (18), the mesh constraint involves all the three-phase currents of the branches belonging to the mesh. Thus, the matrix multiplications needed to create the coefficient matrix $\mathbf{G_{tot}}$ and involving the Jacobian matrix $\mathbf{M}$ for VM (see Equation 10) and the transformation matrix $\mathbf{\Gamma_{zm}}$ for SVR (see Equations 20, 16 and 17) lead to a significant increase of the non-zero elements. At the same time, the presence of the additional branch currents (due to the meshes) lead to an increase of the dimension of the equation system for VM, while no change appears for SVR since such currents are expressed in terms of the reduced state vector. In the case of LM, instead, the explicit expression of the mesh constraints, together with the growth of the dimensions of the equation system (due to both the additional branch currents and the additional constraints) allows to keep the coefficient matrix very sparse.

All these aspects bring direct effects on the efficiency of the different methods. Table 11 reports the obtained results for average iteration number and execution times. It is possible to observe that in this scenario, the VM approach is negatively affected by the presence of the constraints and gives the worst results in terms of execution time. Instead, the proposed method still shows significant benefits on the computational performance with improvements on the execution times around 45%.

Table 10 Numerical properties of the coefficient matrix with weakly meshed network

Method	Coefficient matrix density (%)	Coefficient matrix size
VM	11.42	466 × 466
LM	1.78	716 × 716
SVR	34.43	216 × 216

The same considerations already made for the tests with the radial version of the network hold for the accuracy performance and the conditioning properties.

Conclusions

In this paper, an efficient way to include the equality constraints in a branch current-based state estimator is presented. The method exploits the use of rectangular branch currents as state variables of the system to perform a simple elimination of one of the currents involved in a constraint, expressing it as a linear function of the remaining ones. The method not only is particularly efficient in the management of zero injections but also allows the treatment of mesh constraints. Performed tests prove the goodness of the proposed technique and in particular its capability to significantly improve the computational efficiency of the estimator (with respect to other traditionally used methods). Moreover, full fulfilment of the constraints is guaranteed, and additional benefits can be achieved for the numerical conditioning of the system.

Competing interests

The authors declare that they have no competing interests.

References

1. G Celli, PA Pegoraro, F Pilo, G Pisano, S Sulis, DMS cyber-physical simulation for assessing the impact of state estimation and communication media in smart grid operation. Power Syst. IEEE Trans. **PP**(99), 1–11 (2014)
2. AG Phadke, JS Thorp, *Synchronized Phasor Measurements and Their Applications*. (Springer, New York, 2008)
3. IEEE C37.118.1-2011 - IEEE standard for synchrophasor measurements for power systems. (IEEE, Piscataway, 2011)
4. M Zhou, VA Centeno, JS Thorp, AG Phadke, An alternative for including phasor measurements in state estimators. Power Syst. IEEE Trans. **21**(4), 1930–1937 (2006)
5. S Chakrabarti, E Kyriakides, G Ledwich, A Ghosh, Inclusion of PMU current phasor measurements in a power system state estimator. Generation Transm. Distrib. IET. **4**(10), 1104–1115 (2010)
6. W Jiang, V Vittal, GT Heydt, A distributed state estimator utilizing synchronized phasor measurements. Power Syst. IEEE Trans. **22**(2), 563–571 (2007)
7. ME Baran, AW Kelley, State estimation for real-time monitoring of distribution systems. Power Syst. IEEE Trans. **9**(3), 1601–1609 (1994)

8. ME Baran, AW Kelley, A branch-current-based state estimation method for distribution systems. Power Syst. IEEE Trans. **10**(1), 483–491 (1995)
9. W-M Lin, J-H Teng, S-J Chen, A highly efficient algorithm in treating current measurements for the branch-current-based distribution state estimation. Power Del. IEEE Trans. **16**(3), 433–439 (2001)
10. J-H Teng, Using voltage measurements to improve the results of branch-current-based state estimators for distribution systems. Generation Transm. Distrib. IEE Proc. **149**(6), 667–672 (2002)
11. H Wang, NN Schulz, A revised branch current-based distribution system state estimation algorithm and meter placement impact. Power Syst. IEEE Trans. **19**(1), 207–213 (2004)
12. M Pau, PA Pegoraro, S Sulis, Efficient branch-current-based distribution system state estimation including synchronized measurements. Instrum Meas. IEEE Trans. **62**(9), 2419–2429 (2013)
13. ME Baran, J Jung, TE McDermott, in *IEEE Power Energy Society General Meeting, 2009. PES '09*. Including voltage measurements in branch current state estimation for distribution systems (Calgary, 26–30 July 2009), pp. 1–5
14. M Pau, PA Pegoraro, S Sulis, in *2013 IEEE International Instrumentation and Measurement Technology Conference (I2MTC)*. WLS distribution system state estimator based on voltages or branch currents: accuracy and performance comparison (Minneapolis, 6–9 May 2013), pp. 493–498
15. A Abur, AG Exposito, *Power System State Estimation: Theory and Implementation*. (Marcel Dekker, New York, 2004)
16. W-M Lin, J-H Teng, State estimation for distribution systems with zero-injection constraints. Power Syst. IEEE Trans. **11**(1), 518–524 (1996)
17. Y Guo, W Wu, B Zhang, H Sun, An efficient state estimation algorithm considering zero injection constraints. Power Syst. IEEE Trans. **28**(3), 2651–2659 (2013)
18. P Castello, M Lixia, C Muscas, PA Pegoraro, Impact of the model on the accuracy of synchrophasor measurement. Instrum. Meas. IEEE Trans. **61**(8), 2179–2088 (2012)
19. P Castello, P Ferrari, A Flammini, C Muscas, S Rinaldi, A new IED with PMU functionalities for electrical substations. Instrum. Meas. IEEE Trans. **62**(12), 3209–3217 (2013)
20. Society IEEE Power & Energy, IEEE Test Feeder Specifications. http://ewh.ieee.org/soc/pes/dsacom/testfeeders/. Accessed 25 Feb 2015

7

HOS network-based classification of power quality events via regression algorithms

José Carlos Palomares Salas[1,2*], Juan José González de la Rosa[1,2], José María Sierra Fernández[1,2] and Agustín Agüera Pérez[1,2]

Abstract

This work compares seven regression algorithms implemented in artificial neural networks (ANNs) supported by 14 power-quality features, which are based in higher-order statistics. Combining time and frequency domain estimators to deal with non-stationary measurement sequences, the final goal of the system is the implementation in the future smart grid to guarantee compatibility between all equipment connected. The principal results are based in spectral kurtosis measurements, which easily adapt to the impulsive nature of the power quality events. These results verify that the proposed technique is capable of offering interesting results for power quality (PQ) disturbance classification. The best results are obtained using radial basis networks, generalized regression, and multilayer perceptron, mainly due to the non-linear nature of data.

Keywords: Artificial neural networks (ANN); Power quality (PQ); Cumulants; Higher-order statistics (HOS); Regression algorithms; Smart grid (SG); Spectral kurtosis (SK)

Introduction

With the consequent unstoppable increase of electronic equipments demanding electricity, consumers expect uninterrupted availability and quasi perfect power quality (PQ). For that reason, PQ is being the object of continuous interest by researchers and developers, due to its influence over and from the loads, and the recent potential inclusion in the modern smart grid (SG).

In this frame, an adequate PQ assures the necessary compatibility between all equipment connected to the grid [1]. The terms of this compatibility gather several aspects: sustainable power with low losses and high quality and security of supply and safety, being at the same time economically efficient, reliable, and resilience [2,3].

Certainly, the future SG would introduce transformative technologies to meet these design requirements, integrating intelligence into end-use devices as the key to satisfying the demand response. In parallel, the industry would design ways to incorporate automatic end-use-load participation into the model so that customers are not bothered by these programs and decisions, and so their lifestyles are not inconvenienced [4].

Due to the above arguments, the role of smart meters and sensors is the first being revised in the present and future SG. These automated meters (AM) used a two-way communicating infrastructure and centralized management, as well as new features such as the following: outage management, demand response, automatic load shedding, distribution automation, and the ability to enable and commute alternative energy sources.

Provided with this scenario, this research integrates artificial neural networks (ANNs) and advanced signal processing techniques based in higher-order statistics (HOS) in order to be implemented into an automated smart meter for PQ event detection and classification, within the frame of a SG with a high distribution penetration of renewable sources. Seven regression algorithms are tested based on an hybrid time-frequency battery of characteristics, specially designed to deal with non-stationary measurement time series.

The feature extraction stage from PQ disturbances is based on HOS, idea which has been proven to be efficient in several works. Indeed, since PQ events are sudden changes in the power line, the HOS are potentially useful to characterize each type of electrical anomaly both

*Correspondence: josecarlos.palomares@uca.es
[1] Research Group PAIDI-TIC-168: Computational Instrumentation and Industrial Electronics (ICEI), Av. Ramón Puyol S/N., E-11202 Algeciras-Cádiz, Spain
[2] Area of Electronics, Polytechnic School of Engineering, University of Cádiz, Av. Ramón Puyol S/N., E-11202 Algeciras-Cádiz, Spain

in time and frequency domains. As a novelty, the present research exploits the combination of time and frequency domain features, to deal with the inherent non-stationary associated to the electrical anomalies, with the goal to improve the performance of the ANN and to make feasible the integration in an smart meter.

Regarding backgrounds of HOS applications in this field, in the time domain, several notable works are worthy, e.g., Bollen et al. introduced new advanced statistical features to PQ event detection [5]. In the same direction, Gu and Bollen [6] found relevant characteristics associated to PQ events in the time and frequency domains. The work by Ribeiro et al. is also remarkable [7], which extracted new time-domain features based in cumulants. The same authors performed the classification of single and multiple disturbances using HOS in the time domain and Bayes' theory-based techniques [8]. HOS techniques and estimators have also been implemented to specifically detect sags and swells [9].

The categorization of PQ anomalies had been formerly performed by Nezih and Ece in the work [10], where they proved that HOS and quadratic classifiers improve the second-order-based methods. The same authors previously achieved performance in second-order computing, using 2-D wavelets and compression techniques [11,12]; finding, despite the promising results, the limits of the procedure and quantifying its heavy computational cost. Alienated to this work, the researches of Poisson et al. and Santoso et al. [13,14] also reported a wavelet-based method, finding the potential and the drawbacks of the technique so that to implement it in an intelligent meter.

The direct antecedent of the present research in the work by J.J.G. de la Rosa et al. [15], in which they performed a mixed study involving the time-domain variance, skewness, and kurtosis, and they obtained consequences combined with the spectral kurtosis (SK), and over a set of real-life measurements, some of them with mixed PQ perturbations. The same authors proposed a preliminary criteria for seven types of disturbances based on the former estimators [16]. In a previous work [17], they designed an offline case-based reasoner based on time-domain HOS estimators. Furthermore, the authors used also HOS features in classification techniques for characterization of electrical PQ signals [18].

The present paper is designed as follows. The next section summarizes the main advances in the field, paying special attention at the applications of ANN for PQ analysis, and reasoning the contribution of HOS to the feature extraction stage and the architecture. Then, Section Higher-order statistics for PQ monitoring: an enhancement proposal over the DWT exposes the advantages of HOS for PQ monitoring and the state-of-the art. In Section Proposed methodology: the HOS-based ANN, the procedure is detailed in order to expose the results later in Section Results; finally, conclusions are drawn in Section Conclusions.

ANN for PQ analysis: towards the HOS paradigm

The present paper postulates non-stationary signal processing with higher-order statistics in the time and frequency domains in order to extract a battery of features to be processed via ANNs with regression algorithms. ANNs have been used for classification purposes in myriads of works and have proved the utility for a long time [19]; Figure 1 shows a generic architecture, indicating how the input pattern is processed via the neurons to provide with the output vector, in order to introduce the concrete architecture in the present work.

The performance of the potential PQ monitoring system and consequently the ANN is directly related to the pre-processing and feature extraction techniques used. The main goal of the feature extraction is to represent the data set in a new feature space in which the probability to distinguish classes is higher than the one in the

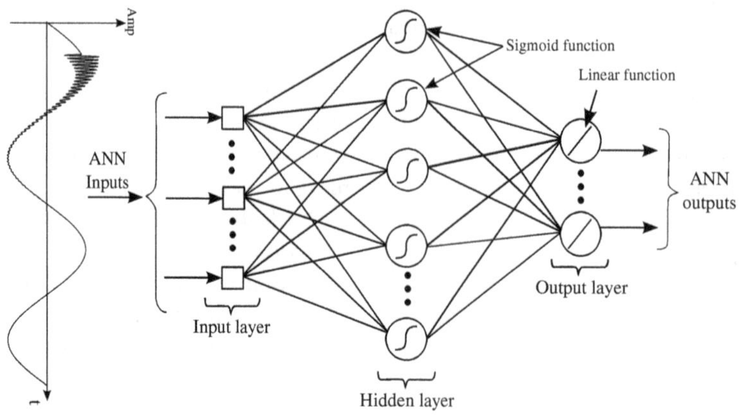

Figure 1 Generic ANN structure which postulates the feature extraction stage previous to the ANN.

original space. Therefore, the identification of efficient pre-processing and feature extraction techniques is a key issue [20].

Some works have shown that the best performance for non-stationary signals is achieved combining time and frequency characteristics obtained in the feature extraction stage via wavelet transforms (WT), on the hypotheses that WT performance is suitable for the analysis of transient signals with easy computation. Three works are worth to be cited as they are directly concerned with our research, involving ANN and PQ analysis. In the first one to mention, Angrisani et al. presented a wavelet network for automated transient detection; basically, it was based in an extended perceptron which incorporated wavelet nodes [21]. They proposed a unique structure based on the introduction of wavelet nodes in the traditional ANN, in the neurons of the first layer, and was applied successfully to typical transients in the PQ analysis. The number of nodes was fixed to 12, and this number was related to the time location of the perturbation. The network has to be pre-conditioned in order to detect concrete transient signals.

In the same former line, using the WT as a preprocessing tool for ANN, the second work by Iñigo Monedero et al. [22] developed a neural perceptron-based (three hidden layers) real-time first system prototype, which was trained using a disturbance generator for pseudo-synthetic waveforms. They obtained a performance of 89% success combining simulated and real-time data. The perturbations were classified in frequency, voltage, and harmonics, and the topology of the ANN (number of hidden neurons and number of outputs) was highly moveable depending on the PQ event under study.

The third and most recent paper, by Martin Valtierra-Rodriguez et al. [23] involves a new dual neural-network-based methodology to detect and classify single and combined PQ disturbances, consisting, on the one hand, of an adaptive linear network for harmonic and inter-harmonic estimation that allowed computing the root-mean-square voltage and THD indices, from which it is possible to detect and classify sags, swells, outages, and harmonics. A complementary feed-forward neural network was used for pattern recognition using the horizontal and vertical histograms of a specific voltage waveform to classify spikes, notching, flicker, and oscillatory transients. The complementary action of these neural networks allows the detection and classification even with simultaneous electrical anomalies, both in noisy and noise less scenarios.

Apart from the difficulties in implementing a real-time processor in the smart grid frame, the main disadvantage of the above procedures lies in the fact that accuracy and repeatability are highly compromised by the second-order estimators (WT and V_{RMS}) used in the feature extraction stage. Drawbacks arise when the data are corrupted by noise; specially when the number of samples of the signal window is reduced, and the resolution and repeatability are degraded. This facts are dramatically increased if the tests are performed over synthetic signals of controlled-lab experiences, where predictability is tacitely supposed. In the real-world experiences, the system should be prepared for unpredictable phenomena, both in the time and in the frequency domains. As explained hereinafter, this goal is accomplished by statistical parameters of an order higher than two.

Higher-order statistics for PQ monitoring: an enhancement proposal over the DWT

During the last decade, some researches [7,15,24] have demonstrated the usage of HOS features for PQ monitoring. The motivation of HOS in PQ analysis is twofold. By one side, as HOS measurements are correlations that involve powers higher than two [25], the HOS information for Gaussian signals is null, as Gaussian processes are described up to the second order. As many measurement noises are Gaussian or symmetrically distributed, the HOS may be less degraded by the background noise than the second-order calculations. Secondly, the capability to reveal non-linear characteristics from the data, which is important for pattern recognition scenarios, in concrete to target PQ events, waveforms which exhibit high time variability, like impulsiveness of peakedness.

The usage of HOS as a feature extraction technique for PQ monitoring systems is very promising, and several recent works presented good results with respect to both detection and classification tasks. The most similar work was recently committed by Liu et al. [26]; transients were classified according to the features extracted via the SK and using ANNs. They worked with simulated signals over the five types of synthetic perturbations.

Formerly, it had been shown that combining techniques allows efficient classification of single and simultaneous disturbances, and more, the usage of the second- and fourth-order HOS features, for a specific lag chosen from Fisher's discriminant ratio (FDR) criterion, has been enough to deal with the majority of the disturbances considered [24].

Regarding specifically signal processing issues, some results show that HOS are capable of detecting disturbances even using short acquisition time windows, which represents an important characteristic for several power system applications such as protection, signal segmentation, and disturbance localization. Being specific, the results shown that the detection of disturbances can be accomplished in less than a quarter of cycle, which is excellent for protection application, where speed and accuracy need to be combined to guarantee selectivity and reliability during the occurrence, for example, of a fault in a system [24,25].

Regarding the HOS that has been used in the present research, it is worth remarking that we have made use of a combination in the time and the frequency domains. The ensemble of Equation 11, whose rigorous mathematical treatment is in the appendix, constitutes indirect measurements of the variance, skewness, and kurtosis that has been used in the present paper, along with the SK, which is used to locate the transients in the frequency domain.

With the aim of motivating the performance of HOS over second-order methods, a comparison has been developed via the discrete wavelet transform (DWT), which is a promising technique for PQ analysis according to the literature (e.g., [21]). Without the further necessity of expanding expressions, in this section, we recall that every finite energy signal $s(t)$ can be decomposed over a wavelet orthogonal basis according to:

$$s(t) = \sum_{j=-\infty}^{+\infty} \sum_{k=-\infty}^{+\infty} \langle s, \psi_{j,k} \rangle \psi_{j,k}. \tag{1}$$

Each partial sum, indexed by k, in Equation 1 represents the detail variations at the scale $a = 2^j$ (at each level j, the scale is increased by a factor of two):

$$d_j(t) = \sum_{k=-\infty}^{+\infty} \langle s, \psi_{j,k} \rangle \psi_{j,k} \qquad s(t) = \sum_{j=-\infty}^{+\infty} d_j(t). \tag{2}$$

The approximation of the signal $s(t)$ can be progressively improved by obtaining more levels, with the aim of recovering the signal selectively. For example, if $s(t)$ varies smoothly, we can obtain an approximation by removing fine scale details, which gather information regarding the high frequencies or rapid variations of the signal. This is done by truncating the sum in Equation 1 at the scale $a = 2^J$:

$$s_J(t) = \sum_{j=J}^{+\infty} d_j(t). \tag{3}$$

Details corresponding to indexes $j < J$ are not considered in Equation 3.

Hereinafter, a simple experience is exposed illustrating the limitations of the method based in the DWT, in case DWT coefficients were selected to be used as features. The wavelet decomposition tree has been expanded up to level 7 using the mother wavelet *sym8*. The idea in the background literature is to study the distribution of the signal energy for each level of the tree, which acts as a filter bank. As stated and proved in myriads of references (e.g., [27]), low index levels point rapid fluctuations of the signal, whereas slow variations are associated to higher (or deeper) decomposition branches. The difference between the energy of the perfect power signal and the energy of the PQ event under test has been studied for each level. The energy expression that we have considered is the traditional 2-norm, according to Equation 4:

$$E(\mathbf{s}) := \sum_{i=1}^{N} \|s_i\|^2, \tag{4}$$

where $\mathbf{s}$ is the N-point vector signal and s_i, their associated components. Figure 2 shows the processing results for six prototypes. The limitations of the DWT can be concluded via a simple analysis of the graphs. Firstly, as the DWT analysis consists of a selective filtering, sag and swells cannot be distinguished. Secondly, mixed PQ events are also difficult to separate because the constant incremental energy does not indicate whether a transient is coupled or not. Finally, an eventual noisy environment would mask results in the case of the oscillatory transient (this has been extensively proved).

At the light of the second-order analysis, a more robust set of features is required in order to process real-life measurements. This is achieved via HOS.

Proposed methodology: the HOS-based ANN

The final goal of this work is the implementation in the future SG to guarantee compatibility between all equipment connected. Data used in this work have been simulated using MATLAB software. The data set generated by the simulation consists of 550 samples including the different studied disturbance kinds, which cover the following disturbances: oscillatory transient, impulsive transient, interruption, harmonic permanent distortion, harmonic temporal distortion, sag, sag plus oscillatory transient, and swell. Each signal comprises a 20K-point synthetic time-domain register with a duration of 1 s (20 KHz sampling frequency). An additive normal noise process (1% of the amplitude of the signal) has been added in order to achieve a more realistic behavior. Figure 3 shows an example of these signals as well as healthy signal that are utilized in this paper.

The classification techniques used to classify the PQ disturbances are based on regression algorithms. These techniques are multiple linear regression (MLR), adaptive linear neuron (LIN), multilayer ANNs (BP1 and BP2), radial basis function (RBF) network, exact radial basis (ERB), and generalized regression network (GRN). Every model has been intelligently adapted to meet the objective of PQ classification because each has different characteristics. Table 1 shows the selected parameters corresponding to the architecture and activation function of each model used. The final design of each model has been obtained by the optimization of the parameters shown in this table.

Data used to realize the classification are based on representative coefficients obtained from the PQ disturbances referred above. These coefficients are acquired

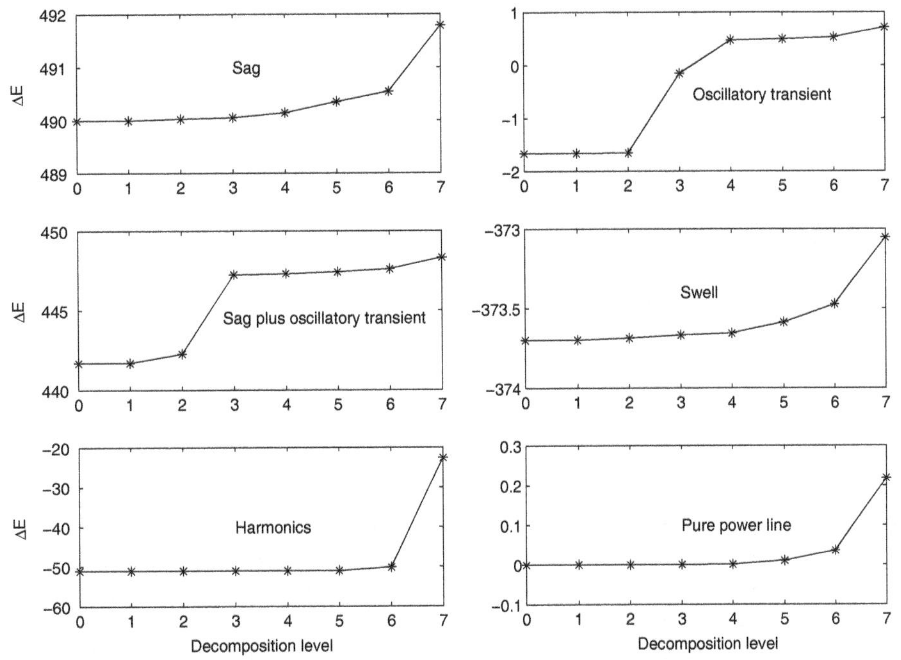

Figure 2 DWT analysis for different PQ events. From top to bottom and left to right: sag, oscillatory transient, sag plus oscillatory transient, swell, harmonics, and pure power signal. The energy difference is depicted vs. the decomposition level. In the case of additive noise, the method does not guarantee success in the feature extraction stage.

by a process of feature extraction which is based on the combination of higher-order statistics in time and frequency domains. The HOS have been computed using a 400-points sliding window (which corresponds to a signal period), with a shift of 10 points over a vector of 20,000 points. After extraction stage, a total of 14 characteristic features are selected, nine of whom correspond to time domain and the remaining five to frequency domain. The coefficients selected in the first one correspond to the maxima and minima and stable in the second-, third-, and fourth-order cumulants at zero lags (directly related to the variance, skewness, and kurtosis).

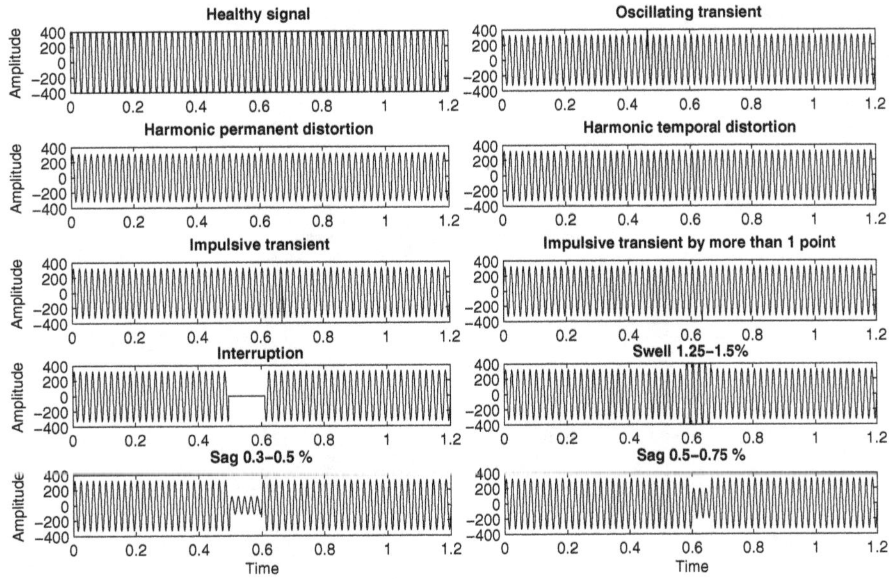

Figure 3 Example of healthy signal and different disturbances studied.

Table 1 Parameters of the ANNs

ANN	LIN	BP1	BP2	RBF	ERB	GRN
Hidden layers	-	1	2	1	1	1
Neurons hidden layer 1	-	[4 to 10]	[4 to 10]	[1 to 150]	[1 to 150]	[1 to 150]
Neurons hidden layer 2	-	-	[2 to 5]	-	-	-
TF	-	S	S	G	G	G
TF output	L	L	L	L	L	L
Training algorithm	WH	LM	LM	*k*	*k*	*k*
Spread	-	-	-	[1 to 20]	[1 to 20]	[1 to 20]

LM, Levenberg Marquardt; WH, Widrow Hoff; S, sigmoid; G, Gaussian; L, linear; *k*, *k*-means; TF, transfer function.

On the other hand, the coefficients selected in the second one correspond to the frequency of extreme value of SK, bandwidth of dome, extreme value of SK, number of peaks in SK, and dome very targeted (between 0 and 1). The abstract graphic of this work with the proposed methodology can be seen in Figure 4 where it specifies the feature extraction module. In order to illustrate the capability of the SK to discriminate PQ disturbances, we have selected a practical example consisting of an oscillatory transient coupled to the power sine wave. The analysis result is depicted in Figure 5. The time-domain variance increases when it bumps into the transient; this behavior is independent of the transient frequency. Similarly, the time-domain skewness and kurtosis detect slight variations. The real detection takes place in the frequency domain; the SK produces a real enhancement in 2,000 Hz, along with the high-resolution bump (narrow peak).

The resulting data after carrying out the feature extraction is a matrix of dimension 550 × 14 (samples × features). Then, the building of the models is performed by following two steps in order to efficiently classify the disturbances. First, data are normalized so that they are in the interval $[-1, 1]$, for a faster computation [28]. And second one, we divide randomly the data set into three subsets: training, evaluation, and test sets. The training and validation sets, with 70% and 15% of data, respectively, were used for ANN model building; and the third set, with the last 15%, was used to test on the out-of-sample set the classification power of a model.

For comparison purposes, the classification criterion is based on the parameter hit rate (HR), that is defined as follows:

$$HR = \frac{N_C}{N_T} \times 100\% \tag{5}$$

where N_T is the number of test samples and N_C is the number of correct disturbance recognition.

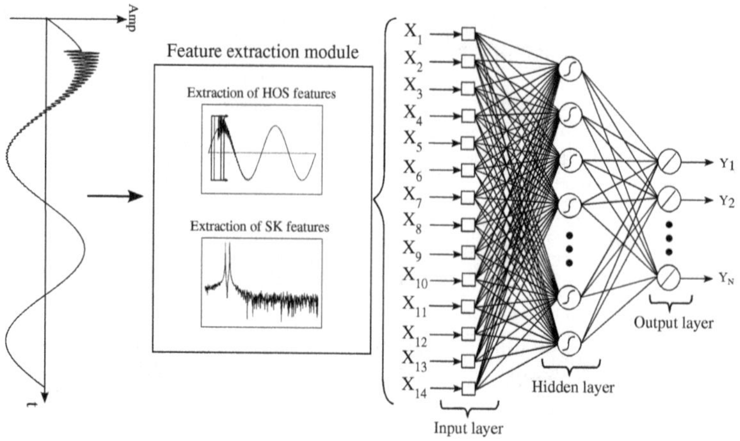

Figure 4 Procedure of feature extraction.

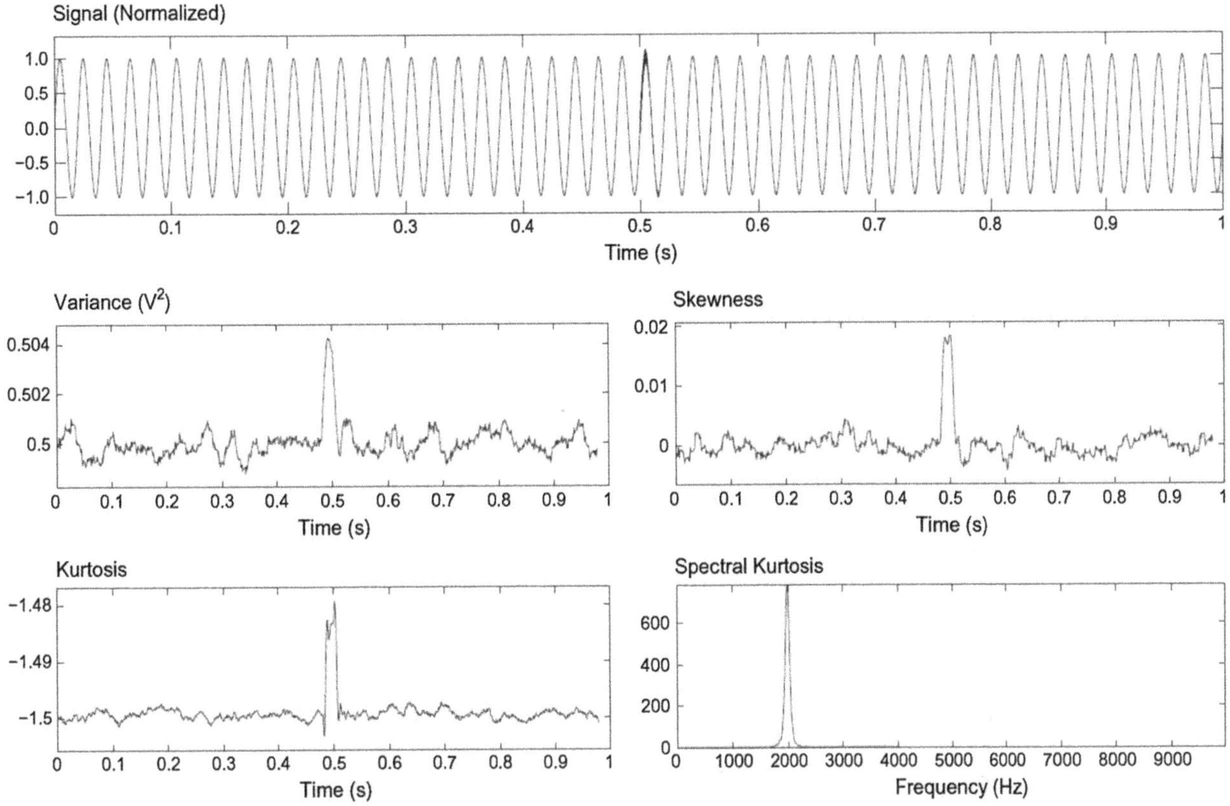

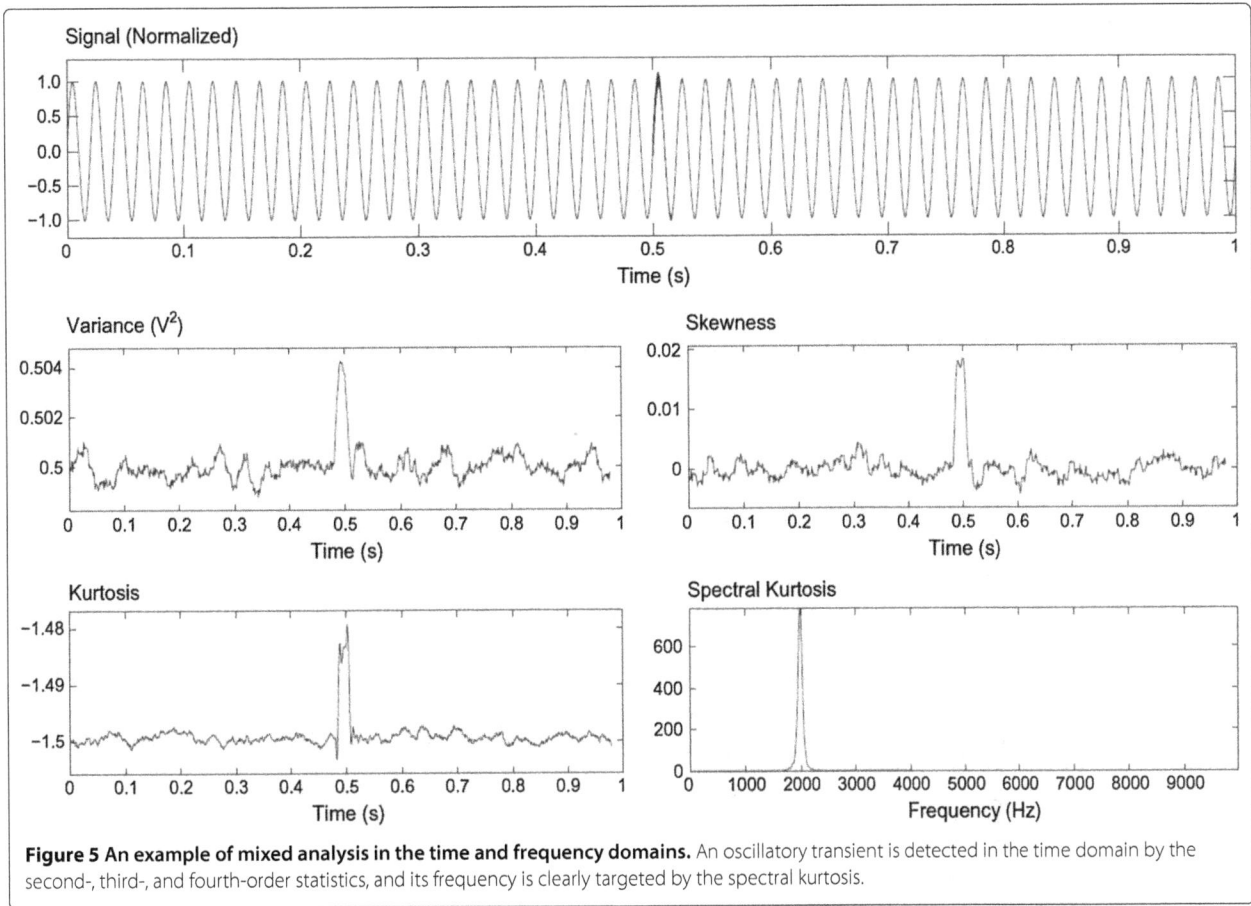

Figure 5 An example of mixed analysis in the time and frequency domains. An oscillatory transient is detected in the time domain by the second-, third-, and fourth-order statistics, and its frequency is clearly targeted by the spectral kurtosis.

Results

Once the assessed models were configured and optimized, they are used in the out-of-sample set. Because the database is small, we choose randomly two test sets, and in each of them, 100 experiments are launched by model. This is realized to achieve statistically meaningful results which rule out the random factors influencing the ANNs.

For each model, we have performed three analysis in function of the features used: HOS in the time domain (HOS_t), the SK, and the mixed (time and frequency domains) analysis ($\mathrm{HOS}_t + \mathrm{SK}$). In all of them, the HR was calculated for each model on the 200 offline tests, observing the arithmetic mean. The obtained results showing the percentage of effectiveness for classifying disturbances are presented in Table 2. In most models, results obtained in the mixed analysis are better than in the other analyses except isolated results. These exceptions represent the 27.27% of the cases presented in Table 2, where the 14.28% of them correspond to (HOS_t) features, and the remaining 12.99% correspond to SK features.

As can be seen, the algorithms based on linear models are worse than those based on non-linear schemes. The best models both individually and collectively are ERB, GRN, and BP1.

Conclusions

In this paper, seven regression algorithms have been applied and compared for PQ disturbances classification. The novel aspect is the introduction of new representative coefficients based on HOS in time and frequency domains. These coefficients are the inputs used in the classification algorithms to verify the occurrence or not of single or multiple disturbances in the electric signals.

The data used to test the proposed method were generated by the MATLAB software. The PQ disturbances considered are the most common on the supply. The best models both individually and collectively are obtained employing radial basis networks, generalized regression, and multilayer perceptron. The overall hit rates obtained are 94.70%, 79.59%, and 74.17%, respectively. This is consistent with non-linearity of the used data and emphases of the non-linearity that provide the HOS.

Once the obtained results demonstrate that the proposed method can effectively classify different kinds of PQ disturbances, is necessary to do more training and tests

Table 2 Regression algorithms for ANNs

	MLR			LIN			BP1			BP2			RBF			ERB			GRN		
	HOS_t + SK	HOS_t	SK	HOS_t + SK	HOS_t	SK	HOS_t + SK	HOS_t	SK	HOS_t + SK	HOS_t	SK	HOS_t + SK	HOS_t	SK	HOS_t + SK	HOS_t	SK	HOS_t + SK	HOS_t	SK
Harmonic permanent distortion	42.45	15.22	0.00	0.00	0.00	0.00	85.89	59.88	88.00	78.30	49.61	70.00	54.94	52.87	100.00	100.00	83.22	100.00	69.39	1.83	0.00
Harmonic temporal distortion	34.14	29.44	4.70	10.54	0.00	1.30	74.56	60.73	17.12	70.00	59.91	24.82	42.17	53.02	1.79	93.00	79.71	80.58	48.41	33.77	0.00
Impulsive transient	65.71	14.18	6.91	53.40	0.00	3.82	82.84	53.23	56.52	81.44	56.37	55.42	38.59	19.85	42.22	96.39	80.38	82.85	100.00	10.00	0.00
Impulsive transient by more than one point	59.50	23.95	43.47	94.36	17.66	57.47	83.79	49.20	68.25	81.75	55.27	73.43	68.02	16.94	77.51	95.97	81.19	79.46	100.00	12.35	100.00
Oscillatory transient	14.88	82.34	4.47	5.99	98.58	3.22	75.23	24.12	52.14	77.29	25.81	58.24	83.60	45.02	39.01	96.93	79.55	79.84	90.16	95.39	2.90
Interruption	25.81	19.14	48.67	18.64	14.60	56.59	84.29	88.81	33.22	81.34	88.20	33.20	95.69	100.00	31.72	91.04	80.26	79.98	96.36	86.74	58.19
Sag 0.3% to 0.5%	21.43	20.74	12.41	40.51	38.88	21.23	67.49	54.49	29.60	69.70	59.02	33.37	38.93	60.64	23.49	90.11	79.76	78.38	64.80	14.52	39.83
Sag 0.5% to 0.75%	36.48	43.36	38.86	44.83	46.79	47.71	53.03	42.57	47.49	58.81	35.85	45.67	33.42	28.32	41.62	89.91	79.61	79.08	87.70	88.14	59.29
Swell 1.25% to 1.5%	31.13	27.31	35.67	18.89	17.92	8.57	84.02	76.86	43.56	78.64	84.36	45.18	40.06	39.42	27.85	99.66	80.36	79.50	94.90	50.42	0.00
Sag + oscillatory	0.00	0.00	0.00	0.00	0.00	0.00	43.93	14.13	24.37	51.66	11.95	25.62	18.64	8.89	100.00	88.68	80.28	80.71	27.40	0.00	0.00
Healthy signal	0.00	0.00	0.00	0.00	0.00	0.00	80.83	12.07	78.50	74.91	25.20	63.00	99.98	0.00	51.00	100.00	95.28	100.00	96.35	0.00	0.00
Global	30.14	25.06	17.74	26.11	21.31	18.17	74.17	48.74	48.98	73.05	50.14	48.00	55.82	38.63	48.75	94.70	81.78	83.64	79.59	35.74	23.66

Number of simulations: 200. HOS_t, higher-order statistics in time domain; SK, spectral kurtosis.

to find an ANN classifier with better characteristics than obtained in this research.

Appendix

Time-domain cumulants and higher-order spectra

Time-domain HOS

Higher-order cumulants are being used extensively to deduce newly statistical features from the data of non-Gaussian measurement time series [29-33]. To reach a compact expression, nomenclature is first introduced.

Let us consider $\{x(t)\}$ be an rth-order stationary real-valued random process; the rth-order cumulant is defined as the joint rth-order cumulant of the random variables $x(t), x(t+\tau_1),\ldots, x(t+\tau_{r-1})$. This compacted notation is expressed via Equation 6:

$$C_{r,x}(\tau_1, \tau_2, \ldots, \tau_{r-1}) = \mathrm{Cum}[x(t), x(t+\tau_1), \ldots, x(t+\tau_{r-1})] \tag{6}$$

where $\tau_1, \tau_2, \ldots, \tau_{r-1}$ are time shifts, and the nth shifting is a multiple of the data acquisition sampling period, T_s, and is usually expressed as $\tau_n = n \cdot T_s$.

Cumulants, defined in the Equation 6, are estimated by using the well-known *Leonov-Shiryaev* formula, which expresses the compact relationship among the cumulants of stochastic signals and their moments [9]. In this sense, the expressions for the second-, third-, and fourth-order cumulants for a real, random, and zero-mean (central cumulants) time series $x(t)$ can be estimated via:

$$C_{2,x}(\tau) = E\{x(t) \cdot x(t+\tau)\} \tag{7a}$$

$$C_{3,x}(\tau_1, \tau_2) = E\{x(t) \cdot x(t+\tau_1) \cdot x(t+\tau_2)\} \tag{7b}$$

$$\begin{aligned} C_{4,x}(\tau_1, \tau_2, \tau_3) = {} & E\{x(t) \cdot x(t+\tau_1) \cdot x(t+\tau_2) \cdot x(t+\tau_3)\} \\ & - C_{2,x}(\tau_1)C_{2,x}(\tau_2 - \tau_3) \\ & - C_{2,x}(\tau_2)C_{2,x}(\tau_3 - \tau_1) \\ & - C_{2,x}(\tau_3)C_{2,x}(\tau_1 - \tau_2) \end{aligned} \tag{7c}$$

where $E\{*\}$ is the expected value operator. Then, looking at Equation 7, each cumulant is easily interpreted as a correlation between the original time series and its associated time-shifted versions, being the computational result of an rth-order cumulant is the rth degree of similarity among the aforementioned time series.

Considering a finite N-sample vector, signal vector $x(n)$, $n = 0, \cdots, N-1$, the following expressions, Equations 8, 9, and 10 describe the three unbiased estimates for the second-, third-, and fourth-order cumulants, respectively:

$$\hat{C}_{2,x}(k) = \frac{1}{N} \sum_{n=0}^{N-1} [x(n)]\,[x(n+\tau)], \tag{8}$$

$$\begin{aligned} \hat{C}_{3,x}(k,l) &= \hat{\mathrm{Cum}}[x(n), x(n+k), x(n+l)] \\ &= \frac{1}{N} \sum_{n=0}^{N-1} x(n)x(n+k)x(n+l) \end{aligned} \tag{9}$$

$$\begin{aligned} \hat{C}_{4,x}(k,l,m) &= \hat{\mathrm{Cum}}[x(n), x(n+k), x(n+l), x(n+m)] \\ &= \frac{1}{N} \sum_{n=0}^{N-1} x(n) \cdot x(n+k)^* \cdot x(n+l)^* \cdot x(n+m)^* \\ &\quad - \frac{1}{N^2} \left[\sum_{n=0}^{N-1} x(n) \cdot x(n+k)^*\right] \left[\sum_{n=0}^{N-1} x(n+l)^* \cdot x(n+m)^*\right] \\ &\quad - \frac{1}{N^2} \left[\sum_{n=0}^{N-1} x(n) \cdot x(n+l)^*\right] \left[\sum_{n=0}^{N-1} x(n+k)^* \cdot x(n+m)^*\right] \\ &\quad - \frac{1}{N^2} \left[\sum_{n=0}^{N-1} x(n) \cdot x(n+m)^*\right] \left[\sum_{n=0}^{N-1} x(n+k)^* \cdot x(n+l)^*\right] \end{aligned} \tag{10}$$

where $k, l, m \in [-\chi, \ldots, -1, 0, 1, \ldots, +\chi]$ and $n = 0, 1, \ldots, N-1$; χ is the index of the maximum time shift (lag) between samples of a record. The biased expressions are estimates over the real terms in the summations of expressions 8, 9, and 10. These expressions establish the correlation between the original signal and its time-shifted versions for the three orders of comparison. The second-order version is the classical auto-correlation, the third-order one account with the symmetry of the signal, and the fourth-order cumulant quantifies the impulsiveness in the time domain.

Avoided time shifting, $\tau_1 = \tau_2 = \tau_3 = 0$ in Equation 7, leads to the simplest computational expressions for cumulants, in Equation 11:

$$\gamma_{2,x} = E\{x^2(t)\} = C_{2,x}(0) \tag{11a}$$

$$\gamma_{3,x} = E\{x^3(t)\} = C_{3,x}(0,0) \tag{11b}$$

$$\gamma_{4,x} = E\{x^4(t)\} - 3(\gamma_{2,x})^2 = C_{4,x}(0,0,0). \tag{11c}$$

The ensemble of Equation 11 constitutes indirect measurements of the variance, skewness, and kurtosis. If $x(t)$ is symmetrically distributed, its skewness is zero (but not *vice versa*, improbable situations); if $x(t)$ is Gaussian distributed, its kurtosis is necessarily zero (but not *vice versa*). Standardization (statistical normalization) makes estimators shift and scale invariant. Standardized quantities are defined as $\gamma_{4,x}/(\gamma_{2,x})^2$ and $\gamma_{3,x}/(\gamma_{2,x})^{3/2}$, for kurtosis and skewness, respectively.

Frequency-domain HOS

Poly-spectra are defined to be the Fourier transforms of the higher-order cumulant sequences. The rth-order spectra are defined as the (r-1)-dimensional Fourier transforms of the rth-order cumulants, according to:

$$S_{r,x}\left(f_1,f_2,\ldots,f_{r-1}\right) = \sum_{\tau_1=-\infty}^{\tau_1=+\infty} \cdots \sum_{\tau_{r-1}=-\infty}^{\tau_{r-1}=+\infty} C_{r,x}(\tau_1,\tau_2,\ldots,\tau_{r-1}) \times \exp\left[-j2\pi\left(f_1\tau_1+f_2\tau_2+\cdots+f_{r-1}\tau_{r-1}\right)\right]. \quad (12)$$

The power spectrum is the decomposition of the signal power in the frequency domain. When this concept is extended to higher orders, as suggested by Equation 12, the result is called a poly-spectrum. Power spectrum, bi-spectrum, and tri-spectrum are specific cases (particular poly-spectra) of Equation 12, with $r = 2, 3$, and 4, respectively. Only power spectrum is real, and the others are complex magnitudes.

The more common higher-order spectra are the bi-spectrum and the tri-spectrum. The first one identifies contributions to a signal's skewness as a function of frequency pairs, meanwhile the tri-spectrum refers to contributions to a signal's kurtosis as a function of frequency triplets. For this reason, poly-spectra output multidimensional data structures which comprise redundant information, distributed in multi-dimensional geometries, often called tensors. As a consequence, their computation may be impractical in many cases, and to extract the desired information, one-dimensional slices of cumulant sequences and spectra and bi-frequency planes are considered [34,35]. To show this, in the following sections, we present two particular cases of the third- and fourth-order spectra, respectively: a 3-D bi-spectrum application and the performance of an estimator of the fourth-order spectrum for zero time lags, the spectral kurtosis (SK).

Ideally, the spectral kurtosis is a representation of the kurtosis of each frequency component of a process (or data from a measurement instrument x_i). For estimation issues, we will consider M realizations of the process; each realization containing N points; i.e., we consequently consider M measurement sweeps, each sweep with N points. The time spacing between points is the sampling period, T_s, of the data acquisition unit. The SK unbiased indirect estimator is given by Equation 13:

$$\widehat{G}_{2,X}^{N,M} = \frac{M}{M-1}\left[\frac{(M+1)\sum_{i=1}^{M}\left|X_N^i(m)\right|^4}{\left(\sum_{i=1}^{M}\left|X_N^i(m)\right|^2\right)^2} - 2\right] \quad (13)$$

where m indicates the frequency index and $\widehat{G}_{2,X}^{N,M}$ indicates the value of the kurtosis for this Fourier frequency. This expression offers an indirect calculation of the SK, as it is obtained directly from the Fourier transforms, and it supposes low computational burden. The graphical representation of the SK allows the identification of non-Gaussian frequency components. The higher the peak the more variable is the amplitude associated to this Fourier component.

Competing interests
The authors declare that they have no competing interests.

Acknowledgements
The authors would like to thank the Spanish Government for funding the research Project *TEC*2010 − 19242 − *C*03 − 03 (SIDER-HOSAPQ). This work is newly supported by the Spanish Ministry of Economy and Competitiveness in the frame of the Statal Plan of Excellency for Research, via the project *TEC*2013 − 47316 − *C*3 − 2 − *P* (SCEMS-AD-TED-PQR). Our unforgettable thanks to the trust we have from the Andalusian Government for funding the Research Group *PAIDI* − *TIC* − 168 in Computational Instrumentation and Industrial Electronics (ICEI).

References
1. MHJ Bollen, S Bahramirad, A Khodaei, in *Proceedings on the 2014 IEEE 16th International Conference on Harmonics and Quality of Power (ICHQP)*. Is there a place for power quality in the smart grid? (University Politehnica of Bucharest Romania, 2014), pp. 713–717
2. EU Expert group 1:, Functionalities of smart grids and smart meters. Technical report, EU Commission Task Force for Smart Grids (2009). http://ec.europa.eu/energy/en/topics/markets-and-consumers/smart-grids-and-meters
3. Y Xiao, *Communication and networking in smart grids*. (CRC Press, Broken Sound Parkway NW, Suite 300, 2012)
4. WK Reder, IEEE smart grid. Part 2: a grand vision for smart grid. Technical report. IEEE (2014). http://smartgrid.ieee.org/education
5. MHJ Bollen, IY-H Gu, PGV Axelberg, E Styvaktakis, Classification of underlying causes of power quality disturbances: deterministic versus statistical methods. EURASIP J. Adv. Signal Process. **2007**(1), 1–17 (2007)
6. YH Gu, MHJ Bollen, Time-frequency and time-scale domain analysis of voltage disturbances. IEEE Trans. Power Deliv. **15**(4), 1279–1283 (2000)
7. MV Ribeiro, CAG Marques, CA Duque, AS Cerqueira, JLR Pereira, Detection of disturbances in voltage signals for power quality analysis using HOS. EURASIP J. Adv. Signal Process. **2007**(1), 1–13 (2007)
8. MV Ribeiro, JLR Pereira, Classification of single and multiple disturbances in electric signals. EURASIP J. Adv. Signal Process. **2007**(1), 1–18 (2007)
9. A Agüera-Pérez, JC Palomares-Salas, JJG de la Rosa, JM Sierra-Fernández, D Ayora-Sedeño, A Moreno-Muñoz, Characterization of electrical sags and swells using higher-order statistical estimators. Measurement. **44**(Issue 8), 1453–1460 (2011)
10. ON Gerek, DG Ece, Power-quality event analysis using higher order cumulants and quadratic classifiers. IEEE Trans. Power Deliv. **21**(2), 883–889 (2006)
11. ON Gerek, DG Ece, 2-D analysis and compression of power quality event data. IEEE Trans. Power Deliv. **19**(2), 791–798 (2004)
12. DG Ece, ON Gerek, Power quality event detection using joint 2D wavelet subspaces. IEEE Trans. Instrumentation Meas. **53**(4), 1040–1046 (2004)
13. O Poisson, P Rioual, M Meunier, Detection and measurement of power quality disturbances using wavelet transform. IEEE Trans. Power Deliv. **15**(3), 1039–1044 (2000)
14. S Santoso, WM Grady, EJ Powers, J Lamoree, SC Bhatt, Characterization of distribution power quality events with fourier and wavelet transforms. IEEE Trans. Power Deliv. **15**(1), 247–254 (2000)
15. JJG de la Rosa, JM Sierra-Fernández, A Agüera-Pérez, JC Palomares-Salas, A Moreno-Muñoz, An application of the spectral kurtosis to characterize power quality events. Electrical Power Energy Syst. **49**, 386–398 (2013)
16. JJG de la Rosa, JM Sierra-Fernández, A Agüera-Pérez, JC Palomares-Salas, A Jiménez-Montero, A Moreno-Muñoz, in *Proceedings on the IEEE International Workshop on the Applied Measurements for Power Systems (AMPS)*, vol. 1. Power quality events' measurement criteria based in higher-order statistics: towards new measurement indices (IEEE Aachen, Germany, 2013), pp. 73–79

17. JJG de la Rosa, A Agüera-Pérez, JC Palomares-Salas, JM Sierra-Fernández, A Moreno-Muñoz, A novel virtual instrument for power quality surveillance based in higher-order statistics and case-based reasoning. Measurement. **45**(7), 1824–1835 (2012)
18. JC Palomares-Salas, JJG de la Rosa, A Agüera-Pérez, A Moreno-Muñoz, Intelligent methods for characterization of electrical power quality signals using higher order statistical features. Przeglad Elektrotechniczny. **2012**(8), 236–243 (2012)
19. S Haykin, *Neural Networks*. (Englewood Cliffs, NJ, 1994)
20. DD Ferreira, CAG Marques, JM de Seixas, AS Cerqueira, MV Ribeiro, CA Duque, *Exploiting Higher-Order Statistics Information for Power Quality Monitoring*. (InTech Open Science, 2011). http://intechopen.com
21. L Angrisani, P Daponte, M D'Apuzzo, Wavelet network-based detection and classification of transients. IEEE Trans. Instrumentation Meas. **50**(5), 1425–1435 (2001)
22. I Monedero, C León, J Ropero, A García, JM Elena, JC Montaño, Classification of electrical disturbances in real time using neural networks. IEEE Trans. Power Deliv. **22**(3), 1288–1296 (2007)
23. M Valtierra-Rodriguez, de Jesus Romero-Troncoso R, RA Osornio-Rios, A Garcia-Perez, Detection and classification of single and combined power quality disturbances using neural networks. IEEE Trans. Ind. Electron. **61**(1), 2473–2482 (2014)
24. DD Ferreira, AS Cerqueira, CA Duque, MV Ribeiro, HOS-based method for classification of power quality disturbances. Electron. Lett. **45**(3), 183–185 (2009)
25. JJG de la Rosa, A Agüera-Pérez, JC Palomares-Salas, A Moreno-Muñoz, Higher-order statistics: discussion and interpretation. Measurement. **46**(8), 2816–2827 (2013)
26. Z Liu, Q Zhang, Z Han, G Chen, A new classification method for transient power quality combining spectral kurtosis with neural network. Neurocomputing. **125**(1), 95–101 (2014)
27. S Mallat, *A wavelet tour of signal processing*. (Academic Press, Burlington, MA 01803, 2009)
28. J Sola, J Sevilla, Importance of input data normalization for the application of neural networks to complex industrial problems. IEEE Transa. Nuclear Sci. **44**(3), 1464–1468 (1997)
29. JM Mendel, Tutorial on higher-order statistics (spectra) in signal processing and system theory: theoretical results and some applications. Proc. IEEE. **79**(3), 278–305 (1991)
30. CL Nikias, JM Mendel, Signal processing with higher-order spectra. IEEE Signal Process. Mag., 10–37 (1993)
31. CL Nikias, AP Petropulu, *Higher-Order Spectra Analysis. A Non-Linear Signal Processing Framework*. (Prentice-Hall, Englewood Cliffs, NJ, 1993)
32. AK Nandi, *Blind Estimation Using Higher-Order Statistics, vol 1*, 1st edn. (Kluwer Academic Publishers, Boston, 1999)
33. JJG de la Rosa, A Moreno-Muñoz, Electrical transients monitoring via higher-order cumulants and competitive layers. Przeglad Elektrotechniczny - Electrical Rev. **85**(10), 284–289 (2009)
34. J Jakubowski, K Kwiatos, A Chwaleba, S Osowski, Higher order statistics and neural network for tremor recognition. IEEE Trans. Biomed. Eng. **49**(2), 152–159 (2002)
35. JJG de la Rosa, I Lloret, CG Puntonet, JM Górriz, Higher-order statistics to detect and characterize termite emissions. Electron. Lett. **40**(20), 1316–1317 (2004)

8

Out-of-band and adjacent-channel interference reduction by analog nonlinear filters

Alexei V Nikitin[1,2*], Ruslan L Davidchack[3] and Jeffrey E Smith[4]

Abstract

In a perfect world, we would have 'brick wall' filters, no-distortion amplifiers and mixers, and well-coordinated spectrum operations. The real world, however, is prone to various types of unintentional and intentional interference of technogenic (man-made) origin that can disrupt critical communication systems. In this paper, we introduce a methodology for mitigating technogenic interference in communication channels by analog nonlinear filters, with an emphasis on the mitigation of out-of-band and adjacent-channel interference.

Interference induced in a communications receiver by external transmitters can be viewed as wide-band non-Gaussian noise affecting a narrower-band signal of interest. This noise may contain a strong component within the receiver passband, which may dominate over the thermal noise. While the total wide-band interference seen by the receiver may or may not be impulsive, we demonstrate that the interfering component due to power emitted by the transmitter into the receiver channel is likely to appear impulsive under a wide range of conditions. We give an example of mechanisms of impulsive interference in digital communication systems resulting from the nonsmooth nature of any physically realizable modulation scheme for transmission of a digital (discontinuous) message.

We show that impulsive interference can be effectively mitigated by nonlinear differential limiters (NDLs). An NDL can be configured to behave linearly when the input signal does not contain outliers. When outliers are encountered, the nonlinear response of the NDL limits the magnitude of the respective outliers in the output signal. The signal quality is improved in excess of that achievable by the respective linear filter, increasing the capacity of a communications channel. The behavior of an NDL, and its degree of nonlinearity, is controlled by a single parameter in a manner that enables significantly better overall suppression of the noise-containing impulsive components compared to the respective linear filter. Adaptive configurations of NDLs are similarly controlled by a single parameter and are suitable for improving quality of nonstationary signals under time-varying noise conditions. NDLs are designed to be fully compatible with existing linear devices and systems and to be used as an enhancement, or as a low-cost alternative, to the state-of-art interference mitigation methods.

Keywords: Adjacent-channel interference; Impulsive noise; Interchannel interference; Spectral density; Nonlinear differential limiters; Nonlinear filters; MANET; Out-of-band interference; Technogenic noise

1 Introduction

In a utopian world, our communication technology would have 'brick wall' filters, no-distortion amplifiers and mixers, and well-coordinated spectrum operations. In the real world, wireless communications are prone to various types of natural and technogenic (man-made) interference. Over the years, engineers developed effective filters and approaches to dealing with natural interference, but the need to transmit more and more data leads to ever-increasing levels of technogenic interference as we saturate the information-carrying capacity of the electromagnetic spectrum. This brings the understanding of the types of technogenic interference and development of effective ways of its mitigation to the forefront of challenges facing modern communication technology.

Technogenic noise comes in a great variety of forms, but it will typically have a temporal and/or amplitude structure which distinguishes it form the natural (e.g., thermal) noise. It will typically also have non-Gaussian amplitude

*Correspondence: avn@avatekh.com
[1] Avatekh Inc., 901 Kentucky Street, Suite 303, Lawrence, KS 66044, USA
[2] Dept. of Electrical and Computer Engineering, Kansas State University, Manhattan, KS 66506, USA
Full list of author information is available at the end of the article

distribution. These features of technogenic noise provide an opportunity for its mitigation by nonlinear filters, especially for the in-band noise, where linear filters that are typically deployed in the communication receiver (see top part of Figure 1) have very little or no effect. Indeed, at any given frequency, a linear filter affects both the noise and the signal of interest proportionally. When a linear filter is used to suppress the interference outside of the passband of interest, the resulting signal quality is affected by the total power and spectral composition, but not by the type of the amplitude distribution of the interfering signal. On the other hand, the spectral density of a non-Gaussian interference in the signal passband can be reduced, without significantly affecting the signal of interest, by introducing an appropriately chosen feedback-based nonlinearity into the response of the linear filter.

In particular, impulsive interference that is characterized by a frequent occurrence of outliers (i.e., relatively high-amplitude, short-duration events) can be effectively mitigated by the nonlinear differential limiters (NDLs) described in [1-4] and in Section 3 of this paper. An NDL can be configured to behave linearly when the input signal does not contain outliers, but when the outliers are encountered, the nonlinear response of the NDL limits the magnitude of the respective outliers in the output signal. As a result, the improvement in signal quality achieved by the NDL exceeds that achievable by the respective linear filter, increasing the capacity of a communications channel. Even if the interference appears nonimpulsive, the non-Gaussian nature of its amplitude distribution enables simple analog pre-processing which can increase its peakedness and thus increases the effectiveness of the NDL mitigation.

Another important consideration is the dynamic nonstationary nature of technogenic noise. When the frequency bands, modulation/communication protocol schemes, power levels, and other parameters of the transmitter and the receiver are stationary and well defined, the interference scenarios may be analyzed in great detail. Then, the system may be carefully engineered (albeit at a cost) to minimize the interference[a]. It is far more challenging to quantify and address the multitude of complicated interference scenarios in nonstationary communication systems such as, for example, software-defined

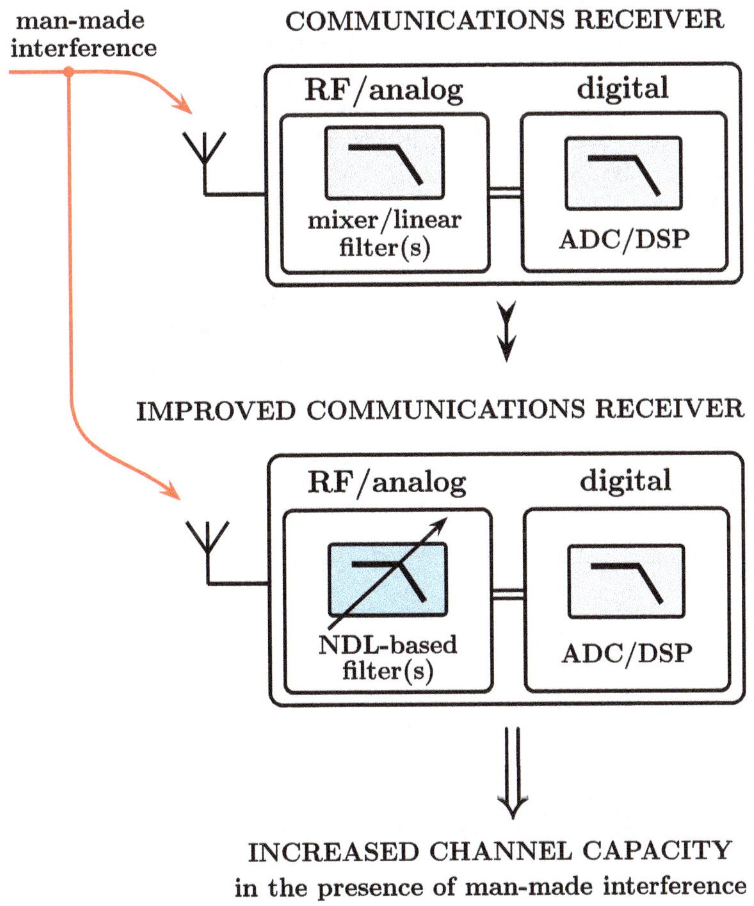

Figure 1 Simplified diagram of improving receiver performance in the presence of technogenic (man-made) interference.

radio (SDR)-based and cognitive *ad hoc* networks comprising mobile transmitters and receivers, each acting as a local router communicating with a mobile *ad hoc* network (MANET) access point [5]. In this scenario, the transmitter positions, powers, and/or spectrum allocations may vary dynamically. In multiple access schemes, the interference is affected by the varying distribution and arrangement of transmitting nodes [6]. In addition, with MANETs, the fading distribution also varies dynamically, and the path loss distribution is unbounded [7]. With spectrum-aware MANETs, frequency allocations could also depend on various criteria, e.g., whitespace and the customer quality of service goals [8]. This is a very challenging situation which requires interference mitigation tools to adapt to the dynamically changing interference. Following the dynamic nature of the *ad hoc* networks, where the networks themselves are scalable and adaptive, and include spectrum sensing and dynamic re-configuration of the network parameters, interference mitigation tools are needed to be scalable and adaptive to the dynamically changing interference. Adaptive NDLs (ANDLs) [2,3] have been developed to address this challenge.

Based on the above considerations, we propose a modification of a communications receiver system as illustrated in the bottom part of Figure 1, where an NDL or ANDL filter is used as a replacement, or in conjunction with the existing mixer/linear filters.

To give more specificity to our presentation, let us consider a single transmitter-receiver system. Figure 2 provides a simplified qualitative illustration of different contributions into the interference which a receiver (RX) experiences from a transmitter (TX). Since real-time 'brick-wall' filters are not physically realizable as they have infinite latency (i.e., their compact support in the frequency domain forces their time responses not to have compact support, meaning that they are everlasting) and infinite order (i.e., their responses cannot be expressed as a linear differential equation with a finite sum), TX emissions would 'leak' outside of the nominal (allocated) passband of a TX channel $[f_1, f_2]$ as out-of-band (OOB) emissions. Likewise, an RX filter would have nonzero response outside of its nominal (allocated) passband $[f_3, f_4]$. As a result, there is nonzero interference from the TX into the RX.

The total power of this interference may be broken into three parts. Part I is the power of the TX signal in its nominal band $[f_1, f_2]$, weighted by the response of the RX filter in this band. Part II is the TX OOB emissions in the RX nominal band $[f_3, f_4]$, weighted by the response of the RX filter in this band. The rest of the interference power comes from the TX emissions outside of the nominal bands of both channels and can be normally ignored in practice since in those frequency regions both the emitted TX power and the RX filter response would be relatively small.

While part I of the interference contributes into the total power in the RX channel and may cause RX overload, it does not normally degrade the quality of the communications in the RX since the frequency content of this part of the interference lies outside of the RX channel. Part II, however, in addition to contributing to overload, also causes degradation in the RX communication signal as it raises the noise floor in the RX channel.

Theoretical [9,10] as well as the experimental [11] data suggests that the TX OOB interference in the RX channel (part II of the interference in Figure 2) can appear impulsive under a wide range of conditions, as will be additionally illustrated in Section 2. While this interference cannot be reduced by the subsequent linear filtering in the RX channel, it may be effectively mitigated by the NDLs introduced in Section 3.

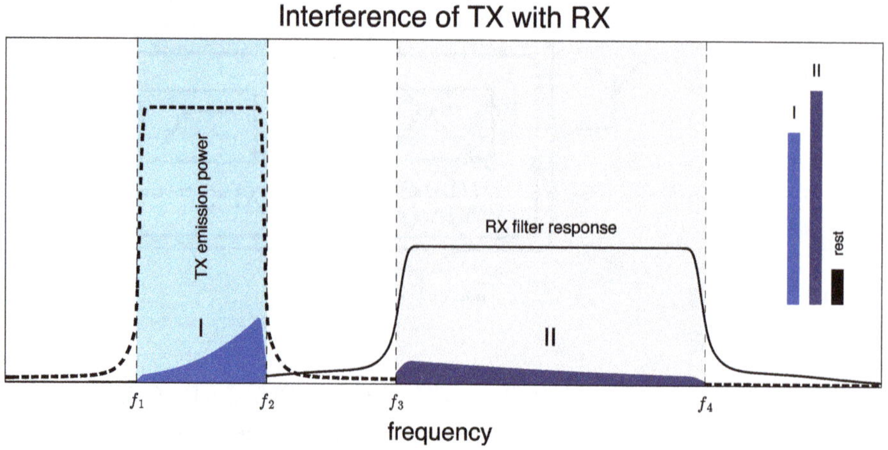

Figure 2 Qualitative illustration of different contributions into the interference which a receiver (RX) experiences from a transmitter (TX).

In Section 4, we show that an NDL deployed in the RX channel can reduce the spectral density of impulsive interference in the signal passband without significantly affecting the signal of interest, thus improving the baseband signal-to-noise ratio and increasing the channel capacity. We also show that, when part I of the interference in Figure 2 dominates over part II and the total interference observed in the receiver does not appear impulsive, one can deploy a bandstop linear filter in the signal chain of the receiver preceding the NDL to suppress part I of the interference without affecting the baseband signal of interest. By suppressing part I of the interference and thus increasing the peakedness of the remaining interference affecting the baseband signal, this additional filter can greatly improve the effectiveness of interference mitigation by the subsequently deployed NDL.

In Section 5, we provide some concluding remarks and comment on the possibility of digital implementations and deployment of the NDLs.

2 Impulsive nature of interchannel interference

As shown in more detail in [9,10], with additional experimental evidence presented in [11], the signal components induced in a receiver by out-of-band communication transmitters can appear impulsive under a wide range of conditions. For example, in the transmitter-receiver pair schematically shown at the top of Figure 3, for a sufficiently large absolute value of the difference between the transmit and receive frequencies $\Delta f = f_{\mathrm{RX}} - f_{\mathrm{TX}}$, the instantaneous power $I^2(t, \Delta f) + Q^2(t, \Delta f)$ of the in-phase and quadrature components of the receiver signal may appear as a train of pulses consisting of a linear combination of pulses originating at discrete times and shaped

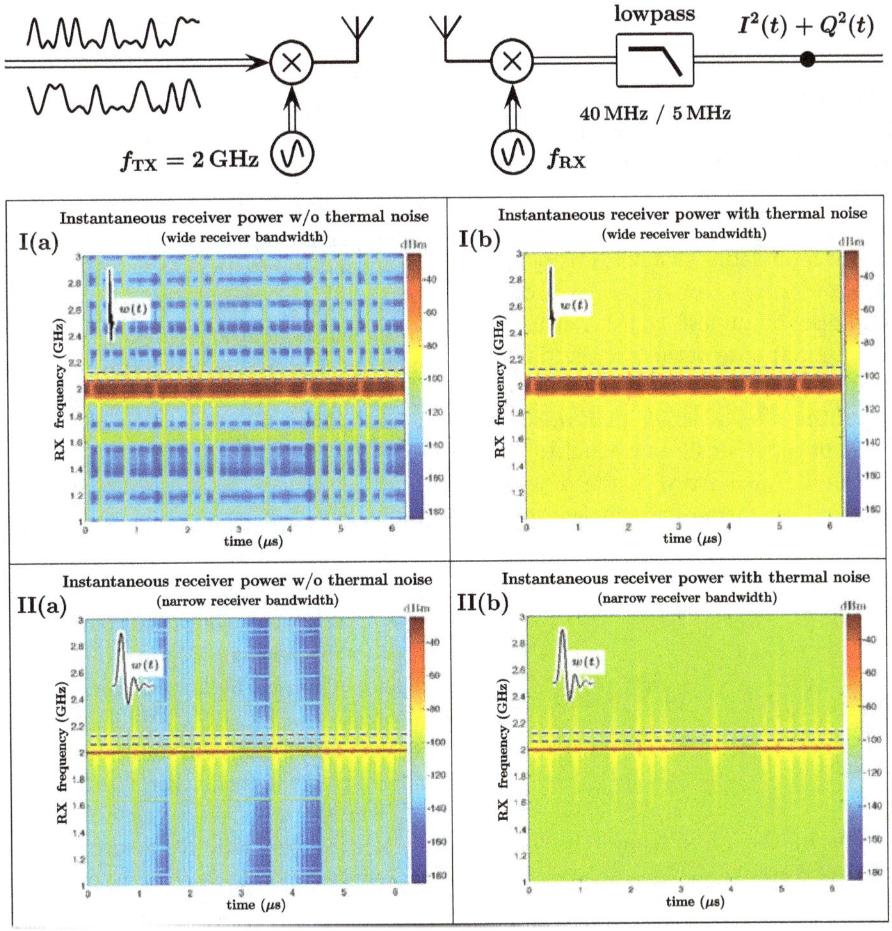

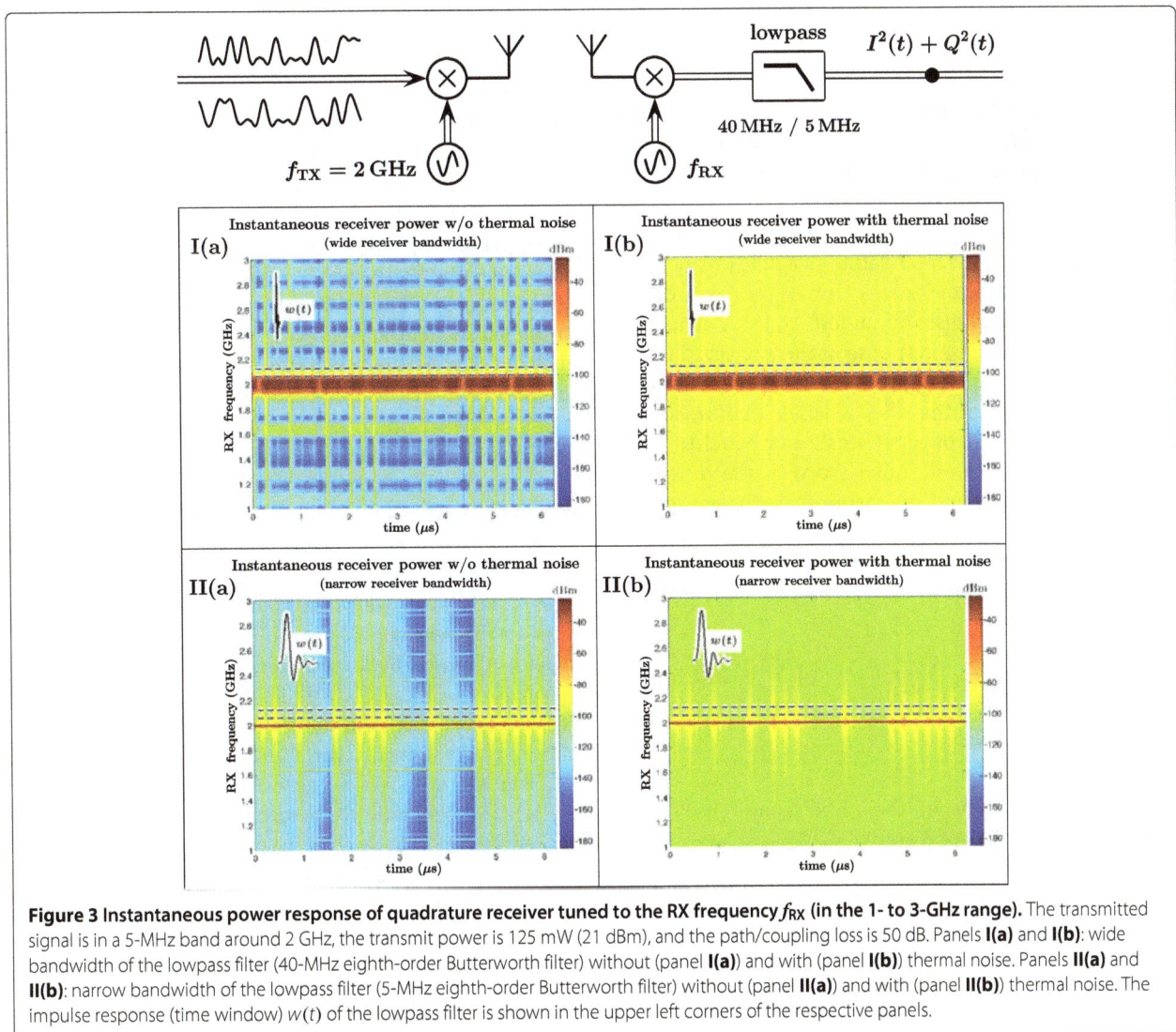

Figure 3 Instantaneous power response of quadrature receiver tuned to the RX frequency f_{RX} (in the 1- to 3-GHz range). The transmitted signal is in a 5-MHz band around 2 GHz, the transmit power is 125 mW (21 dBm), and the path/coupling loss is 50 dB. Panels **I(a)** and **I(b)**: wide bandwidth of the lowpass filter (40-MHz eighth-order Butterworth filter) without (panel **I(a)**) and with (panel **I(b)**) thermal noise. Panels **II(a)** and **II(b)**: narrow bandwidth of the lowpass filter (5-MHz eighth-order Butterworth filter) without (panel **II(a)**) and with (panel **II(b)**) thermal noise. The impulse response (time window) $w(t)$ of the lowpass filter is shown in the upper left corners of the respective panels.

as the squared (i.e., raised to the power of 2) impulse response of the receiver lowpass filter.

For a single transmitter, the typical intervals between those discrete times are multiples of the symbol duration of the transmitted signal (or other discrete time intervals used in the designed modulation scheme, for example, chip and guard intervals). The nonidealities in hardware implementation of designed modulation schemes, such as the nonsmooth behavior of the modulator around zero, and/or nonlinearities in the power amplifier, can also lead to the appearance of additional discrete origins for the pulses and exacerbate the OOB emissions. If the typical value of those discrete time intervals is large in comparison with the inverse bandwidth of the lowpass filter in the receiver, this pulse train may be highly impulsive.

A key mathematical argument leading to this conclusion can be briefly recited as follows [9,10]. The total emission from various digital transmitters can be written as a linear combination of the terms of the following form:

$$x(t) = A_T(\bar{t})\, e^{i\omega_c t}, \tag{1}$$

where ω_c is the frequency of a carrier, $\bar{t} = \frac{2\pi}{T} t$ is the *nondimensionalized time*, and $A_T(\bar{t})$ is the desired (or designed) complex-valued modulating signal representing a data signal with symbol duration (unit interval) T. Let us assume that, for some integer n, all derivatives of order smaller than $n-1$ of the modulating signal $A_T(\bar{t})$ are finite, but the derivative of order $n-1$ of $A_T(\bar{t})$ has a countable number of step discontinuities[b] at $\{\bar{t}_i\}$. Let us now assume that the impulse response of the lowpass filters in both channels of a quadrature receiver is $w(t) = \frac{2\pi}{T} h(\bar{t})$, and that the order of the filter is larger than n, so that all derivatives of $w(t)$ of order smaller or equal to $n-1$ are continuous.[c] Then, if $\Delta\omega = 2\pi\Delta f$ is the difference between the receiver and the carrier frequencies, and the bandwidth of the lowpass filter $w(t)$ in the receiver is much smaller than $|\Delta f|$, the instantaneous power in the quadrature receiver due to $x(t)$ can be expressed as[d]:

$$P_x(t,\Delta f) = \frac{1}{(T\,\Delta f)^{2n}} \sum_i \alpha_i\, h(\bar{t}-\bar{t}_i) \sum_j \alpha_j^*\, h(\bar{t}-\bar{t}_j) \quad \text{for} \quad T\Delta f \gg 1, \tag{2}$$

where α_i is the value of the ith discontinuity of the order $n-1$ derivative of $A_T(\bar{t})$:

$$\alpha_i = \lim_{\varepsilon\to 0} \left[A_T^{(n-1)}(\bar{t}_i+\varepsilon) - A_T^{(n-1)}(\bar{t}_i-\varepsilon) \right] \neq 0. \tag{3}$$

The detailed derivation of Equation 2 can be found in [9,10].

When viewed as a function of both time and frequency, the interpretation of Equation 2 for the instantaneous power in a quadrature receiver is a *spectrogram* ([12], for example) in the time window $w(t)$ of the term $x(t)$ of the transmitted signal. Figure 3 provides an illustrative example of such spectrograms for the $I^2(t,\Delta f) + Q^2(t,\Delta f)$ receiver signal in the transmitter-receiver pair schematically shown at the top of the figure.

The spectrograms displayed in the panels of the figure show the instantaneous power response of a quadrature receiver tuned to the RX frequency f_{RX}, where f_{RX} is in the 1- to 3-GHz range. The transmitted signal is in a 5-MHz band around 2 GHz, the transmit power is 125 mW (21 dBm), and the path/coupling loss is 50 dB. A more detailed description of the simulation parameters used in Figure 3 and the subsequent examples can be found in Appendix A.

In panels I(a) and I(b) of Figure 3, the bandwidth of the lowpass filter (eighth-order Butterworth) is 40 MHz (wide), while in panels II(a) and II(b), it is 5 MHz (narrow). Panels I(a) and II(a) show the receiver power due to the transmitter signal only (without thermal noise), while panels I(b) and II(b) show the receiver power with additive white Gaussian noise (AWGN) taken as the thermal noise multiplied by the noise figure of the receiver (assumed 5 dB). The shape of the impulse response (time window) $w(t) = \frac{2\pi}{T} h(\bar{t})$ of the lowpass filters is shown in the upper left corners of the respective panels. The dashed horizontal lines in the panels indicate the specific receiver offset frequencies $\Delta f = 65$ MHz and $\Delta f = 125$ MHz used in the subsequent examples. To make the OOB interference induced by the transmitter less idealized, moderate intermodulation (resulting from 'clipping' of the carrier signal at high amplitudes) was added to the simulation. This results in the intermodulation distortion (IMD) that appears as horizontal bands at frequencies different from the carrier frequency in panels I(a) and II(a).

The upper panels of Figure 4 show the instantaneous receiver power averaged over time, for both wide (blue lines) and narrow (red lines) bandwidths of the lowpass filter in the receiver. These would be akin to the power spectra obtained by a spectrum analyzer with the resolution bandwidth (RBW) filters of 5 MHz (red line) and 40 MHz (blue line), without (left panels) and with (right panels) thermal noise taken into account.

Obviously, the *average* receiver power as a function of the RX frequency does not provide information on the peakedness of the receiver signal. The lower panels of Figure 4 quantify such peakedness of the receiver signal $z(t) = I(t) + iQ(t)$ in terms of the measure K_{dBG} found in [2,3]:

$$K_{dBG}(z) = 10\lg\left(\frac{\langle |z|^4\rangle - |\langle zz\rangle|^2}{2\langle |z|^2\rangle^2}\right), \tag{4}$$

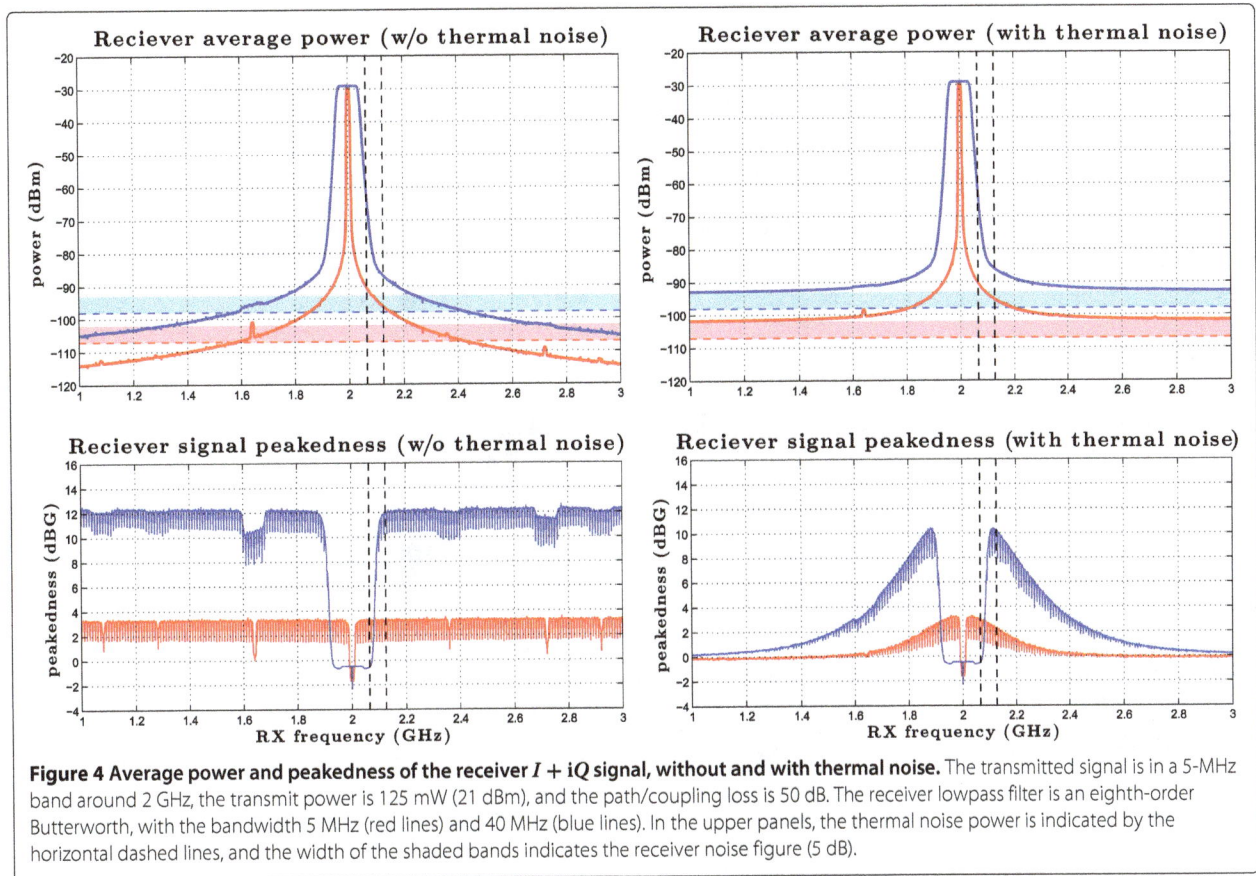

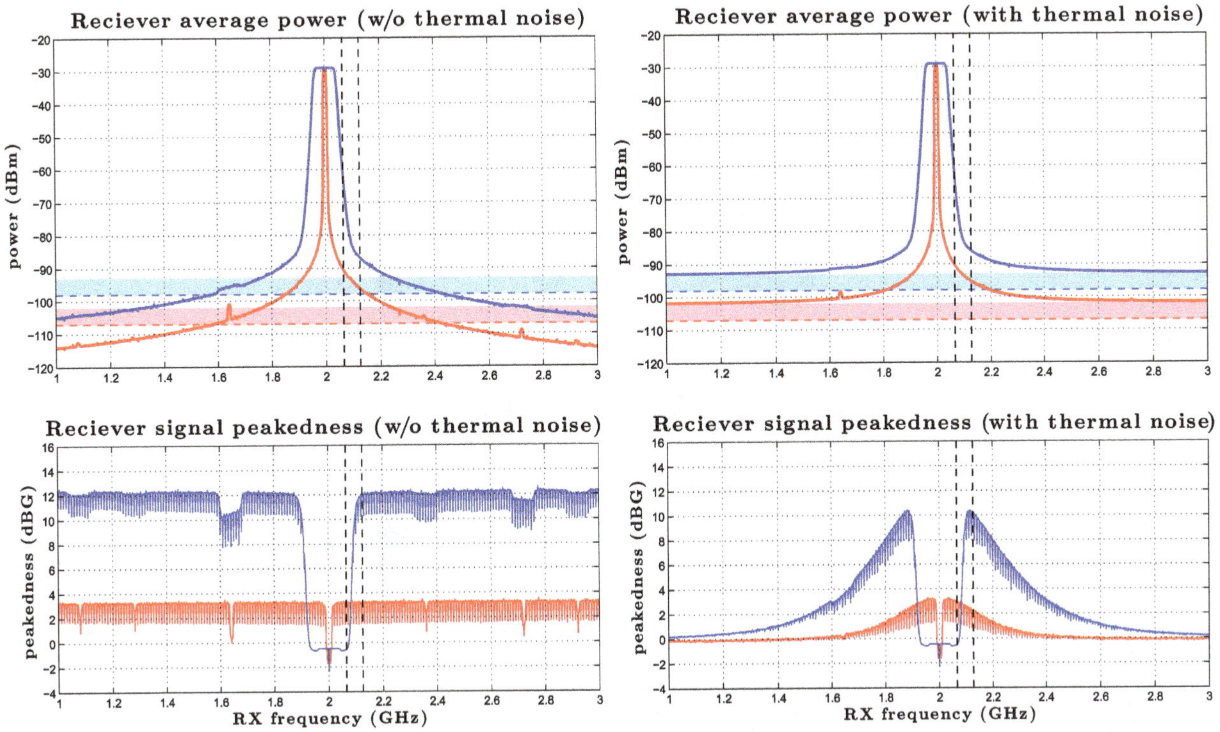

Figure 4 Average power and peakedness of the receiver $I + iQ$ signal, without and with thermal noise. The transmitted signal is in a 5-MHz band around 2 GHz, the transmit power is 125 mW (21 dBm), and the path/coupling loss is 50 dB. The receiver lowpass filter is an eighth-order Butterworth, with the bandwidth 5 MHz (red lines) and 40 MHz (blue lines). In the upper panels, the thermal noise power is indicated by the horizontal dashed lines, and the width of the shaded bands indicates the receiver noise figure (5 dB).

where the angular brackets denote time averaging. This measure of peakedness is based on an extension of the classical definition of *kurtosis* ([13], for example) to complex variables ([14], for example). According to this definition, the peakedness is measured in units of 'decibels relative to Gaussian' (dBG) (i.e., in relation to the kurtosis of the Gaussian (aka normal) distribution). Gaussian distribution has zero dBG peakedness, while sub-Gaussian and super-Gaussian distributions have negative and positive dBG peakedness, respectively.

As can be seen in the lower left panel of Figure 4, the peakedness of the receiver $I + iQ$ signal at large values of $|\Delta f|$ is much higher for the wide-bandwidth receiver (blue line) than for the narrow-band receiver (red line). As follows from the linearity property of kurtosis, adding a Gaussian (zero dBG) signal to a super-Gaussian (positive dBG) signal would lower the peakedness of the mixture. This can be seen in the lower right panel of Figure 4, where the peakedness remains high while the power of the OOB interference dominates over the thermal noise, asymptotically approaching zero as the OOB interference decays at large values of $|\Delta f|$.

Figure 5 provides time (upper panel) and frequency (lower panel) domain quantification of the receiver $I + iQ$ signal without thermal noise for $\Delta f = 125$ MHz and wide (blue lines) and narrow (red lines) bandwidths of the lowpass filter.

As discussed in Section 2.1, for a receiver lowpass filter of a given type and order, the amplitude ('height') of the interference pulses would be proportional to the bandwidth of the filter. This can be seen in the upper panel of Figure 5, where the peak amplitude of the pulses shown by the blue lines (40-MHz filter) is eight times the peak amplitude of the pulses shown by the red lines (5-MHz filter).

Since the duration of the pulses is inversely proportional to the lowpass filter bandwidth, the time average of the squared amplitudes of the pulses would be proportional to the bandwidth, while the average of the amplitudes raised to the fourth power would be proportional to the bandwidth raised to the third power. As a result, the measure of peakedness given by Equation 4 would be approximately proportional to a logarithm of the bandwidth. Thus, the increase in the bandwidth of the receiver lowpass filter from 5 to 40 MHz (by 9 dB) would result in a 9-dB increase of peakedness. This is confirmed by the measured values of peakedness indicated in the lower panel of Figure 5 for the 5-MHz bandwidth filter (3.2 dBG, red text) and the 40-MHz bandwidth filter (12 dBG, blue text).

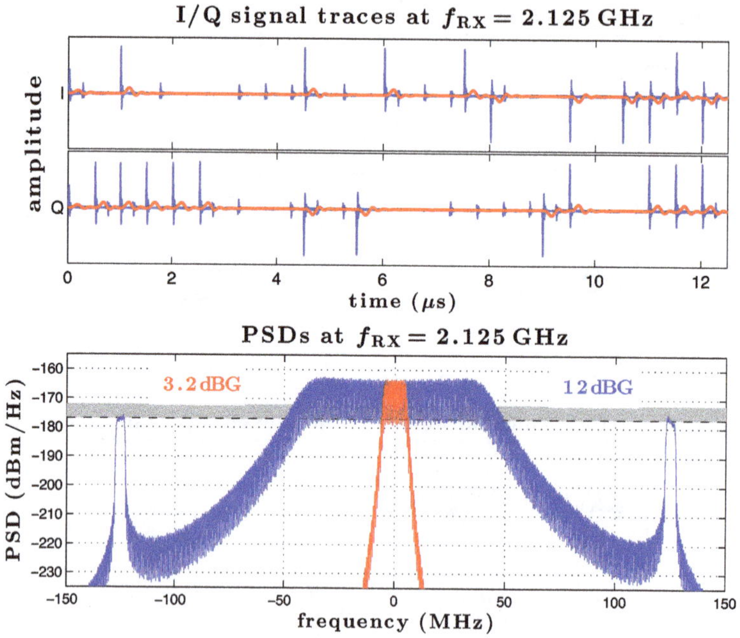

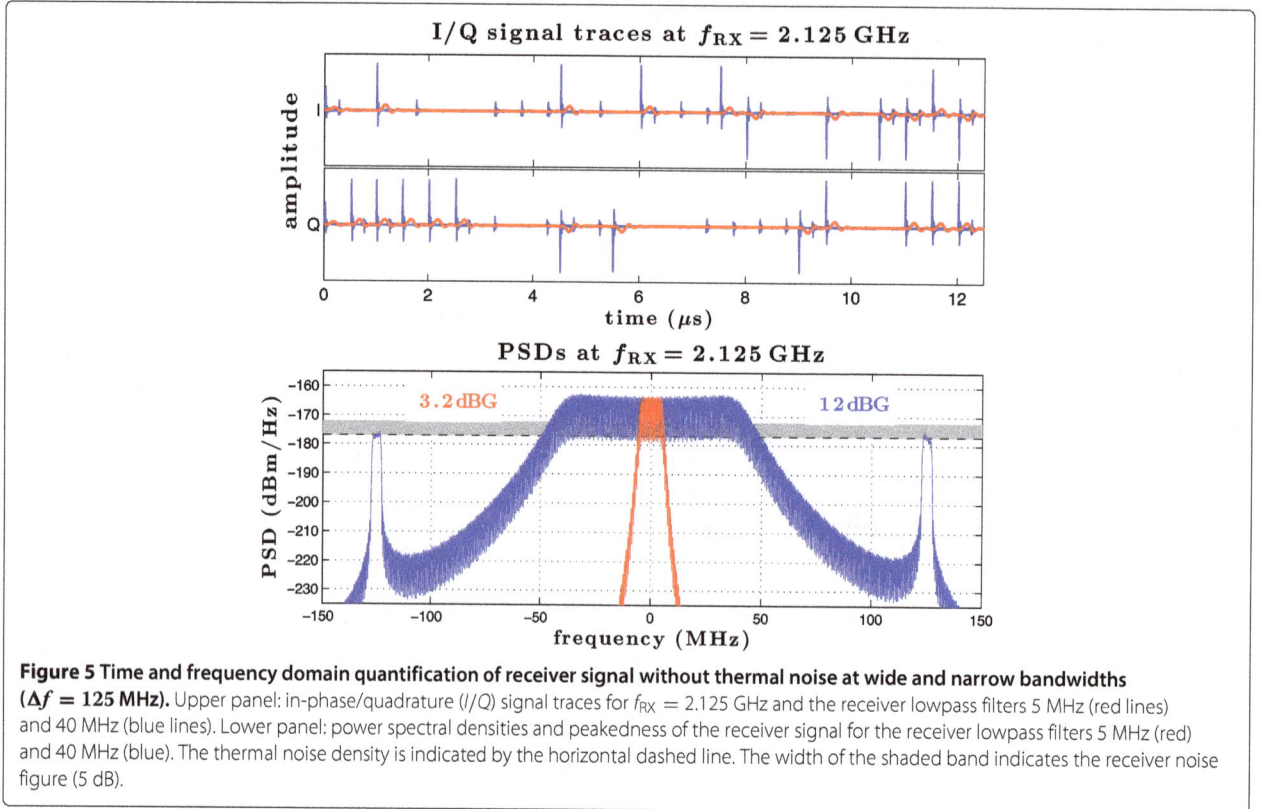

Figure 5 Time and frequency domain quantification of receiver signal without thermal noise at wide and narrow bandwidths (Δf = 125 MHz). Upper panel: in-phase/quadrature (I/Q) signal traces for $f_{RX} = 2.125$ GHz and the receiver lowpass filters 5 MHz (red lines) and 40 MHz (blue lines). Lower panel: power spectral densities and peakedness of the receiver signal for the receiver lowpass filters 5 MHz (red) and 40 MHz (blue). The thermal noise density is indicated by the horizontal dashed line. The width of the shaded band indicates the receiver noise figure (5 dB).

Not surprisingly, as can be seen in the lower panel of Figure 5, the power spectral density (PSD) of the interference around $\Delta f = 0$ (in baseband) is identical for both wide- and narrow-bandwidth receiver filters, and if subsequent linear processing is used (e.g., the signal is digitized and filtered with a matching digital filter), the resulting signal quality is independent of the bandwidth of the lowpass filter in the receiver. However, while the increase in the bandwidth of the receiver lowpass filter does not affect the baseband PSD of either the interference or the thermal noise, widening this bandwidth increases the peakedness of the interference, enabling its more effective mitigation by the NDLs introduced in Section 3.

Figure 6 provides time (upper panel) and frequency (lower panel) domain quantification of the receiver $I + \mathrm{i}Q$ signal without thermal noise for $\Delta f = 65$ MHz, for a 40-MHz lowpass filter (green lines) and for a 40-MHz lowpass filter cascaded with a 65-MHz notch filter (black lines).

The response of the receiver 40-MHz lowpass filter at 65 MHz is relatively large, and as can be seen in both panels of Figure 6 (green lines and text), the contribution of the TX signal in its nominal band (part I of the interference in Figure 2) into the total interference becomes significant, reducing the peakedness of the total interference and making it sub-Gaussian (−0.5 dBG peakedness). However, since the sub-Gaussian part of the interference lies outside of the baseband, cascading a 65-MHz notch filter with the lowpass filter would reduce this part of the interference without affecting either the signal of interest or the PSD of the impulsive interference around the baseband. Then, as shown by the black lines and text in Figure 6, the interference becomes super-Gaussian (10.8-dBG peakedness), enabling, as illustrated further in Section 4, its effective mitigation by the NDLs.

2.1 Effects of symbol rates and pulse shaping on the interference power

When the origins of the OOB interference lie in the finite duration of the finite impulse response (FIR) filters used for pulse shaping, an average value of $t_{i+1} - t_i$ in Equation 2 is of the same order of magnitude as the symbol duration (unit interval) T (in the range from $T/2$ to T, and equal to T if the group delay is a multiple of T). If the reciprocal of this value (the symbol rate) is small in comparison with the bandwidth of the receiver, the contribution of the terms $\alpha_i \alpha_j^* h(\bar{t} - \bar{t}_i)\, h(\bar{t} - \bar{t}_j)$ for $i \neq j$ is negligible, and Equation 2 describes an impulsive pulse train consisting of a linear combination of pulses shaped as $w^2(t)$ and originating at $\{t_i\}$, namely:

$$P_x(t, \Delta f) = \frac{1}{(T\,\Delta f)^{2n}} \sum_i |\alpha_i|^2\, h^2(\bar{t} - \bar{t}_i) \tag{5}$$

for sufficiently large T and Δf.

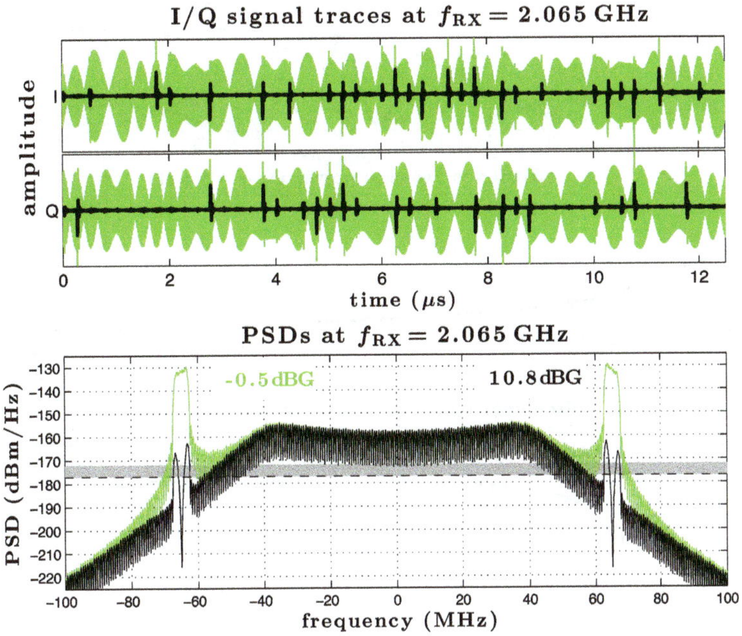

Figure 6 Quantification of receiver signal without thermal noise at wide bandwidth, with and without notch filter (Δf = 65 MHz). Upper panel: in-phase/quadrature (*I*/*Q*) signal traces for f_{RX} = 2.065 GHz, for a 40-MHz lowpass filter (green lines) and for a 40-MHz lowpass filter cascaded with a 65-MHz notch filter (black lines). Lower panel: power spectral densities and peakedness of the receiver signal for a 40-MHz lowpass filter (green) and for a 40-MHz lowpass filter cascaded with a 65-MHz notch filter (black). The thermal noise density is indicated by the horizontal dashed line. The width of the shaded band indicates the receiver noise figure (5 dB).

In Equation 5, the terms under the summation sign are functions of the nondimensionalized time $\bar{t} = \frac{2\pi}{T} t$. Then, for a given transmitter power and the modulation pulse shape, if the discontinuities are due to the modulation pulse shape only and thus the time intervals between t_i and t_{i+1} are proportional to the unit interval T, the differences between $\bar{t}_i$ and $\bar{t}_{i+1}$ and thus the time average of the sum in Equation 5 are independent of the symbol rate. As a result, provided that the conditions for Equation 5 are met, for the given offset frequency Δf, transmitter power, and modulation pulse shape, the average interference power is proportional to the symbol rate raised to the power of $2n$.

This is illustrated in Figure 7, where the upper panel shows the (highly oversampled) FIR root-raised-cosine filters ([15], for example) used for pulse shaping. All four filters have group delays equal to three times the unit interval T, two with roll-off factor one fourth (red and blue lines), and two with roll-off factor zero (black and green lines). The nonzero end values of the filters shown by the red and blue lines lead to discontinuities in the modulation signal ($n = 1$). The ratio of the unit intervals for these filters is equal to two (for the symbol rates 2 and 1 Mbit/s, respectively), and thus, the ratio of the respective interference powers at high Δf is $2^{2n} = 2^2$, or 6 dB, as can be seen in the middle panel of Figure 7.

While for the filters shown by the black and green lines the modulating signal itself is continuous (the end values are zero), the first time derivative of the modulation signal is discontinuous ($n = 2$). The ratio of the unit intervals for these filters is equal to four (for the symbol rates 8 and 2 Mbit/s, respectively), and thus, the ratio of the respective interference powers at high Δf is $4^{2n} = 4^4$, or 24 dB, as can be seen in the middle panel of Figure 7. As can be seen in the lower panel of Figure 7, as the offset frequency Δf increases, the impulsive component of the OOB interference becomes dominant, leading to a high peakedness of the interference.

As can also be seen from Equation 5, the average interference power depends on the impulse response $w(t)$ of the receiver lowpass filter and, for a filter of a given type and order, is proportional to its bandwidth. On the other hand, the thermal noise power is also proportional to the bandwidth of the receiver lowpass filter, and thus, the ratio of the powers of the interference and the thermal noise is independent of this bandwidth. This can be seen in the upper panels of Figure 4, where the increase in the bandwidth of the receiver lowpass filter from 5 to 40 MHz (by 9 dB) results in a 9-dB increase of both the interference and the thermal noise powers.

While the increase in the bandwidth does not affect the baseband PSD of either the interference or the thermal noise, widening of this bandwidth increases the

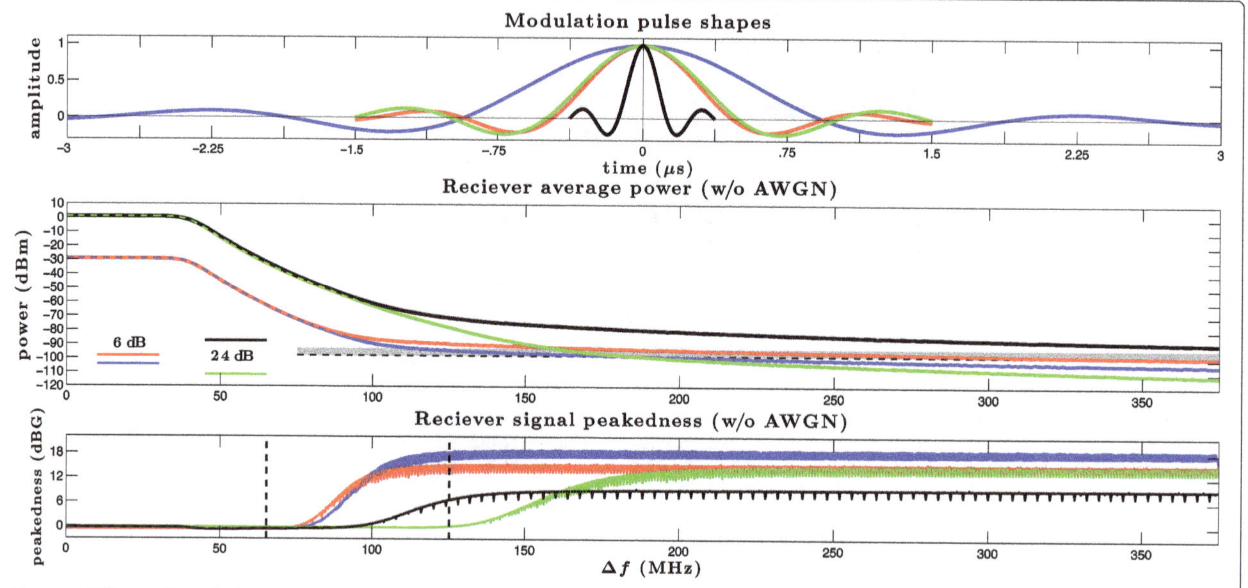

Figure 7 Effects of symbol rates and pulse shaping on the interference power and peakedness.

peakedness of the interference, enabling its more effective mitigation by the NDLs.

2.2 Limitations of mainstream approaches to impulsive noise mitigation

Since a signal of interest typically occupies a different and/or narrower frequency range than the noise, linear filters are typically applied to the incoming mixture of the signal and the noise in order to reduce the frequency range of the mixture to that of the signal. This reduces the power of the interference to a fraction of the total, limited to the frequency range of the signal. However, the noise having the same frequency power spectrum may have various peakedness and be impulsive (super-Gaussian) or non-impulsive (sub-Gaussian). For example, white shot noise is much more impulsive than white thermal noise, while both have identically flat power spectra. Linear filtering cannot improve the signal-to-noise ratio (SNR) in a passband of interest, does not discriminate between impulsive and nonimpulsive noise contributions, and does not allow mitigation of the impulsive noise relative to the nonimpulsive. In addition, as can be justified by the central limit theorem [16], reduction in the bandwidth of an initially impulsive noise by linear filtering typically reduces the peakedness and makes the noise less impulsive (more 'Gaussian-like'), decreasing the ability to separate the signal from the noise based on the peakedness.

Effective suppression of impulsive interferences in the signal path typically requires nonlinear means, for example, digital processing based on order statistics, and various approaches to design of nonlinear receivers with improved performance in the presence of impulsive interference have been proposed. Many of these are *model-based approaches*, which rely on theoretical or empirical assumptions and models of interference distributions. For example, the α-stable [17,18] and Middleton class A, B, and C [19,20] distributions are commonly used to model the interference in wireline [21] and wireless [22,23] communications. Such approaches, designed under specific interference model assumptions, are often limited by parameter estimation schemes (e.g., are sensitive to inaccuracies in obtaining derivatives) and may not be robust under a model mismatch. Alternative methods that do not explicitly rely on noise distribution models have also been proposed. Those include receiver designs based on flexible classes of distributions (e.g., myriad filter [24-26], normal inverse Gaussian (NIG) [27]) or directly on the log-likelihood ratio shape (e.g., soft limiter [28], hole puncher [29], p-norm [27,30,31]).

While linear filters cannot increase the passband SNR, they can be optimal if the noise is purely Gaussian (e.g., thermal). Nonlinear filters, on the other hand, can improve quality of a signal if the latter is affected by a non-Gaussian interference ([32], for example). When the noise is Gaussian, however, nonlinear filters would be typically inferior to linear filters. For example, a *median filter* [33] can be significantly more effective than a linear averaging filter in suppression of impulsive noise but less effective in removal of Gaussian noise. In addition, nonlinear filters are generally not compatible with the existing linear systems as they introduce various types of nonlinear distortions to the signal of interest. As will be demonstrated later in this paper, one of the significant advantages of NDLs is that, while being nonlinear filters, their nonlinear behavior is intermittent and is controlled by a single parameter in a manner that makes them fully compatible with existing linear devices and systems.

Another key methodological distinction of NDLs is that they are deployed (in their either analog or digital implementations) sufficiently early in the signal chain to combine bandwidth reduction with interference mitigation. Numerical filtering algorithms (e.g., digital nonlinear filters) are typically deployed after the analog-to-digital conversion (ADC), when the bandwidth of the signal + interference mixture is reduced to below half of the sampling rate and it is often already 'too late' to deal with non-Gaussian interference effectively. Indeed, the non-Gaussian nature of interference can be viewed as a result of the 'coupling' of various interference components in a wide frequency band. Bandwidth reduction destroys this coupling, making the interference appear more 'Gaussian-like' and reducing our ability to distinguish it from the thermal noise and to effectively remove it. Thus, insufficient processing bandwidth often severely limits the effectiveness of state-of-art nonlinear interference mitigation techniques. While this can be overcome by increasing the sampling rate (and thus the acquisition bandwidth), this further exacerbates the memory and DSP intensity of numerical algorithms, making them unsuitable for real-time implementation and treatment of nonstationary noise.

3 Nonlinear differential limiters

In this section, we provide a brief introduction to NDLs. More comprehensive descriptions of NDLs, with detailed analysis and examples of various NDL configurations, nonadaptive as well as adaptive, can be found in [1-3].

3.1 Theoretical foundation of NDLs

For the optimal mitigation of non-Gaussian interference by nonlinear filters, it is imperative that the distributional properties of the interference are known, either *a priori* or through measurements. The 'blind' NDL-based approach proposed in this paper arises from the methodology introduced in [34], which relies on the transformation of discrete or continuous signals into normalized continuous scalar fields with the mathematical properties of distribution functions. This methodology enables a variety of nonlinear signal processing techniques that naturally incorporate the consideration of such distributional properties, including those which have no digital counterparts.

For example, as detailed in [34], the time-dependent amplitude distribution $\Phi(D,t)$ of a continuous signal $x(t)$ obtained in a time window $w(t)$ can be expressed as:

$$\Phi(D,t) = w(t) * \mathcal{F}_{\Delta D}\left[D - x(t)\right] , \quad (6)$$

where D is a *threshold* value, asterisk denotes convolution, and $\mathcal{F}_{\Delta D}(D)$ is a *discriminator function* that changes monotonically from zero to one in such a way that most of this change occurs over some characteristic range of threshold values ΔD around zero.

Since $\Phi(D,t)$ can be viewed as a surface in the three-dimensional space (t, D, Φ), the expression:

$$\Phi\left(D_q(t),t\right) = q, \qquad 0 < q < 1 , \quad (7)$$

defines $D_q(t)$ as a level (or contour) curve obtained from the intersection of the surface $\Phi = \Phi(D,t)$ with the plane $\Phi = q$, as illustrated in Figure 8.

As is well known in differential geometry, an explicit (albeit differential) equation of the level curve $D_q(t)$ can be obtained by differentiating Equation 7 with respect to time (see, for example, [35], p. 551, equation (4.29)), leading to:

$$\frac{\mathrm{d}D_q}{\mathrm{d}t} = -\frac{\partial\Phi(D_q,t)/\partial t}{\phi(D_q,t)} + \nu\left[q - \Phi(D_q,t)\right], \qquad \nu > 0 . \quad (8)$$

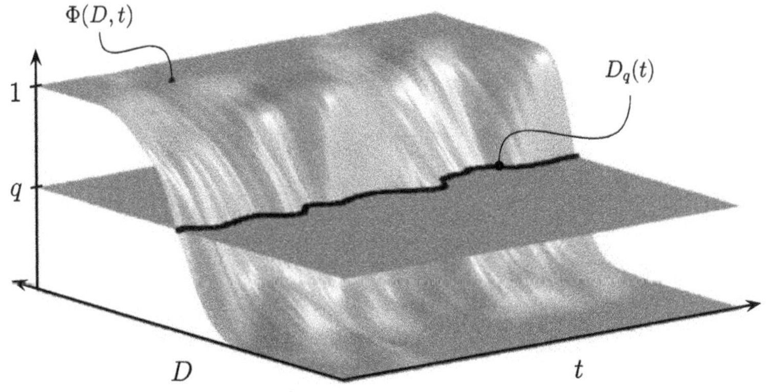

Figure 8 $D_q(t)$ as a level curve of the distribution function $\Phi(D,t)$.

In Equation 8, $\phi(D,t) = \partial\Phi(D,t)/\partial D$ is the amplitude *density* of $x(t)$ in the time window $w(t)$, and since $\Phi(D,t)$ is a monotonically increasing function of D for all t, the added term in the right-hand side ensures the convergence of the solution to the chosen quantile order q regardless of the initial condition.

It can be shown that, depending on the shape of the discriminator function $\mathcal{F}_{\Delta D}(D)$, Equation 8 corresponds to a variety of nonlinear filters with desired characteristics. For example, as demonstrated in our earlier work [34], in the limit $\Delta D \to 0$, Equation 8 describes an analog *rank* filter (e.g., a *median filter* for $q = 1/2$) in an arbitrary time window $w(t)$, leading, as illustrated below, to the introduction of NDLs.

3.2 First-order canonical differential limiter

The digital median filter introduced in the early 1970s [33] is a widely recognized tool for removing outlier (i.e., impulsive) noise. In our prior work [34], we introduce *analog rank filters* in arbitrary time windows and derive differential and integro-differential equations that enable their implementation in analog feedback circuits. In particular, from equation (4.6) in [34], an expression for the output $\chi(t)$ of an 'exact' (or 'true') analog median filter in an exponential time window with the time constant τ_0 can be written as:

$$\dot{\chi}(t) = \lim_{\alpha\to 0} \frac{\frac{1}{2} - \mathcal{F}_{2\alpha}\left[\chi(t) - x(t)\right]}{\int_{-\infty}^{t} ds \, \exp\left(\frac{s-t}{\tau_0}\right) f_{2\alpha}\left[\chi(t) - x(s)\right]}, \quad (9)$$

where $x(t)$ is the input signal, $\mathcal{F}_{2\alpha}(x)$ is a discriminator function with a characteristic width 2α, and $f_{2\alpha}(x) = \mathrm{d}\mathcal{F}_{2\alpha}(x)/\mathrm{d}x$ is its respective *probe*. In Equation 9, $\mathcal{F}_{2\alpha}(x)$ and $f_{2\alpha}(x)$ are such that $\lim_{\alpha\to 0}\mathcal{F}_{2\alpha}(x) = \theta(x)$ and $\lim_{\alpha\to 0} f_{2\alpha}(x) = \delta(x)$, where $\theta(x)$ is the Heaviside unit step function [36] and $\delta(x)$ is the Dirac δ-function [37]. In Equation 9, the parameter α can be called the *resolution parameter*.

Let us now choose a particularly simple discriminator function with a 'ramp' transition, such that the respective probe will be a boxcar function, as illustrated in the left-hand panel of Figure 9. In this case, since the main contribution to the integral in the denominator of Equation 9 will come from a relatively close proximity to the point $s = t$, for a finite and sufficiently large α such that $|\chi(t) - x(t)|$ generally remains smaller than the resolution parameter α, except for relatively rare outliers with a typical duration much smaller than τ_0, the denominator in Equation 9 can be approximated by a constant value equal to $\tau_0/(2\alpha)$. Then, for a finite α, Equation 9 becomes:

$$\chi = x - \tau(|x - \chi|)\,\dot{\chi}, \quad (10)$$

where the time parameter $\tau = \tau(|x - \chi|)$ is given by:

$$\tau(|x - \chi|) = \tau_0 \times \begin{cases} 1 & \text{for} \quad |x - \chi| \le \alpha \\ \frac{|x-\chi|}{\alpha} & \text{otherwise} \end{cases}, \quad (11)$$

as illustrated in the right-hand panel of Figure 9.

We shall call a filter described by Equations 10 and 11 a *first-order canonical differential limiter (CDL)*. Note that when the time parameter τ is a constant (e.g., in the limit $\alpha \to \infty$), Equation 10 describes a first-order linear analog filter (RC integrator), wherein the rate of change of the output is proportional to the *difference signal* $x - \chi$. When the magnitude of the difference signal $|x - \chi|$ exceeds the resolution parameter α, however, the rate of change of the output is proportional to the *sign function* of the difference signal and no longer depends on the magnitude of the incoming signal $\chi(t)$, providing an output insensitive to outliers with a characteristic amplitude determined by the resolution parameter.

3.3 Higher order NDLs

A high-order analog linear lowpass filter would be typically constructed as a first- (for odd-order filters) or

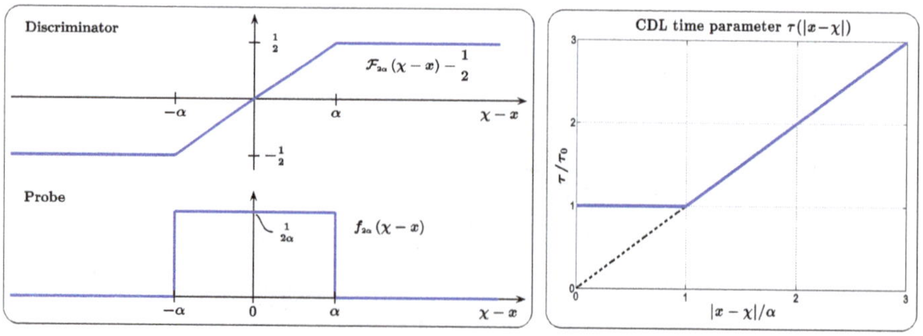

Figure 9 'Ramp' discriminator and CDL time parameter. *Left panel:* 'Ramp' discriminator and its respective boxcar probe. *Right panel:* CDL time parameter $\tau = \tau(|x - \chi|)$.

second- (for even-order filters) order stage followed by cascaded second-order stages, typically arranged from the lowest to the highest quality factor. A similar approach can be taken to extend the previous example to higher order NDLs. For example, a third-order NDL can be constructed as a first-order CDL followed by a second-order linear filter, and a fourth-order NDL as a second-order NDL (introduced below) followed by a second-order linear filter. Such general constructions are of theoretical interest as it may be practically unnecessary to cascade the NDL stages. Indeed, the main burden of removing outliers will be carried out by the first stage, and the subsequent stages would be needed only to provide a desired frequency and phase response for the linear-regime NDL operation.

For even-order NDLs, a second-order NDL stage can be introduced as follows. Let us consider a second-order lowpass stage that can be described by the differential equation:

$$\zeta(t) = z(t) - \tau\,\dot{\zeta}(t) - (\tau Q)^2\,\ddot{\zeta}(t)\,, \tag{12}$$

where $z(t)$ and $\zeta(t)$ are the input and the output signals, respectively (which can be real-, complex-, or vector-valued), τ is the *time parameter* of the stage, Q is the quality factor, and the dot and the double dot denote the first and the second time derivatives, respectively. Note that, when written in such a form, Equation 12 with $Q = 0$ describes a first-order lowpass filter.

For a linear time-invariant filter, the time parameter τ and the quality factor Q in Equation 12 are constants, so that, when the input signal $z(t)$ is increased by a factor of K, the output $\zeta(t)$ is also increased by the same factor, as is the difference between the input and the output. For convenience, we will call the difference between the input and the output $z(t) - \zeta(t)$ *the difference signal*. A transient outlier in the input signal would result in a transient outlier in the difference signal of a filter, and an increase in the input outlier by a factor of K would result, for a linear filter, in the same factor increase in the respective outlier of the difference signal. If a significant portion of the frequency content of the input outlier is within the passband of the linear filter, the output will typically also contain an outlier corresponding to the input outlier, and the amplitudes of the input and the output outliers will be proportional to each other. A reduction (limiting) of the output outliers, while preserving the relationship between the input and the output for the portions of the signal not containing the outliers, can be achieved by proper dynamic modification of the filter parameters τ and Q in Equation 12 based on the magnitude (for example, the absolute value) of the difference signal. A filter comprising such dynamic modification of the filter parameters based on the magnitude of the difference signal will be called an NDL.

Since at least one of the filter parameters depends on the instantaneous magnitude of the difference signal, the differential equation describing such a filter is nonlinear. However, even though in general an NDL is a nonlinear filter, if the parameters remain constant as long as the magnitude of the difference signal remains within a certain range, the behavior of the NDL will be linear during that time. Thus, an NDL can be configured to behave linearly as long as the input signal does not contain outliers. By specifying a proper dependence of the NDL filter parameters on the difference signal, it can be ensured that, when the outliers are encountered, the nonlinear response of the NDL limits the magnitude of the respective outliers in the output signal.

A comprehensive discussion and illustrative examples of various dependencies of the NDL parameters on the difference signal can be found in [1-3]. For example, one can set the quality factor in Equation 12 to a constant value and allow the time parameter τ to be a *nondecreasing* function of the absolute value of the difference signal satisfying the following equation:

$$\tau(|z-\zeta|) = \tau_0 \times \begin{cases} 1 & \text{for} \quad |z-\zeta| \le \alpha \\ > 1 & \text{otherwise} \end{cases}\,, \tag{13}$$

where $\alpha > 0$ is the *resolution* parameter. A particular example can be given by

$$\tau(|z-\zeta|) = \tau_0 \times \begin{cases} 1 & \text{for} \quad |z-\zeta| \le \alpha \\ \left(\frac{|z-\zeta|}{\alpha}\right)^{\beta} & \text{otherwise} \end{cases} \tag{14}$$

with $\beta > 0$. Parameter β in Equation 14 controls the behavior of the NDL in the presence of outliers - the larger its value, the stronger the suppression of outliers. From practical considerations, the value $\beta = 1$ is convenient, so we refer to the NDL with $\beta = 1$ as a CDL. If stronger suppression is desirable, then we can use $\beta > 1$, which we call a differential over-limiter (DoL).

It should be easily seen from Equations 13 or 14 that in the limit of a large resolution parameter, $\alpha \to \infty$, an NDL becomes equivalent to the respective linear filter with $\tau = \tau_0 =$ const. This is an important property of the proposed NDL, enabling its full compatibility with linear systems. At the same time, when the noise affecting the signal of interest contains impulsive outliers, the signal quality (e.g., as characterized by a SNR, a throughput capacity of a communication channel, or other measures of signal quality) exhibits a global maximum at a certain finite value of the resolution parameter $\alpha = \alpha_0$. As illustrated in the next section, this property of an NDL enables its use for improving the signal quality in excess of that achievable by the respective linear filter, effectively reducing the spectral density of the interference in the signal passband without significantly affecting the signal of interest.

4 Examples of mitigation of out-of-band interference by NDLs

The incoming 'native' (in-band) RX signal used in the examples of Figures 10, 11, 12, and 13 was a QPSK signal with the I/Q modulating signals as two independent random bit sequences with the rate 4.8 Mbit/s. An FIR root-raised-cosine (RRC) filter with a roll-off factor of one fourth and group delay $3T$ was used for the RX incoming signal pulse shaping, and the same FIR filter was used for matched filtering in the baseband. In all examples, the signal-to-noise ratio for the RX signal was measured in the baseband, after applying the matched FIR filter. The PSD of the RX signal without noise was approximately -167 dBm/Hz in the baseband, leading to the S/N ratio without interference of approximately 5 dB, as indicated by the upper horizontal dashed lines in Figures 10 and 12. The OOB interference was created by the TX signal used in the examples of Section 2. In all the examples of this section, the NDLs were fourth-order 'Butterworth-like' filters constructed as a second-order constant-Q CDL with the pole quality factor $Q = 1/\sqrt{2+\sqrt{2}}$ and the initial cutoff frequency $f_0 = 5.25$ MHz, followed by a second-order linear lowpass filter with $Q = 1/\sqrt{2-\sqrt{2}}$ and the same cutoff frequency.

In Figures 10 and 11, the receiver with the 40-MHz lowpass filter was tuned to $f_{RX} = 2.125$ GHz. Figure 10 shows the SNR in the receiver baseband as a function of the NDL resolution parameter α. In the limit of a large-resolution parameter, an NDL is equivalent to the respective linear filter (in this example, the fourth-order Butterworth lowpass filter with the cutoff frequency $f_0 = 5.25$ MHz), resulting in the same signal quality of the filtered output as provided by the linear filter (indicated by the lower horizontal dashed line). When viewed as a function of the resolution parameter, however, the signal quality of the NDL output exhibits a global maximum at some $\alpha = \alpha_0$. This property of an NDL enables its use for improving the signal quality in excess of that achievable by the respective linear filter, effectively reducing the in-band impulsive interference. As can be seen in Figure 10, when linear processing is used, the OOB interference reduces the SNR by approximately 6.7 dB. The NDL with $\alpha = \alpha_0$ improves the SNR by approximately 4.8 dB, suppressing the OOB interference by approximately a factor of three.

For the resolution parameter of the NDL set to $\alpha = \alpha_0$, Figure 11 shows the time domain I/Q traces and PSDs of

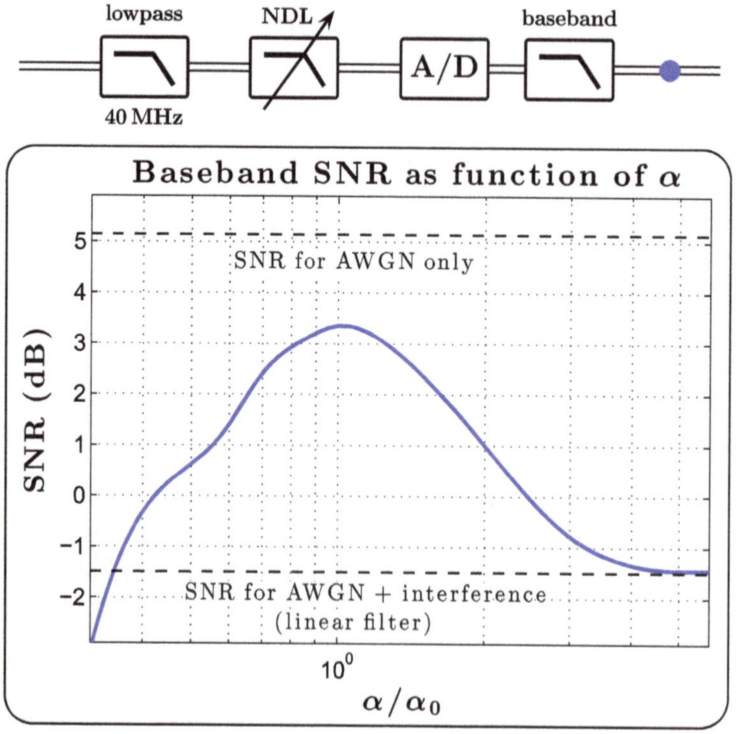

Figure 10 SNR in the receiver baseband as a function of the NDL resolution parameter α. The RX frequency is $f_{RX} = 2.125$ GHz, and the NDL follows the 40-MHz lowpass filter.

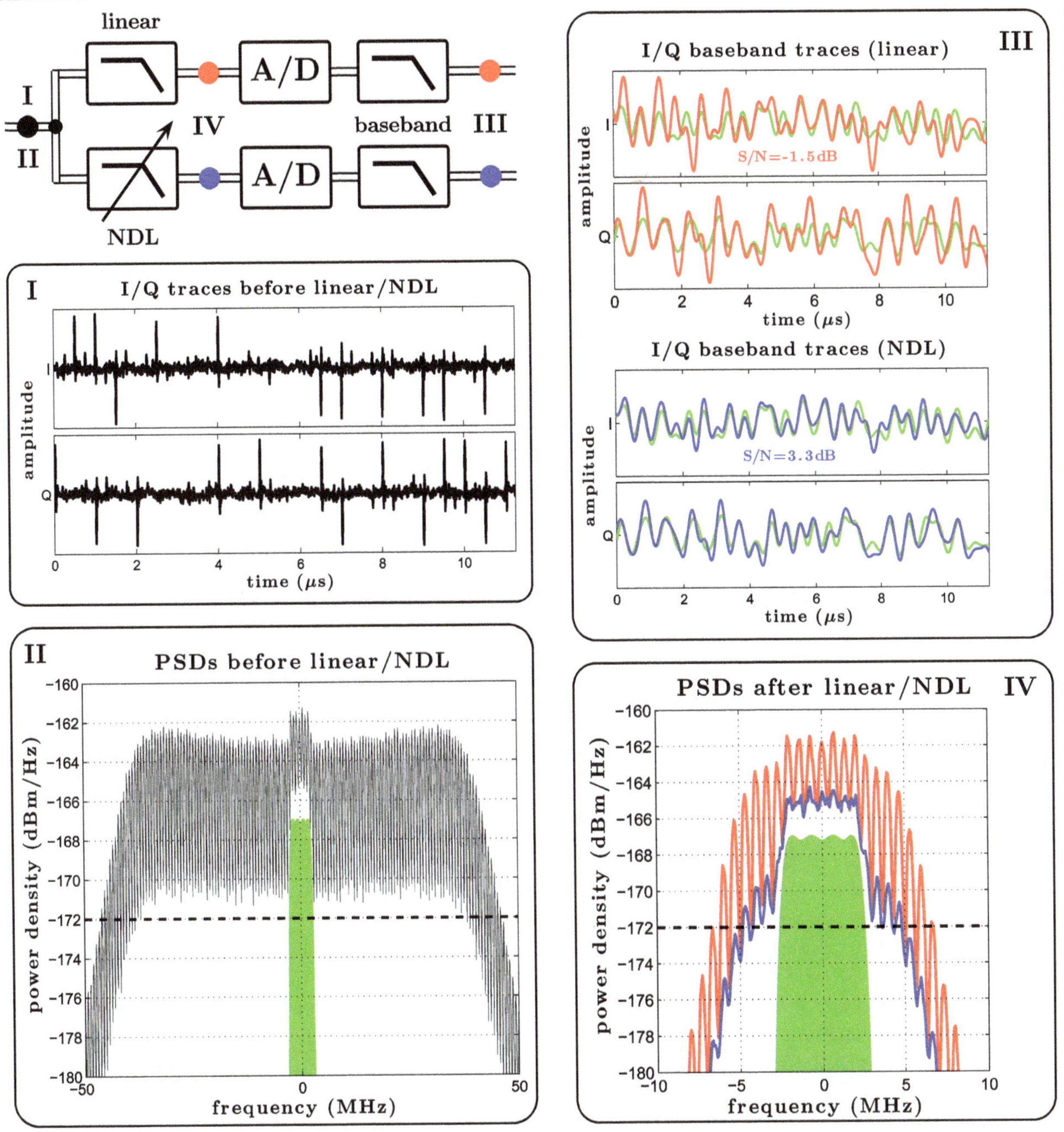

Figure 11 Time domain I/Q traces and PSDs of the signals measured at the test points. The RX frequency is $f_{RX} = 2.125$ GHz. In panels **II** and **IV**, the AWGN level is shown by the horizontal dashed lines, and the PSD of the RX signal without noise is shown by the green shading. In panel **III**, the green lines show the I/Q traces of the baseband RX signal without noise.

the signals measured at the test points indicated by the fat colored dots on the signal path diagram outlined in the upper left of the figure. In panels II and IV, the AWGN level is shown by the horizontal dashed lines, and the PSD of the RX signal without noise is shown by the green shading. In panel III, the green lines show the I/Q traces of the baseband RX signal without noise. In panel IV of Figure 11, the fact that the NDL indeed reduces the spectral density of the interference without significantly affecting the signal of interest can be deduced and gauged from observing how the quasiperiodic structure of the PSD is affected by the NDL in comparison with the linear filter.

If the Shannon formula [38] is used to calculate the capacity of a communication channel, the baseband SNR increase from −1.5 to 3.3 dB provided by the NDL in

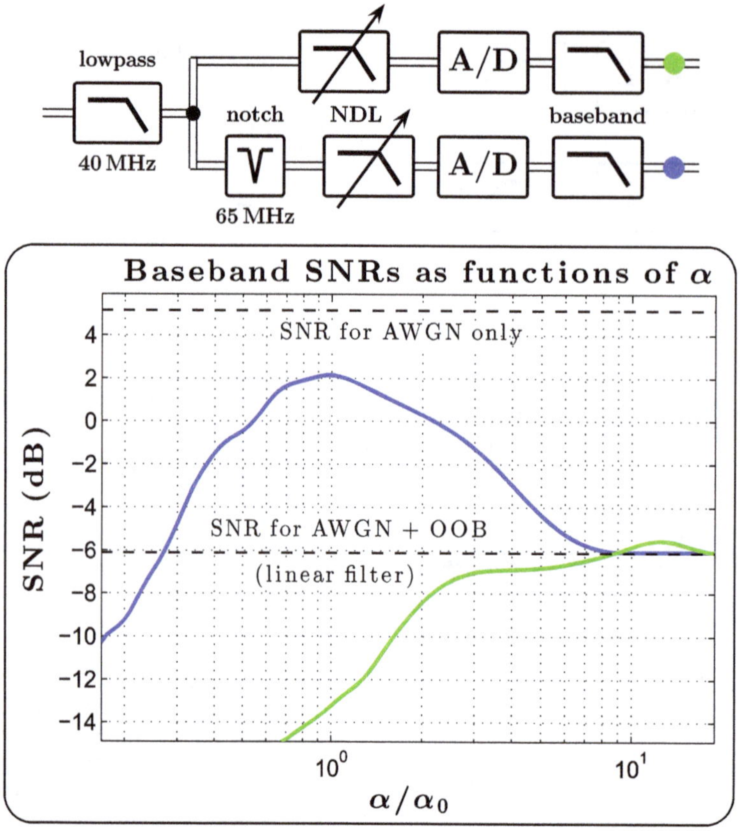

Figure 12 SNRs in the receiver baseband as functions of the NDL resolution parameter α. The RX frequency is $f_{RX} = 2.065$ GHz. *Green line:* the NDL is applied directly to the output of the 40-MHz lowpass filter. *Blue line:* a 65-MHz notch filter precedes the NDL.

the examples of Figures 10 and 11 results in a 114% (2.14 times) increase in the channel capacity.

Figure 12 shows the SNRs in the receiver baseband as functions of the NDL resolution parameter α for the RX frequency $f_{RX} = 2.065$ GHz, when the NDL is applied directly to the output of the 40-MHz lowpass filter (green line) and when a 65-MHz notch filter precedes the NDL (blue line). As can be seen in Figure 12 from the distance between the horizontal dashed lines, when linear processing is used, the OOB interference reduces the SNR by approximately 11 dB.

As was discussed in Section 2 (see, in particular, the description of Figure 6), the response of the receiver 40-MHz lowpass filter at 65 MHz is relatively large, and the contribution of the TX signal in its nominal band (part I of the interference in Figure 2) into the total interference is significant, which makes the total interference sub-Gaussian (−0.5-dBG peakedness). Thus, an NDL deployed immediately after the 40-MHz lowpass filter will not be effective in suppressing the interference, as can be seen from the SNR curve shown by the green line in Figure 12. However, a 65-MHz notch filter preceding the NDL attenuates the nonimpulsive part of the interference without affecting either the signal of interest or the PSD of the impulsive interference, making the interference impulsive and enabling its effective mitigation by the subsequent NDL. This can be seen from the SNR curve shown by the blue line in Figure 12, where the NDL with $\alpha = \alpha_0$ improves the SNR by approximately 8.2 dB, suppressing the OOB interference by approximately a factor of 6.6.

For the resolution parameter of the NDL set to $\alpha = \alpha_0$, Figure 13 shows the time domain I/Q traces and the PSDs of the signals measured at the test points indicated by the fat colored dots on the signal path diagram in the upper left of the figure. The RX frequency is $f_{RX} = 2.065$ GHz, and the notch is at 65 MHz. In panels II and IV, the AWGN level is shown by the horizontal dashed lines, and the PSD of the RX signal without noise is shown by the green shading. In panel III, the green lines show the I/Q traces of the baseband RX signal without noise. In panel IV of Figure 13, the fact that the NDL indeed reduces the spectral density of the interference without significantly affecting the signal of interest can be deduced and gauged from observing how the quasiperiodic structure of the PSD is affected by the NDL in comparison with the linear filter.

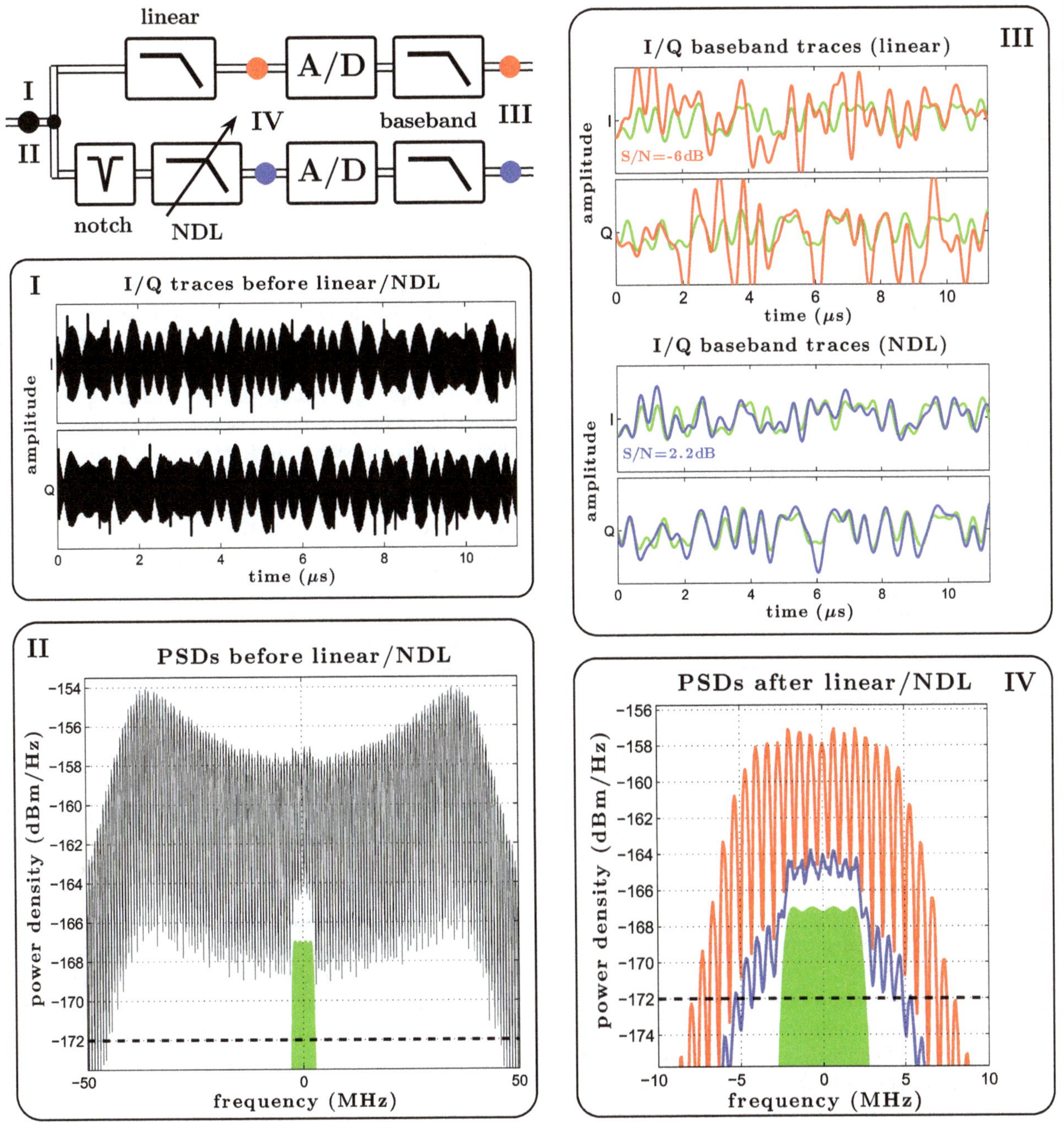

Figure 13 Time domain I/Q traces and PSDs of the signals measured at the test points. The RX frequency is $f_{RX} = 2.065$ GHz, and the notch is at 65 MHz. In panels **II** and **IV**, the AWGN level is shown by the horizontal dashed lines, and the PSD of the RX signal without noise is shown by the green shading. In panel **III**, the green lines show the I/Q traces of the baseband RX signal without noise.

If the Shannon formula [38] is used to calculate the capacity of a communication channel, the baseband SNR increase from −6 to 2.2 dB provided by the NDL in the examples of Figures 12 and 13 results in a 337% (4.37 times) increase in the channel capacity.

The value of α_0 that maximizes the signal quality may vary in a wide range depending on the composition of the signal + noise mixture, for example, on the SNR and the relative spectral and temporal structures of the signal and the noise. ANDL configurations (see [1-3]) contain a sub-circuit (characterized by a *gain* parameter) that monitors a chosen measure of the signal + noise mixture and provides a time-dependent resolution parameter $\alpha = \alpha(t)$ to the main NDL circuit, making it suitable for improving the quality of nonstationary signals under time-varying noise conditions.

5 Conclusions

Interference from various technogenic (man-made) sources, unintentional as well as intentional, typically has temporal and/or amplitude structure, and its amplitude distribution is usually non-Gaussian. A simplified explanation of the non-Gaussian (and often impulsive) nature of a technogenic noise produced by digital electronics and communication systems can be as follows. An idealized discrete-level (digital) signal can be viewed as a linear combination of Heaviside unit step functions ([36], for example). Since the derivative of the Heaviside unit step function is the Dirac δ-function ([37], for example), the derivative of an idealized digital signal is a linear combination of Dirac δ-functions, which is a limitlessly impulsive signal with zero interquartile range and infinite peakedness. The derivative of a 'real' (i.e., no longer idealized) digital signal can thus be viewed as a convolution of a linear combination of Dirac δ-functions with a continuous kernel. If the kernel is sufficiently narrow (for example, the bandwidth is sufficiently large), the resulting signal will appear as an impulse train protruding from a continuous background signal. Thus, impulsive interference occurs naturally in digital electronics as "$\mathrm{d}i/\mathrm{d}t$" (inductive) noise or as the result of the coupling (for example, capacitive) between various circuit components and traces, leading to the so-called 'platform noise' ([39], for example).

In this paper, we focus on particular illustrative mechanisms of impulsive interference in digital communication systems resulting from the nonsmooth nature of any physically realizable modulation scheme for transmission of a digital (discontinuous) message. Even modulation schemes designed to be 'smooth,' e.g., continuous-phase modulation, are, in fact, not smooth because their higher order time derivatives still contain discontinuities.

The non-Gaussian nature of technogenic interference provides an opportunity for its mitigation by nonlinear filtering that is more effective than the mitigation achievable by linear filters. When a linear filter is used to suppress interference outside the passband of interest, the filtered signal quality is not influenced by the type of the amplitude distribution of the interfering signal, as long as the total power and the spectral composition of the interference is the same. It may be possible to reduce the spectral density of the *in-band* technogenic interference (that is, in the signal's passband) without significantly affecting the signal of interest by introducing an appropriately chosen feedback-based nonlinearity into the response of a filter. As a result, the signal quality can be improved in excess of that achievable by the respective linear filter.

In this paper, we describe such nonlinear filters (NDLs), outline a methodology for mitigation of technogenic interference in communication channels by NDLs, and provide several examples of such mitigation. We demonstrate that an NDL replacing a linear filter in the receiver channel can improve the receiver by increasing the signal quality in the presence of man-made noise and thus the capacity of a communication channel.

5.1 Comment on use methodology

As was stated in our prior work [1-3] and illustrated in Section 2 of this paper, the distributions of non-Gaussian signals are generally modifiable by linear filtering, and non-Gaussian interference can often be converted from sub-Gaussian into super-Gaussian, and *vice versa*, by linear filtering that does not affect the signal of interest. As a result, employing appropriate linear filtering preceding an NDL in a signal chain can greatly improve effectiveness of NDL-based interference mitigation. While we have previously outlined several approaches to such distribution modification by linear front end (LFE) filtering, and to identifying non-Gaussian components in an interfering signal [1,2], the development of systematic procedures for identification of non-Gaussian interference components and for design of appropriate LFE filtering remains a challenging task that is a subject of ongoing research.

5.2 Comment on digital NDLs

Conceptually, NDLs are analog filters that combine bandwidth reduction with mitigation of interference. For the interference to appear strongly impulsive, the bandwidth of the receiver lowpass filter needs to be much larger than a typical value of $(t_{i+1} - t_i)^{-1}$ in Equation 2, and effective use of an NDL may require that its input signal has a bandwidth much larger than the bandwidth of the RX signal of interest. Thus, the best conceptual placement for an NDL is in the analog part of the signal chain, for example, as part of the antialiasing filter preceding an ADC, as shown in panel (a) of Figure 14. However, digital NDL implementations may offer many advantages typically associated with digital processing, including simplified development and testing, configurability, and reproducibility.

While near-real-time finite-difference implementations of the NDLs described in Section 3 would be relatively simple and computationally inexpensive, their use would still require a digital signal with a sampling rate much higher than the Nyquist rate of the signal of interest. Increasing the sampling rate of a high-resolution converter in order to enable the use of an NDL would be impractical for many reasons, including the ADC cost and its saturation by high-amplitude impulsive outliers. Instead, as illustrated in panel (b) of Figure 14, a low-bit high-rate A/D converter should be used to provide the input to a digital NDL. Then, the NDL output can be

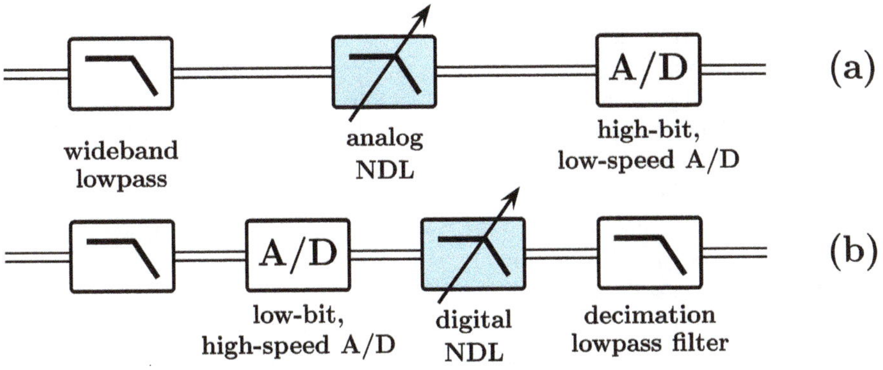

Figure 14 Analog (panel (a)) and digital (panel (b)) NDL deployment.

downsampled (after appropriate digital lowpass filtering) to provide the desired high-resolution signal at a lower sampling rate.

Endnotes

[a] For example, the out-of-band (OOB) emissions of a transmitter may be greatly reduced by employing a high-quality bandpass filter in the antenna circuit of the transmitter. Such an additional filter, however, may negatively affect other properties of a system, for example, by increasing its cost and power consumption (due to the insertion loss of the filter).

[b] One will encounter discontinuities in a derivative of some order in the modulating signal sooner or later, since any physical pulse shaping is implemented using causal filters of finite order.

[c] In general, if n is the order of a causal analog filter, then $n-1$ is the order of the first discontinuous derivative of its impulse response.

[d] Equation (2) will still accurately represent the instantaneous power in the quadrature receiver if the 'real' (physical) modulating signal can be expressed as $A(t) = \psi(t) * A_T(t)$, where the convolution kernel $\psi(t)$ is a lowpass filter of a bandwidth much larger than $|\Delta f|$.

Appendix

A Simulation parameters

The transmitter signal used in all simulations was a QPSK signal with the I/Q modulating signals as two independent random bit sequences. In all simulations except those shown in Figure 7, the symbol rate was 4 Mbit/s (unit interval $T = 250$ ns), and an FIR RRC filter ([15], for example) with the roll-off factor one fourth and the group delay $3T$ was used for pulse shaping. The average TX signal power in all simulations was set to 125 mW (21 dBm), and it was assumed that the additional path/coupling loss at any RX frequency was 50 dB, except for the TX signals shaped with the filters shown by black and green lines in Figure 7, where it was 20 dB.

A rather small transmission power (6 dB below a typical cellular phone power of 27 dBm) and a relatively large 50-dB loss were chosen as somewhat of a 'safety margin' to ensure that, even if the OOB emissions are significantly (e.g., by 20 to 30 dB) reduced by a carefully selected combination of the roll-off factor and the group delay of the shaping filter, and/or by reducing the bandwidth (symbol rate) of the TX signal, the spectral density of the interference may still be comparable with or dominate over the spectral density of the thermal noise for a respectively smaller path/coupling loss and/or larger transmit power.

The quasiperiodic time domain structures of the spectrograms in Figure 3, and of the time domain traces seen in the upper panels of Figures 5 and 6, and in panel I of Figure 11, are related to the unit interval T and have a period $T = 250$ ns. The quasiperiodic structures that can be seen in the PSDs shown in the lower panels of Figures 5 and 6, and in panels II and IV of Figures 11 and 13, are related to the group delay of the FIR pulse shaping filters and have a period equal to half of the inverse group delay or $(6T)^{-1} = 2/3$ MHz.

As can be seen in the lower panels of Figure 4, the peakedness of the interference also exhibits a quasiperiodic structure (quasiperiodic local minima). This period is related to the symbol rate T^{-1} as $(2\,T)^{-1} = 2$ MHz or half of the symbol rate. In those simulations, the f_{RX} sampling interval of 1.25 MHz was used. This is why only one out of four local minima is visible, and the apparent period of peakedness in Figure 4 is 10 MHz. The local minima in the peakedness plots shown in the lower panel of Figure 7 also occur at halves of the respective symbol rates. For example, the peakedness shown by the black line has a structure with the period 4 MHz.

A constant 5-dB noise figure of the receiver was assumed at all receiver frequencies f_{RX}. This, combined with the −177-dBm/Hz two-sided PSD of the thermal noise at room temperature, leads to the total AWGN noise level of −172 dBm/Hz. The incoming RX signal used in Figures 10, 11, 12, and 13 was a QPSK signal with the I/Q modulating signals as two independent random bit sequences with the rate of 4.8 Mbit/s. An FIR RRC filter with roll-off factor one fourth and group delay $3T$ was used for the RX incoming signal pulse shaping, and the same FRI filter was used for the matched filtering in the baseband. The PSD of the RX signal without noise was approximately −167 dBm/Hz in the baseband, leading to the S/N ratio without interference of approximately 5 dB.

Abbreviations

ACI, adjacent-channel interference; ADC, analog-to-digital converter; AMC, adaptive modulation and coding; ANDL, adaptive nonlinear differential limiter; AWGN, additive white Gaussian noise; A/D, analog-to-digital; dBm, dB per milliwatt; DoL, differential over-limiter; DSP, digital signal processing; EMC, electromagnetic compatibility; EMI, electromagnetic interference; FIR, finite impulse response; IMD, intermodulation distortion; I/Q, in-phase/quadrature; LFE, linear front end; MANET, mobile *ad hoc* network; Mbit/s, megabit per second; NDL, nonlinear differential limiter; OOB, out-of-band; PSD, power spectral density; QAM, quadrature amplitude modulation; QPSK, quadrature phase-shift keying (4-QAM); RBW, resolution bandwidth (electronic signal term used in spectrum analyzers and EMI/EMC testing); RF, radio frequency; RRC, root-raised-cosine; RX, receiver; SDR, software-defined radio; SNR, signal-to-noise ratio; S/N, signal-to-noise; TX, transmitter.

Competing interests

This work was supported in part by the National Science Foundation under Grant Number 1314790. The authors have no other relationships or activities that could appear to have influenced the submitted work.

Acknowledgements

The authors would like to thank Malcolm D Macleod of QinetiQ and Bala Natarajan of Kansas State University for their valuable suggestions and critical comments.

Author details

[1]Avatekh Inc., 901 Kentucky Street, Suite 303, Lawrence, KS 66044, USA. [2]Dept. of Electrical and Computer Engineering, Kansas State University, Manhattan, KS 66506, USA. [3]Dept. of Mathematics, University of Leicester, Leicester LE1 7RH, UK. [4]BAE Systems Technology Solutions, 6 New England Executive Park, Burlington, MA 01803, USA.

References

1. AV Nikitin, Method and Apparatus for Signal Filtering and for Improving Properties of Electronic Devices. *US patent 8,489,666* (2013)
2. Nikitin, AV, Method and Apparatus for Signal Filtering and for Improving Properties of Electronic Devices. *US Patent Application Publications 2013/0297665 (Nov. 7, 2013), 2013/0339418 (Dec. 19, 2013), and 2014/0195577 (July 10, 2014)*
3. AV Nikitin, RL Davidchack, TJ Sobering, in *Proceedings of IEEE Military Communications Conference 2013*. Adaptive Analog Nonlinear Algorithms and Circuits for Improving Signal Quality in the Presence of Technogenic Interference (San Diego, CA) p. 2013
4. AV Nikitin, RL Davidchack, JE Smith, in *Proceedings of 3rd IMA Conference on Mathematics in Defence*. Out-of-band and adjacent-channel interference reduction by analog nonlinear filters (Malvern, UK) p. 2013
5. EM Royer, CK Toh, A review of current routing protocols for ad-hoc mobile wireless networks. IEEE Pers. Commun. **6**(2), 46–55 (1999)
6. S Weber, JG Andrews, N Jindal, The Effect of Fading, Channel Inversion, and Threshold Scheduling on Ad Hoc Networks. IEEE Trans. Inf. Theory. **53**(11), 4127–4149 (2007)
7. H Inaltekin, M Chiang, HV Poor, SB Wicker, On Unbounded Path-Loss Models: Effects of Singularity on Wireless Network Performance. IEEE J. Selected Areas Commun. **27**(7), 1078–1092 (2009)
8. S Basagni, M Conti, S Giordano, (eds.) Stojmenovic I, *Mobile Ad Hoc Networking*. (Wiley-IEEE Press, New York, 2004)
9. AV Nikitin, in *Proc. IEEE Radio and Wireless Symposium*. On the Impulsive Nature of Interchannel Interference in Digital Communication Systems (Phoenix, AZ, 2011), pp. 118–121
10. AV Nikitin, On the interchannel interference in digital communication systems, its impulsive nature, and its mitigation. EURASIP J Adv. Signal Process. **2011**(137) (2011)
11. AV Nikitin, M Epard, JB Lancaster, RL Lutes, EA Shumaker, Impulsive interference in communication channels and its mitigation by SPART and other nonlinear filters. EURASIP J. Adv. Signal Process. **2012**(79) (2012)
12. L Cohen, *Time-frequency analysis*. (Prentice-Hall, Englewood, NJ, 1995)
13. M Abramowitz, IA Stegun (eds.), *Handbook of Mathematical Functions with Formulas, Graphs, and Mathematical Tables*, 9th printing, (Dover, New York, 1972)
14. A Hyvärinen, J Karhunen, E Oja. Independent component analysis (Wiley, New York, 2001)
15. JG Proakis, DG Manolakis, *Digital signal processing: principles, algorithms, and applications*, 4th edn. (Prentice Hall, 2006)
16. R Durrett, *Probability: Theory and examples*, 4th edn. (Cambridge University Press, 2010)
17. G Samorodnitsky, MS Taqqu, *Stable Non-Gaussian Random Processes: Stochastic Models with Infinite Variance*. (Chapman and Hall, New York, 1994)
18. GA Tsihrintzis, CL Nikias, Fast estimation of the parameters of alpha-stable impulsive interference. IEEE Trans. Signal Process. **44**(6), 1492–1503 (1996)
19. D Middleton, Statistical-Physical Models of Electromagnetic Interference. IEEE Trans. Electromagnetic Compatibility. **EMC-19**(3), 106–127 (1977)
20. D Middleton, Non-Gaussian noise models in signal processing for telecommunications: New methods and results for class A and class B noise models. IEEE Trans. Inf. Theory. **45**(4), 1129–1149 (1999)
21. M Zimmermann, K Dostert, Analysis and modeling of impulsive noise in broad-band powerline communications. IEEE Trans. Electromagnetic Capability. **44**, 249–258 (2002)
22. X Yang, AP Petropulu, Co-channel interference modeling and analysis in a Poisson field of interferers in wireless communications. IEEE Trans. Signal Process. **51**, 64–76 (2003)
23. K Gulati, BL Evans, JG Andrews, KR Tinsley, Statistics of co-channel interference in a field of Poisson and Poisson-Poisson clustered interferers. IEEE Trans. Signal Proc. **58**(12), 6207–6222 (2010)
24. JG Gonzalez, GR Arce, Optimality of the myriad filter in practical impulsive-noise environments. IEEE Trans. Signal Process. **49**(2), 438–441 (2001)
25. P Zurbach, JG Gonzalez, GR Arce, Weighted myriad filters for image processing. IEEE Int Symp. Circuits Syst. **2**, 726–729 (1996)
26. JG Gonzalez, GR Arce, Statistically-efficient filtering in impulsive environments: weighted myriad filters. EURASIP J. Appl. Signal Process. **2002**, 4–20 (2002)
27. W Gu, G Peters, L Clavier, F Septier, I Nevat, in *2012 International Symposium on Wireless Communication Systems (ISWCS)*. Receiver study for cooperative communications in convolved additive E±-stable interference plus Gaussian thermal noise, (2012), pp. 451–455
28. GL Stuber, Soft-limiter receivers for coded DS/DPSK systems. IEEE Trans. Commun. **38**, 46–53 (1990)
29. S Ambike, J Ilow, D Hatzinakos, Detection for binary transmission in a mixture of Gaussian noise and impulsive noise modeled as an alpha-stable process. IEEE Signal Process. Lett. **1**(3), 55–57 (1994)
30. NC Beaulieu, H Shao, J Fiorina, P-order metric UWB receiver structures with superior performance. IEEE Trans. Commun. **56**(10), 1666–1676 (2008)
31. W Gu, L Clavier, Decoding metric study for turbo codes in very impulsive environment. IEEE Commun. Lett. **16**(2), 256–258 (2012)
32. HV Poor, *An Introduction to signal detection and estimation*. (Springer, 1998)
33. JW Tukey, *Exploratory Data Analysis*. (Addison-Wesley, 1977)
34. AV Nikitin, RL Davidchack, Signal analysis through analog representation.Proc. R. Soc. Lond. A. **459**(2033), 1171–1192 (2003)

35. IN Bronshtein, KA Semendiaev, *Handbook of Mathematics*, 3rd edn. (Springer, 1997)
36. R Bracewell, *The Fourier Transform and Its Applications, 2000 chap. "Heaviside's Unit Step Function, H(x)"*, 3rd edn. (McGraw-Hill, New York) pp. 61–65
37. PAM Dirac, *The Principles of Quantum Mechanics*, 4th edn. (Oxford University Press, London, 1958)
38. CE Shannon, Communication in the presence of noise. Proc. Inst. Radio Eng. **37**, 10–21 (1949)
39. K Slattery, H Skinner, *Platform Interference in Wireless Systems*. (Elsevier, 2008)

Impacts of frequency increment errors on frequency diverse array beampattern

Kuandong Gao*, Hui Chen†, Huaizong Shao, Jingye Cai and Wen-Qin Wang†

Abstract

Different from conventional phased array, which provides only angle-dependent beampattern, frequency diverse array (FDA) employs a small frequency increment across the antenna elements and thus results in a range angle-dependent beampattern. However, due to imperfect electronic devices, it is difficult to ensure accurate frequency increments, and consequently, the array performance will be degraded by unavoidable frequency increment errors. In this paper, we investigate the impacts of frequency increment errors on FDA beampattern. We derive the beampattern errors caused by deterministic frequency increment errors. For stochastic frequency increment errors, the corresponding upper and lower bounds of FDA beampattern error are derived. They are verified by numerical results. Furthermore, the statistical characteristics of FDA beampattern with random frequency increment errors, which obey Gaussian distribution and uniform distribution, are also investigated.

Keywords: Frequency diverse array (FDA); Frequency increment errors; Beampattern error; FDA beampattern; FDA radar

1 Introduction

Beampattern is widely used to assess the performance of phased arrays [1]. However, a limitation of phased array is that the beam steering is fixed at one angle for all the ranges [2]. Recently, a flexible array called frequency diverse array (FDA) has been proposed [3-5]. Different from phased array, a small frequency increment, as compared to the carrier frequency, is applied across the array elements [6]. This small frequency increment results in a range angle-dependent beampattern [7,8]. Several investigations have been carried on FDA radars. The time and angle periodicity of FDA beampattern was analyzed in [9]. A linear FDA was proposed in [10] for forward-looking radar ground moving target indication. The multipath characteristics of FDA radar over a ground plane were investigated and compared with phased array in [11]. FDA radar full-wave simulation and implementation with linear frequency-modulated continuous waveform were presented in [12,13]. In [14], we have investigated the FDA Cramér-Rao lower bounds for estimating direction, range, and velocity. More recent work about the applications of FDA in MIMO radars can be found in [15-19].

Perfect frequency increments are often assumed in existing literatures [20]. However, in an actual array system, there will have imperfect errors including element position errors, mutual coupling, phase errors, and frequency increment errors [21-24]. Some results have been reported about the impacts of element position error, mutual coupling and phase error on beampattern, and direction-of-arrival (DOA) estimation performances. Since FDA beampattern has similar properties with conventional phased array for the impacts of element position errors, mutual coupling, and phase errors, this paper considers only the impacts of frequency increment errors on FDA beampattern. Since FDA beampattern is dependent on the angle and range, it has a potential for target localization , which is different from traditional time-of-arrival (TOA) and angle-of-arrival (AOA)-based localization [25-28]. The contributions can be summarized as follows. (i) More tighter bounds of FDA beampattern deviation are derived. (ii) Statistical analysis of FDA beampattern in terms of expectation value, variance, and probability density function (PDF) are provided.

*Correspondence: kuandonggao@gmail.com
†Equal contributors
School of Communication and Information Engineering, University of Electronic Science and Technology of China, 2008, Road XiYuan, Chengdu, 611731 Sichuan, China

The rest of this paper is organized as follows. Section 2 formulates the data models of FDA radar without and with frequency increment errors, respectively. Section 3 analyzes the impacts of deterministic frequency increment errors on FDA beampattern. Thereafter, the impacts of random frequency increment errors are investigated in Section 4. Finally, simulation results are provided in Section 5, and conclusions are drawn in Section 6.

2 FDA beampattern without frequency increment errors

Suppose an N-element uniform linear FDA with inter-element spacing denoted as d. The radiated frequency from the nth element is as follows:

$$f_n = f_0 + n\Delta f, n = 0, 1, 2, \ldots, N-1 \tag{1}$$

where f_0 and Δf are the carrier frequency and frequency increment, respectively. Taking the first element as the reference for the array, under far-field condition, one might express the direct wave component of the electric field emitted from the FDA at the observation point (θ, r) as [17]:

$$A(\theta, r, t) = \sum_{n=0}^{N-1} a_n \varsigma_n(\theta\,|w_n) \frac{e^{jw_n\left(t-\frac{r+d_n\sin\theta}{c_0}\right)}}{r} \tag{2}$$

where N is the number of FDA elements, a_n represents the complex excitation coefficient for the nth element, $\varsigma_n(\theta\,|w_n)$ stands for the far-field vector radiation pattern for the nth element at range r and angular frequency $w_n = 2\pi f_n$, c_0 is the light speed, d_n is the element position of the nth element reference relative to the first element, and t is the time parameter. In accordance with the far-field assumption, $e^{jw_n\left(t-\frac{r+d_n\sin\theta}{c_0}\right)}/r$ corresponds to the delayed carrier with free space loss.

To interpret the effect of frequency diversity within the scope of an array factor, we should factor the vector element pattern out of Equation 2. This can indeed be done under certain conditions. Assuming all elements in the FDA are identical, we can eliminate the frequency dependence in the element factor. Since $r \gg d_n$, we have:

$$\varsigma_n(\theta|w_n) \approx \varsigma(\theta|w_0). \tag{3}$$

where w_0 is the carrier angular frequency. So Equation 2 can be rewritten as:

$$A(\theta, r, t) = \varsigma(\theta\,|w_0) \sum_{n=0}^{N-1} a_n \frac{e^{jw_n\left(t-\left(\frac{r+d_n\sin\theta}{c_0}\right)\right)}}{r}. \tag{4}$$

Further simplification becomes possible by considering particular FDA arrangements that are simple to handle and yet able to provide valuable insight. A uniform linear FDA utilizing discrete, linear frequency increments is such a practical configuration, and it is examined in this section as a special case. By definition, the elements are excited with uniform amplitude, but they are allowed to have a phase progression across the array. These specifications translate to the following expressions for d_n and a_n:

$$d_n = nd \text{ and } a_n = e^{-jn\phi_a} \tag{5}$$

where ϕ_a stands for the phase progression. Submitting Equations 1 and 5 into Equation 4 yields:

$$\begin{aligned} A(\theta, r, t) &= \varsigma(\theta\,|w_0) \sum_{n=0}^{N-1} e^{-jn\phi_a} \frac{e^{j2\pi(f_0+n\Delta f)\left(t-\left(\frac{r+nd\sin\theta}{c_0}\right)\right)}}{r} \\ &= \frac{\varsigma(\theta\,|w_0)}{r} e^{j\varphi_0} \sum_{n=0}^{N-1} e^{-jn\phi_a} e^{j2\pi n\Delta f\left(t-\left(\frac{r+nd\sin\theta}{c_0}\right)\right)-j2\pi f_0\frac{nd\sin\theta}{c_0}} \end{aligned} \tag{6}$$

where $\varphi_0 = 2\pi f_0 t - 2\pi f_0 \frac{r}{c_0}$. For notation convenience, we define $\bar{\varsigma} = \varsigma(\theta\,|w_0)\, e^{j2\pi f_0 t - j2\pi f_0 \frac{r}{c_0}}$, Equation 6 can then be rewritten as:

$$A(\theta, r, t) = \frac{\bar{\varsigma}}{r} \sum_{n=0}^{N-1} e^{-jn\phi_a} e^{j2\pi n\left(t\Delta f - \frac{r+nd\sin\theta}{c_0}\Delta f - f_0\frac{d\sin\theta}{c_0}\right)} \tag{7}$$

Since $nd\sin\theta \ll r$, Equation 7 can be reformulated as:

$$\begin{aligned} A(\theta, r, t) &\approx \frac{\bar{\varsigma}}{r} \sum_{n=0}^{N-1} e^{-jn\phi_a} e^{j2\pi n\left(t\Delta f - \frac{r}{c_0}\Delta f - f_0\frac{d\sin\theta}{c_0}\right)} \\ &= \frac{\bar{\varsigma}}{r} \mathbf{w}^H \mathbf{v}(\theta, r, t) \end{aligned} \tag{8}$$

where

$$\mathbf{w} = \left[\, 1 \;\; e^{j\phi_a} \;\; \ldots \;\; e^{j(N-1)\phi_a} \,\right]^T \tag{9}$$

and

$$\begin{aligned} \mathbf{v}(\theta, r, t) = \Big[1 \;\; & e^{j2\pi\left(t\Delta f - \frac{r}{c_0}\Delta f - f_0\frac{d\sin\theta}{c_0}\right)} \;\; \ldots \\ & e^{j2\pi(N-1)\left(t\Delta f - \frac{r}{c_0}\Delta f - f_0\frac{d\sin\theta}{c_0}\right)} \Big]^T. \end{aligned} \tag{10}$$

with $[\cdot]^T$ and $[\cdot]^H$ being the transpose operator and hermitian transpose operator, respectively. For simplicity, $\bar{\varsigma} = 1$ is assumed in the following discussions. Figure 1 shows the ideal linear uniform FDA beampattern with $t = 1/\Delta f$.

2.1 FDA beampattern with frequency increment errors

Suppose the frequency increment errors being $\rho = [\, 0 \;\; \rho_1 \;\; \ldots \;\; \rho_{N-1} \,]$, Equation 8 can be rewritten as:

$$A(\theta, r) = \frac{\mathbf{w}^H \tilde{\mathbf{v}}(\theta, r, t)}{r} \tag{11}$$

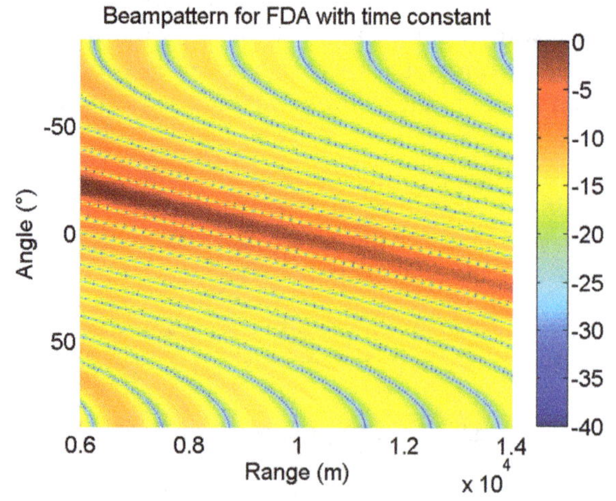

Figure 1 Ideal two-dimension FDA beampattern with $N = 16$, $\Delta f = 30\ KHz, t = 1/(3 \times 10^4), \phi_a = 0$, and $w = 1_{N\times 1}$.

where

$$\tilde{\mathbf{v}}(\theta,r,t) = \Big[1 \quad e^{j2\pi\left(t(\Delta f+\rho_1)-\frac{r}{c_0}(\Delta f+\rho_1)-f_0\frac{d\sin\theta}{c_0}\right)} \dots \\ e^{j2\pi(N-1)\left(t(\Delta f+\rho_{N-1})-\frac{r}{c_0}(\Delta f+\rho_{N-1})-f_0\frac{d\sin\theta}{c_0}\right)}\Big]^T \\ = \Big[1 \quad e^{j2\pi\left(t\Delta f-\frac{r}{c_0}\Delta f-f_0\frac{d\sin\theta}{c_0}\right)}e^{j2\pi\left(t\rho_1-\frac{r}{c_0}\rho_1\right)} \dots \\ e^{j2\pi(N-1)\left(t\Delta f-\frac{r}{c_0}\Delta f-f_0\frac{d\sin\theta}{c_0}\right)}e^{j2\pi\left(N-1)(t\rho_{N-1}-\frac{r}{c_0}\rho_{N-1}\right)}\Big]^T. \tag{12}$$

Then Equation 11 can be reformulated as:

$$\begin{aligned} \tilde{A}(\theta,r) &= \frac{\mathbf{w}^H\tilde{\mathbf{v}}(\theta,r,t)}{r} \\ &= \frac{\mathbf{w}^H(\mathbf{v}(\theta,r,t)\odot\Delta\mathbf{v}(r,t))}{r} \\ &= \frac{\mathbf{w}^H diag(\Delta\mathbf{v}(r,t))\mathbf{v}(\theta,r,t)}{r} \\ &= \frac{\mathbf{w}^H\mathbf{C}(r,t)\mathbf{v}(\theta,r,t)}{r} \end{aligned} \tag{13}$$

where

$$\Delta\mathbf{v}(r,t) = \left[1 \quad e^{j2\pi\left(t\rho-\frac{r}{c_0}\rho\right)} \quad \dots \quad e^{j2\pi(N-1)\left(t\rho-\frac{r}{c_0}\rho\right)}\right]^T$$

and

$$\mathbf{C} = \text{diag}\ (\Delta\mathbf{v}(r,t))$$

and $\odot$ denotes Hadamard product. Note that diag$(\Delta\mathbf{v}(r,t))$ denotes a diagonal matrix with $\Delta\mathbf{v}(r,t)$ being its diagonal elements. For small frequency increment errors, Taylor series expansion about $\mathbf{C}$ is performed as follows:

$$\mathbf{C}(r,t) = \mathbf{I} + \mathbf{C}_+(r,t). \tag{14}$$

According to Equation 13, it can be known that:

$$\begin{aligned} \mathbf{C}_+(r,t) &= \text{diag}\left(j2\pi\left(-\frac{(\mathbf{N}\odot\rho)r}{c_0}+t(\mathbf{N}\odot\rho)\right)\right. \\ &\quad \left.\odot\left(e^{j2\pi\left(-\frac{(\mathbf{N}\odot\rho)r}{c_0}+t(\mathbf{N}\odot\rho)\right)}\right)\right)+O(\rho) \\ &\approx \text{diag}\left(j2\pi\left(-\frac{(\mathbf{N}\odot\rho)r}{c_0}+t(\mathbf{N}\odot\rho)\right)\right. \\ &\quad \left.\odot\left(e^{j2\pi\left(-\frac{(\mathbf{N}\odot\rho)r}{c_0}+t(\mathbf{N}\odot\rho)\right)}\right)\right). \end{aligned} \tag{15}$$

where $\mathbf{N} = [0,1,2,\dots,N-1]$ denotes element number of FDA radar, and $O(\rho)$ represents for the high order terms about ρ, which can be ignored for Equation 15. Using Equation 15, we can rewrite the beampattern (13) as:

$$\begin{aligned} \tilde{A}(\theta,r,t) &= \frac{\mathbf{w}^H\mathbf{C}(r,t)\mathbf{v}(\theta,r,t)}{r} \\ &= \frac{\mathbf{w}^H\mathbf{v}(\theta,r,t)+\mathbf{w}^H\mathbf{C}_+(r,t)\mathbf{v}(\theta,r,t)}{r} \\ &= A(\theta,r,t)+\frac{\mathbf{w}^H\mathbf{C}_+(r,t)\mathbf{v}(\theta,r,t)}{r} \\ &= A(\theta,r,t)+\Delta A(r,t) \end{aligned} \tag{16}$$

where $\Delta A(r,t) = \frac{\mathbf{w}^H\mathbf{C}_+(r,t)\mathbf{v}(\theta,r,t)}{r}$ denotes the FDA beampattern deviation. Figure 2 shows the FDA beampattern with frequency increment errors. The array parameters are the same as that for Figure 1. Due to the frequency increment errors, the FDA beampattern sidelobes are different from Figure 1. Stronger sidelobe peaks will make the FDA energy scattering and consequently degrade the array performance.

3 Impacts of deterministic frequency increment errors

Firstly, we consider uniform frequency increment errors, i.e., $\rho = [0\ \rho_1 \dots \rho_{N-1}] = [0\ \rho \dots \rho]$. In this case, Equation 13 can be rewritten as:

$$\begin{aligned} \tilde{A}(\theta,r) &= \frac{\mathbf{w}^H\text{diag}\ (\Delta\mathbf{v}(r,t))\,\mathbf{v}(\theta,r,t)}{r} \\ &= \sum_{n=1}^{N}\frac{e^{j2\pi(n-1)\left(t(\Delta f+\rho)-\frac{r}{c_0}(\Delta f+\rho)-f_0\frac{d\sin\theta}{c_0}\right)}}{r} \\ &= \frac{k}{r}\frac{\sin\left(\frac{\pi N}{2}\left(t\left(\Delta f+\rho\right)-\frac{r}{c_0}(\Delta f+\rho)-f_0\frac{d\sin\theta}{c_0}\right)\right)}{\sin\left(\pi\frac{t(\Delta f+\rho)-\frac{r}{c_0}(\Delta f+\rho)-f_0\frac{d\sin\theta}{c_0}}{2}\right)} \\ &= \frac{k}{r}\frac{\sin\left(\frac{\pi N}{2}\left(\Delta f\left(t+\frac{\rho}{\Delta f}t\right)-\frac{\Delta f}{c_0}\left(r+\frac{\rho}{\Delta f}r\right)-f_0\frac{d\sin\theta}{c_0}\right)\right)}{\sin\left(\pi\frac{\Delta f\left(t+\frac{\rho}{\Delta f}t\right)-\frac{\Delta f}{c_0}\left(r+\frac{\rho}{\Delta f}r\right)-f_0\frac{d\sin\theta}{c_0}}{2}\right)} \end{aligned} \tag{17}$$

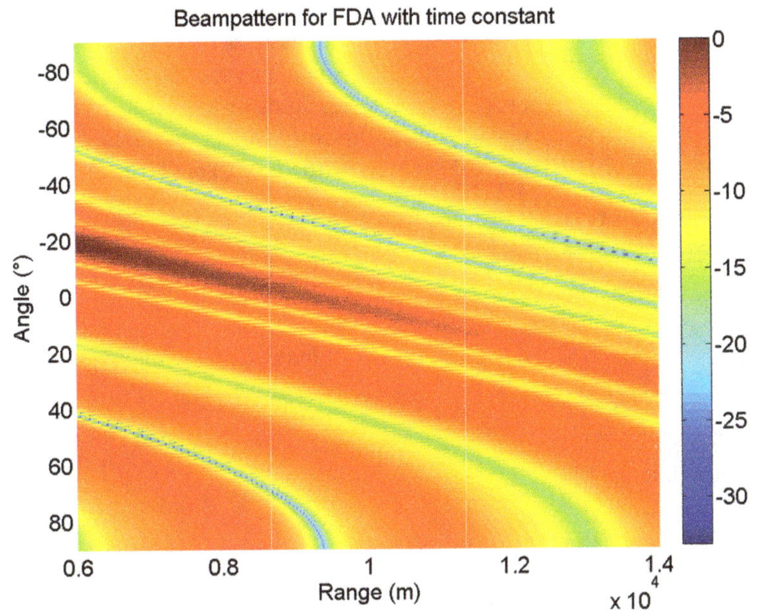

Figure 2 Two-dimension FDA beampattern with frequency increment error ρ_n and $\rho_n \sim N(0, \rho^2)$ with $\rho = 500$.

where $k = e^{j\pi N\left(t(\Delta f+\rho)-\frac{r}{c_0}(\Delta f+\rho)-f_0\frac{d\sin\theta}{c_0}\right)}$ and $\mathbf{w} = \mathbf{1}_{N\times 1}$. Equation 17 arrives the maximum value when:

$$\Delta f\left(t+\frac{\rho}{\Delta f}t\right)-\frac{\Delta f}{c_0}\left(r+\frac{\rho}{\Delta f}r\right)-\left(f_0\frac{d\sin\theta}{c_0}\right)=2m \quad m=0,1,2,\ldots. \tag{18}$$

If no frequency increment error exists, the FDA mainlobe will pass the location $\left(0°, \frac{c_0}{\Delta f}\right)$ at $t = \frac{1}{\Delta f}$. However, when there are frequency increment errors, the location $\left(0°, \frac{c_0}{\Delta f}\right)$ may be not at the beampattern maximum point. According to Equation 18, the corresponding range error is as follows:

$$\Delta r = \frac{\rho}{\Delta f}r. \tag{19}$$

Since the phase error caused by time error can be equivalently regarded as range error and angle error, Equation 18 can be reformulated as:

$$\begin{aligned}&\Delta f\left(t+\frac{\rho}{\Delta f}t\right)-\frac{\Delta f}{c_0}\left(r+\frac{\rho}{\Delta f}r\right)-\left(f_0\frac{d\sin\theta}{c_0}\right)\\&=\Delta ft+\frac{\rho}{\Delta f}t-\frac{\Delta f}{c_0}\left(r+\frac{\rho}{\Delta f}r\right)-\left(f_0\frac{d\sin\theta}{c_0}\right)\\&=\Delta ft-\frac{\Delta f}{c_0}\left(r+\frac{\rho}{\Delta f}r-\frac{\rho c_0}{2\Delta f^2}t\right)-f_0\frac{d}{c_0}\left(\sin\theta-\frac{\rho c_0}{2\Delta f f_0 d}t\right)\\&=2m, \quad \mathrm{m}=0,1,2,\ldots\end{aligned} \tag{20}$$

Therefore, the corresponding range error is as follows:

$$\Delta r = \frac{\rho}{\Delta f}r - \frac{\rho c_0}{2\Delta f^2}t \tag{21}$$

and the angle error is as follows:

$$\Delta\theta = \arcsin\left(\frac{\rho c_0}{2\Delta f f_0 d}t\right). \tag{22}$$

If the range error and angle error are required to be smaller than 50 m and 0.5°, respectively, the frequency increment error should be smaller than 150 Hz.

Figure 3a,b shows the FDA beampattern with uniform frequency increment errors ρ, frequency increment $\Delta f =$ 30 KHz, and $t = 1/\Delta f$. It can be noticed that the phase error for $\theta = 0.5°$ is $\rho/\Delta f = 5\times 10^{-3}$. Note that since the range attenuation exists, the range error has a small deviation from the calculation of Equation 21.

4 Impacts of stochastic frequency increment errors

In this section, we investigate the statistical properties of FDA beampattern with random frequency increment errors.

4.1 Error boundary of FDA beampattern

Suppose the frequency increment errors are random. In this case, it is not possible to derive a deterministic beampattern expression like Equation 17. Here, we are interested in the maximum beampattern deviation, namely:

$$\max_{t,r,\theta}|\Delta A(t,r,\theta)|. \tag{23}$$

According to Equation 16 and using the Cauchy-Schwarz inequality, we have:

$$\max_{t,r,\theta}|\Delta A(\theta,r,t)| \le \frac{1}{r}\|w_{\max}\mathbf{C}_+\|_2\|\mathbf{v}(\theta,r)\|_2$$

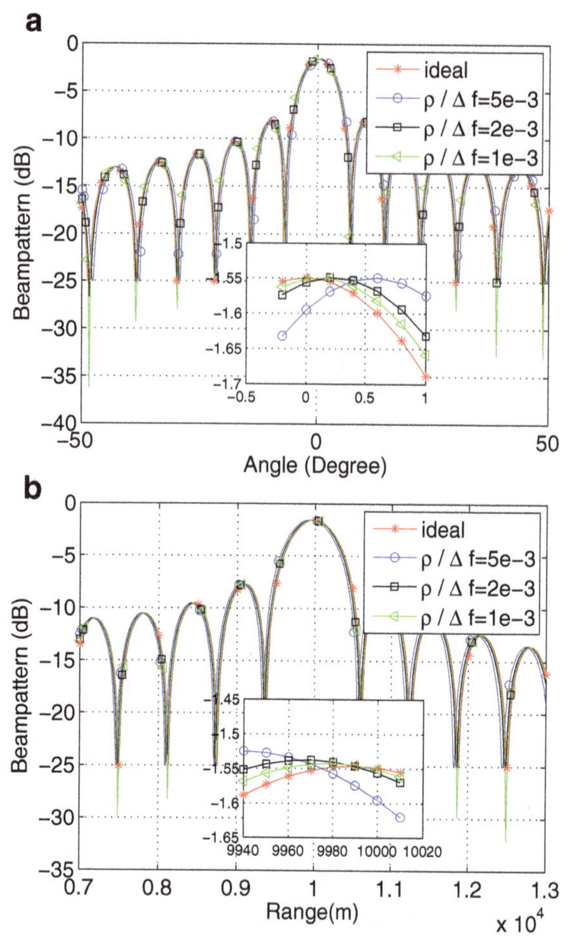

Figure 3 FDA beampattern profiles with frequency increment errors. **(a)** In angle dimension. **(b)** In range dimension.

$$\begin{aligned} &= \frac{1}{r}\|w_{\max}\mathbf{C}_+\|_2\sqrt{N} \\ &\leq \frac{1}{r}|w_{\max}|\|\mathbf{C}_+\|_2\sqrt{N} \\ &= \Delta A_{\max} \end{aligned} \tag{24}$$

where $|\cdot|$ and $\|\cdot\|_2$ denote the absolute value and Euclidean norm, respectively, and $w_{\max}$ denotes the maximum element of $\mathbf{w}$. Note that compared with [29], $|w_{\max}| \leq |\mathbf{w}|_2$, the bound expressed in Equation 24 is much tighter. According to Equation 16, $\mathbf{C}_+$ can be rewritten as:

$$\begin{aligned} \|\mathbf{C}_+\|_2 &= \left\| \mathrm{diag}\left(\left(j2\pi\left(-\frac{(\mathbf{N}\odot\rho)r}{c_0} + t(\mathbf{N}\odot\rho)\right)\right)\right.\right. \\ &\qquad \left.\left.\odot\left(e^{j2\pi(-\frac{(\mathbf{N}\odot\rho)r}{c_0}+t(\mathbf{N}\odot\rho))}\right)\right)\right\|_2 \\ &= \left\|\mathrm{diag}\left(j2\pi\left(-\frac{r}{c_0}+t\right)(\mathbf{N}\odot\rho)\right)\right\|_2 \\ &= 2\pi\left|-\frac{r}{c_0}+t\right|\cdot\left\|\mathrm{diag}(\mathbf{N}\odot\rho)\right\|_2. \end{aligned} \tag{25}$$

The maximum deviation, in a linear scale, is the same over the whole beampattern. When only $\mathbf{C}_+$ (and thus $\mathbf{C}$) is known, or when the influence of unknown factors on $\mathbf{C}_+$, e.g., aging or temperature, comes into play, it makes sense to consider as random matrix, with some statistical model for the entries of $\mathbf{C}_+$. We then have to calculate:

$$\max \quad \|\mathbf{C}_+\|_2 \tag{26}$$

for the random matrix $\mathbf{C}_+$ to find an upper bound.

$$\begin{aligned} \max\|\mathbf{C}_+\|_2 &= \max 2\pi\left(-\frac{r}{c_0}+t\right)\left\|\mathrm{diag}(\mathbf{N}\odot\rho)\right\|_2 \\ &= \max 2\pi\left(-\frac{r}{c_0}+t\right)\|\mathbf{N}\odot\rho\|_2 \\ &= 2\pi\left|-\frac{r}{c_0}+t\right|\cdot\|\mathbf{N}\|_2\,|\rho_{\max}|. \end{aligned} \tag{27}$$

When applying this result to an array with uniform weighting $w_{\max} = 1$, Equation 24 leads to:

$$\Delta A_{\max} = \frac{2\pi}{r}\left|-\frac{r}{c_0}+t\right|\cdot\|\mathbf{N}\|_2\,|\rho_{\max}|\sqrt{N}. \tag{28}$$

For a constant time $t = 1/\Delta f$, the time error caused by frequency increment error will become amplitude error of FDA beampattern, as shown follows:

$$\Delta A_{\max} = \frac{2\pi}{r}\left|-\frac{r}{c_0}+\frac{1}{\Delta f}\right|\cdot\|\mathbf{N}\|_2\,|\rho_{\max}|\sqrt{N}. \tag{29}$$

It can be noticed from Equation 29 that the errors caused by time and by range counterbalance the total beampattern error. Therefore, we delete $1/\Delta f$. Then, Equation 29 can be reformulated as:

$$\begin{aligned} \Delta A_{\max} &\leq \frac{2\pi}{r}\left|-\frac{r}{c_0}\right|\|\mathbf{N}\|_2\,|\rho_{\max}|\sqrt{N} \\ &= \frac{2\pi}{c_0}\|\mathbf{N}\|_2\,|\rho_{\max}|\sqrt{N}. \end{aligned} \tag{30}$$

Equation 30 gives FDA beampattern error bound which indicates its worst case. Since it is caused by the frequency increment errors, the maximum device errors which produce the FDA frequency increment are regulated by the bound. This gives guideline about device selections and predicts the possible FDA beampattern derivation bound.

4.2 Statistical properties of FDA beampattern

In most cases, we know the statistical properties of frequency increment errors and would like to derive the respective properties of impacted beampattern. In this subsection, we will give the formulas of expectation value, variance, and PDF about FDA beampattern based on different PDFs of the frequency increment errors.

4.2.1 *Expected value*

Using Equations 8 and 13, it is straightforward to show that:

$$E\{\Delta A(\theta,r,t)\} = E\{\mathbf{w}^H\mathbf{C}(r,t)\mathbf{v}(\theta,r,t)\} - \mathbf{w}^H\mathbf{v}(\theta,r,t) = \mathbf{w}^H E\{\mathbf{C}(r,t)\}\mathbf{v}(\theta,r,t) - \mathbf{w}^H\mathbf{v}(\theta,r,t) \tag{31}$$

where $E\{\cdot\}$ denotes the expected value.

Assume that the frequency increment error ρ_n of the nth element satisfies the Gaussian statistical model $\rho_n \sim N(0,\sigma^2)$ and the PDF is as follows:

$$f_n(\rho) = \frac{1}{\sqrt{2\pi}\sigma}e^{-\frac{\rho^2}{2\sigma^2}}. \tag{32}$$

Therefore, the expected value of the nth element in diag(**C**) can be calculated as [30]:

$$\begin{aligned} E\{\mathbf{C}_{nn}(r,t)\} &= E\left\{e^{j2\pi(n-1)\left(t\rho-\frac{\rho r}{c_0}\right)}\right\} \\ &= \frac{1}{\sqrt{2\pi}\sigma}\int_{-\infty}^{+\infty} e^{j2\pi(n-1)\left(t\rho-\frac{\rho r}{c_0}\right)}e^{-\frac{\rho^2}{2\sigma^2}}d\rho \\ &= e^{-2\pi^2(n-1)^2\left(t-\frac{r}{c_0}\right)^2\sigma^2}. \end{aligned} \tag{33}$$

where $\mathbf{C}_{nn}$ denotes the nth element of diag(**C**). When applying Equation 33 onto 31, we can get the expected value of FDA beampattern deviation with Gaussian distribution frequency increment errors as follows:

$$E\{\Delta A(\theta,r,t)\} = \mathbf{w}^H E\{\mathbf{C}(r,t)\}\mathbf{v}(\theta,r,t) - \mathbf{w}^H\mathbf{v}(\theta,r,t) = \mathbf{w}^H\left(\text{diag}\left(\mathbf{b_g}\right) - \mathbf{I}\right)\mathbf{v}(\theta,r,t) \tag{34}$$

where

$$\mathbf{b_g} = \left[1, e^{-2\pi^2\left(t-\frac{r}{c_0}\right)^2\sigma^2}, \ldots, e^{-2\pi^2(N-1)^2\left(t-\frac{r}{c_0}\right)^2\sigma^2}\right]. \tag{35}$$

and **I** denotes $N\times N$ unit matrix. In this case, the expected value of the FDA beampattern deviation is dependent on the variance of frequency increment errors and is not 0 for the Gaussian distribution model.

For another case, we assume that ρ_n is uniform distribution, i.e., $\rho_n \sim u(-\rho_{\max},\rho_{\max})$ with minimum and maximum values $-\rho_{\max}$ and $\rho_{\max}$. We can get the expected value of nth element in diag(**C**) as follows:

$$\begin{aligned} E\{\mathbf{C}_{nn}(r,t)\} &= E\left\{e^{j2\pi(n-1)\left(t\rho-\frac{\rho r}{c_0}\right)}\right\} \\ &= \int_{-\rho_{\max}}^{\rho_{\max}} e^{j2\pi(n-1)\left(t-\frac{r}{c_0}\right)\rho}\frac{1}{2\rho_{\max}}d\rho \\ &= \frac{\sin\left(2\pi(n-1)\left(t-\frac{r}{c_0}\right)\rho_{\max}\right)}{2\pi(n-1)\left(t-\frac{r}{c_0}\right)\rho_{\max}}. \end{aligned} \tag{36}$$

By using Equation 36 onto 31, the expected value of FDA beampattern deviation with uniform distribution frequency increment error can be reformulated as:

$$E\{\Delta A(\theta,r,t)\} = \mathbf{w}^H E\{\mathbf{C}(r,t)\}\mathbf{v}(\theta,r,t) - \mathbf{w}^H\mathbf{v}(\theta,r,t) = \mathbf{w}^H\left(\text{diag}\left(\mathbf{b_u}\right) - \mathbf{I}\right)\mathbf{v}(\theta,r,t) \tag{37}$$

where

$$\mathbf{b_u} = \left[1, \frac{\sin\left(2\pi\left(t-\frac{r}{c_0}\right)\rho_{\max}\right)}{2\pi\left(t-\frac{r}{c_0}\right)\rho_{\max}}, \ldots, \frac{\sin\left(2\pi(N-1)\left(t-\frac{r}{c_0}\right)\rho_{\max}\right)}{2\pi(N-1)\left(t-\frac{r}{c_0}\right)\rho_{\max}}\right]. \tag{38}$$

Similarly, the expected value of the FDA beampattern deviation is dependent on the maximum value of frequency increment error and not 0 for the uniform distribution model. From Equations 34 and 37, the expected values of FDA beampattern have range offset and angle offset, which is caused by time dependent error on the two distributions.

4.2.2 *Variance*

The beampattern deviation variance $\text{var}\{\Delta A(\theta,r,t)\}$ equals to the beampattern variance:

$$\text{var}\{\Delta A(\theta,r,t)\} = \text{var}\{A(\theta,r,t)\} = \text{var}\left\{\mathbf{w}^H\mathbf{C}(r,t)\mathbf{v}(\theta,r,t)\right\}. \tag{39}$$

As we deal with vectors and matrices, we utilize the covariance matrix $\text{cov}\{\cdot\}$, and $\text{var}\{\cdot\} = \text{cov}\{\cdot\}$ for any scalar. We thus have:

$$\text{var}\{\Delta A(\theta,r,t)\} = \text{cov}\left\{\mathbf{w}^H\mathbf{C}(r,t)\mathbf{v}(\theta,r,t)\right\} \tag{40}$$

Using the Kronecker product $\otimes$, the vectorization transformation $\text{vec}(\cdot)$ and the identity:

$$\text{vec}\{\mathbf{ABC}\} = \mathbf{C}^T \otimes \mathbf{A}\text{vec}\{\mathbf{B}\}. \tag{41}$$

Since

$$\mathbf{w}^H\mathbf{C}(r,t)\mathbf{v}(\theta,r,t) = \mathbf{v}^T(\theta,r,t)\otimes\mathbf{w}^H\text{vec}\{\mathbf{C}(r,t)\} \tag{42}$$

we use $\mathbf{c} = \text{vec}\{\mathbf{C}(r,t)\}$ and the vector $\mathbf{t}(\theta,r,t) = \mathbf{v}^T(\theta,r,t) \otimes \mathbf{w}^H$, which utilizing:

$$\text{cov}\{\mathbf{AX}\} = \mathbf{A}\text{cov}\{\mathbf{X}\}\mathbf{A}^H \tag{43}$$

to yield:

$$\text{var}\{\Delta A(\theta,r,t)\} = \mathbf{t}(\theta,r,t)\text{cov}\{\mathbf{c}(r,t)\}\mathbf{t}^H(\theta,r,t). \tag{44}$$

Since $\mathbf{C}$ is a diagonal matrix and its entries are independent random variables, cov($\mathbf{c}$) is a diagonal matrix and has non-zero value with the nNth entry. We then have:

$$\text{var}\{\Delta A(\theta,r,t)\} = \sum_k \mathbf{t}_k(\theta,r,t)\mathbf{t}_k^H(\theta,r,t)\text{cov}\{\mathbf{c}(r,t)\}_{kk} \tag{45}$$

where $k = (n-1)N + 1$ denotes the kth entry of vector. So calculating the var$\{\Delta A(\theta,r)\}$ can be equivalently obtained by calculating cov$\{\mathbf{c}(r)\}_{kk}$. This again is a very general result that we can evaluate and review for different statistical models for cov$\{\mathbf{c}(r)\}_{kk}$.

$$\begin{aligned}\text{cov}\{\mathbf{c}(r,t)\}_{kk} &= \int_{-\infty}^{+\infty}\left[\mathbf{c}(r,t)_{kk} - E(\mathbf{c}(r,t)_{kk})\right]^2 f(\rho)d\rho \\ &= \int_{-\infty}^{+\infty}\left[\mathbf{C}(r,t)_{nn} - E(\mathbf{C}(r,t)_{nn})\right]^2 f(\rho)d\rho\end{aligned} \tag{46}$$

Assume that all the random frequency increment errors have the same distribution. For the first case, frequency increment error ρ_n of the nth element satisfies the Gaussian statistical model $\rho_n \sim N(0,\sigma^2)$. According to the Equations 13 and 36, 46 can be rewritten as:

$$\begin{aligned}\text{cov}\{\mathbf{c}(r,t)\}_{kk} &= \int_{-\infty}^{+\infty}\left[\mathbf{C}(r,t)_{nn} - E(\mathbf{C}(\theta,r,t)_{nn})\right] \\ &\quad \times \left[\mathbf{C}(r,t)_{nn} - E(\mathbf{C}(r,t)_{nn})\right]^* f(\rho)d\rho \\ &= \int_{-\infty}^{+\infty}\left[\mathbf{C}(r,t)_{nn}\mathbf{C}(r,t)^*_{nn} - E(\mathbf{C}(r,t)_{nn})\mathbf{C}(r,t)^*_{nn}\right. \\ &\quad \left. - \mathbf{C}(r,t)_{nn}E(\mathbf{C}(r,t)^*_{nn}) + E(\mathbf{C}(r,t)_{nn})E(\mathbf{C}(r,t))^*_{nn}\right]f(\rho)d\rho \\ &= \int_{-\infty}^{+\infty}\mathbf{C}(r,t)_{nn}\mathbf{C}(r,t)^*_{nn}f(\rho)d\rho \\ &\quad - \int_{-\infty}^{+\infty}E(\mathbf{C}(r,t))_{nn}\mathbf{C}((r,t)^*_{nn})f(\rho)d\rho \\ &\quad - \int_{-\infty}^{+\infty}\mathbf{C}(r,t)_{nn}E(\mathbf{C}(r,t))^*_{nn}f(\rho)d\rho \\ &\quad + \int_{-\infty}^{+\infty}E(\mathbf{C}(r,t))_{nn}E(\mathbf{C}(r,t))^*_{nn}f(\rho)d\rho \\ &= \int_{-\infty}^{+\infty}\mathbf{C}(r,t)_{nn}\mathbf{C}(r,t)^*_{nn}f(\rho)d\rho - E(\mathbf{C}(r,t))^2_{nn}\end{aligned} \tag{47}$$

where $[\cdot]^*$ denotes the conjugate operator. Here, the $E(\mathbf{C}(r,t)_{nn}) = e^{-2\pi^2(n-1)^2\left(t-\frac{r}{c_0}\right)^2\sigma^2} = E\left(\mathbf{C}(r,t)^*_{nn}\right)$ is utilized. Applying Equation 32, the first term of Equation 47 can be rewritten as:

$$\begin{aligned}&\int_{-\infty}^{+\infty}\mathbf{C}(r,t)_{nn}\mathbf{C}(r,t)^*_{nn}f(\rho)d\rho = \frac{1}{\sqrt{2\pi}\sigma} \\ &\int_{-\infty}^{+\infty} e^{j2\pi(n-1)\left(t\rho-\frac{\rho r}{c_0}\right)}e^{-j2\pi(n-1)\left(t\rho-\frac{\rho r}{c_0}\right)}e^{-\frac{\rho^2}{2\sigma^2}}d\rho = 1\end{aligned} \tag{48}$$

Applying it to Equation 47 yields:

$$\begin{aligned}\text{cov}\{\mathbf{c}(r,t)\}_{kk} &= 1 - E(\mathbf{C}(r,t))^2_{nn} \\ &= 1 - e^{-4\pi^2(n-1)^2\left(t-\frac{r}{c_0}\right)^2\sigma^2}\end{aligned} \tag{49}$$

where k is the kth element of $\mathbf{c}(r,t)$, which is used to distinguish from n. Consequently, we can get the FDA beampattern deviation variance:

$$\begin{aligned}\text{var}\{\Delta A(\theta,r,t)\} &= \sum_k \mathbf{t}_k(\theta,r,t)\mathbf{t}_k^H(\theta,r,t) \\ &\quad \times \left(1 - e^{-4\pi^2(n-1)^2\left(t-\frac{r}{c_0}\right)^2\sigma^2}\right).\end{aligned} \tag{50}$$

Similarly, we can get the FDA beampattern variance with the random frequency increment errors, which are uniformly distributed as aforementioned:

$$\begin{aligned}\text{var}\{\Delta A(\theta,r,t)\} &= \sum_k \mathbf{t}_k(\theta,r,t)\mathbf{t}_k^H(\theta,r,t) \\ &\quad \times \left(1 - \left(\frac{\sin(2\pi(n-1)\left(t-\frac{r}{c_0}\right)\rho_{\max})}{2\pi(n-1)\left(t-\frac{r}{c_0}\right)\rho_{\max}}\right)^2\right).\end{aligned} \tag{51}$$

Compared with Equations 50 and 51, we can see that the FDA beampattern variances at both kinds of random frequency increment error distributions are dependent on the statistics properties. Given the same frequency increment errors, the variances will decrease when the number of elements is increased.

4.2.3 PDF

We have shown how to calculate the expectation value and the variance of the beampattern deviation and the beampattern itself. A remained question is: What form will their respective PDF have? In this subsection, we will derive the PDF of beampattern based on the PDF of the frequency increment errors.

For the first case, we investigate the PDF of beampattern when the frequency increment errors obey the Gaussian distribution as aforementioned. According to Equation 13, we derive the PDF of **C**. It is known that:

$$\begin{aligned} C_{nn} &= e^{-j2\pi(n-1)\left(t-\frac{r}{c_0}\right)\rho} \\ &= \cos\left(2\pi(n-1)\left(t-\frac{r}{c_0}\right)\rho\right) \\ &\quad - j\sin\left(2\pi(n-1)\left(t-\frac{r}{c_0}\right)\rho\right) \\ &= \text{real}(C_{nn}) - j\left(\text{imag}(C_{nn})\right) \end{aligned} \tag{52}$$

where $\text{real}(C_{nn}) = \cos\left(2\pi(n-1)\left(t-\frac{r}{c_0}\right)\rho\right)$ and $\text{imag}(C_{nn}) = \sin\left(2\pi(n-1)\left(t-\frac{r}{c_0}\right)\rho\right)$ denote the real part and imaginary part of C_{nn}, respectively. Consequently, we have:

$$\begin{aligned} \rho &= \frac{\arcsin(\text{imag}(C_{nn}))}{2\pi(n-1)\left(t-\frac{r}{c_0}\right)} \\ &= \frac{\arcsin(g_n)}{2\pi(n-1)\left(t-\frac{r}{c_0}\right)} \\ &= h(g_n) \end{aligned} \tag{53}$$

where $g_n = \text{imag}(C_{nn})$ denotes the imaginary part of C_{nn}. As we known that if a variable x has PDF $f_x(x)$, the PDF of $y = hg(x)$ is as follows:

$$f_y(y) = f_x[h(y)]\left|h'(y)\right| \tag{54}$$

where $x = h(y)$ is the inverse function of $hg(x)$, $f_y(\cdot)$ denotes the PDF responding to y, and $[\cdot]'$ denotes the derivation operation. We have:

$$\begin{aligned} f_{g_n}(g_n) &= f_\rho[h(g_n)]\left|h'(g_n)\right| \\ &= \frac{1}{\sqrt{2\pi}\sigma}e^{-\frac{h^2(g_n)}{2\sigma^2}}\left|h'(g_n)\right|, -1 < g_n < 1 \end{aligned} \tag{55}$$

where

$$h'(g_n) = \frac{1}{2\pi(n-1)\sqrt{1-g_n^2}\left(t-\frac{r}{c_0}\right)}$$

denotes the derivation of $h(g_n)$. Utilizing g_n, Equation 13 can be rewritten as:

$$\begin{aligned} \tilde{A}(\theta,r,t) &= \frac{\mathbf{w}^H\mathbf{C}(\theta,r,t)\mathbf{v}(\theta,r,t)}{r} \\ &= \sum_{n=2}^{N}\left(\sqrt{1-g_n^2}+jg_n\right)\frac{\mathbf{v}_n(\theta,r,t)}{r} + \frac{\mathbf{v}_1(\theta,r,t)}{r} \\ &= \sum_{n=2}^{N}\left(\sqrt{1-g_n^2}+jg_n\right)v_n + v_1 \end{aligned} \tag{56}$$

where $\mathbf{v}_n$ denotes the nth element of $\mathbf{v}$. Defining $v_n = \frac{\mathbf{v}_n(\theta,r,t)}{r}$ for notation convenience, we can find

$$\sqrt{1-g_2^2}+jg_2 = \left(\tilde{A}(\theta,r,t) - \sum_{n=3}^{N}\left(\sqrt{1-g_n^2}+jg_n\right)v_n - v_1\right)/v_2. \tag{57}$$

Let

$$\begin{aligned} a_2 &= \text{imag}\left(\tilde{A}(\theta,r) - \sum_{n=3}^{N}\left(\sqrt{1-g_n^2}+jg_n\right)v_n - v_1\right) \\ &= \text{imag}(\tilde{A}(\theta,r)) - \text{imag}\left(\sum_{n=3}^{N}\left(\sqrt{1-g_n^2}+jg_n\right)v_n - v_1\right) \end{aligned}$$

and $v_2 = \cos\phi + j\sin\phi$, where ϕ is the phase of v_2, utilizing Equation 57, we can get:

$$g_2\cos\varphi + \sqrt{1-g_2{}^2}\sin\varphi = a_2. \tag{58}$$

Solving Equation 58, it yields:

$$\begin{aligned} g_2 &= a_2\cos\varphi + \sin\varphi\sqrt{1-a_2{}^2} \\ &= \xi_2(\tilde{A}_i) \end{aligned} \tag{59}$$

where $\tilde{A}_i = \text{imag}\left(\tilde{A}(\theta,r)\right)$ denotes the imaginary part of $\tilde{A}$. And we define:

$$\begin{aligned} g_3 &= g_3 = \xi_3(g_3) \\ &\cdots \\ g_N &= g_N = \xi_N(g_N). \end{aligned} \tag{60}$$

The joint PDF of the variables above is as follows:

$$\begin{aligned} & f_{\tilde{A}_i g_3 \cdots g_N}(\tilde{A}_i g_3 \cdots g_N) \\ &= f_{g_1 g_3 \cdots g_N}\left[\xi_2\left(\tilde{A}_i(\theta,r,t)\right), \xi_3(g_3), \ldots, \xi_N(g_N)\right]|J| \end{aligned} \tag{61}$$

According to Equations 59 and 60, we can get that:

$$\begin{aligned} J &= \begin{vmatrix} \frac{\partial\xi_2}{\partial\tilde{A}_i} & \frac{\partial\xi_2}{\partial g_3} & \cdots & \frac{\partial\xi_2}{\partial g_N} \\ \frac{\partial\xi_3}{\partial\tilde{A}_i} & \frac{\partial\xi_3}{\partial g_3} & \cdots & \frac{\partial\xi_3}{\partial g_N} \\ \cdots & \cdots & \cdots & \cdots \\ \frac{\partial\xi_N}{\partial\tilde{A}_i} & \frac{\partial\xi_N}{\partial g_3} & \cdots & \frac{\partial\xi_N}{\partial g_N} \end{vmatrix} = \begin{vmatrix} \frac{\partial\xi_2}{\partial\tilde{A}_i} & \frac{\partial\xi_2}{\partial g_3} & \cdots & \frac{\partial\xi_2}{\partial g_N} \\ 0 & 1 & \cdots & 0 \\ \cdots & \cdots & \cdots & \cdots \\ 0 & 0 & \cdots & 1 \end{vmatrix} \\ &= \cos\phi - \frac{\sin\phi}{\sqrt{1-a_2^2}} \end{aligned} \tag{62}$$

where $\frac{\partial\xi_1}{\partial\tilde{A}_i} = \cos\phi - \frac{\sin\phi}{\sqrt{1-a_2^2}}$. Since the each element of **C** is responding to only one frequency increment error which is random and independent, g_n is the independent with other g_m with $m = 1, 2, \ldots, N-1$ and $m \neq n$. Therefore, Equation 61 can be rewritten as:

$$\begin{aligned} & f_{g_2 g_3 \cdots g_N}\left[\xi_2(\tilde{A}_i(\theta,r,t)), \xi_3(g_3), \ldots, \xi_N(g_N)\right] \\ &= f_{g_2}\left[\xi_2(\tilde{A}_i(\theta,r,t))\right] f_{g_3}\left[g_3\right] \cdots f_{g_N}\left[g_N\right] \end{aligned} \tag{63}$$

Moreover, we can get the PDF of FDA beampattern as:

$$f_{\tilde{A}_i}(\tilde{A}_i) = \int_{-1}^{1} \cdots \int_{-1}^{1} f_{g_2 g_3 \cdots g_N}\left[\xi_2(\tilde{A}_i(\theta, r, t)), \xi_3(g_3), \ldots, \xi_N(g_N)\right] \times \left(\cos\phi - \frac{\sin\phi}{\sqrt{1-a_2^2}}\right) d_{g_3} \ldots d_{g_N}. \tag{64}$$

Utilizing the same approach, we can get the PDF of real part of FDA beampattern. Since the relationship between real part and imaginary part is complex, so we cannot give the PDF of the whole FDA beampattern.

Similarly, we can derive the FDA beampattern variance with the random frequency increment errors, which are uniformly distributed as aforementioned. The PDF of g_n is as follows:

$$\begin{aligned} f_{g_n}(g_n) &= f_\rho[h(g_n)]\,|h'(g_n)| \\ &= \frac{1}{2\rho_{\max}} \frac{1}{2\pi(n-1)\left(t - \frac{r}{c_0}\right)\sqrt{1-g_n^2}}, \\ &-1 < g_n < 1 \end{aligned} \tag{65}$$

and the PDF of FDA beampattern imaginary part is as follows:

$$f_{\tilde{A}_i}(\tilde{A}_i) = \int_{-1}^{1} \cdots \int_{-1}^{1} f_{g_2 g_3 \cdots g_N}\left[\xi_2(\tilde{A}_i(\theta, r, t)), \xi_3(g_3), \ldots, \xi_N(g_N)\right] \times \left(\cos\phi - \frac{\sin\phi}{\sqrt{1-a_2^2}}\right) d_{g_3} \ldots d_{g_N}. \tag{66}$$

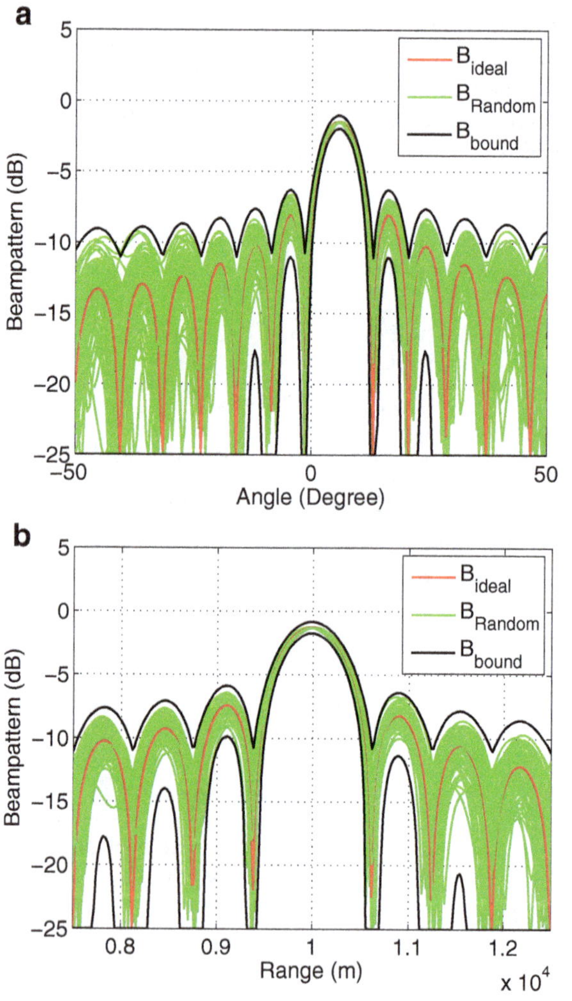

Figure 4 Profiles for the ideal, the bounds, and the errors FDA beampattern. **(a)** In angle dimension. **(b)** In range dimension.

5 Simulation results

5.1 Example 1: FDA beampattern bound

Consider a 16-element uniform linear FDA with half of wavelength λ spacing between neighbor elements. The center frequency f_0 is 10 GHz, and the increment frequency Δf is 30 KHz. The target is located at the range 10 km, angle 0°, and the time is on $1/\Delta f$. Figure 4a,b shows the comparisons of ideal FDA beampattern B_{ideal}, FDA beampattern upper and lower bound B_{bound}, and FDA beampattern with random frequency increment errors B_{Random}. In the FDA beampattern with random frequency increment errors, the frequency increment errors are Gaussian distribution, i.e., $\rho_n \sim N(0, 90^2)$ and the random FDA beampatterns are based on 50 independent Monte Carlo simulation runs. It can be shown that the present bounds hold for all the simulated beampattern realizations for both Figure 4a,b. Simulations results show that the bounds are tighter in range dimension than that in the angle dimension which is caused by range counterbalance of the total error in beampattern.

Furthermore, we give two figures about expectation and variance about FDA beampattern errors caused by frequency increment errors, shown as in Figures 5 and 6, respectively. Figure 5 compares the expectation value of theoretical result and empirical result. The theoretical result is based on the method of Section 4.2.1. The empirical result is based on the 10,000 independent Monte Carlo simulation runs. The comparisons show that when the FDA beampattern errors have smaller value, two results are close to each other. The variances of theoretical result and empirical result are shown in Figure 6. Similar with Figure 5, the theoretical result is based on the method of Section 4.2.2. The empirical result is based on the 1,000 independent Monte Carlo simulation runs. Different from the expectation, the empirical result fluctuates around the theoretical result along with increase of σ.

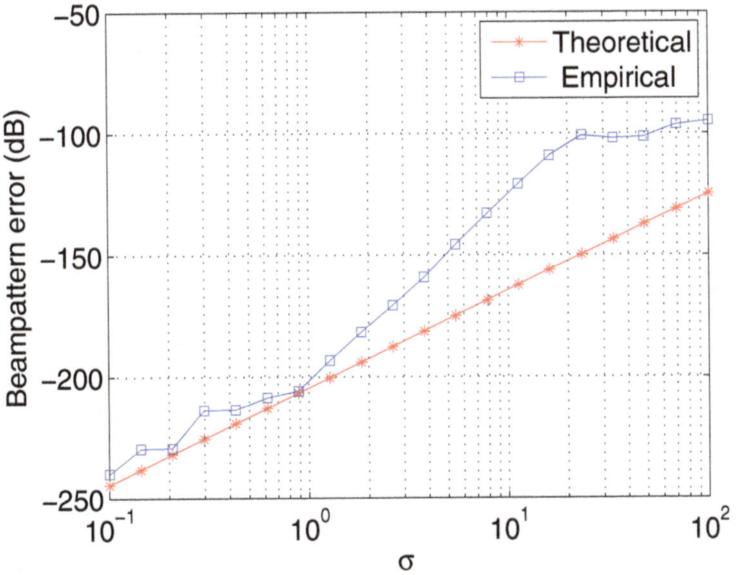

Figure 5 The comparison of Gaussian random FDA beampattern error expectation value.

5.2 Example 2: FDA beampattern PDF

Consider a uniform linear FDA with four elements. In this example, FDA beampattern amplitude error does not divide into the factor r_0, and its value is very large compared with that of Figures 5 and 6. Other array parameters are the same to that of example 1. Figure 7 shows PDFs of theoretical and empirical results, and Figure 8 shows the PDFs of imaginary part of random FDA beampattern with different σs. It can be known that lower σ enjoys the smaller beampattern errors, and PDF curves are centrosymmetric with the center about $A_i = 0$. One might use this figure and Equation 64 to specify tolerance requirements of frequency increment errors to fulfill a given beampttern requrement with certain probability. For instance, if probability of the beampttern errors at the domain (−0.5 to 0.5) is not smaller than 0.95, the standard deviations of ρ for every elements of FDA array will be required to be no larger than 60.

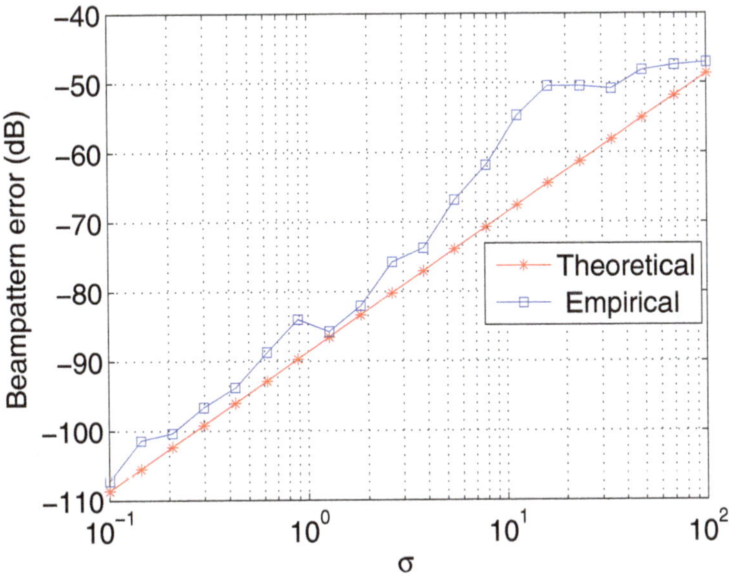

Figure 6 The comparison of Gaussian random FDA beampattern error variance value.

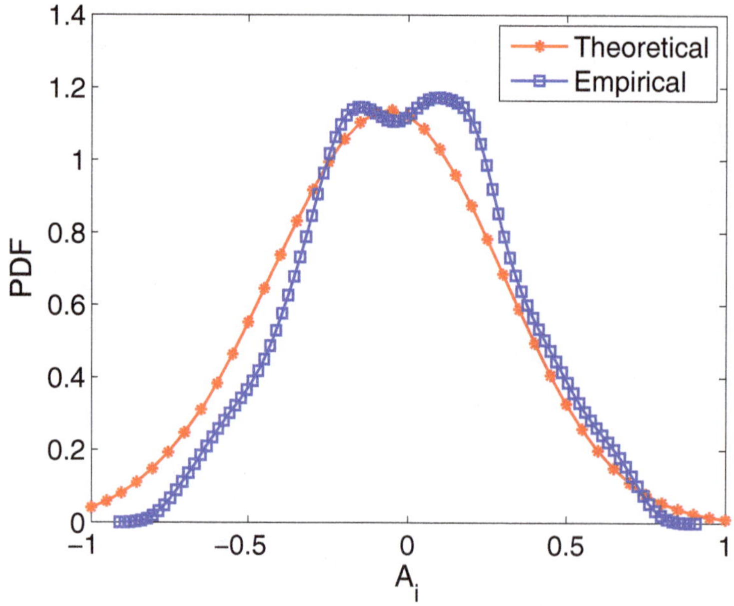

Figure 7 The PDF of Gaussian random FDA beampattern error.

6 Conclusions

In this paper, we have investigated the impacts of frequency increment errors on FDA beampattern based on deterministic errors and random errors. For uniform and linear deterministic frequency increment errors, the specific beampattern error formulations are provided, which gives guideline for device selection. For the stochastic frequency increment errors, we have derived a very tight worst-case boundary of the FDA beampattern. Simulation results show that all the random beampatterns are held for the derived bounds, and the worst-case boundary is helpful to the FDA system design. At last, we derived the statistical properties of the expectation, variance, and PDF. They can be used to analyze the probability of FDA beampattern fluctuations for the given distribution frequency increment errors.

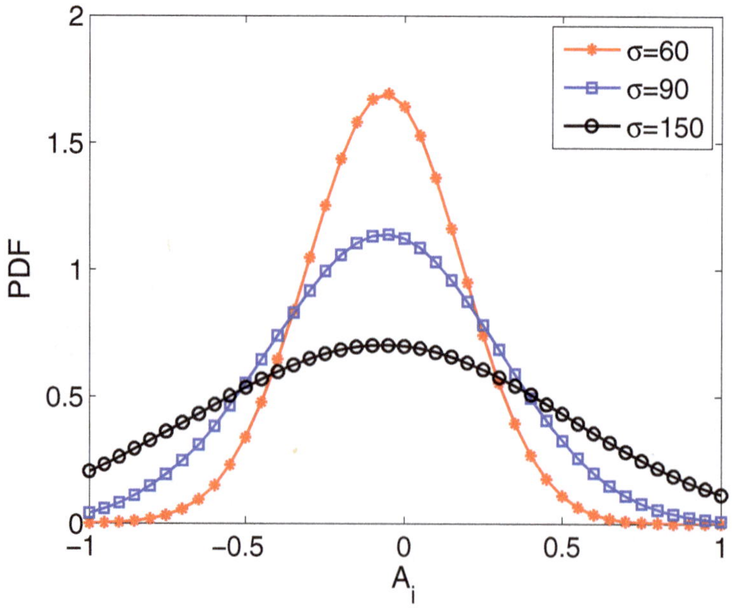

Figure 8 The PDF of imaginary part of random FDA beampattern.

Competing interests
The authors declare that they have no competing interests.

Authors' contributions
All authors formulated and discussed the idea together. Additionally, KG wrote the paper. All authors read and approved the final manuscript.

Acknowledgements
This work was supported in part by the Program for New Century Excellent Talents in University under grant no. NCET-12-0095.

References
1. BDV Veen, KM Buckley, Beamforming: a versatile approach to spatial filtering. IEEE ASSP Mag. **5**(2), 4–24 (1988)
2. J Xie, H Li, Z He, C Li, A robust adaptive beamforming method based on the matrix reconstruction against a large doa mismatch. EURASIP J. Adv. Signal Process. **2014**(91), 1–10 (2014)
3. P Antonik, MC Wicks, HD Griffiths, CJ Baker, in *Proceedings of the IEEE Radar Conference*. Frequency diverse array radars (IEEE, Verona, NY, 2006), pp. 215–217
4. MC Wicks, P Antonik, *Frequency diverse array with independent modulation of frequency, amplitude, and phase. in January 15, 2008*. (U.S.A Patent 7,319,427, USA)
5. P Antonik, MC Wicks, *Frequency diverse array with independent modulation of frequency, amplitude, and phase. in June 5, 2008*. U.S.A Patent 7, Application 20080129584, USA
6. P Antonik, MC Wicks, *Method and apparatus for a frequency diverse array. in March 31, 2009*. U.S.A Patent 7.511, 665B2, USA
7. P Antonik, An investigation of a frequency diverse array. PhD thesis, University College London (2009)
8. S Brady, Frequency diverse array radar: signal characterization and measurement accuacy. PhD thesis, Air Force Institute of Technology (2010)
9. M Secmen, S Demir, A Hizal, T Eker, in *Proceedings of the IEEE Radar Conference*. Frequency diverse array antenna with periodic time modulated pattern in range and angle (IEEE, Boston, MA, 2007), pp. 427–430
10. P Baizert, TB Hale, MA Temple, MC Wicks, Forward-looking radar GMTI benefits using a linear frequency diverse array. Electron. Lett. **42**(22), 1311–1312 (2006)
11. C Cetintepe, S Demir, Multipath characteristics of frequency diverse arrays over a ground plane. IEEE Trans. Antennas Propagation. **62**(7), 3567–3574 (2014)
12. J Shin, J-H Choi, J Kim, J Yang, W Lee, J So, C Cheon, in *Microwave Conference Proceedings (APMC)*. Full-wave simulation of frequency diverse array antenna using the FDTD method (IEEE, Seoul, Korea, 2013), pp. 1070–1072
13. T Eker, S Demir, A Hizal, Exploitation of linear frequency modulated continuous waveform (LFMCW) for frequency diverse arrays. IEEE Trans. Antennas Propagation. **61**(7), 3546–3553 (2013)
14. YB Wang, WQ Wang, HZ Shao, Frequency diverse array Cramér-Rao lower bounds for estimating direction, range and velocity. Int J Antennas Propagation. **2014**, 1–10 (2014)
15. V Ravenni, in *Proceedings of the 4th European Radar Conference*. Performance evaluations of frequency diversity radar system (IEEE, Munich, Germany, 2009), pp. 436–439
16. L Zhuang, XZ Liu, Application of frequency diversity to suppress grating lobes in coherent MIMO radar with separated subapertures. EURASIP J. Adv. Signal Process. **2009**, 1–10 (2009)
17. WQ Wang, Phased-MIMO radar with frequency diversity for range-dependent beamforming. EURASIP J. Adv. Signal Process. **13**(4), 1320–1328 (2013)
18. W-Q Wang, HZ Shao, Range-angle localization of targets by a double-pulse frequency diverse array radar. IEEE J. Sel. Top. Signal Process. **8**(1), 106–114 (2014)
19. W-Q Wang, HC So, Transmit subaperturing for range and angle estimation in frequency diverse array radar. IEEE Trans. Signal Process. **62**(8), 2000–2011 (2014)
20. E Yazdian, S Gazor, MH Bastani, Limiting spectral distribution of the sample covariance matrix of the windowed array data. EURASIP J. Adv. Signal Process. **2013**(42), 1–15 (2013)
21. J Xie, Z He, H Li, J Li, 2D DOA estimation with sparse uniform circular arrays in the presence of mutual coupling. EURASIP J. Adv. Signal Process. **2011**(127), 1–18 (2011)
22. M Khodja, A Belouchrani, K Abed-Meraim, Performance analysis for time-frequency music algorithm in presence of both additive noise and array calibration errors. EURASIP J. Adv. Signal Process. **2012**(94), 1–11 (2012)
23. Y Han, D Zhang, A recursive Bayesian beamforming for steering vector uncertainties. EURASIP J. Adv. Signal Process. **1**(108), 1–10 (2014)
24. S Henault, SK Podilchak, SM Mikki, YMM Antar, A methodology for mutual coupling estimation and compensation in antennas. IEEE Trans. Antennas Propagation. **61**(3), 1119–1132 (2013)
25. H Chen, B Liu, P Huang, J Liang, Y Gu, Mobility-assisted node localization based on TOA measurements without time synchronization in wireless sensor networks. MONET. **17**(1), 90–99 (2012)
26. H Chen, G Wang, Z Wang, H-C So, HV Poor, Non-line-of-sight node localization based on semi-definite programming in wireless sensor networks. IEEE Trans. Wireless Commun. **11**(1), 108–116 (2012)
27. H Chen, Q Shi, R Tan, HV Poor, K Sezaki, Mobile element assisted cooperative localization for wireless sensor networks with obstacles. IEEE Trans. Wireless Commun. **9**(3), 956–963 (2010)
28. W Zhang, Q Yin, H Chen, F Gao, N Ansari, Distributed angle estimation for localization in wireless sensor networks. IEEE Trans. Wireless Commun. **12**(2), 527–537 (2013)
29. CM Schmid, S Schuster, R Feger, A Stelzer, On the effects of calibration errors and mutual coupling on the beam pattern of an antenna array. IEEE Trans. Antennas Propagation. **61**(8), 4063–4071 (2013)
30. H Chen, F Gao, MH Martins, JL P Huang, Accurate and efficient node localization for mobile sensor networks. MONET. **18**(1), 141–147 (2013)

Bit-depth scalable lossless coding for high dynamic range images

Masahiro Iwahashi[1*], Taichi Yoshida[1], Norrima Binti Mokhtar[2] and Hitoshi Kiya[3]

Abstract

In this paper, we propose a bit-depth scalable lossless coding method for high dynamic range (HDR) images based on a reversible logarithmic mapping. HDR images are generally expressed as floating-point data, such as in the OpenEXR or RGBE formats. Our bit-depth scalable coding approach outputs base layer data and enhancement layer data. It can reconstruct the low dynamic range (LDR) image from the base layer data and reconstructs the HDR image by adding the enhancement layer data. Most previous two-layer methods have focused on the lossy coding of HDR images. Unfortunately, the extension of previous lossy methods to lossless coding does not significantly compress the enhancement layer data. This is because the bit depth becomes very large, especially for HDR images in floating-point data format. To tackle this problem, we apply a reversible logarithmic mapping to the input HDR data. Moreover, we introduce a format conversion to avoid any degradation in the quality of the reconstructed LDR image. The proposed method is effective for both OpenEXR and RGBE formats. Through a series of experiments, we confirm that the proposed method decreases the volume of compressed data while maintaining the visual quality of the reconstructed LDR images.

Keywords: High dynamic range imaging; Lossless coding; Bit-depth scalable coding

1 Introduction

Image data compression technologies, such as the JPEG 2000 international standard [1,2], allow high quality images to be transmitted via worldwide digital communication networks. Digital cinema and 4K images are remarkable examples of such technology [3,4]. These images require a huge number of pixels to express fine textures at high spatial resolutions.

Recently, high dynamic range (HDR) images have attracted considerable attention [5]. These images have a high resolution of pixel values, i.e., numerous pixel tones. Compared with the current standard for low dynamic range (LDR) images, which are expressed in 8 bits, HDR images have an extremely long bit depth and high dynamic range of pixel values. To fully utilize this dynamic range under limited memory space, the pixel values are expressed as floating-point data, such as in OpenEXR or RGBE format [6,7]. This paper focuses on the compression of HDR images in these data formats. Moreover, the proposed method, referred to as bit-depth scalable coding, is backward compatible with a standard coding method for LDR images.

Bit-depth scalable coding outputs compressed data in two layers, a base layer and an enhancement layer. From the bit stream in the base layer, the LDR image is decoded with a standard lossy decoder. By adding the bit stream in the enhancement layer, the original HDR image can be decoded without any loss. This scalable coding system has the advantage that it can directly accommodate both HDR and LDR users. Therefore, the system has attracted many researchers, and a number of variations have been reported [8-15].

Ward et al. [8] proposed a backward compatible bit-depth scalable coding method in which the original HDR color image is tone mapped in the base layer to produce an LDR image that is compressed by the JPEG international standard encoder. The enhancement layer then embeds the luminance ratio of the LDR and HDR images. The original HDR color image is decoded by multiplying the luminance ratio in the enhancement layer and the LDR color image in the base layer. This method has been extended to video signals and has attracted attention as

*Correspondence: iwahashi@vos.nagaokaut.ac.jp
[1] Department of Electrical, Electronics and Information Engineering, Nagaoka University of Technology, 1603-1 Kamitomioka, Nagaoka, Niigata, Japan
Full list of author information is available at the end of the article

a bit-depth scalable video coding method in international standardization activities [9-11,16]. For still images, Khan introduced a piecewise linear model of a tone mapping [12]. Jinno et al. improved the coding efficiency in the enhancement layer by replacing the ratio with a low-pass-filtered HDR image [15]. However, these reports focused on 'lossy' coding for HDR images.

Unlike these previous reports, we discuss the 'lossless' coding of HDR images under a scalable coding scheme that is compatible with lossy LDR image coding. The lossless coding of HDR images is especially important for storing and archiving original visual data such as medical, artistic, and astrograph images. Such data can be used for diagnosis based on medical images, analysis of astrograph images, art preservation, and bio-medical detections [17,18].

First, we discuss a baseline method [19] that was simply extended to lossless scalable coding from a non-scalable HDR image coding method [20]. Although the baseline method is straightforward and easy to implement, the coding efficiency in the enhancement layer is not satisfactory. To cope with this problem, we introduced a reversible logarithmic mapping and reduced the dynamic range of the HDR images [19,21]. This approach was shown to be effective for compressing data in the enhancement layer. However, the method was limited to the OpenEXR format [6]. Another representative format, referred to as RGBE [7], has been ignored.

In this paper, we improve on our previous conference papers [19,21] and add some theoretical analysis. First, we show that a simple extension of the reversible logarithmic mapping (Rev) to the RGBE format degrades the visual quality of the decoded LDR images. To avoid this problem, we introduce a format conversion (Cnv) to the system. We demonstrate that simply extending Rev magnifies the quantization error added by the lossy coding in the base layer. Second, we analyze the theoretical basis for why our method improves the coding efficiency of the system. We estimate how the bit depth of the residual image to be encoded in the enhancement layer is reduced by Rev. We also explain why the simply extended Rev degrades the LDR images, and why Cnv improves their quality in the RGBE format.

This paper is organized as follows. In Section 2, we describe two floating-point data formats and a non-scalable HDR image coding method. A baseline scalable coding method that simply extends the non-scalable coding approach is then summarized in Section 3, and the concept and implementation of the proposed method are introduced in Section 4. The theoretical analysis is described in Section 5, and our experimental results are summarized in Section 6. Finally, we present our conclusions in Section 7.

2 Data format and non-scalable coding

We first describe two floating-point data formats for HDR images. A non-scalable lossy coding method, which is extended to scalable lossless coding of HDR images in the next section, is also summarized.

2.1 Type A format of HDR images

To date, there are two well-known representative data formats for HDR images. One is the OpenEXR floating-point data format [6] and the other is the RGBE data format [7].

In the OpenEXR data format, a pixel value $x_{H,c}$ of an HDR image is described by an exponent value $x_{E,c}$, mantissa value $x_{M,c}$, and sign value $x_{S,c}$ as

$$\begin{cases} x_{H,c} = (-1)^{x_{S,c}} \left(1 + 2^{-10} x_{M,c}\right) 2^{-15+x_{E,c}} \\ \text{if } x_{E,c} \neq 0 \end{cases} \quad (1)$$

and

$$\begin{cases} x_{H,c} = (-1)^{x_{S,c}} \left(0 + 2^{-10} x_{M,c}\right) 2^{-14} \\ \text{if } x_{E,c} = 0 \end{cases} \quad (2)$$

for a color component $c \in \{R, G, B\}$. The exponent, mantissa, and sign values are given as integers in the ranges

$$x_{M,c} \in \left[0, 2^{10} - 1\right], \; x_{E,c} \in \left[0, 2^5 - 1\right], \; x_{S,c} \in [0, 1]. \quad (3)$$

The mantissa, exponent, and sign have depths of 10 bits, 5 bits, and 1 bit, respectively. Therefore, a pixel value of an HDR image is expressed in $10 + 5 + 1 = 16$ bits for each color component. Note that in certain special cases, $x_{E,c} = 31$ [6].

In the remainder of this paper, we denote the exponent, mantissa, and sign of each color component as a vector

$$\begin{cases} \mathbf{x}_E = \left[x_{E,R}, \; x_{E,G}, \; x_{E,B}\right]^T \\ \mathbf{x}_M = \left[x_{M,R}, \; x_{M,G}, \; x_{M,B}\right]^T \\ \mathbf{x}_S = \left[x_{S,R}, \; x_{S,G}, \; x_{S,B}\right]^T \end{cases} \quad (4)$$

and define the HDR image data $\mathbf{x}_D$ as

$$\mathbf{x}_D = \left[\mathbf{x}_E, \; \mathbf{x}_M, \; \mathbf{x}_S\right]. \quad (5)$$

Using these vectors, we denote Equations 1 and 2 as

$$\mathbf{x}_H = \mathrm{Flt}_A\left(\mathbf{x}_D\right), \quad (6)$$

where the pixel value of the HDR image $\mathbf{x}_H$ is

$$\mathbf{x}_H = \left[x_{H,R}, \; x_{H,G}, \; x_{H,B}\right]^T. \quad (7)$$

Hereafter, we refer to OpenEXR as the 'type A' format.

2.2 Type B format of HDR images

In the RGBE data format, a pixel value of an HDR image $x_{H,c}$ is given as

$$x_{H,c} = \begin{cases} \frac{x_{M,c}+0.5}{256} 2^{x_{E,0}-128} & \text{if } x_{E,0} \neq 0 \\ 0 & \text{if } x_{E,0} = 0 \end{cases} \quad (8)$$

for a color component $c \in \{R, G, B\}$. Both the mantissa and exponent have depths of 8 bits, i.e.,

$$x_{M,c} \in \left[0, 2^8 - 1\right], \ x_{E,0} \in \left[0, 2^8 - 1\right]. \tag{9}$$

The exponent $x_{E,0}$ is commonly used among three color components. In this format, a pixel value is expressed with a total of 32 bits [7]. Using the vectors, we denote Equation 8 as

$$\mathbf{x}_H = \mathrm{Flt}_B\left(\mathbf{x}_D\right). \tag{10}$$

Hereafter, we refer to RGBE as the 'type B' format. Note that, for a type B image, $\mathbf{x}_H$ in Equation 10 is non-negative. In contrast, $\mathbf{x}_H$ in Equation 6 for a type A image can be negative, zero, or positive.

2.3 Non-scalable lossy coding

Figure 1 illustrates the 'HDR image coding in JPEG 2000' reported in [20]. At the encoder, the HDR image data $\mathbf{x}_D$ is converted into the pixel value $\mathbf{x}_H$ by Flt, where Flt denotes Flt_A in Equation 6 for type A images and Flt_B in Equation 10 for type B images. The logarithmic function $\log_e$ is applied to each color component of $\mathbf{x}_H$. Note that pixel values that are less than or equal to zero are first clipped to the minimum positive pixel value in the image. In terms of the signal-to-noise ratio (SNR) of the variances, the effect on the LDR images is almost zero. At worst, of the nine test images considered in this paper, the SNR is less than 10^{-2} [%] for a type A 'still life' image. The effect on HDR images is also limited, with an SNR of less than 10^{-10} [%] for the same input image.

The pixel values are normalized to the range [0, 255] by

$$\mathrm{Nrm}(x) = (x - \min X) \cdot \frac{255}{\max X - \min X} \tag{11}$$

for $X = \{x | x \in \textit{image}\}$, where $\min X$ and $\max X$ are the minimum and maximum pixel values in the set X, respectively. Because the input values to the encoder must be integers, the results are rounded to be integers. Namely,

$$\mathbf{x}_B = \mathrm{Rnd}(\mathrm{Nrm}(\log_e(\mathrm{Clp}(\mathbf{x}_H)))) \tag{12}$$

is fed into the encoder, where Rnd and Clp are the rounding and clipping operations, respectively. In the decoder, the HDR pixel values $\mathbf{y}_H$ are recovered from the decoded image $\mathbf{y}_B$ with the inverse of each Nrm and $\log_e$.

In this paper, we extend this method to the scalable lossless coding of HDR images. The tone mapping operator Tmo described in Section 2.4 is added to this procedure as 'part A' to display color LDR images with better quality.

2.4 Tone mapping operation

We now summarize the tone mapping operator for color images based on the Hill function [5]. A pixel value of the HDR image $y_{H,c}$ is tone mapped to $y_{L,c}$ of the LDR image as

$$y_{L,c} = \mathrm{Rnd}\left(255 y_{H,c} \cdot y_{L,Y} / y_{H,Y}\right) \tag{13}$$

for $c \in \{R, G, B\}$, where

$$\begin{cases} y_{H,Y} = 0.27 y_{H,R} + 0.67 y_{H,G} + 0.06 y_{H,B} \\ y_{L,Y} = \mathrm{Hill}\left(y_{H,Y} / \bar{Y}_{H,Y}\right), \end{cases} \tag{14}$$

and the Hill function is defined as

$$\mathrm{Hill}(x) = \frac{x^a}{x^a + b^a}. \tag{15}$$

In (14), $\bar{Y}_{H,Y}$ is defined as

$$\bar{Y}_{H,Y} = \exp\left(\mathrm{Ens}\left(\log_e(y_{H,Y})\right)\right), \tag{16}$$

where $\mathrm{Ens}(\cdot)$ denotes the ensemble average over all positive values of $y_{H,Y}$ in the image. a and b are user-set parameters. In our experiments, we use $(a, b) = (1, 1)$. We denote the tone mapping in Equation 13 as

$$\mathbf{y}_L = \mathrm{Rnd}\left(\mathrm{Clp'}\left(\mathrm{Tmo}\left(\mathbf{y}_H\right)\right)\right), \tag{17}$$

where

$$\begin{cases} \mathbf{y}_H = \left[y_{H,R}, \ y_{H,G}, \ y_{H,B}\right]^T \\ \mathbf{y}_L = \left[y_{L,R}, \ y_{L,G}, \ y_{L,B}\right]^T \end{cases} \tag{18}$$

for color components. Because the output values of Tmp exceed 8-bit integers for color images, we clip the output values to the range [0, 255] with Clp'.

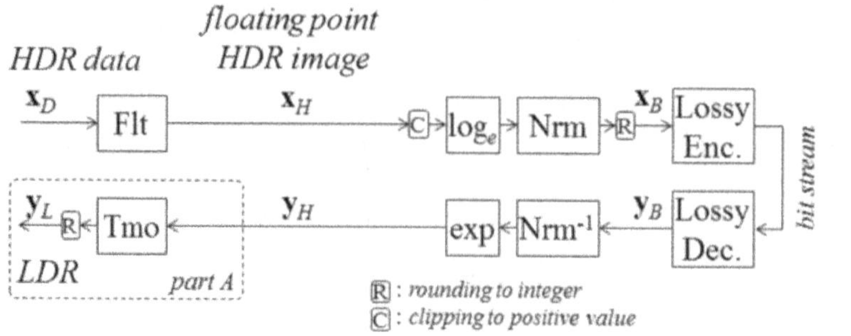

Figure 1 HDR image coding in JPEG 2000. The logarithmic function is applied and normalized to 8-bit depth before lossy encoding.

3 Baseline method

The baseline scalable lossless coding method is simply an extension of the non-scalable lossy coding method. We now summarize this baseline method, as well as the problem considered in this paper.

3.1 Scalable lossless coding

Figure 2 illustrates the baseline method, which we use as a reference in this paper. This is a simple extension of the non-scalable lossy coding in Figure 1 to the scalable lossless coding of HDR images. 'Part B' denotes the processes that have been added.

To achieve the lossless coding of HDR images, $\mathbf{x}_H$ is converted into the integer value $\mathbf{x}_I$ by the reversible integer mapping Int detailed in Section 3.2. Note that the inverse mapping Int^{-1} reconstructs the original value without any loss. The procedure for generating the LDR images is almost the same as the method in Figure 1. The bit stream needed to reconstruct the LDR image is embedded in the base layer. In the enhancement layer, the integer value $\mathbf{y}_I$ is reconstructed from the decoded LDR image $\mathbf{y}_B$ with the inverse normalization Nrm^{-1}, the exponential function exp, and the rounding operation

$$\mathbf{y}_I = \text{Rnd}\left(\exp\left(\text{Nrm}^{-1}(\mathbf{y}_B)\right)\right). \tag{19}$$

Finally, the residual

$$\mathbf{e}_I = \mathbf{y}_I - \mathbf{x}_I \tag{20}$$

is encoded with a lossless coding method to generate the bit stream in the enhancement layer.

3.2 Reversible integer mapping

The reversible integer mapping Int from the real value $\mathbf{x}_H$ to the integer value $\mathbf{x}_I$ was introduced in [19]. It is defined as

$$x_{I,c} = \begin{cases} (-1)^{x_{S,c}}\left(x_{H,c}2^{25-\min X_E} - 2^{10}\right) & \text{if } \min X_E \neq 0 \\ (-1)^{x_{S,c}} x_{H,c}2^{24} & \text{if } \min X_E = 0 \end{cases} \tag{21}$$

where $c \in \{R, G, B\}$ and $X_E = \{x_{E,c} | x_{E,c} \in \text{image}\}$ for type A images. This is a simple scaling applied to the rational number $x_{H,c}$ in Equation 1 so that it becomes an integer. In other words, we shift the decimal point to the right. Note that the minimum $\min X_E$ of all the pixel values $\mathbf{x}_E$ in the image is stored and embedded into the bit stream. We denote the mapping in Equation 21 as

$$\mathbf{x}_I = \text{Int}_A(\mathbf{x}_H) \tag{22}$$

for

$$\mathbf{x}_I = \left[x_{I,R},\ x_{I,G},\ x_{I,B}\right]^T. \tag{23}$$

Similarly, a mapping for type B images can be defined as

$$\mathbf{x}_{I,c} = \left(256x_{H,c}2^{128-\min X_E^+} - 0.5\right)2 + 1 \tag{24}$$

where $c \in \{R, G, B\}$ and $X_E^+ = \left\{x_{E,c} | x_{E,c} > 0\right\}$. Note that the minimum $\min X_E^+$ of all the positive pixel values $\mathbf{x}_E > 0$ in the image is stored and embedded into the bit stream. We denote this mapping as

$$\mathbf{x}_I = \text{Int}_B(\mathbf{x}_H) \tag{25}$$

for type B images.

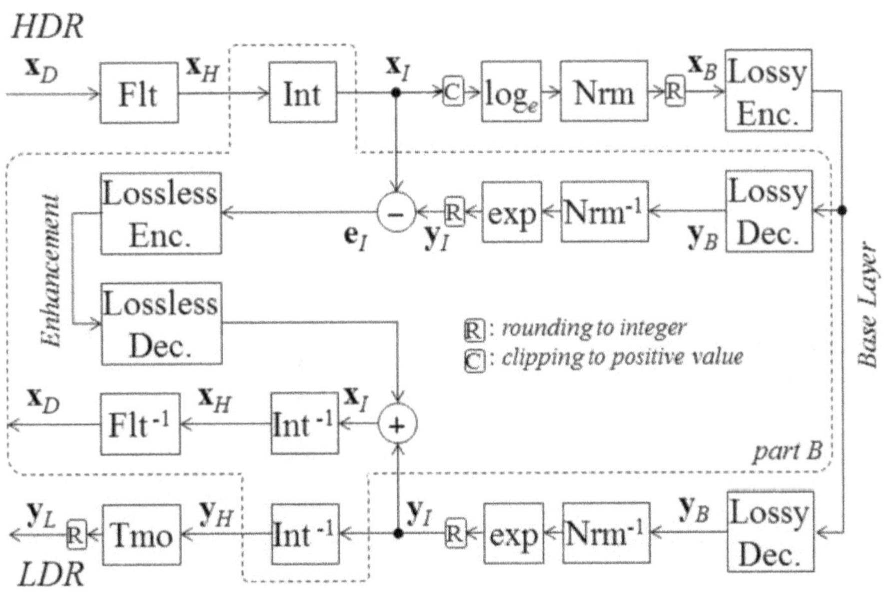

Figure 2 The baseline method. The HDR image is reconstructed without any loss.

Note that the inverse of this mapping recovers the original value without any loss. Therefore, the baseline method becomes lossless for the original HDR image.

3.3 Problem setting

In this paper, we tackle the following limitation of the baseline method. As a result of the reversible integer mapping, the residual $\mathbf{e}_I$ in Equation 20 requires a very large bit depth. It is somewhat difficult to compress this data volume in the high bit rate coding of the LDR image. This is because $\mathbf{e}_I$ is a magnified version of the coding noise $\mathbf{e}_B = \mathbf{y}_B - \mathbf{x}_B$. In lossy coding, the noise $\mathbf{e}_B$ is added in the base layer and is magnified by Nrm^{-1} and exp as indicated in Equation 19. Because this noise tends to have a weak correlation, the difference $\mathbf{e}_I$ also has a weak correlation. Therefore, the data volume of the enhancement layer becomes huge. Note that the correlation of $\mathbf{e}_I$ increases in the low bit rate coding of the LDR image. This is investigated in Section 6.

To cope with this problem, we previously introduced the reversible logarithmic mapping (Rev) to reduce the bit depth of the residual image [19]. However, in this previous report, we only presented experimental results without any theoretical endorsement. In this paper, we theoretically compare Int and Rev in respect of the bit depth of the residual image in the enhancement layer.

In addition, Rev has been limited to the type A format, ignoring type B. In this paper, we show that a simple extension of Rev to the type B format degrades the LDR images. To avoid this problem, we introduce a format conversion (Cnv) from type B to type A in the base layer. We present a theoretical justification for why the simply extended Rev degrades the LDR images and Cnv improves its quality for type B images.

4 Proposed method

The reversible logarithmic mapping (Rev) is introduced to reduce the data volume of the enhancement layer. In particular, for type B format images, the format conversion (Cnv) is introduced to maintain the visual quality of the LDR images.

4.1 Type I method for type A format images

Figure 3 illustrates the proposed type I method. Instead of Flt and Int in the baseline method (Figure 2), the reversible logarithmic mapping Rev defined in Section 4.2 is applied to the HDR data $\mathbf{x}_D$ to produce $\mathbf{x}_R$. This is converted to an 8-bit depth integer $\mathbf{x}_B$ as

$$\mathbf{x}_B = \text{Rnd}(\text{Nrm}(\text{Clp}(\mathbf{x}_R))) \tag{26}$$

and fed into the lossy encoder, which outputs the bit stream in the base layer. The reconstructed pixel $\mathbf{y}_B$ given

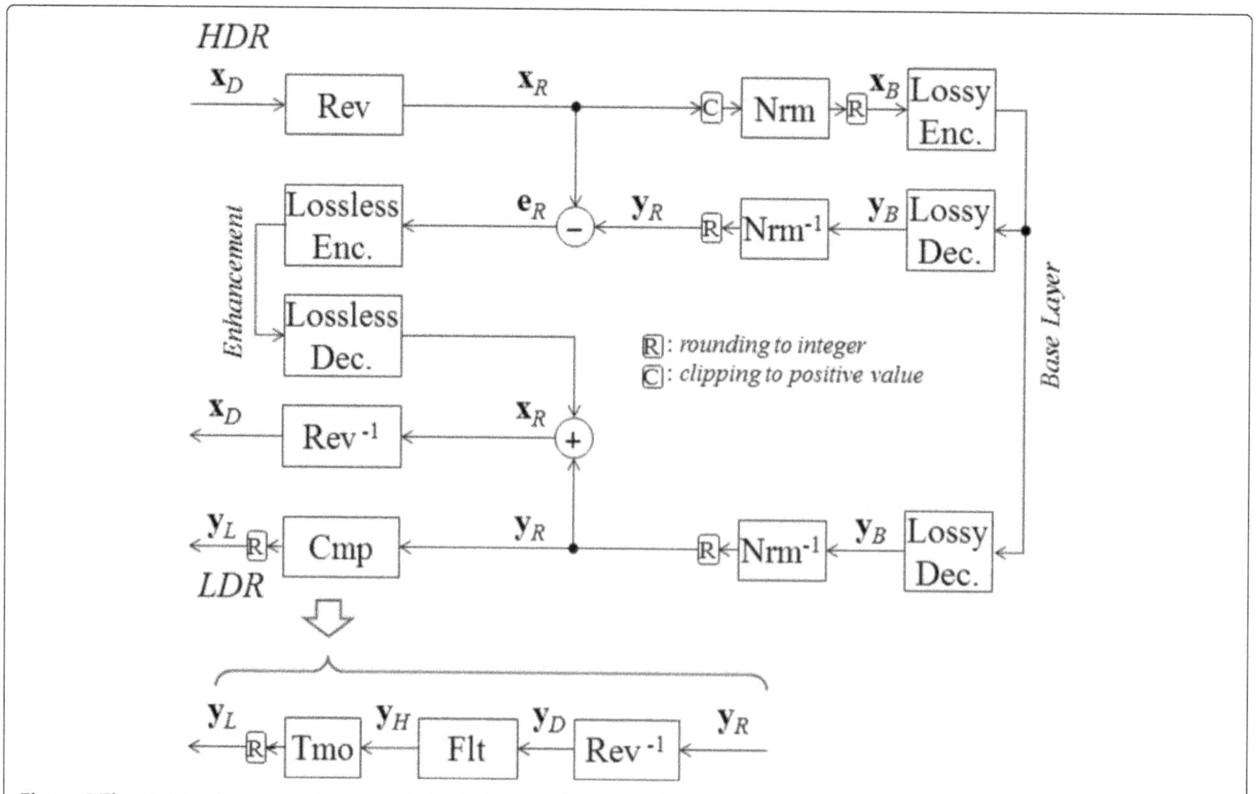

Figure 3 The proposed type I method. The bit depth of the residual $\mathbf{e}_R$ is reduced by the reversible logarithmic mapping Rev.

by the decoder is inversely normalized and rounded to an integer as

$$\mathbf{y}_R = \mathrm{Rnd}\left(\mathrm{Nrm}^{-1}\left(y_B\right)\right). \quad (27)$$

Then, the difference

$$\mathbf{e}_R = \mathbf{x}_R - \mathbf{y}_R \quad (28)$$

is encoded with the lossless encoder to generate the bit stream in the enhancement layer. In the decoder, $\mathbf{y}_R$ is added to $\mathbf{e}_R$ to recover $\mathbf{x}_R$. Applying the inverse of Rev, the original HDR data $\mathbf{x}_D$ are retrieved without any loss. Namely, they are recovered as

$$\mathbf{x}_D = \mathrm{Rev}^{-1}(\mathbf{e}_R + \mathbf{y}_R). \quad (29)$$

The LDR image $\mathbf{y}_L$ is reconstructed from the decoded image $\mathbf{y}_B$ in a similar way to the baseline method with a compensation factor (Cmp). This recovers the HDR pixel value $\mathbf{y}_H$, and then applies the tone mapping operation Tmo as

$$\begin{aligned}\mathbf{y}_L &= \mathrm{Rnd}\left(\mathrm{Tmo}\left(\mathrm{Flt}\left(\mathrm{Rev}^{-1}\left(\mathbf{y}_R\right)\right)\right)\right)\\ &= \mathrm{Rnd}\left(\mathrm{Cmp}\left(\mathbf{y}_R\right)\right).\end{aligned} \quad (30)$$

It is also possible to display $\mathbf{y}_B$ as an LDR image without using Cmp. In this case, $\mathbf{y}_B$ in the proposed method is almost the same as that of the baseline method as illustrated in Figure 4. There are two approaches that use the Hill function in Equation 15 to generate the LDR image $\mathbf{y}_L$ exampled in Figure 5. The first introduces Cmp in the encoding process, and the second introduces Cmp in the decoding process. The former case is convenient for data receivers, because it is not necessary to add Cmp to a standard decoder. However, this increases the data volume of the enhancement layer. In this paper, we employ the latter approach.

4.2 Reversible logarithmic mapping

In the proposed type I method illustrated in Figure 3, the reversible logarithmic mapping is applied to generate the integer value $\mathbf{x}_R$. This technique was originally introduced in [22]. The mapping for type A images is defined as

$$x_{R,c} = (-1)^{x_{S,c}}\left(\left(x_{E,c} - \min X_E\right)2^{10} + x_{M,c}\right) \quad (31)$$

for $c \in \{R, G, B\}$. We denote this mapping as

$$\mathbf{x}_R = \mathrm{Rev}_A(\mathbf{x}_D) \quad (32)$$

for

$$\mathbf{x}_R = \left[x_{R,R},\ x_{R,G},\ x_{R,B}\right]^T. \quad (33)$$

This mapping approximates the logarithm of an HDR image $\mathbf{x}_H$. Substituting $\mathbf{x}_{E,c}$ from Equation 1, i.e.,

$$x_{E,c} = \log_2 x_{H,c} - \log_2\left(1 + 2^{-10}x_{M,c}\right) + 15, \quad (34)$$

for positive values in Equation 31, we have

$$x_{R,c} = \left(\log_2 x_{H,c} + 15 - \min X_E - \epsilon_A\right)2^{10} \quad (35)$$

where

$$\begin{aligned}\epsilon_A &= \log_2\left(2^{-10}x_{M,c} + 1\right) - 2^{-10}x_{M,c}\\ &= \log_2 \frac{\delta_A + 1}{2^{\delta_A}}\end{aligned} \quad (36)$$

for $\delta_A = 2^{-10}x_{M,c}$. As indicated in Equation 35, Rev_A generates a good approximation of the logarithm of the HDR image [23,24]. The approximation error is relatively small, as ϵ_A fluctuates around 0.06 depending on the mantissa. Therefore, Nrm $(\mathbf{x}_R)$ becomes close to $\mathbf{x}_B$ in Equation 12. This is encoded with a standard lossy encoder to generate the bit stream in the base layer.

The reversible logarithmic mapping is suitable for lossless scalable coding because it one-to-one maps an integer to an integer. Therefore, its inverse mapping reconstructs the original integer values without any loss, i.e., Rev is 'reversible'. This property also reduces the dynamic range of the mapped integer values. We have experimentally confirmed [21] that the residual in the enhancement layer $\mathbf{e}_R$ has a lower bit depth than that of $\mathbf{e}_I$ in the baseline method. We provide the theoretical basis for this observation in Section 5.1.

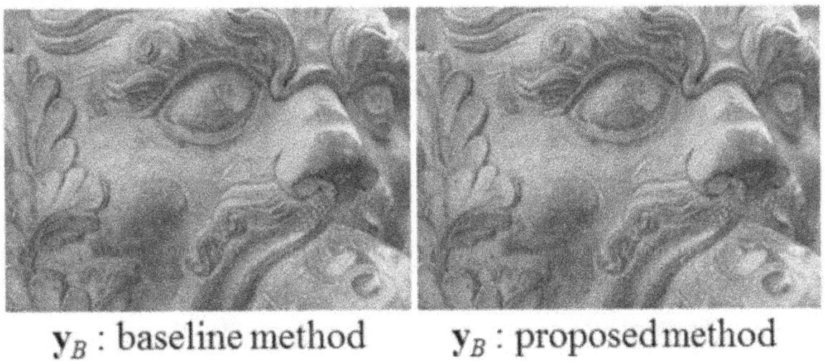

Figure 4 Images decoded in the base layer.

$\mathbf{x}_L$ 'Cannon' $\mathbf{x}_L$ 'Tree'

Figure 5 LDR images tone mapped with the Hill function.

4.3 Type I method for type B format images

For type A images, Rev_A in Equation 32 is applied in Figure 3. For type B images, a direct extension of Rev_A can be defined as

$$x_{R,c} = \begin{cases} x^*_{E,0}2^8 + x_{M,c} + 1 & \text{if } x_{E,0} \neq 0 \\ 0 & \text{if } x_{E,0} = 0 \end{cases} \tag{37}$$

for $c \in \{R, G, B\}$ and

$$x^*_{E,0} = x_{E,0} - \min X^+_E + 1. \tag{38}$$

We denote this mapping as

$$\mathbf{x}_R = \text{Rev}_B\left(\mathbf{x}_D\right), \tag{39}$$

and apply this to type B images in the type I method.

Note that the depth of $\mathbf{x}_R$ from Rev_B is a maximum of 16 bits for each color component. Therefore, it costs 48 bits for all the color components, which exceeds the original 32-bit data. However, using the reversible color transform (RCT) in JPEG 2000 lossless coding reduces the cost by 16 bits. The RCT is defined as

$$\begin{cases} x_1 = \lfloor (x_R + 2x_G + x_B)/4 \rfloor \\ x_2 = x_B - x_G \\ x_3 = x_R - x_G \end{cases} \tag{40}$$

Because the second and third row of this RCT take the difference between the color components, the exponent term $x^*_{E,0}$ in Equation 37 disappears. As a result, the bit depth becomes $48-16 = 32$ bits in total. Furthermore, the exponent term is less than 5 bits in the type B images tested in our experiment. Therefore, in practice, the system can compress the data volume.

In this paper, we show that the quality of LDR images is degraded in this directly extended Rev_B for type B images. Figure 6 shows some LDR images produced by the proposed type I and type II methods. The former has lower quality than the latter, with a peak SNR (PSNR) of 20.79 dB compared with 29.08 dB at the same bit rate of 5.23 bppc in the base layer. The reason for this is analyzed

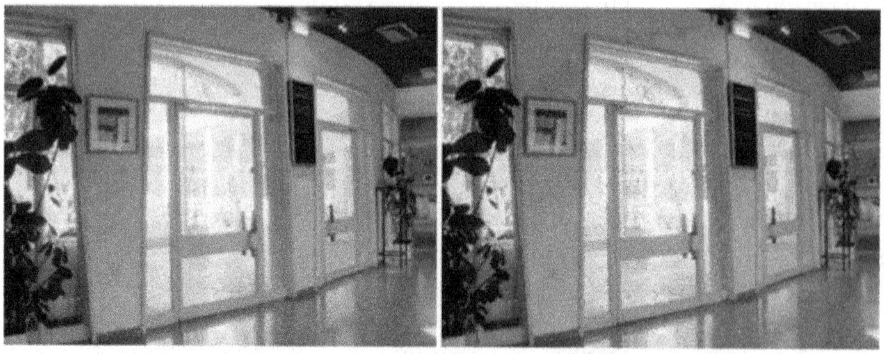

type I type II

Figure 6 LDR images given by the proposed type I and type II methods. The LDR image from the type I method is degraded in type B format.

in Section 5.2. To solve this problem, we introduce the following format conversion.

4.4 Type II method for type B format images

Figure 7 illustrates the modified method for type B images. We refer to this as the type II method. The type II approach includes the format conversion (Cnv). First, Rev_B converts the HDR data $\mathbf{x}_D$ into type B $\mathbf{x}_R$. In the figure, this is denoted as $\mathbf{x}_R^{(B)}$ to clearly indicate the type. The conversion introduced in this paper is defined as

$$\begin{cases} \text{Cnv}(\mathbf{x}) = \text{Rev}_A\left(\text{Flt}_A^{-1}\left(\text{Flt}_B\left(\text{Rev}_B^{-1}(\mathbf{x})\right)\right)\right) \\ \mathbf{x} = \text{Clp}\left(\mathbf{x}_R^{(B)}\right) \end{cases} \quad (41)$$

as illustrated in Figure 8. In the proposed type II method,

$$\mathbf{x}_B = \text{Rnd}\left(\text{Nrm}\left(\text{Cnv}\left(\text{Clp}\left(\mathbf{x}_R^{(B)}\right)\right)\right)\right) \quad (42)$$

is encoded with the lossy encoder.

As a result of this conversion, the type B image is temporarily converted into a type A image in the base layer. Therefore, the problem caused by Rev_B can be avoided, and the quality of the LDR image is improved compared to that given by applying the type I method to type B images. This assertion is theoretically endorsed in Section 5.2. This conversion is reversible as far as a large enough bit depth is assigned to values inside the process. However, reversion is not always necessary, as the system becomes lossless for HDR images in as much as the $\mathbf{y}_R$ are exactly the same in the encoder and the decoder, even though Cnv is not perfectly reversible.

5 Theoretical analysis

We now present a theoretical analysis of why the proposed method reduces the bit depth of the enhancement layer. The rationale for introducing the format conversion is also explained.

5.1 Bit depth of the enhancement layer

We estimate the bit depth of the residual $\mathbf{e}_I$ in the baseline method and $\mathbf{e}_R$ in the proposed method and theoretically demonstrate that the bit depth of the proposed method is smaller than that of the baseline method.

The bit depth of pixel values in an image $\mathbf{x}$ is defined as

$$B_{dp}(\mathbf{x}) = \log_2(\max X - \min X + 1), \quad (43)$$

where $\max X$ and $\min X$ denote the maximum and minimum pixel values in the image. We must calculate the maximum of $\mathbf{e}_I$ in the baseline method to estimate its bit depth. In Figure 2, the relations

$$\begin{cases} \mathbf{x}_B = \text{Rnd}\left(\text{Nrm}\left(\log_e \text{Clp}\left(\mathbf{x}_I\right)\right)\right) \\ \mathbf{y}_I = \text{Rnd}\left(\exp(\text{Nrm}^{-1}(\mathbf{y}_B))\right) \end{cases} \quad (44)$$

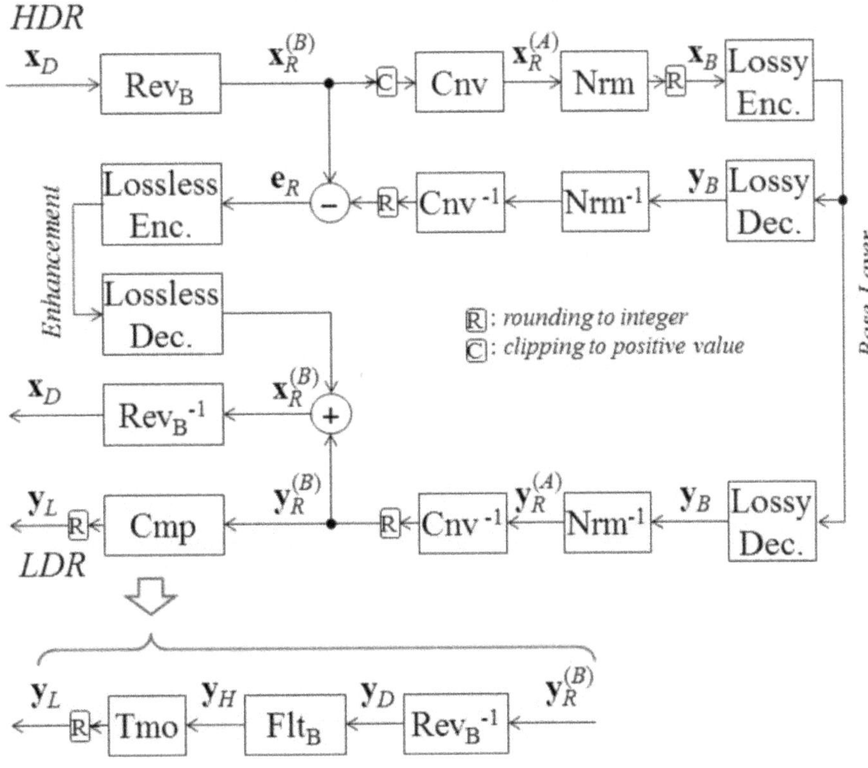

Figure 7 The proposed type II method. Cnv converts type B to type A in the base layer.

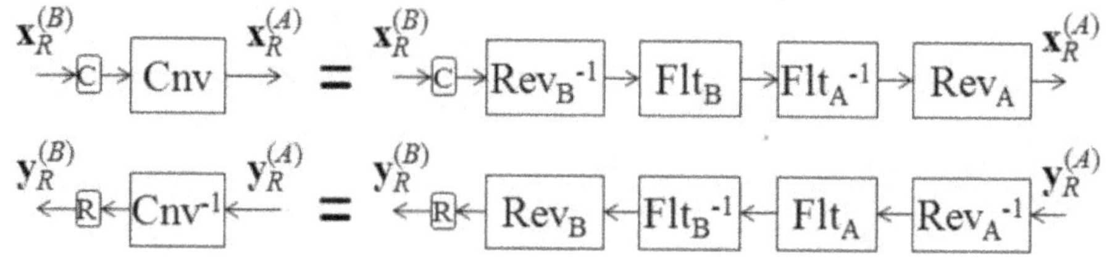

Figure 8 The format conversion Cnv in the proposed type II method.

are modeled as

$$\begin{cases} \mathbf{x}_B = \mathrm{Nrm}(\log_e \mathbf{x}_I) + e_1 \\ \mathbf{y}_I = \exp(\mathrm{Nrm}^{-1}(\mathbf{y}_B)) + e_2 \end{cases} \tag{45}$$

for positive values of $\mathbf{x}_I$, where $e_1, e_2 \in [-0.5, 0.5]$ are rounding errors due to Rnd in Equation 44. Therefore, the maximum of

$$\begin{aligned} \mathbf{e}_I &= \mathbf{y}_I - \mathbf{x}_I \\ &= \exp\left(\mathrm{Nrm}^{-1}\left(\mathbf{y}_B\right)\right) - \mathbf{x}_I + e_2 \\ &= \exp\left(\mathrm{Nrm}^{-1}\left(\mathbf{x}_B + \mathbf{e}_B\right)\right) - \mathbf{x}_I + e_2 \\ &= \exp\left(\mathrm{Nrm}^{-1}\left(\mathrm{Nrm}\left(\log_e \mathbf{x}_I\right)\right.\right. \\ &\quad \left.\left. +\ \mathbf{e}_B + e_1\right)\right) - \mathbf{x}_I + e_2 \end{aligned} \tag{46}$$

is estimated as

$$\max E_I = \max E_B \cdot \max X_I \cdot C/255 \tag{47}$$

for

$$C = \log_e(\max X_I) - \log_e\left(\min X_I^+\right), \tag{48}$$

as detailed in Appendix A. Substituting

$$\min E_I = -\max E_I \tag{49}$$

and Equation 47 into Equation 43, the bit depth of $\mathbf{e}_I$ is estimated as

$$\begin{aligned} B_{dp}(\mathbf{e}_I) &= \log_2(2 \cdot \max E_I + 1) \\ &\approx \log_2(\max E_B \cdot \max X_I \cdot C/255) + 1, \end{aligned} \tag{50}$$

giving the bit depth of the residual of the baseline method. Similarly, using the model

$$\begin{cases} \mathbf{x}_B = \mathrm{Nrm}'(\mathbf{x}_R) + e_1' \\ \mathbf{y}_R = \mathrm{Nrm}'^{-1}(\mathbf{y}_B) + e_2' \end{cases} \tag{51}$$

in the proposed method, the maximum of

$$\begin{aligned} \mathbf{e}_R &= \mathbf{y}_R - \mathbf{x}_R \\ &= \mathrm{Nrm}'^{-1}\left(\mathbf{x}_B + \mathbf{e}_B\right) - \mathbf{x}_R + e_2' \\ &= \mathrm{Nrm}'^{-1}\left(\mathrm{Nrm}'(\mathbf{x}_R) + \mathbf{e}_B + e_1'\right) \\ &\quad - \mathbf{x}_R + e_2' \end{aligned} \tag{52}$$

is given as

$$\max E_R = \max E_B \cdot \max X_R/255, \tag{53}$$

as shown in Appendix B. Substituting

$$\min E_R = -\max E_R. \tag{54}$$

and Equation 53 into Equation 43, the bit depth of $\mathbf{e}_R$ can be estimated as

$$\begin{aligned} B_{dp}\left(\mathbf{e}_R\right) &= \log_2(2 \cdot \max E_R + 1) \\ &\approx \log_2\left(\max E_B \cdot \max X_R\right)/255) + 1, \end{aligned} \tag{55}$$

giving the bit depth of the residual of the proposed method.

We can now compare $\mathbf{e}_I$ and $\mathbf{e}_R$ in terms of bit depth. The error in the base layer $\mathbf{e}_B$ is composed of errors due to the rounding before applying the lossy encoder, as well as quantization errors added by the lossy coding. Therefore, the maximum and minimum of

$$\mathbf{e}_B = \mathbf{y}_B - \mathbf{x}_B \tag{56}$$

are

$$\max E_B = -\min E_B = Q, \tag{57}$$

where Q is determined by the quantization step size of the lossy coding in the base layer. Taking the difference between Equation 50 and Equation 55, we have

$$\begin{aligned} \Delta B_{dp} &= B_{dp}\left(\mathbf{e}_I\right) - B_{dp}\left(\mathbf{e}_R\right) \\ &= \log_2\left(\max E_B \cdot \max X_I/255 \cdot C\right) \\ &\quad - \log_2\left(\max E_B \cdot \max X_R/255\right), \end{aligned} \tag{58}$$

and therefore

$$\Delta B_{dp} = \log_2 \frac{\max X_I \cdot C}{\max X_R} \tag{59}$$

is the difference in bit depth. From Equations 1, 21, and 31, the maxima of $\mathbf{x}_I$ and $\mathbf{x}_R$ are expressed as

$$\begin{cases} \max X_I = \left(2^{C^*} - 1\right) 2^{10} \approx 2^{C^*} \cdot 2^{10} \\ \max X_R = C^* \cdot 2^{10} \end{cases} \tag{60}$$

for

$$C^* = \max X_E - \min X_E + \gamma, \tag{61}$$

where $\gamma \in [0, 1)$ is determined according to the mantissa $\in \left[0, 2^{10}\right)$. Substituting Equation 60 into Equation 59, we have the difference as

$$\Delta B_{dp} = \log_2 \frac{2^{C^*} \cdot C}{C^*} > 0, \tag{62}$$

which is always a positive value. This indicates that the bit depth of the proposed method is smaller than that of the baseline method. Thus, we have theoretically shown that the proposed method achieves bit-depth reduction in the enhancement layer.

5.2 Difference between type I and type II for type B format

Next, we show that the format conversion introduced in Section 4.4 alleviates the degradation of LDR images in the base layer. The output LDR image $\mathbf{y}_L$ is tone mapped from the decoded HDR image $\mathbf{y}_H$, which is generated from $\mathbf{y}_R$ in the proposed method. Therefore, we analyze the relation between $\mathbf{y}_H$ and $\mathbf{y}_R$ for the type I and type II methods.

As illustrated in Figure 3, the proposed type I method produces $\mathbf{y}_H$ as

$$\mathbf{y}_H = \mathrm{Flt}_B\left(\mathrm{Rev}_B^{-1}\left(\mathbf{y}_R\right)\right). \quad (63)$$

for type B images. For example, the exponent of the type B image data becomes

$$x^*_{E,0} = \left(x_{R,c} - x_{M,c} - 1\right)/256 \quad (64)$$

from the inverse of Equation 37. Substituting this equation into Equation 8, we have

$$x_{H,0} = f_{Ia}(x_{R,c}) \cdot f_{Ib}(x_{M,c}) \cdot 2^{\min X_E^+ - 127}, \quad (65)$$

where

$$\begin{cases} f_{Ia}(x_{R,c}) &= \exp(x_{R,c} \cdot 2^{-8} \log_e 2), \\ f_{Ib}(x_{M,c}) &= \frac{\delta_B + 2^{-9}}{2^{\delta_B + 2^{-8}}} \in [0.002, 0.499], \\ \delta_B &= \frac{x_{M,c}}{256} \in [0, 1). \end{cases}$$

This is the relation between $\mathbf{x}_H$ and $\mathbf{x}_R$ and includes a function f_{Ia} that is proportional to the exponent of $\mathbf{x}_R$. However, note that this is chopped by the function f_{Ib}. This is confirmed by Figure 9, which indicates the relation between $\mathbf{x}_B$ and $\mathbf{x}_H$ for the type B 'tree' image. Note that $\mathbf{x}_B$ is a scaled version of $\mathbf{x}_R$. The points 'o' indicate where the mapping given by the type I method becomes discontinuous.

In contrast, the proposed type II method in Figure 7 produces $\mathbf{y}_H$ as

$$\begin{aligned} \mathbf{y}_H &= \mathrm{Flt}_B\left(\mathrm{Rev}_B^{-1}\left(\mathrm{Rnd}\left(\mathrm{Cnv}^{-1}\left(\mathbf{y}_R^{(A)}\right)\right)\right)\right) \\ &\approx \mathrm{Flt}_A\left(\mathrm{Rev}_A^{-1}\left(\mathbf{y}_R^{(A)}\right)\right) \end{aligned} \quad (66)$$

neglecting the effect of Rnd. This means that the image is converted to type A in the base layer. Therefore, taking the inverse of Equation 35, we have

$$x_{H,c} = f_{IIa}(x_{R,c}) \cdot f_{IIb}(x_{M,c}) \cdot 2^{\min X_E - 15}, \quad (67)$$

where

$$\begin{cases} f_{IIa}(x_{R,c}) &= \exp(x_{R,c} \cdot 2^{-10} \log_e 2), \\ f_{IIb}(x_{M,c}) &= \frac{\delta_A + 1}{2^{\delta_A}} \in [1, 1.06), \\ \delta_A &= \frac{x_{M,c}}{1024} \in [0, 1). \end{cases}$$

Similar to the type I method, the function f_{IIa} is proportional to the exponent of $\mathbf{x}_R$. Note that the function f_{IIb} is close to one. Therefore, unlike the type I method, the type II method gives an HDR image $\mathbf{x}_H$ that is approximately proportional to the exponent of $\mathbf{x}_R$. This is confirmed by the points marked 'x' in Figure 9.

Next, we investigate how the mappings in Equations 65 and 67 magnify the quantization error $\mathbf{e}_B$. Denoting the mapping as $\mathbf{x}_H = f(\mathbf{x}_B)$, the error is magnified as

$$\mathbf{y}_H - \mathbf{x}_H = f(\mathbf{x}_B + \mathbf{e}_B) - f(\mathbf{x}_B) \quad (68)$$

$$\approx \frac{\partial f(\mathbf{x}_B)}{\partial \mathbf{x}_B} \cdot \mathbf{e}_B. \quad (69)$$

Figure 10 illustrates the absolute value of

$$\frac{\partial f(\mathbf{x}_B)}{\partial \mathbf{x}_B} \approx \frac{\mathbf{y}_H - \mathbf{x}_H}{\mathbf{y}_B - \mathbf{x}_B} = \frac{\Delta \mathbf{x}_H}{\Delta \mathbf{x}_B} \quad (70)$$

for the type I method (marked 'o') and the type II method (marked 'x'). In the figure, a larger value signifies greater

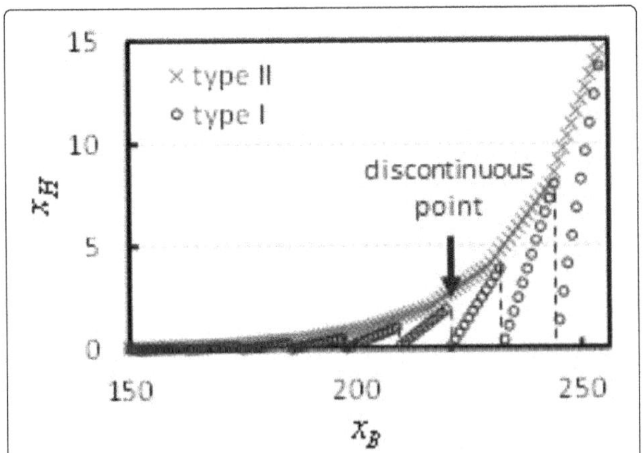

Figure 9 Mapping in the systems for the type B image 'tree'.

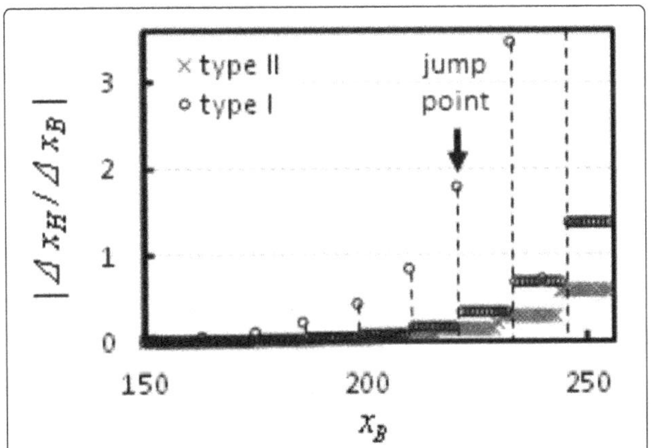

Figure 10 Magnification of coding errors in the systems for the type B image 'tree'.

amplification of the error. We can see that the type I method has larger values, especially at the jump points of Figure 10 which come from the discontinuous points of Figure 9. This implies that the degradation in LDR image quality produced by the type I method is alleviated by the type II method, which uses the format conversion in Section 4.4.

6 Experiments

We now describe a series of experiments that tested nine HDR images, including five type A images and four type B images. For the lossy coding in the base layer and the lossless coding in the enhancement layer, we used the JPEG 2000 international standard [1] in lossy mode and lossless mode, respectively.

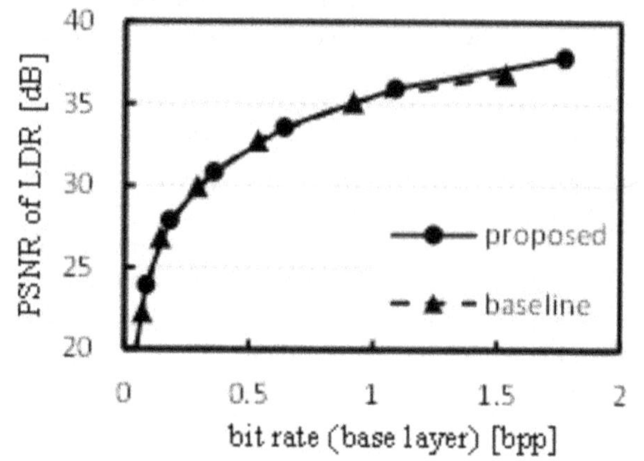

Figure 12 Coding performance in the base layer for type B image 'Belgium'.

6.1 Base layer

We compared the coding performance in the base layer of the baseline method and the proposed method. In this section, the proposed method denotes the type I procedure in Figure 3 for type A images and the type II procedure in Figure 7 for type B images. Figure 11 compares the baseline and proposed methods for the type A 'cannon' image. The horizontal axis records the data volume of the base layer in bits per pixel per color component (bppc). The vertical axis indicates the LDR image quality in terms of PSNR, which is defined by

$$\mathrm{PSNR} = 10\log_{10}\frac{255^2}{\mathrm{Ens}\left((\mathbf{y}_L-\mathbf{x}_L)^2\right)} \quad [dB] \tag{71}$$

for

$$\mathbf{x}_L = \mathrm{Rnd}\left(\mathrm{Clp'}\left(\mathrm{Tmo}(\mathbf{x}_H)\right)\right), \tag{72}$$

where Ens(·) denotes the ensemble average over all pixels in the image. The results indicate that the proposed method is slightly worse (by 0.46 dB at 3.1 bpp) than the baseline approach. Figure 12 indicates the rate distortion curves for the type B 'Belgium' image. The results are very similar to those for 'cannon'. The 'tree' image was investigated in different formats. Figures 13 and 14 indicate the curves for this image in type A and type B formats, respectively. Figure 15 summarizes the PSNR at 1.5 bppc in the base layer. This indicates that the proposed method is slightly better than the baseline technique. It can be concluded that the proposed method is comparable to or slightly better than the baseline method. This is considered to be because of the similarity of $\mathbf{x}_B$ in the baseline method and the proposed method. Both quantities represent the logarithm of the original HDR image $\mathbf{x}_H$.

6.2 Enhancement layer

Figure 16 compares the output from the proposed and baseline methods for the type A 'cannon' image. The horizontal axis indicates the PSNR of the reconstructed

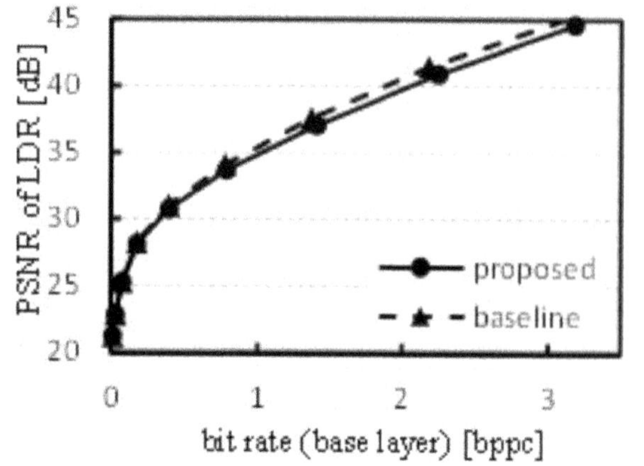

Figure 11 Coding performance in the base layer for type A image 'cannon'.

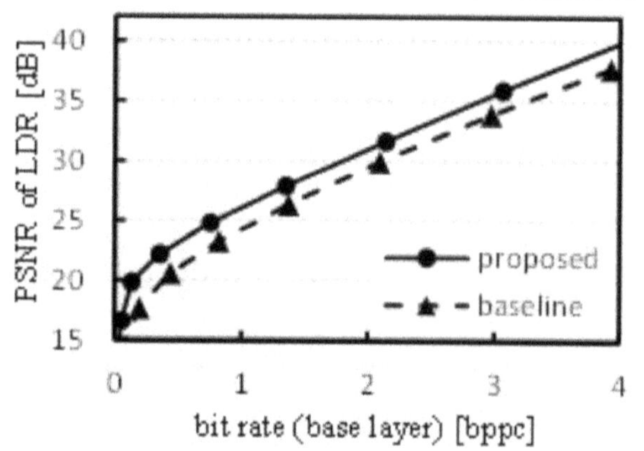

Figure 13 Coding performance in the base layer for type A image 'tree'.

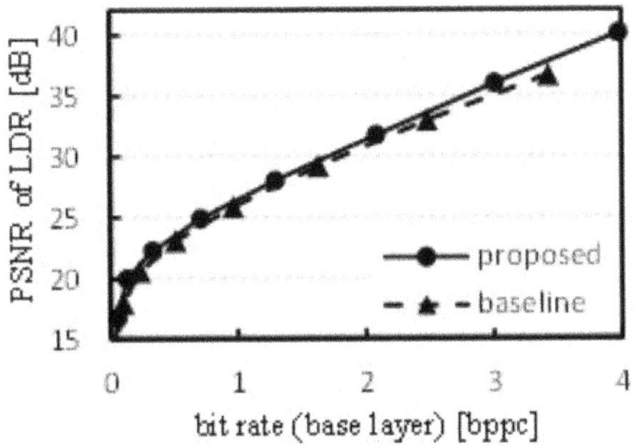

Figure 14 Coding performance in the base layer for type B image 'tree'.

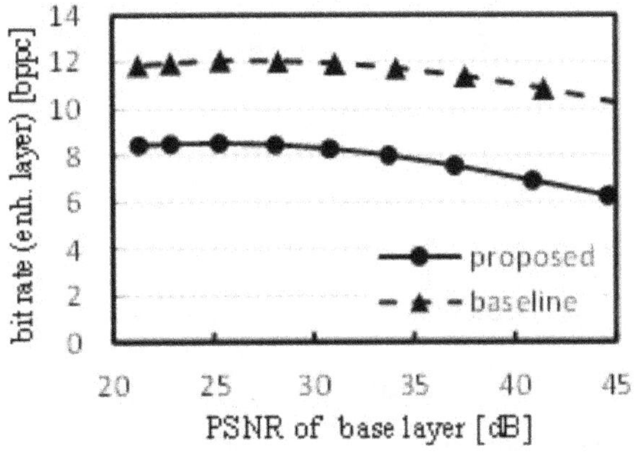

Figure 16 Bit rate of the enhancement layer for type A image 'cannon'.

LDR images, and the vertical axis indicates the bit rate of the bit stream in the enhancement layer. Note that, because the decoded HDR images are lossless, the PSNR is infinite. This figure indicates that the proposed method reduces the bit rate by more than 3.4 bppc for this image. As indicated in the figure, the bit rate in the enhancement layer decreases as the PSNR increases. However, the bit rate in the base layer increases with PSNR, which means that there is a trade-off in the bit rate in these layers.

Figure 17 shows the results for the type B 'Belgium' image. We can observe that the bit rate decreases by 8.03 bppc at 35 dB LDR image quality. Unlike the case in Figure 16, the bit rate increases with the PSNR. This is because the correlation among neighboring pixels in $\mathbf{e}_I$ increases in low PSNR (low bit rate) coding of the LDR image as indicated in Figure 18. For this input image, the correlation is observed to be 0.14 at a PSNR of 36.9

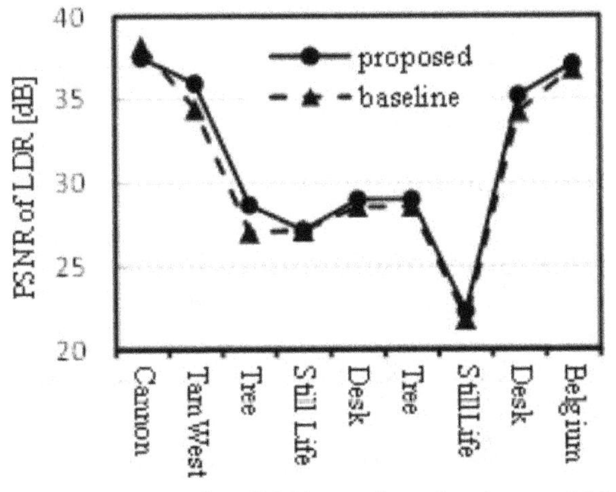

Figure 15 Image quality of LDR images for various images at 1.5 bppc enhancement layer bit stream.

dB. The correlation monotonically increases as PSNR decreases, reaching 0.80 at 18.8 dB. Because $\mathbf{e}_I$ is encoded with a transform that uses this correlation, a higher correlation serves to lower the bit rate. This is why the curve of the baseline method in the figure increases monotonically. The bit depth of the enhancement layer decreases monotonically from 26.7 bits at a PSNR of 18.8 dB to 24.1 bits at 36.9 dB as indicated in Figure 19. Furthermore, the logarithm of the variance of $\mathbf{e}_I$ is also monotonically decreasing as indicated in Figure 20.

The 'tree' image was again investigated in different formats. Figures 21 and 22 show the bit rate for the type A and type B image formats, respectively. We can see that better PSNR in the LDR images brings about a lower bit rate in the enhancement layer. This suggests that a higher data volume in the base layer will lead to a lower volume in the enhancement layer. Figure 23 summarizes the bit rate at 35 dB LDR image quality. This figure indicates that the proposed method reduces the data volume of type

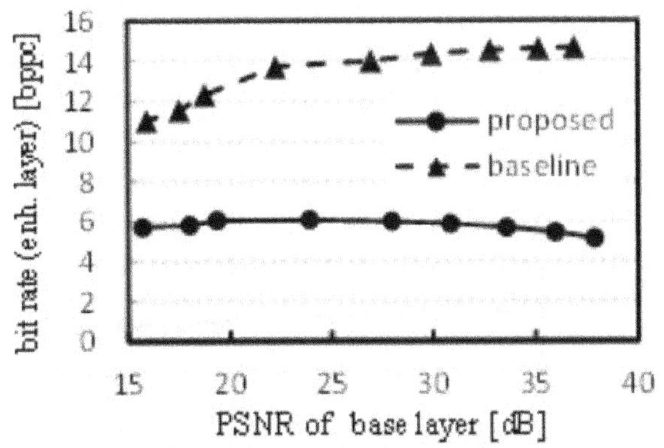

Figure 17 Bit rate of the enhancement layer for type B image 'Belgium'.

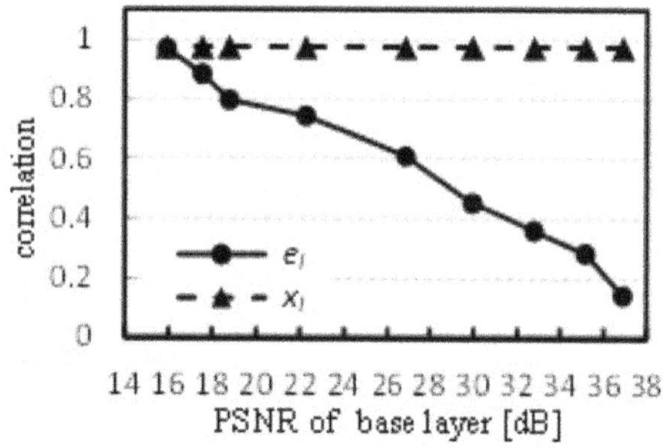

Figure 18 Correlation of the difference for type B image 'Belgium'.

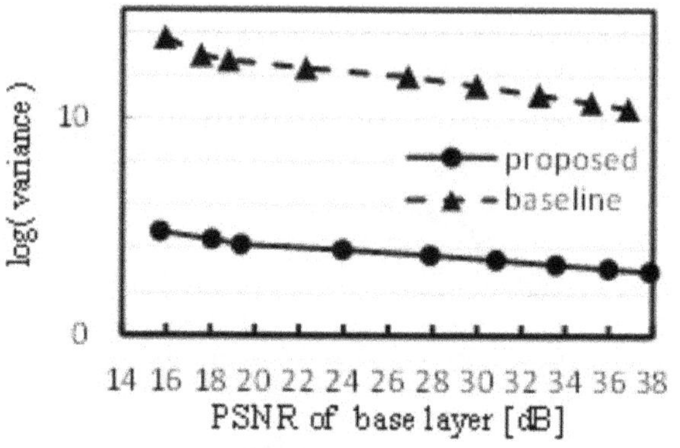

Figure 20 Log of variance of the difference for type B image 'Belgium'.

A images by a minimum of 3.82 bppc (for the 'cannon' image) and by a maximum of 8.82 bppc (for 'still life'). For type B images, the data volume is reduced by a minimum of 7.8 bppc for 'desk'. It was confirmed that the proposed method significantly reduces the data volume of the enhancement layer for both type A and type B format images.

7 Conclusions

In this paper, we have presented a bit-depth scalable lossless coding for HDR images in floating-point data formats. Unlike most conventional scalable coding methods, the proposed method reconstructs the original HDR image without any loss. Introducing a reversible logarithmic mapping and format conversion technique, it was confirmed that the proposed method reduces the bit depth as well as the bit rate in the enhancement layer. It was also confirmed that the proposed method maintains the LDR image quality and coding performance of the baseline method in the base layer for both the OpenEXR and RGBE formats.

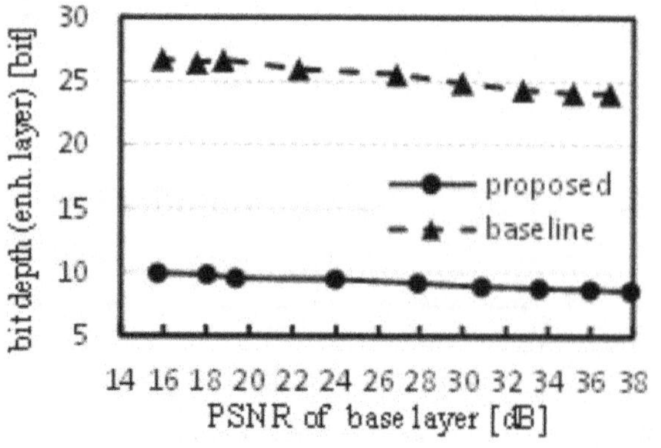

Figure 19 Bit depth of the difference for type B image 'Belgium'.

As our investigation has been limited to a difference-based approach, it is necessary to include ratio-based approaches, such as [8].

Appendix A

Substituting

$$\begin{cases} \mathrm{Nrm}^{-1}(\mathbf{x}_B) = \mathbf{x}_B \cdot C/255 + C_1 \\ C = C_2 - C_1 \\ C_1 = \log_e\left(\min X_I^+\right), \; C_2 = \log_e\left(\max X_I\right) \end{cases}$$

into

$$\mathbf{e}_I = \exp\left(\mathrm{Nrm}^{-1}\left(\mathrm{Nrm}\left(\log_e \mathbf{x}_I\right) + \mathbf{e}_B + e_1\right)\right) - \mathbf{x}_I + e_2,$$

we have

$$\mathbf{e}_I = \exp(x + \epsilon) - \mathbf{x}_I + e_2$$

where

$$\begin{cases} x = \log_e \mathbf{x}_I, \\ \epsilon = (\mathbf{e}_B + e_1) \cdot C/255. \end{cases}$$

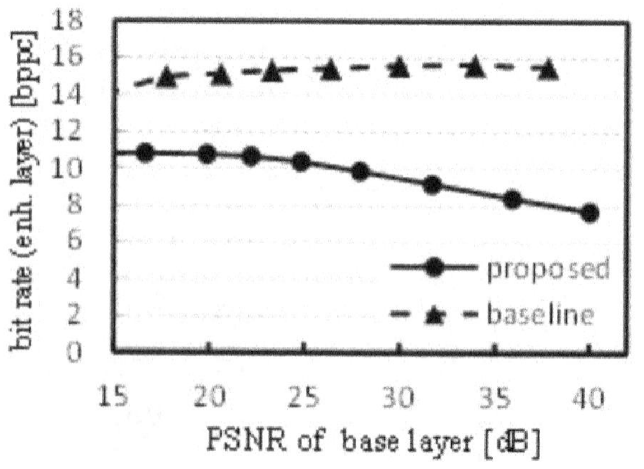

Figure 21 Bit rate of the enhancement layer for type A image 'tree'.

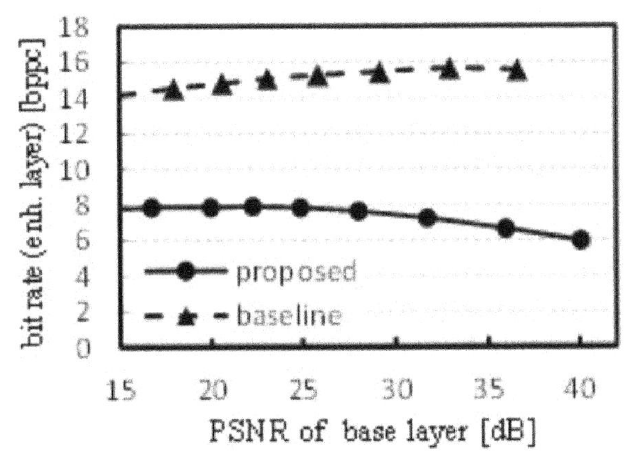

Figure 22 Bit rate of the enhancement layer for type B image 'tree'.

When x takes its maximum value, $\epsilon \ll x$ holds. For example, the value of ϵ/x for all images tested in this paper is less than 10^{-4}. In this case,

$$\begin{aligned} \mathbf{e}_I &= \exp(x+\epsilon) - \mathbf{x}_I + e_2 \\ &\approx \frac{\partial \exp(x)}{\partial x}\epsilon + \exp(x) - \mathbf{x}_I + e_2 \end{aligned}$$

holds. Therefore, we have

$$\begin{aligned} \mathbf{e}_I &= \exp(x)\epsilon + \exp(x) - \mathbf{x}_I + e_2 \\ &= \exp\left(\log_e \mathbf{x}_I\right)(\mathbf{e}_B + e_1)\cdot C/255 \\ &\quad + \exp\left(\log_e \mathbf{x}_I\right) - \mathbf{x}_I + e_2 \\ &= \mathbf{x}_I\,(\mathbf{e}_B + e_1)\cdot C/255 + \mathbf{x}_I - \mathbf{x}_I + e_2 \\ &= \mathbf{x}_I\,(\mathbf{e}_B + e_1)\cdot C/255 + e_2 \end{aligned}$$

namely,

$$\max E_I = \max X_I\,(\max E_B + e_1)\cdot C/255 + e_2.$$

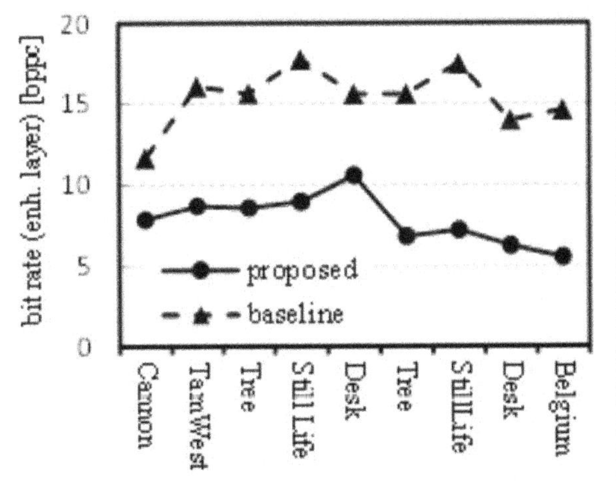

Figure 23 Bit rate of the enhancement layer for various images at 35 dB LDR image quality.

According to our experiments, $\max E_I$ is 7.23×10^3 in 'cannon' at minimum. Therefore e_2 is negligible compared to $\max E_I$ since the maximum of e_2 is 0.5. Similarly, $\max E_B$ takes value between 3 and approximately 2^7 depending on the bit rate of the base layer. Therefore, e_1 is negligible in low bit rate compared to $\max E_B$ and we have

$$\max E_I = \max X_I \cdot \max E_B \cdot C/255.$$

Note that precision of this estimation slightly decreases in high bit rate coding in which $\max E_B$ takes small value such as 3.

Appendix B

Substituting

$$\begin{cases} \mathrm{Nrm}'^{-1}(\mathbf{x}_B) = \mathbf{x}_B \cdot C'/255 + C'_1 \\ \mathrm{Nrm}'(\mathbf{x}_R) = \left(\mathbf{x}_R - C'_1\right)\cdot 255/C' \\ C' = C'_2 - C'_1 \\ C'_1 = \min X_R,\ C'_2 = \max X_R \end{cases}$$

into

$$\mathbf{e}_R = \mathrm{Nrm}'^{-1}\left(\mathrm{Nrm}'(\mathbf{x}_R) + \mathbf{e}_B + e'_1\right) - \mathbf{x}_R + e'_2,$$

we have

$$\begin{aligned} \mathbf{e}_R &= \mathrm{Nrm}'^{-1}\left(\left(\mathbf{x}_R - C'_1\right)\cdot 255/C' + \mathbf{e}_B + e'_1\right) \\ &\quad - \mathbf{x}_R + e'_2, \\ &= \mathbf{x}_R - C'_1 + \left(\mathbf{e}_B + e'_1\right)\cdot C'/255 + C'_1 \\ &\quad - \mathbf{x}_R + e'_2, \\ &= \left(\mathbf{e}_B + e'_1\right)\cdot C'/255 + e'_2. \end{aligned}$$

According to our experiments, $\max E_R$ is 94 in 'cannon' at minimum. Therefore $e'_2 \in [-0.5, 0.5]$ is negligible compared to $\max E_R$. Similarly to Appendix A, e_1 is negligible compared to $\max E_B$. As a result, we have

$$\max E_R = \max E_B \cdot \max X_R/255.$$

Competing interests
The authors declare that they have no competing interests.

Acknowledgements
This work was supported by JSPS KAKENHI Grant Number 26289117.

Author details
[1]Department of Electrical, Electronics and Information Engineering, Nagaoka University of Technology, 1603-1 Kamitomioka, Nagaoka, Niigata, Japan. [2]Department of Electrical Engineering, University of Malaya, 50603 Kuala Lumpur, Malaysia. [3]Department of Information and Communication Systems, Faculty of System Design, Tokyo Metropolitan University, 6-6 Asahigaoka, Hino, Tokyo, Japan.

References
1. ISO/IEC 15444-1, *Information Technology - JPEG, 2000 Image Coding System, Core coding system, ITU-T Recommendation, T800*, (Geneva, 2002), pp. 1–194
2. A Skodras, C Christopoulos, T Ebrahimi, The JPEG, 2000, still image compression standard. IEEE Signal Process. Mag. **18**, 36–58 (2001)

3. A Descampe, F-O Devaux, G Rouvroy, J Legat, J-J Quisquater, B Macq, A flexible hardware JPEG, 2000 decoder for digital cinema. IEEE Trans. Circuits Syst. Video Technol. **16**(11), 1397–1410 (2006)
4. K Kaneko, N Ohta, in *Proc. Int. Conf. Virtual Syst. Multimedia*. 4K applications beyond digital cinema (Seoul, 2010), pp. 133–136
5. E Reinhard, G Ward, S Pattanaik, P Debevec, *High Dynamic Range Imaging: Acquisition, Display, and Image-Based Lighting*. (Morgan Kaufmann, San Francisco, CA, USA, 2010)
6. R Bogart, F Kainz, D Hess, *OpenEXR image file format. ACM SIGGRAPH, Sketches & Applications*, (San Diego, 2003)
7. G Ward, Real Pixels Graphics Gems II (1991)
8. G Ward, M Simmons, in *Proc. ACM SIGGRAPH 2006 Courses*. JPEG-HDR: a backwards-compatible, high dynamic range extension to JPEG (Boston, 2006)
9. R Mantiuk, A Efremov, K Myszkowski, H-P Seidel, Backward compatible high dynamic range MPEG video compression. ACM Trans. Graph. **25**(3), 2006
10. M Winken, D Marpe, H Schwarz, T Wiegand, in *Proc. IEEE Int. Conf. Image Process*. Bit-depth scalable video coding, vol. 1 (San Antonio, 2007), pp. 5–8
11. S Park, KR Rao, Bit-depth scalable video coding based on H.264/AVC. IEICE Trans. Fundamentals. **E91-A**(6), 1541–1544 (2008)
12. IR Khan, in *Proc. IEEE Int. Conf. Acoustics Speech Signal Process*. Two layer scheme for encoding of high dynamic range images (Las Vegas, 2008), pp. 1169–1172
13. H Kikuchi, W Otake, M Iwahashi, in *Proc. Int. Symp. Intell. Signal Process. Commun. Syst.* Bit rate reduction of enhancement layer in bit-depth scalable coding (Kanazawa, 2009), pp. 264–267
14. A Boschetti, N Adami, R Leonardi, M Okuda, in *Proc. IEEE Int. Conf. Image Process*. Flexible and effective high dynamic range image coding (Hong Kong, 2010), pp. 3145–3148
15. T Jinno, M Okuda, N Adami, in *Proc. IEEE Int. Conf. Acoustics Speech Signal Process*. New local tone mapping and two-layer coding for HDR images (Kyoto, 2012), pp. 765–768
16. J-C Chiang, W-T Kuo, P-H Kao, Bit-depth scalable video coding with new inter-layer prediction. EURASIP J. Adv. Signal Process. **2011**(1), 1–9 (2011)
17. Y Zhang, DR Bull, E Reinhard, in *Proc. Int. Conf. Image Process*. Perceptually lossless high dynamic range image compression with JPEG, 2000 (Kyoto, 2012), pp. 1057–1060
18. YS Shih, WC Zhang, H Sheng, YH Yang, ST Sun, CC Wu, YC Lee, SS Lee, SC Huang, in *Proc. Int. Comp. Symp*. Bio-inspired JPEG XR CODEC design for lossless HDR biomedical image, (2010), pp. 148–153
19. M Iwahashi, H Kiya, in *Proc. Asia-Pacific Signal Inf. Process. Association Annual Summit Conf.* Efficient lossless bit depth scalable coding for HDR images (Hollywood, 2012), pp. 1–4
20. R Xu, SN Pattanaik, CE Hughes, High-dynamic-range still-image encoding in JPEG 2000. IEEE Comput. Graphics Appl. **25**(6), 57–64 (2005)
21. CY Ping, M Iwahashi, H Kiya, in *Proc. Int. Workshop Adv. Image Technol.* Lossless bit depth scalable coding for floating point images (Nagoya, 2013), pp. 169–174
22. JF Blinn, Floating-point tricks. IEEE Comput. Graph. Appl. **17**(4), 80–84 (1997)
23. M Iwahashi, H Kiya, in *Proc. Int. Conf. Acoust. Speech Signal Process*. Two layer lossless coding of HDR images (Vancouver, 2014), pp. 1340–1344
24. ML Pendu, C Guillemot, D Thoreau, in *Proc. Int. Conf. Acoust. Speech Signal Process*. Adaptive re-quantization for high dynamic range video compression (Vancouver, 2014), pp. 7367–7371

11

Some advanced parametric methods for assessing waveform distortion in a smart grid with renewable generation

Luisa Alfieri

Abstract

Power quality (PQ) disturbances are becoming an important issue in smart grids (SGs) due to the significant economic consequences that they can generate on sensible loads. However, SGs include several distributed energy resources (DERs) that can be interconnected to the grid with static converters, which lead to a reduction of the PQ levels. Among DERs, wind turbines and photovoltaic systems are expected to be used extensively due to the forecasted reduction in investment costs and other economic incentives. These systems can introduce significant time-varying voltage and current waveform distortions that require advanced spectral analysis methods to be used. This paper provides an application of advanced parametric methods for assessing waveform distortions in SGs with dispersed generation. In particular, the Standard International Electrotechnical Committee (IEC) method, some parametric methods (such as Prony and Estimation of Signal Parameters by Rotational Invariance Technique (ESPRIT)), and some hybrid methods are critically compared on the basis of their accuracy and the computational effort required.

Keywords: Smart grid; Dispersed generation; Power quality; Waveform distortion; DFT; Parametric methods

1 Introduction

Currently, significant modifications are taking place in distribution systems as they move toward the future use of smart grids (SGs). The main objectives of SGs are the efficient use of energy, the reduction of losses in the system, improvement in the power quality (PQ), and to encourage the use of distributed generation, in particular renewable energy sources [1-4]. In this context, the increasing use of controllable and non-linear loads and the new needs of liberalized markets impose new PQ requirements in order to avoid dangerous effects [5-7].

Among the disturbances mentioned above, the distortions of the voltage and current waveforms are of great interest, and they have been discussed extensively in the literature. Both types of distortions are due mainly to the extensive use of electronic power converters to supply loads and to connect dispersed generation (DG) units or electrical storage systems to the grid.

Photovoltaic systems (PVSs) and wind turbine systems (WTSs) are the fastest growing systems for meeting the requirements of dispersed generation. Their growth is due to the expected cost reduction and to active government policies in many countries that have encouraged their use in power grids in the last few decades. These DG units can be interfaced to the grid through either partially rated power converters or full-scale power electronic devices that can inject harmonic currents that cause voltage distortions [6-8]. Therefore, there is a pressing need to study the impacts of such DG systems on waveform distortions in distribution networks [9]. Thus, the main objective of this paper is to apply some of the advanced methods for the assessment of the waveform distortions that are caused by different configurations of PVSs and WTSs [10-20].

PVSs and WTSs generate distorted, time-varying waveforms. As a result, great attention is required to evaluate the PQ indices to acquire information that must be deduced from the analyses of the spectral components of the waveforms and their locations as a function of time [5,21].

The International Electrotechnical Committee's (IEC) standard recommendations for signal processing use the Discrete Fourier Transform (DFT) over successive, rectangular time windows with the time duration set to 10

Correspondence: luisa.alfieri@unina.it
Department of Electrical Engineering and Information Technology, University of Naples Federico II, Via Claudio, 21, 80125 Napoli, Italy

or 12 cycles of the fundamental period for 50-Hz systems or 60-Hz systems, respectively, to evaluate the spectral components [22,23]. However, even though the standard method can provide a global quantification of the waveform distortion, it has limited ability to obtain more detailed information in the analysis of a single spectral component. This is due to some well-known problems that characterize the DFT, i.e., the spectral leakage that arises when the duration of the time window is not correctly synchronized with the fundamental period of the power system [5].

Many solutions have been proposed in the relevant literature to overcome the spectral leakage problems by using DFT-based methods or parametric methods, such as Prony and Estimation of Signal Parameters by Rotational Invariance Technique (ESPRIT) [24-30]. In particular, in [27,28], the Sliding-Window Prony method and the Sliding-Window ESPRIT method were used to provide an accurate estimation of both the harmonic and interharmonic components with high-frequency resolution, but the computational burden was excessive. Other authors have proposed hybrid methods that include DFT and parametric methods to analyze the different frequency bands of a signal separately to reduce the computational burden and provide acceptable accuracy [29,30].

Recently, some authors [31,32] proposed two new, modified sliding-window parametric methods based on modifications of the Prony and ESPRIT signal models. In fact, these methods are based on the observation of the reduced time variability of the frequencies of the spectral components. Then, the frequency values of the spectral components were assumed to be constant over time, thereby reducing the number of unknown parameters of the signal model and the dimension of related equation systems to be solved.

The main aims of this paper can be summarized as follows:

- To provide a depth application of some methods for the assessment of waveform distortion caused by PVS and WTS generators. In particular, the DFT method, parametric methods (Prony and ESPRIT), hybrid methods, and modified sliding-window parametric methods are presented and critically compared on the basis of the accuracy of the results they produce and their computational burdens, taking into account both test and measured waveforms.
- To explore and compare the theoretical waveform distortions introduced by PVS and WTS generators and the distortions detected by the considered methods in some current waveforms measured at the point of common coupling of PVS and WTS.

The paper is organized as follows. Section 1.1 describes waveform distortions in detail that result from the different configurations of the PVS and the WTS schemes. In Section 1.2, the Sliding-Window ESPRIT and Prony methods are exposed, and in Section 1.3, detailed descriptions of the sliding-window hybrid methods and the sliding-window modified parametric methods are presented. In Section 1.4, we describe several case studies that were based on synthetic and measured waveforms of PVSs and WTSs. This article's conclusions are given in Section 2, and the DFT method is discussed in Appendix A.

1.1 Waveform distortion in PV systems and WT systems

It is well known that solar energy and wind energy currently are the most diffuse sources of renewable energy. The following sub-sections provide an overview of the waveform distortions caused by the most common PVS and WTS configurations. The overview deals with the primary spectral emissions that result from disturbances introduced by the specific nature of the system that is being considered, and it also deals with the secondary spectral emissions that are due to disturbances caused by other sources near the system, such as non-linear loads and power communication signals [18]. Note that the distortions introduced by the PVSs and WTSs generally correspond to spectral components that are included in a wide range of frequencies.

For the sake of clarity, in the following, the spectral components up to 2 kHz are referred as 'low-frequency components', and the spectral components over 2 kHz are referred as 'high-frequency components'.

1.1.1 Photovoltaic systems

Photovoltaic panels are connected to the grid through an inverter that converts DC to AC and ensures that particular specifications for voltage and frequency are met and that other important tasks are performed, such as maximum power point tracking (MPPT) [10].

Three-phase or single-phase inverters are used, with the former ensuring no zero-sequence emission in the spectra of the waveforms. Multi-string inverters also are available to obtain the combined benefit of the previous configurations [19].

Figure 1 shows the most diffuse topology solutions for the inverters, i.e., (i) inverters with line frequency isolation transformers (Figure 1a), (ii) inverters with high-frequency isolation transformers (Figure 1b), and (iii) inverters without isolation transformers (Figure 1c) [10,19].

Note that the different topologies and operating conditions of plants produce different waveform distortions. However, in general, it has been observed that every PVS has a low level of distorting spectral components

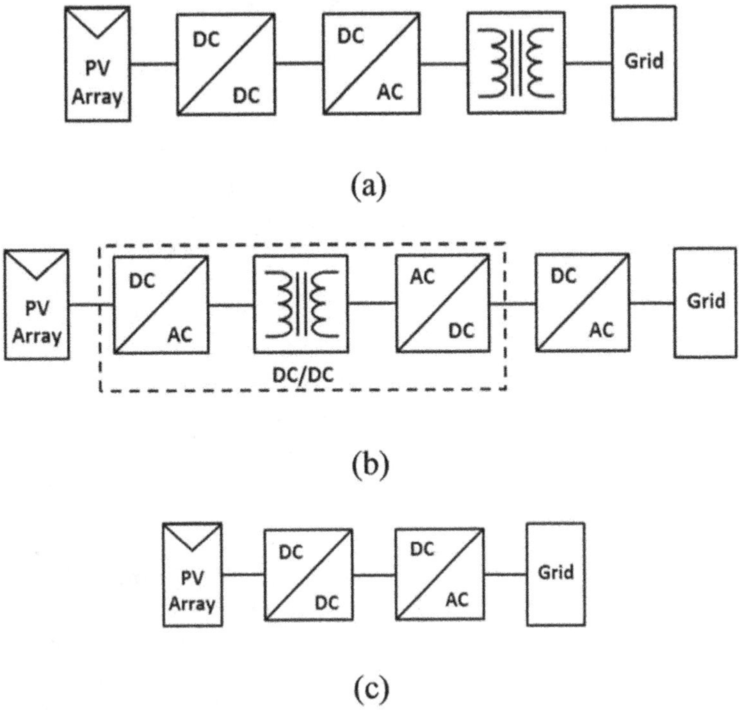

Figure 1 Schemes of single-phase PVSs. (a) Inverter with line-frequency isolation transformer; **(b)** inverter with high-frequency isolation transformer; **(c)** inverter without an isolation transformer.

that increases slightly as the amount of power being produced increases.

Practical experience has indicated that PVSs constituted by multiple inverters produce lower levels of emissions than PVSs that have a unique, large-size inverter [20]. The spectral emissions of PVSs at the point of common coupling (PCC) determine both the low- and high-frequency spectral components.

The low-frequency emissions can be due to both background distortion and over-modulation by the inverter. The amplitudes of these low-frequency spectral components in a single photovoltaic plant are generally characterized by a current total harmonic distortion (THDi) smaller than 10%, but in resonance conditions, significant voltage distortions and problems for the electric network have been documented [19].

The high-frequency emissions basically are due to the PVS inverter and they are specifically due to the pulse-width modulation (PWM) technique that is used. These emissions are always present during the power production of the system, and they are null when the inverter is turned off; moreover, for the current waveform of a single photovoltaic plant in ideal operating conditions, the amplitudes of these high-frequency spectral components are higher than those of the low-frequency spectral components [18,19]. These spectral components depend on the type of inverter and on its switching frequency, and they are mostly harmonics that appear as sidebands that are centered around integer multiples of the switching frequency [19,33]. Since the switching frequency is generally in a range between 10 and 20 kHz, actually, there still are no adequate and consolidated standards for these high-frequency emissions. Currently, however, the increasing number of spectral components above 2 kHz has resulted in an increased intensity of research activities on this issue [19]. Note that, also at high-frequency, there are often spectral components due to the background voltage, which could become relevant for the series resonance effect.

1.1.2 Wind turbine systems

At the current time, the main large wind turbine schemes are (i) the fixed-speed wind turbine system, (ii) the doubly fed induction generator (DFIG), and (iii) the full-converter wind turbine generator [17,34,35]. Figure 2 shows the block schemes of the three configurations.

Figure 2a shows a fixed-speed wind turbine system in which a squirrel-cage induction generator (SCIG) works in a speed-limited range above the synchronous speed. An electronic soft starter generally is provided to avoid the increase of current during the start-up phase. Also, a bank of capacitors is used to compensate for the reactive power.

The spectral emissions of this scheme primarily are caused by the action of the soft starter, which introduces odd harmonic components at low frequency. Triple

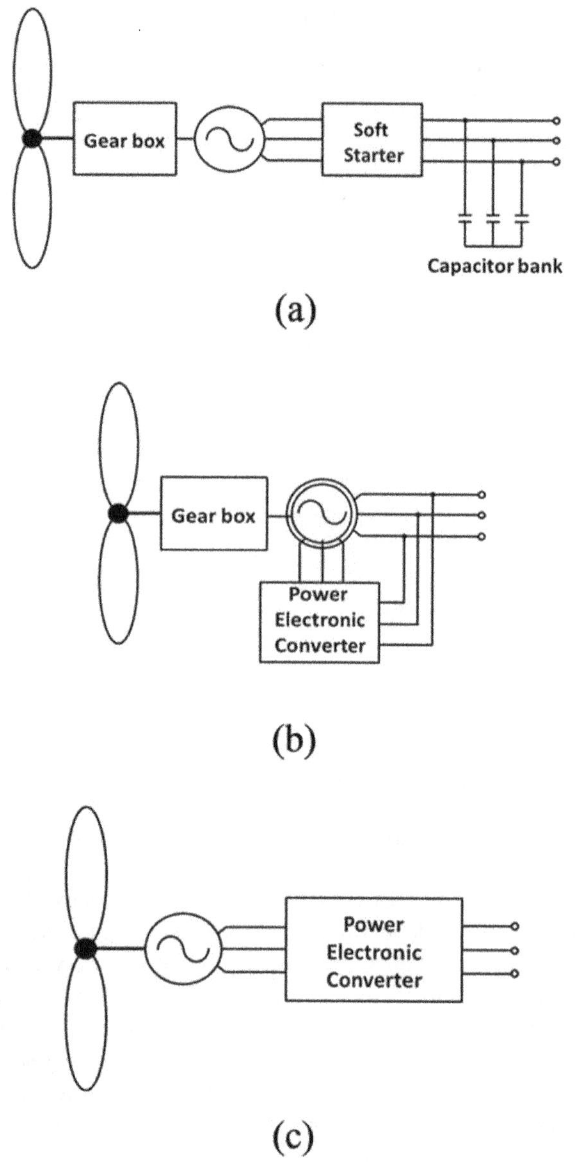

Figure 2 Scheme of the most-diffuse WTSs. (a) Fixed-speed wind turbine; **(b)** DFIG; **(c)** full-converter wind turbine generator.

components are usually due to voltage unbalances. Generally, for this configuration, the most significant components detected at the PCC are the 3rd, 5th, 7th, 9th, 11th, and 13th harmonic orders [17].

In a DFIG, the windings of the stator are connected directly to the grid, whereas, on the rotor circuit, there is a back-to-back, partial-scale static converter with a rated power that is roughly equal to 30% of the generator's power.

For this scheme, the spectral emissions in the grid are mainly due to the static converter. Three main types of spectral emissions can arise at the PCC, i.e., (i) inherent components, (ii) switching spectral components, and (iii) spectral components due to the unbalance conditions and/or background voltages [17,36].

The first category includes mostly low-frequency components, and it is due to a non-sinusoidal air gap flux, which produces distorted voltages and currents at the frequencies $f_k = |6\,k(1 - s) \pm 1|f_{0,s}$, with $k = 1, 2, \ldots,$ and where s is the generator slip, and $f_{0,s}$ is the fundamental frequency of the stator voltage. These spectral components are time-varying with the speed of the DFIG rotor.

The spectral components that belong to the second class are due to both the rotor-side and grid-side converters and are linked to the PWM technique that was used. They are mostly high-frequency spectral components, but in over-modulation conditions, low-frequency components also can be detected. Under ideal conditions, the aforesaid harmonics and interharmonics are as follows: (i) for the grid side, in correspondence with the frequencies $f_{k,m}^{\mathrm{PWM},g} = \left|kf_{sw,g} \pm mf_{0,s}\right|$, with $k, m = 1, 2, \ldots$ and where $f_{0,s}$ is the nominal frequency of the system, and $f_{\mathrm{sw},g}$ is the PWM switching frequency on the grid side; (ii) for the rotor side, in correspondence with the frequencies $f_{k,m}^{\mathrm{PWM},r} = \left|kf_{sw,r} \pm mf_{0,r}\right|$, with $k, m = 1, 2, \ldots$ and where $f_{0,r} = sf_{0,s}$ is the fundamental frequency of the waveform to generate for the rotor winding, and $f_{\mathrm{sw},r}$ is the PWM switching frequency on the rotor side.

Note that the rotor-side converter works at a frequency that is linked to the generator slip and that the aforesaid spectral components could be shifted because of the air-gap coupling between the rotor current and the stator circuit [36].

The third type of spectral emissions is the spectral components that are introduced by not-ideal operating conditions and by the WTS's auxiliary, non-linear loads, such as controllers and motors [12,17,36].

Figure 2c shows the full-converter wind-turbine generator scheme, which consists of a permanent-magnet synchronous generator connected to the grid by means of a full-scale, power electronic converter with a rated power that is 110% of the generator's power.

Also in this scheme, the emission of spectral components is caused mainly by the static converter, but unlike the previous case, there is no direct influence on the air-gap flux, so the spectral content of the waveforms at the PCC is less. The most significant components are harmonics of $6\,k \pm 1$ order (5th, 7th, 11th, 13th,...) at low frequency, typical of a six-pulse, three-phase bridge. However, for the high frequency, the typical spectral components of the PWM technique are detected. The Unbalanced condition and background voltage can introduce additional distortions [12-17,34,35].

WTS spectral emissions appear to vary significantly when they are observed over a large period of time; this is because they depend on wind conditions and, as a result, on electrical production. In addition, the harmonic emissions increase as the amount of power generated increases.

1.2 Basic parametric methods for the assessment of time-varying waveform distortion

This section deals with two of the most popular parametric methods used to assess waveform distortion in power systems, i.e., the ESPRIT and the Prony methods, which are used as the basis of the other hybrid methods presented in Section 1.3.

1.2.1 The SW ESPRIT method

The ESPRIT method is one of the most well-known subspace methods that model waveform samples by means a linear combination of M complex exponentials added to white noise $r(n)$. In more detail, a given sequence of sampled data $x(n)$ of size N is approximated by [37]:

$$\hat{x}(n) = \sum_{k=1}^{M} A_k e^{j\psi_k} e^{(\alpha_k + j2\pi f_k) n T_s} + r(n), \quad n = 0, 1, \ldots, N-1 \tag{1}$$

where T_s is the sampling time, and A_k, ψ_k, f_k, and α_k are the amplitude, the initial phase, the frequency, and the damping factor of the kth complex exponential, respectively, which are the unknown parameters to be evaluated.

Model (1) can be written in a matrix form as:

$$\hat{x} = V\Phi^n H + r, \tag{2}$$

where:

$$\hat{x} = [\hat{x}(n) \ldots \hat{x}(n+N_1-1)]^T, H = [h_1 \ldots h_M]^T,$$

$$r = [r(n) \ldots r(n+N_1-1)]^T,$$

$$V = \begin{bmatrix} 1 & 1 & \cdots & 1 \\ e^{(\alpha_1+j2\pi f_1)T_s} & e^{(\alpha_2+j2\pi f_2)T_s} & \cdots & e^{(\alpha_M+j2\pi f_M)T_s} \\ \vdots & \vdots & \ddots & \vdots \\ e^{(\alpha_1+j2\pi f_1)(N_1-1)T_s} & e^{(\alpha_2+j2\pi f_2)(N_1-1)T_s} & \cdots & e^{(\alpha_M+j2\pi f_M)(N_1-1)T_s} \end{bmatrix},$$

$$\Phi = \begin{bmatrix} e^{(\alpha_1+j2\pi f_1)T_s} & 0 & \cdots & 0 \\ 0 & e^{(\alpha_2+j2\pi f_2)T_s} & \cdots & 0 \\ \vdots & \vdots & \ddots & \vdots \\ 0 & 0 & \cdots & e^{(\alpha_M+j2\pi f_M)T_s} \end{bmatrix},$$

$h_k = A_k e^{j\psi_k}$ and with $N_1 < N$ the selected order of the correlation matrix. The symbol Φ represents the rotation matrix, and its properties can be used to determine a solution for the system (2). In particular, problem (2) can be solved by introducing the matrix $\hat{S}$ of the eigenvector associated with the first M eigenvalues of the correlation matrix R_x and a matrix Ψ that has the same eigenvalues of Φ and verifies the relationship:

$$\hat{S}_2 = \hat{S}_1 \Psi, \tag{3}$$

where $\hat{S}_1 = [I_{N-1} \quad 0]\hat{S}, \hat{S}_2 = [0 \quad I_{N-1}]\hat{S}$ and I_{N-1} is an identity matrix of order $N-1$. Then, by using the least squares approach, the matrix Ψ can be estimated as:

$$\hat{\Psi} = \left(\hat{S}_1^* \hat{S}_1\right)^{-1} \hat{S}_1^* \hat{S}_2 \tag{4}$$

Note that the eigenvalues $\hat{\lambda}_i$ of the matrix Ψ are the elements of the main diagonal of Φ, so $\hat{\lambda}_i = e^{(\alpha_i + j2\pi f_i)T_s}$, and the unknown damping factors and frequencies can be computed as real and imaginary parts, respectively, of the natural logarithm of these eigenvalues. For the amplitudes, it is necessary to use another least squares approach in which the theoretical definition of the correlation matrix R_x [28] is:

$$R_x = VAV^H + \sigma_w^2 I, \tag{5}$$

where A is the diagonal matrix of the squares of the unknown amplitudes, I is an identity matrix, and σ_w^2 is the variance of the white noise. In this way, by assuming that the noise r is negligible, the initial phases also can be evaluated by making appropriate substitutions in Equation 2 and solving the modified equation.

If the analyzed waveform is time-varying, the Sliding-Window ESPRIT is used with a window that slides forward successively over time in order to obtain the time-dependent estimates of the parameters of the ESPRIT model [5,28].

The reliability of the results and the computational burden of the ESPRIT method are dependent significantly on the number of exponentials M, the order N of the correlation matrix, and the sampling frequency [38].

1.2.2 The SW Prony method

The Prony method is another parametric method for spectral analysis. This method models the waveform samples by means of a linear combination of M complex exponentials. Specifically, a given sequence of sampled data $x(n)$ of size N is approximated by [5]:

$$x(n) = \sum_{k=1}^{M} h_k z_k^n \quad n = 0, 1, \ldots, N-1, \tag{6}$$

where $h_k = A_k e^{j\psi_k}$ and $z_k = e^{(\alpha_k + j\omega_k)T_s}$. To achieve this estimation of the unknown parameters in a traditional way, a severely non-linear problem should be solved [5], but the Prony's intuition allows another approach to be used, i.e., the initial problem is divided into two systems of linear equations, which can be solved easily.

In the first step of the aforesaid approach, the following system of linear equations is solved to determine the damping factors and the frequencies:

$$\sum_{m=0}^{M} a(m)x(n-m) = 0, n = M, M+1, \ldots, N-1 \tag{7}$$

System (7) is comprised of $(N-M)$ linear equations, and the coefficients $a(m)$ are the M unknowns to be computed. By imposing $a(0) = 1$, system (7) can be written in matrix form as:

$$\begin{bmatrix} x(M) & x(M-1) & \dots & x(1) \\ x(M+1) & x(M) & \dots & x(2) \\ \vdots & \vdots & \vdots & \vdots \\ x(N-1) & x(N-2) & \dots & x(N-M) \end{bmatrix} \cdot \begin{bmatrix} a(1) \\ a(2) \\ \vdots \\ a(M) \end{bmatrix} = - \begin{bmatrix} x(M+1) \\ x(M+2) \\ \vdots \\ x(N) \end{bmatrix} \tag{8}$$

After the coefficients $a(m)$ have been determined, the polynomial $F(z)$ can be obtained:

$$F(z) = \sum_{m=0}^{M} a(m) z^{M-m} \tag{9}$$

The roots z_k of $F(z)$ are used to calculate the damping factors and the frequencies of each exponential by means of simple relationships.

The second step provides the amplitude and phase of each exponential by calculating the h_k. This is possible by replacing the obtained z_k in system (6) and, so, by solving another system of linear equations that in matrix form is [5]:

$$\begin{bmatrix} z_1^0 & z_2^0 & \dots & z_M^0 \\ z_1^1 & z_2^1 & \dots & z_M^1 \\ \vdots & \vdots & \vdots & \vdots \\ z_1^{M-1} & z_2^{M-1} & \dots & z_M^{M-1} \end{bmatrix} \cdot \begin{bmatrix} h_1 \\ h_2 \\ \vdots \\ h_M \end{bmatrix} = \begin{bmatrix} x(1) \\ x(2) \\ \vdots \\ x(M) \end{bmatrix} \tag{10}$$

If the analyzed waveform is time-varying, the Sliding-Window Prony is used once again with sliding windows in order to obtain the time-dependent estimates of the parameters of the Prony model [5,27]. In [27], an adaptive technique also was proposed in order to achieve an optimal and adaptive duration of the sliding window. The accuracy and computational burden of the Prony method depend significantly on the number of exponentials M, the number N of samples for analysis window, and the sampling frequency [38].

1.3 Advanced parametric methods for the assessment of time-varying waveform distortion

In this section, two types of advanced parametric methods for spectral analysis are analyzed: the Three-step Sliding-Window Hybrid method and the Sliding-Window Modified Parametric method. These methods are based on the parametric methods considered in Section 1.2 and are able to provide both accurate results and reduced computational burden.

1.3.1 The Three-step sliding-window hybrid method

The Three-step Sliding-Window Hybrid method is a joint parametric-DFT scheme that was developed in three stages in which either the sliding-window (SW) ESPRIT, the SW Prony, or the SW DFT can be used alternatively to estimate (i) the fundamental and the interharmonic components in the 0 to 100 Hz band, (ii) the harmonics, and (iii) the interharmonics at frequency $f > 100\,Hz$ of power system waveforms [29,30], reducing the typical problems that characterize the SW DFT and the SW parametric method (i.e., spectral leakage problems and high computational efforts). Figure 3 shows the block diagram of the three-step method.

In more detail, in the first step, the SW parametric method (Prony or ESPRIT) is applied to the output of a low-pass band filter of 0 to 200 Hz, in order to estimate the power system's fundamental and the interharmonics in the frequency range from 0 to 100 Hz. Then, the filtered waveform is approximated with a linear combination of exponentials according to either the (1) (SW ESPRIT) or (6) (SW Prony).

In the second step, the SW DFT is used to evaluate the harmonic components; in fact, there is also an estimation of the fundamental frequency among the outputs of the first step, so it is possible to set the duration of the DFT analysis window equal to an integer multiple of the fundamental period $\hat{T}_{\text{fund}}$, greatly limiting the spectral leakage.

As shown in Figure 3, the waveform analyzed by the synchronized DFT is obtained by subtracting from the original signal $x(n)$ the reconstructed waveform $\hat{x}_{\text{l.f.}}(n)$ that was obtained by the first step. The duration of the analysis window in this step can be set to 10 cycles

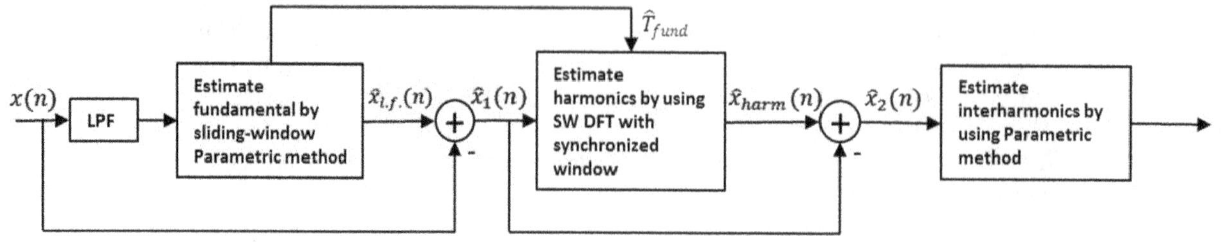

Figure 3 Three-step sliding-window method.

(12 cycles) of the fundamental period $\hat{T}_{\text{fund}}$, for 50-Hz systems (60-Hz systems), according to the IEC recommendations, but if it is required, it also is possible to select a different number of cycles [29]. We note that the aforesaid synchronization of the DFT time window guarantees a very accurate estimation of the harmonic components with a reduced computational effort, and this is attributable to the significant limitation of the spectral leakage due only to the possible presence of interharmonics at frequencies $f > 100\,Hz$ in $x_1(n)$, which yields relatively low errors [5,29].

The waveform $\hat{x}_{\text{harm}}(n)$ in Figure 3 was obtained summing the estimated harmonics and using the results to compute the residual $x_2(n)$ by subtracting $\hat{x}_{\text{harm}}(n)$ from $x_1(n)$. The waveform $x_2(n)$ is analyzed in the third step by the same parametric method used in the beginning of the scheme.

Note that, in the first and third steps, the number of exponentials required in the model was considerably lower than the number needed in the application of SW ESPRIT or SW Prony to the full band signal. This positive outcome resulted from the explanation of the decomposition of the waveform in different frequency ranges separately analyzed in the first and third steps. As a result, the Three-step SW method has a significantly lower computational burden than SW ESPRIT or SW Prony while still providing highly accurate results.

In order to improve the global performances of the Three-step SW method, each step could have a different duration of the analysis window, and then, the results could be processed and associated with a common time interval.

1.3.2 The sliding-window modified parametric method

The Sliding-Window Modified Parametric methods are based on a reduction of the unknown parameters in the signal model in the original parametric methods with obvious improvements in terms of computational efforts. In fact, starting from the consideration that the damping factors and the frequencies of spectral components in power system applications are slightly variable versus time [25,39-43], the estimation of frequencies is conducted only a few times, and the damping factors are assumed to be constant along the entire signal to be analyzed [31,32].

Specifically, in [32], the frequencies of the spectral components were computed initially by the analysis of the first sliding window, the *basis window*, and then, the values that were obtained were assumed to be constant with time, and they were imposed as known quantities in the successive sliding windows (*no-basis windows*). The damping factors were assumed to be null in the SW M-Prony and fixed to previously calculated values during the analysis of the '*basis window*' in the SW M-ESPRIT [32].

We note that, to elude masking effects due to significant variation of the frequencies, it is expected that the estimation of the frequencies would be repeated periodically. If the deviation between the values obtained was greater than a fixed threshold, the frequencies are updated and the analysis restarts with these new values. We also note that a different choice is required for the damping factors in the Prony and ESPRIT methods. The damping factors, in fact, are able to take into account temporal variations of the amplitudes of the spectral components and to link two consecutive sliding windows smoothly.

When the duration of the windows is very short (one to two times the fundamental period), the damping factors have very low values, and their contribution is negligible. This is the case for the SW Prony, so for this method, the damping factors are set equal to zero. However, for the SW ESPRIT, the duration of the analysis window is significantly larger (four to five times the fundamental period), so in this case, the damping factors are assumed to be constant for the no-basis windows and equal to the damping factors estimated for the basis window [32].

Figure 4 shows the block diagram of the Sliding-Window Modified Parametric methods.

In the first window (basis window for $k = 0$), a traditional parametric algorithm (TPA) is used to estimate all of the unknown parameters $\left(A_k^{\text{BW}}, f_k^{\text{BW}}, \psi_k^{\text{BW}}, \alpha_k^{\text{BW}}\right)$, which, then, are stored; in this window, the research for the optimal values for M and N_1 or N (respectively for the ESPRIT-based and Prony-based method) is effected, updating them if the reconstruction error overcomes a prefixed threshold. The same algorithm is also used after k_f windows, where another basis window is generated in order to avoid the masking effect. The choice of the value k_f is related to the particular nature of the analyzed waveform. In the *no-basis windows*, a modified parametric algorithm (MPA) is used only for the computation of the unknown amplitudes A_k and initial phases ψ_k of the spectral components, since the frequencies are equal to f_k^{BW}, and the damping factors are forced to be null or equal to α_k^{BW} for the Prony and ESPRIT methods, respectively. Specifically, the equations solved in the MPA when ESPRIT is used are as follows:

$$\text{i)}\left\{\begin{array}{l} \boldsymbol{R}_x = \boldsymbol{V}_{\text{BW}}\boldsymbol{A}\boldsymbol{V}_{\text{BW}}^H + \sigma_w^2\boldsymbol{I} \\ \hat{\boldsymbol{x}} = \boldsymbol{V}_{\text{BW}}\boldsymbol{\Phi}_{\text{BW}}^n\boldsymbol{H} \end{array}\right. \tag{11}$$

When Prony is used, the equation system solved in MPA is as follows:

$$\text{i)}\, x(n) = \sum\nolimits_{k=1}^{M} h_k\left(\hat{z}_k^{\text{BW}}\right)^n, \quad \text{n} = 0, 1, \ldots, \text{N}-1 \tag{12}$$

The assumption of constant frequencies and damping factors inside the no-basis windows decreases the number of unknown parameters to be estimated, and the

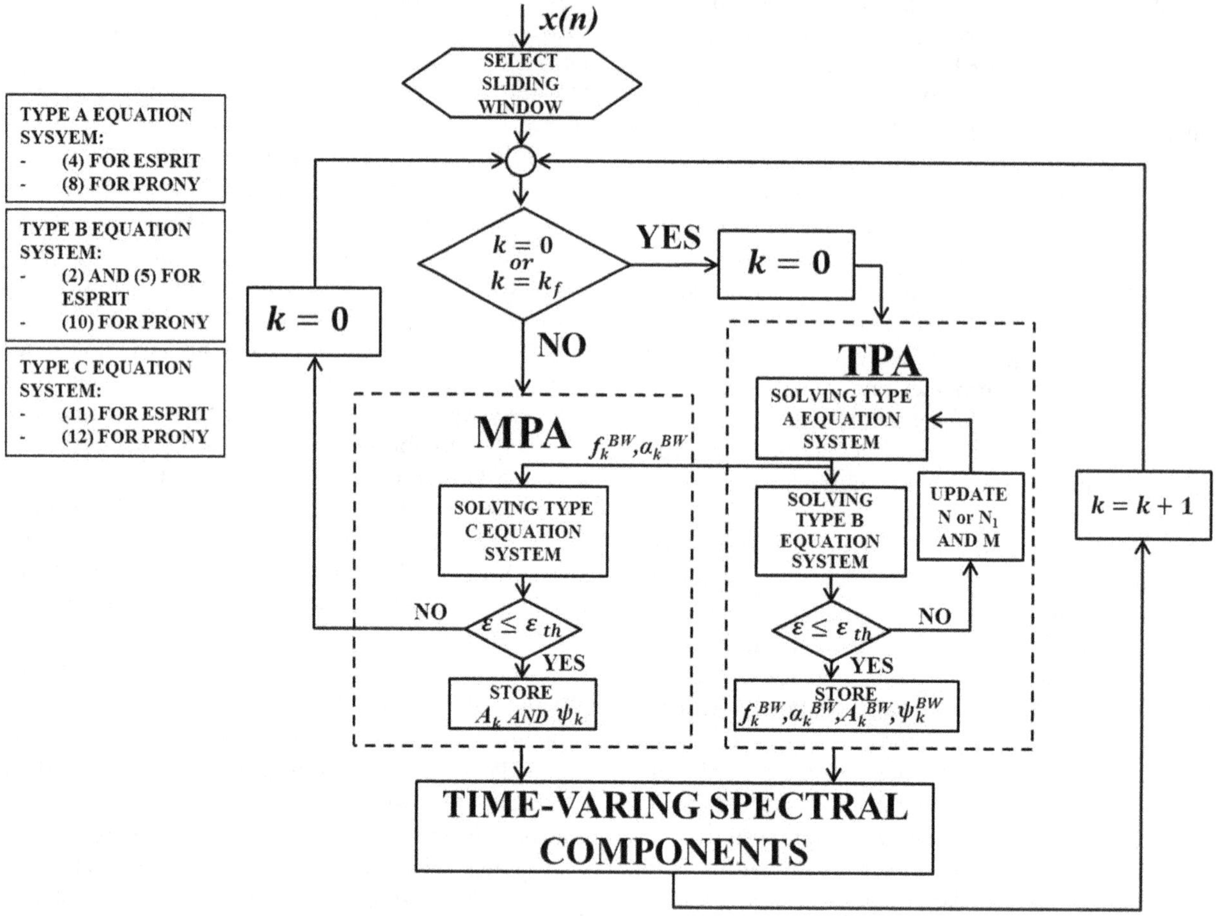

Figure 4 sliding-window modified parametric methods.

computational effort is reduced significantly compared to the classical Prony or ESPRIT approach [32].

1.4 Numerical applications

The basic and advanced parametric methods shown in Sections 1.2 and 1.3, respectively, were used to analyze the waveform distortion due to PVSs and WTSs, and the accuracy of their results and the computational burden of the methods were compared. Several numerical experiments were conducted, analyzing both synthetic and measured waveforms in many operating conditions of DG units, but for sake of brevity, only four case studies are reported in this section, and all of them are referred to the analysis of the current waveforms.

Specifically, the synthetic and measured currents of PVSs and WTSs were analyzed by using the following:

- the IEC standard method (IECM);
- the Sliding-Window ESPRIT method (SWEM);
- the Sliding-Window Prony method (SWPM);
- the Three-step Sliding-Window ESPRIT-DFT-ESPRIT method (SWEDEM);
- the Three-step Sliding-Window Prony-DFT-Prony method (SWPDPM);
- the Sliding-Window Modified ESPRIT method (SWMEM);
- the Sliding-Window Modified Prony method (SWMPM).

The results of each case study were referred to the same duration of signal, and the IECM computational burden was considered as a reference for the other methods. It is important to emphasize that, for the IECM, since its frequency resolution was fixed at 5 Hz, the fundamental component was always detected in correspondence with 50 Hz, so the harmonic components are estimated also as multiples of 50 Hz.

All of the programs were conducted in MATLAB, and they were not optimized for computational speed because we were interested only in obtaining a rough and relative quantification of the efficiency of the different methods. The MATLAB programs were developed and tested on a Windows PC with an Intel i7-3770 3.4 GHz and 16 GB of RAM.

1.4.1 Case study 1: test signal of a photovoltaic system

The synthetic 3-s waveform emulated a current at the PCC of a PVS that had a full-bridge, unipolar inverter. It was a sort of an 'acid test,' since our aim was to observe the behavior of each of the methods in the analysis of a waveform with a very wide spectrum, i.e., exceeding a frequency of 20 kHz. Specifically, the waveform was assembled assuming a frequency modulation index m_f equal to 200 for the inverter PWM technique and the presence of all odd harmonics up to the 27th order for the low-frequency components. The fundamental component was fixed at 50.02 Hz, and it had an amplitude of 9 A. Also, a white noise with a standard deviation of 0.001 was added to the aforesaid components.

The sampling rate was 100 kHz in order to provide the most appropriate operating conditions for the parametric methods that were used so that they could provide estimates of the spectral components around the order $2m_f$, which are the most significant introduced by the aforesaid type of PWM and whose amplitudes were fixed up to 12% of the fundamental[a]. For the parametric methods, the error threshold was fixed equal to 10^{-5}, and for all of the methods, the window of analysis slides of 0.04 s was used. In the second steps of both SWPDPM and SWEDEM, the analysis window was fixed equal to 5 cycles of the fundamental period.

Table 1 shows the average percentage errors in the estimates of the frequencies and amplitudes of five spectral components, particularly interesting in the comparison of the methods that were used. Specifically, the components that were examined were the fundamental and the harmonics of order 3rd, 11th, 401st, and 405th, with amplitudes equal to 1.6%, 6%, 12%, and 3% of the fundamental amplitude, respectively.

It was evident that SWEM, SWPM, SWMEM, and SWMPM always provided negligible percentage errors, with values almost of the same order of magnitude. As the first part of Table 1 shows, the percentage errors obtained using the SWEDEM and the SWPDEPM also were very low; however, the second part of Table 1 shows SWEDEM and SWPDEPM with higher errors for the amplitude than with the other basic and advanced parametric methods.

This behavior of the three-step methods in estimating the amplitudes of the fundamental and third harmonic was due to the presence of the low-pass band filter of 0 to 200 Hz, which introduces a little attenuation in the amplitude of the components in the frequency band of 0 to 100 Hz and slight distortions for the components near the filter cut-off frequency. However, it is important to emphasize that, for the other harmonic components (far from the filter cut-off frequency) both at low and high frequency, the errors in the amplitude were practically negligible. Table 1 shows that the IECM errors were globally higher, both in terms of frequency and amplitude, and that, for the high-frequency components, the average percentage error in the estimation of amplitudes increased, reaching values greater than 90%.

This was due to the desynchronization between the duration of the time window and the fundamental period,

Table 1 Case study 1: average percentage estimation errors

	Fundamental	3rd	11th	401st	405th
(a) Average errors by frequency [%]					
SWEM	1.08×10^{-4}	0.0028	1.74×10^{-4}	2.93×10^{-6}	1.15×10^{-5}
SWPM	5.77×10^{-5}	0.0012	8.47×10^{-5}	1.18×10^{-6}	4.81×10^{-6}
SWEDEM	6.86×10^{-4}	1.58×10^{-5}	1.58×10^{-5}	1.58×10^{-5}	1.58×10^{-5}
SWPDPM	1.55×10^{-5}	1.60×10^{-5}	1.60×10^{-5}	1.60×10^{-5}	1.60×10^{-5}
SWMEM	1.12×10^{-4}	0.0023	1.50×10^{-4}	3.01×10^{-6}	1.89×10^{-6}
SWMPM	4.43×10^{-5}	0.0013	2.21×10^{-5}	2.27×10^{-6}	5.54×10^{-7}
IECM	0.04	0.04	0.04	0.04	0.04
(b) Average errors of amplitude [%]					
SWEM	9.85×10^{-4}	0.067	0.016	0.0081	0.031
SWPM	4.29×10^{-4}	0.026	0.0078	0.0037	0.017
SWEDEM	0.11	0.09	0.064	0.073	0.070
SWPDPM	0.11	0.66	0.073	0.074	0.071
SWMEM	5.27×10^{-4}	0.019	0.0056	0.0024	0.014
SWMPM	3.86×10^{-4}	0.023	0.0065	0.0032	0.014
IECM	0.060	0.70	0.28	82.31	91.51

Average percentage estimation errors of (a) frequencies and (b) amplitudes of the fundamental, 3rd, 11th, 401st, and 405th harmonic components by means of SWEM, SWPM, SWEDEM, SWPDEPM, SWMEM, SWMPM, and IECM.

which produces spectral leakage problems that increase as the harmonic order increases. Table 2 shows the computational times obtained by using all of the methods to analyze the 3-s waveform per unit of computational time required by IECM. It is evident that the basic parametric methods require greater computational time in order to provide accurate results; however, the advanced parametric methods reduced the computational effort significantly while simultaneously providing highly accurate results. Note that the SWEM (SWMEM) required less computational effort than the SWPM (SWPEM); this was due to under-sampling the signal before analysis [44].

Tables 3 and 4 are provided to make it clearer what occurred in terms of the accuracy and computational efforts when the sampling rate in SWEM and SWMEM remained at 100 kHz. Table 3 shows that there was an irrelevant gain in the accuracy of the results compared to the values shown in Table 1.

However, the computational burdens of SWEM and SWMEM (Table 3) were considerably greater than those observed for the same methods in Table 2. This proves that the optimal sampling rate for each parametric method guarantees the best performance with respect to accuracy and computational time. It also proves that the accuracy of the results was not affected by exceeding the optimal sampling rate, but the computational effort increased significantly, especially at the very high-sampling rates.

Table 2 shows that the computational time of SWEDEM was greater than that of SWPDPM. This was because the presence of the filter in the first step resulted in some negligible spectral components at low frequency that made it impossible to reduce the signal sampling rate in the first step of SWEDEM and maintain an acceptable accuracy.

It was interesting to observe that, in this case study, the advanced parametric methods required computational times that were, in the worst case (i.e., the SWEDEM and the SWMPM), an order of magnitude greater than that needed by the IECM. SWPDPM and SWMEM had computational times that were the same order of magnitude as that of the IECM, although these methods provided better results than the method recommended by the IEC standards.

Table 2 Case study 1: computational time of all of the methods

	Computational time [p.u.]
SWEM	45.56
SWPM	618.03
SWEDEM	12.86
SWPDPM	4.24
SWMEM	2.87
SWMPM	10.27
IECM	1

The values are per unit of computational time required by IECM.

1.4.2 Case study 2: measurement of the current of the photovoltaic system

A 1-s current waveform measured at the PCC of a 10-kW, three-phase inverter without an isolation transformer, which, together with another twin-inverter, is included in a PVS. The sampling rate was 10 kHz, but in order to provide better operating conditions for the detection of spectral components by the parametric methods that were used, a resampling to 20 kHz was used [44].

The error threshold for the parametric methods was fixed at 10^{-3}, and the window of analysis slides of 0.04 s was used for all of the methods. In the second steps of the SWPDPM and the SWEDEM, the analysis window was fixed equal to 5 cycles of the fundamental period. Figure 5 shows the time trend of the measured current.

All of the methods that were used detected mainly low-frequency spectral components. In fact, in the frequency band from 5 to 10 kHz, the advanced methods that used the ESPRIT method and the DFT method detected the same components, but the values of amplitude were lower than the 0.05% of the fundamental, so they are not discussed here.

Table 5 shows the average values of frequency and amplitude of some spectral components detected by all of the methods that were used. Among the most significant spectral components, Table 5 reports the results related to the fundamental, a harmonic at about 1,000 Hz, and an interharmonic at about 2,915 Hz. They are good representatives that provide a perception of the behaviors of the methods that were used in the evaluation of the various types of spectral components.

First, as expected, it was evident that both harmonic and interharmonic components had low amplitudes and that each method was able to estimate them adequately, since the values that were obtained for both frequency and amplitude were in a narrow range.

The IECM provided an amplitude value of about 1 order of magnitude lower than the other methods only for the interharmonic detection, although the component was almost a multiple of 5 Hz (the IECM frequency resolution). This phenomenon was compatible with the observation in the previous case study; in fact, it confirmed that spectral leakage increases at high frequencies.

Table 6 shows the computational time, per unit of the computational time required by IECM, obtained by analyzing the current waveform by all of the methods. The advanced parametric methods always required less computational time than the basic parametric methods. In particular, all of the advanced parametric methods required a computational time that was 2 orders of magnitude less than those of SWEM and SWPM.

Table 3 Case study 1: average percentage estimation errors

	Fundamental	3rd	11th	401st	405th
(a) Average errors of frequency [%]					
SWEM	7.80×10^{-5}	0.0013	1.10×10^{-4}	1.53×10^{-6}	6.02×10^{-6}
SWMEM	6.07×10^{-5}	7.39×10^{-4}	9.75×10^{-5}	3.22×10^{-6}	2.29×10^{-7}
(b) Average errors of amplitude [%]					
SWEM	5.30×10^{-4}	0.026	0.0092	0.0046	0.018
SWMEM	2.58×10^{-4}	0.029	0.0036	0.0034	0.014

Average percentage estimation errors of (a) frequency and (b) amplitude of spectral components detected by SWEM and SWMEM.

Also, in this case, the SWMEM was the method that came the closest to matching the performance of IECM with respect to computational time. Note that, for the analysis of the measured current, all of the advanced parametric methods required computational times that were only 1 order of magnitude greater than that required by IECM.

1.4.3 Case study 3: test signal of wind turbine system

The synthetic 1-s waveform emulated a current at the PCC of a WTS that had a doubly fed induction generator. Also, this case study was a sort of 'acid test,' since our aim was to observe the behavior of each of the methods in the analysis of a waveform with a spectrum that exceeded a frequency of 2 kHz and that had embedded successive harmonic and interharmonic components.

The waveform was assembled assuming a frequency modulation index m_f of 40, and the fundamental frequency and amplitude were fixed at 50.02 Hz and 5.4 A, respectively. Figure 6 shows the low-frequency and the high-frequency spectrum of the test signal. White noise with a standard deviation of 0.001 was added to the signal in order to make the waveform more realistic and to stress the performance of the spectral analysis methods that were used.

The sampling rate was 20 kHz. For the parametric methods, the error threshold was fixed at 5×10^{-8}, and for all of the methods, the window of analysis slides was 0.04 s. In the second steps of both the SWPDPM and the SWEDEM, the analysis window was equal to 10 cycles of the fundamental period.

Table 7 shows the average percentage errors in the estimates of the frequency and amplitude of the five spectral components, which are particularly interesting for comparing the methods that were used. Specifically, the components that were examined were the fundamental, the 5th, and the 38th order of harmonic and two interharmonics at 74.79 and 382.35 Hz.

Table 4 Case study 1: computational time for the oversampled waveform

	Computational time [p.u.]
SWEM	107.01
SWMEM	58.57

Relative computational time in p.u. of IECM time, when the analyzed waveform was over-sampled at 100 kHz.

In this case study, it was observed that the SWPM, in order to provide reasonably accurate results, requires a very high number of exponentials M, and as a result, it also requires a high value of N for the optimal window of analysis. This is due to the contemporaneous presence of noise interference and many small spectral components in the signal. In fact, since the Prony signal model does not include noise, Prony-based methods require an increasing number of M in the presence of so many small spectral components and noise. Basically, the SWPM adds many spectral components to the real spectrum, since it also must individuate the noise spectrum to approach the analyzed waveform adequately.

However, it is easier for the SWEM to detect the spectrum, since noise is accounted for in its model.

However, as shown in Table 7, both the SWPM and the SWEM provide better results in terms of accuracy, with average percentage errors of the same order of magnitude.

Obviously, noise interference also was a problem for the SWMPM, which produced less accurate results than the SWMEM; however, it performed well in detecting the spectral components, with errors of the same order of magnitude as the basic parametric methods (error of less than 0.021% for the estimation of frequency and less than 0.9% for the estimation of the amplitude). The second part of Table 7 shows that the SWPDPM and the SWEDEM had some difficulties in estimating the amplitudes of the components up to the fifth harmonic; as underlined in case study 1, this was due to the effect of the low-pass band filter. However, the first part of Table 7 shows that the average percentage errors for these methods were comparable to the best results obtained by the basic methods.

However, in the estimations of both the frequency and the amplitude, the IECM had the largest errors, especially for the interharmonic components, which were detected with average percentage errors in the frequency

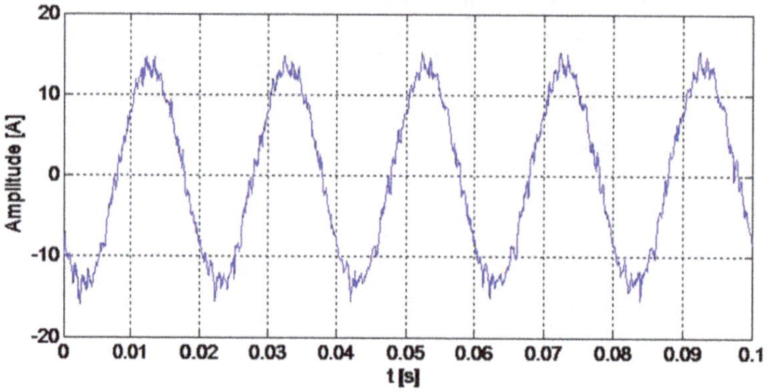

Figure 5 Case study 2: time trend of the analyzed current.

of 0.2% and 0.6% for the components near 74.79 and 382.35 Hz, respectively. For the same components, the average errors of amplitude were even worse, at 3% and 31%, respectively.

Table 8 shows the computational time required to analyze the test signal by all of the methods. Also in this case, the SWEM and the SWPM appear to be the slowest methods, whereas the advanced parametric methods had computational times that were significantly less than those of SWEM and SWPM. In fact, in the best case, they were only 1 order of magnitude greater than the time required by IECM. In particular, once again, SWMEM is the method that required a computational time that was the closest to that of IECM.

Table 5 Case study 2: average values

	Fundamental	Harmonic	Interharmonic
(a) Average values of frequency [Hz]			
SWEM	49.98	1001.38	2915.63
SWPM	49.97	1001.61	2914.50
SWEDEM	49.98	999.55	2911.02
SWPDPM	49.98	999.51	2912.03
SWMEM	49.98	1001.15	2912.28
SWMPM	49.97	1000.73	2912.00
IECM	50.00	1000.00	2915.00
(b) Average values of amplitude [A]			
SWEM	13.44	0.18	0.15
SWPM	13.39	0.24	0.15
SWEDEM	13.38	0.23	0.14
SWPDPM	13.39	0.23	0.14
SWMEM	13.44	0.27	0.15
SWMPM	13.44	0.24	0.14
IECM	13.40	0.20	0.052

Average values of (a) frequency and (b) amplitude of the fundamental, a harmonic, and an interharmonic detected by SWEM, SWPM, SWEDEM, SWPDEPM, SWMEM, SWMPM, and IECM.

1.4.4 Case study 4: measured signal of wind turbine system

A 6-s current waveform measured during the soft starting of a fixed-speed wind turbine was analyzed. The original sampling rate was 2,048 Hz, so a resampling at 10 kHz was conducted.

For the parametric methods, the error threshold was fixed equal to 10^{-7}, and the window of analysis slides was 0.02 s for all of the methods. In the second steps of both the SWPDPM and the SWEDEM, the analysis window was fixed equal to 3 cycles of the fundamental period. Figure 7 shows the highly time-varying trend of this waveform, which allowed us to test and compare methods also in the analysis of a non-stationary waveform.

As was predictable based on the theoretical considerations reported in Section 1.1, all of the methods identified the 3rd, 5th, 7th, and 11th harmonics as significant components, in addition to the fundamental. The amplitudes of the aforesaid components were characterized initially by an increasing trend up to a maximum value, after which the descending phase began and continued until each component achieved its steady-state value.

Table 6 Case study 2: computational time

	Computational time [p.u.]
SWEM	9447.06
SWPM	5227.87
SWEDEM	21.59
SWPDPM	15.58
SWMEM	11.75
SWMPM	21.35
IECM	1

Computational time, in p.u. of the computational time required by the IECM, of all of the methods.

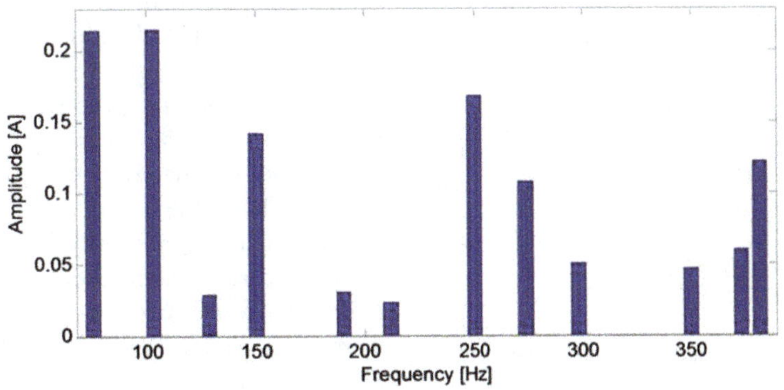

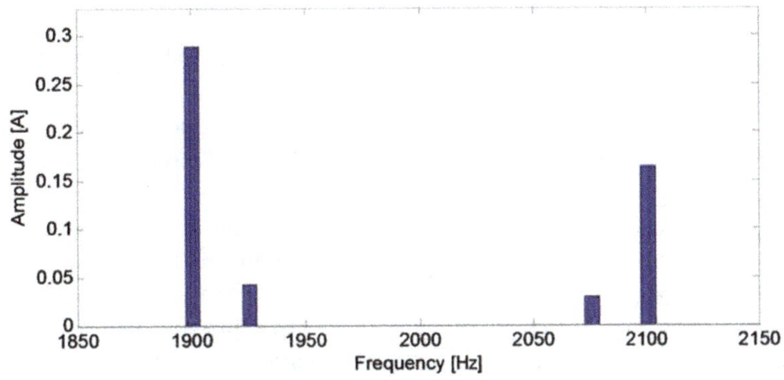

Figure 6 Case study 3: spectrum of the WTS test signal. (a) Low-frequency spectrum; **(b)** high-frequency spectrum.

Table 7 Case study 3: average percentage errors

	Fundamental	74.79 Hz	5th	382.35 Hz	38th
(a) Average errors of frequency [%]					
SWEM	1.18×10^{-4}	0.0036	9.24×10^{-4}	0.0041	3.75×10^{-5}
SWPM	7.84×10^{-5}	0.0013	4.35×10^{-4}	0.0015	2.39×10^{-5}
SWEDEM	0.0056	0.0063	0.0042	0.0028	0.0042
SWPDPM	9.37×10^{-5}	0.0025	1.60×10^{-5}	0.0019	1.60×10^{-5}
SWMEM	3.88×10^{-4}	0.0041	0.0028	0.021	5.18×10^{-5}
SWMPM	0.0016	0.011	0.11	0.12	3.69×10^{-4}
IECM	0.04	0.28	0.04	0.64	0.04
(b) Average errors of amplitude [%]					
SWEM	0.0019	0.059	0.066	0.48	0.026
SWPM	0.0012	0.039	0.035	0.22	0.028
SWEDEM	0.19	2.80	2.01	0.43	0.25
SWPDPM	0.19	2.77	2.09	0.36	0.077
SWMEM	0.0014	0.027	0.20	0.89	0.011
SWMPM	0.0058	1.16	1.50	7.34	0.041
IECM	0.10	3.41	2.15	31.48	3.12

Estimation of the average percentage errors of the fundamental, harmonic components, and interharmonics obtained by SWEM, SWPM, SWEDEM, SWPDEPM, SWMEM, SWMPM, and IECM: (a) frequencies; (b) amplitudes.

Table 8 Case study 3: computational time

	Computational time [p.u.]
SWEM	2116.47
SWPM	6507.25
SWEDEM	79.79
SWPDPM	180.81
SWMEM	11.84
SWMPM	30.02
IECM	1

Computational time, in p.u. of the computational time required by the IECM, of all of the methods.

Table 9 reports the maximum peak values of amplitude for the fundamental, the fifth, and the seventh harmonics that were detected by the methods that were used.

The results presented in Table 9 provide evidence that the IECM had significant spectral leakage since the estimated values always were considerably lower than those obtained by the parametric methods. These lasts, in fact, estimate, for each component, values that differ slightly from each other.

Note that, as expected, the behaviors of the time trends of the frequencies were different; in fact, all of the parametric methods had spectral-component frequencies with negligible time variations.

Table 10 shows the computational time required by all of the methods to analyze the current's waveform. The results were similar to those in the previous case study. It is important to observe that the SWMEM required a computational time that was only double that of IECM.

2 Conclusions

In this paper, we analyzed some methods used to estimate waveform distortions caused by photovoltaic and wind turbine systems. An overview of the waveform distortions caused by the most common configurations of PVSs and WTS schemes was also provided.

The theoretical aspects of the basic and advanced parametric methods proposed in literature were presented, focusing on their high accuracy while, at the same time, emphasizing how the advanced parametric methods have the advantage of having computational times that are significantly less than those required by the basis parametric methods.

The basic parametric methods that were considered were the Sliding-Window ESPRIT method and the Sliding-Window Prony method. The advanced parametric methods were the following:

the Three-step Sliding-Window ESPRIT-DFT-ESPRIT method;
the Three-step Sliding-Window Prony-DFT-Prony method;
the Sliding-Window Modified ESPRIT method;
the Sliding-Window Modified Prony method.

These methods were used in the spectral analysis of synthetic and measured currents of PVSs and WTSs in order to compare them in terms of the accuracy of the results they produced and computational efforts for both stationary and non-stationary waveforms. The waveforms were analyzed also by the IEC standard method, which we used as a reference for comparisons of computational times.

The case studies that were considered highlighted the importance of an adequate sampling rate for the parametric methods and the importance of the effect of noise on the detection of spectra by the Prony-based method when the signal components have very small amplitudes. In fact, in these cases, the ESPRIT-based methods can estimate the spectral components more easily, requiring a lower computational burden than the methods that use the Prony model.

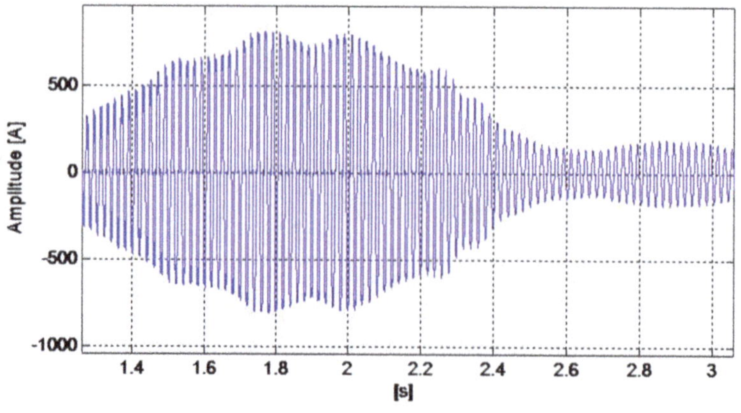

Figure 7 Case study 4: time trend of the analyzed current waveform.

Table 9 Case study 4: estimated peak values

Maximum peak value [A]			
	Fundamental	5th	7th
SWEM	866.92	139.96	66.97
SWPM	857.84	139.79	64.56
SWEDEM	867.06	139.31	64.14
SWPDPM	867.68	139.26	63.90
SWMEM	864.24	143.03	70.99
SWMPM	867.52	140.25	67.19
IECM	814.20	133.40	58.50

Peak values of the fundamental, 5th, and 7th harmonics detected by SWEM, SWPM, SWEDEM, SWPDEPM, SWMEM, SWMPM, and IECM.

All of the parametric methods were shown to have high levels of accuracy, exceeding that provided by the IECM, especially when the waveform spectrum was very wide. The IECM, in fact, failed in the detection of both interharmonic components and high-frequency harmonics.

Conversely, the advanced parametric methods took less computational effort than the basic parametric methods, with computational times of the same order of magnitude or no more than 1 order or magnitude greater than that of the IECM. The aforesaid numerical experiments also demonstrated that the SWMEM generally provided the best compromise between high-accuracy results and low computational effort.

It is important to note that, for very wide spectra, the three-step solution could inspire a new method for spectral analysis in which separating the waveform into low- and high-frequency components could result in the rapid and accurate detection of the spectral components of both the aforesaid bands by adapting the duration of the analysis window to the lowest frequency of each band.

3 Appendix A - IEC method

The standards IEC 61000-4-7 [23] and IEC 61000-4-30 [22] propose to use, for the spectral analysis of the non-stationary waveforms, the Sliding-Window Discrete Fourier Transform, which consists of applying the DFT to consecutive windows that slide forward successively over time.

Table 10 Case study 4: computational time

	Computational time [p.u.]
SWEM	145.44
SWPM	367.38
SWEDEM	25.34
SWPDPM	57.26
SWMEM	2.34
SWMPM	4.29
IECM	1

Computational time, in p.u. of the computational time required by the IECM, of all of the methods.

If $x(n)$ is a waveform sampled data, it is possible to obtain the corresponding N-point DFT, $X(k)$, as:

$$X(k) = \sum_{n=0}^{N-1} x(n)e^{-j2\pi\frac{k}{N}n}, \quad k = 0, 1, ..., N-1 \tag{13}$$

The SW DFT, $X_m(k)$, by definition, follows from the (13), introducing, for the sampled waveform $x(n)$ with length L, a window function $w(n)$, generally rectangular, with size $N < L$:

$$X_m(k) = \sum_{n=0}^{N-1} x(n)w(n-m)e^{-j2\pi\frac{k}{N}n}, \quad k = 0, 1, ..., N-1 \tag{14}$$

where m is the starting time instant.

It is worth to observe that the aforesaid window function is very important both in terms of analysis resolution and of spectral leakage. Specifically, it is important to choose adequately the time duration T_w of the analysis window; in fact, for a sampling interval T_s, it is $T_w = NT_s$, and the frequency resolution of the spectrum is $f = 1/T_w$, so the selected value N has to be a compromise to obtain a good resolution both in time and in frequency. According to the IEC standard, for a rectangular window, the compromise is a time duration T_w equal to 10 and 12 cycles of fundamental period, respectively, for 50-Hz systems and 60-Hz systems [22,23].

Moreover, selecting a synchronized window with a T_w equal to an integer multiple of the waveform fundamental period, it is also possible to avoid the spectral leakage phenomenon, which is the most significant problem of the DFT method, because it impacts negatively on the analysis results accuracy. In this regard, the time duration recommended by the IEC standards becomes generally inadequate, especially in the presence of interharmonic components.

3.1 Endnote

[a]Note that, as shown in [44], if fs is the chosen sampling rate, the Prony's method is able to detect a maximum component at a frequency of fs/4, whereas the ESPRIT method can estimate components up to fs/2. Then, the signal was under-sampled at 50 kHz before the application of SWEM and SWMEM.

Abbreviations

DER: distributed energy resources; DFIG: doubly fed induction generator; DFT: Discrete Fourier Transform; DG: dispersed generation; IEC: International Electrotechnical Committee; IECM: IEC standard method; MPPT: maximum power point tracking; PCC: point of common coupling; PQ: power quality; PVS: photovoltaic system; PWM: pulse-width modulation; SCIG: squirrel-cage induction generator; SG: smart grid; SW: sliding-window; ESPRIT: estimation of signal parameters by rotational invariance technique; SWEDEM: Three-step

Sliding-Window ESPRIT-DFT-ESPRIT method; SWEM: Sliding-Window ESPRIT method; SWMEM: Sliding-Window Modified ESPRIT method; SWMPM: Sliding-Window Modified Prony method; SWPDPM: Three-step Sliding-Window Prony-DFT-Prony method; SWPM: Sliding-Window Prony method; THDi: current total harmonic distortion; WTS: wind turbine system.

Competing interests
The author declares that she has no competing interests.

References

1. E Santacana, G Rackliffe, X Feng, Getting smart. IEEE Power Energ Mag **8**(2), 41–48 (2010)
2. H Farhangi, The path of the smart grid. IEEE Power Energ Mag **8**(1), 18–28 (2010)
3. RH Lasseter, Smart distribution: coupled microgrids. Proc IEEE **99**(6), 1074–1082 (2011)
4. A Vojdani, Smart integration. IEEE Power Energ. Mag. **6**(6), 71–79 (2008). November-December
5. P Caramia, G Carpinelli, P Verde, *Power quality indices in liberalized markets* (Wiley-IEEE Press, New Jersey, 2009)
6. F Blaaiberg, Z Chen, SB Kjaer, Power electronics as efficient interface in dispersed power generation systems. IEEE Trans Power Electron **19**(5), 1184–1194 (2004)
7. R Angelino, A Bracale, G Carpinelli, M Mangoni, D Proto, Dispersed generation units providing system ancillary services in distribution networks by a centralized control. IET Renew Power Gener **5**(4), 311–321 (2011)
8. CIGRE WG 37–23: Impact of increasing contribution of dispersed generation on the power system (1999)
9. F Gronwald, Frequency versus time domain immunity testing of smart grid components. Adv Radio Sci **12**, 149–153 (2014)
10. RC Variath, MAE Andersen, ON Nielsen, A Hyldgard, A review of module inverter topologies suitable for photovoltaic systems. in IPEC, 2010 Conference Proc., Singapore, 27–29 October 2010
11. SA Papathanassiou, MP Papadopoulos, Harmonic analysis in a power system with wind generation. IEEE Trans Power Delivery **21**(4), 2006–2016 (2006)
12. ST Tentzerakis, SA Papathanassiou, An investigation of the harmonic emissions of wind turbines. IEEE Trans Energy Convers **22**(1), 150–158 (2007)
13. AE Feijoo, J Cidras, Modeling of wind farms in the load flow analysis. IEEE Transactions on Power Systems **15**(1), 110–115 (2000)
14. JI Herrera, TW Reddoch, JS Lawler, Harmonics generated by two variable speed wind generating systems. IEEE Trans Energy Convers **3**, 267–273 (1988)
15. MHJ Bollen, K Yang, Harmonic aspects of wind power integration. J Modern Power Syst Clean Energy **1**(1), 14–21 (2013)
16. IEC 61400–21: Wind turbine generator systems, part 21: measurement and assessment of power quality characteristics of grid connected wind turbines (2001)
17. L Alfieri, A Bracale, P Caramia, G Carpinelli, Advanced methods for the assessment of time varying waveform distortions caused by wind turbine systems. Part I: theoretical aspects. in 13th Int. Conf. on Environment and Electrical Engineering, EEEIC 2013, Poland, November 2013
18. SK Rönnberg, *MH Bollen* (EOA Larsson, Emission from small scale PV-installations on the low voltage grid. Renewable Energy and Power Quality Journal, 2014)
19. D Gallo, R Langella, A Testa, JC Hernandez, I Papic, B Blazic, J Meyer, Case studies on large PV plants: harmonic distortion, unbalance and their effects. in Power and Energy Society General Meeting (PES), IEEE , 21–25 July 2013
20. SK Ronnberg, K Yang, MHJ Bollen, AG de Castro, Waveform distortion - a comparison of photovoltaic and wind power. in 16th Int. Con. on Harmonics and Quality of Power, Bucharest, Romania, 25–28 May 2014
21. A Andreotti, A Bracale, P Caramia, G Carpinelli, Adaptive Prony method for the calculation of power quality indices in the presence of non-stationary disturbance waveforms. IEEE Trans Power Delivery **24**(2), 874–883 (2009)
22. IEC standard 61000-4-30: Testing and measurement techniques – power quality measurement methods (2008)
23. IEC standard 61000-4-7: General guide on harmonics and interharmonics measurements, for power supply systems and equipment connected thereto (2009)
24. P Ribeiro, *Time-Varying Waveform Distortions in Power Systems* (John Wiley & Sons, New York, 2009)
25. GW Chang, C-I Chen, Measurement techniques for stationary and time-varying harmonics. in IEEE Power & Energy Society General Meeting 2010, PES 2010, Minneapolis, USA, July 2010
26. PM Silveira, C Duque, T Baldwin, PF Ribeiro, Sliding window recursive DFT with dyadic downsampling — a new strategy for time-varying power harmonic decomposition. in IEEE Power & Energy Society General Meeting, Canada, 2009
27. A Bracale, P Caramia, G Carpinelli, Adaptive Prony method for waveform distortion detection in power systems. Int J Electrical Power Energ Systems **29**(5), 371–379 (2007)
28. I Gu, MHJ Bollen, Estimating interharmonics by using sliding-window ESPRIT. IEEE Trans on Pow Del **23**(1), 13–23 (2008)
29. A Bracale, G Carpinelli, IYH Gu, MHJ Bollen, A new joint sliding-window ESPRIT and DFT scheme for waveform distortion assessment in power systems. Electr Power Syst Res **88**, 112–120 (2012)
30. L Alfieri, A Bracale, P Caramia, G Carpinelli, Waveform distortion assessment in power systems with a new three-step sliding-window method. in 12th Int. Conf. EEEIC, EEEIC 2013, Wroclaw, Poland, 5–8 May 2013
31. J Zygarlicki, M Zygarlicka, J Mroczka, KJ Latawiec, A Reduced, Prony's method in power-quality analysis—parameters selection. IEEE Trans Power Delivery **25**(2), 979–986 (2010)
32. L Alfieri, G Carpinelli, A Bracale, New ESPRIT-based method for an efficient assessment of waveform distortions in power systems. Electr Power Syst Res **122**, 130-139 (2015)
33. N Mohan, *WP Robbins, TM Undeland, Power Electronics: Converters, Applications, and Design* (Wiley, New York, 1995)
34. H Li, Z Chen, Overview of different wind generator systems and their comparisons. IET Ren. Pow. Gen. 2(2), 123–138
35. E Muljadi, N Samaan, V Gevorgian, Jun Li, S Pasupulati, Short circuit current contribution for different wind turbine generator types. in IEEE PES General Meeting, Minneapolis, United States, 25–29 July 2010
36. GL Xie, BH Zhang, Y Li, CX Mao, Harmonic propagation and interaction evaluation between small-scale wind farms and nonlinear loads. Energies **6**, 3297–3322 (2013)
37. R Roy, T Kailath, ESPRIT - estimation of signal parameters via rotational invariance techniques. IEEE Trans. on Ac., Sp., and Sig. Proc. 37(7), 984–995 (1989)
38. A Bracale, P Caramia, G Carpinelli, Optimal evaluation of waveform distortion indices with Prony and RootMusic methods. Int J of Pow & En Syst **27**(4), 360–370 (2007)
39. T Lobos, Z Leonowicz, J Rezmer, P Schegner, High-resolution spectrum estimation methods for signal analysis in power systems. IEEE Trans Instrum Meas **55**(1), 219–225 (2006)
40. A Bracale, G Carpinelli, Z Leonowicz, T Lobos, J Rezmer, Measurement of IEC groups and subgroups using advanced spectrum estimation methods. IEEE Trans on Instr Meas **57**(4), 672–681 (2008)
41. A Bracale, P Caramia, G Carpinelli, A Rapuano, A new, sliding-window prony and DFT scheme for the calculation of power-quality indices in the presence of non-stationary waveforms. Int. J. of Emerging Electric Power Systems. 13(5), 6 (December 2012)
42. C-I Chen, GW Chang, An efficient Prony-based solution procedure for tracking of power system voltage variations. IEEE Trans Ind Electron **60**(7), 2681–2688 (2013)
43. A Testa, MF Akram, R Burch, G Carpinelli, G Chang, V Dinavahi et al., Interharmonics: theory and modeling. IEEE Trans Power Del **22**(4), 2335–2348 (2007)
44. L Alfieri, A Bracale, P Caramia, G Carpinelli, Advanced methods for the assessment of time varying waveform distortions caused by wind turbine systems. Part II: experimental applications. in 13th International Conference on Environment and Electrical Engineering, EEEIC 2013, Wroclaw, Poland, 1–3 November 2013

A distributed approach to the OPF problem

Tomaso Erseghe

Abstract

This paper presents a distributed approach to optimal power flow (OPF) in an electrical network, suitable for application in a future smart grid scenario where access to resource and control is decentralized. The non-convex OPF problem is solved by an augmented Lagrangian method, similar to the widely known ADMM algorithm, with the key distinction that penalty parameters are constantly increased. A (weak) assumption on local solver reliability is required to always ensure convergence. A certificate of convergence to a local optimum is available in the case of bounded penalty parameters. For moderate sized networks (up to 300 nodes, and even in the presence of a severe partition of the network), the approach guarantees a performance very close to the optimum, with an appreciably fast convergence speed. The generality of the approach makes it applicable to any (convex or non-convex) distributed optimization problem in networked form. In the comparison with the literature, mostly focused on convex SDP approximations, the chosen approach guarantees adherence to the reference problem, and it also requires a smaller local computational complexity effort.

Keywords: Alternating direction method of multipliers; Augmented Lagrangian methods; Convergence guarantee; Distributed processing; Optimal power flow; Smart grid

Introduction

One of the key aspects of the current research trends for the future smart grid is the possibility of devising distributed algorithms for solving a global problem. This corresponds to the idea of a decentralized access to generation/storage resources, as well as to the much more challenging task of decentralized control.

The typical smart grid problem taken into consideration for distributed optimization is that of optimal power flow (OPF), that is, the optimal management of electrical power throughout the grid under a number of (electrical) constraints (e.g., the satisfaction of a power request from a load, the presence of a dispatchable/non dispatchable renewable generator or of a storage system). The OPF problem, being non-convex in nature in both the target function and the constraints, is very difficult to solve. For this reason, a widely used approach is to *map* it into a (somehow) close convex problem, and then solve the convex counterpart by means of distributed methods, e.g., the alternating direction method of multipliers. In this context, semi definite programming (SDP) relaxations have emerged as a common option, e.g., see Lavaei and Sojoudi et al. [1-4], Lam, Tse, and Zhang et al. [5,6], Dall'Anese and Giannakis et al. [7,8], Gayme and Topcu [9], and Erseghe and Tomasin [10]. One of the limits of this approach lies in the lack of adherence to the original problem, and in fact, optimality of the solution can only be ensured for very specific networks. But complexity is also an issue, since the number of variables involved in the local processing is squared with respect to its natural size. A few other worth mentioning approaches are available from the literature. Šulc et al. [11] exploit the (convex) LinDistFlow approximation as a lower complexity alternative to SDP relaxation. Magnusson et al. [12] avoid SDP relaxation and propose a sequential convex approximation approach, which, however, is known to imply slow convergence speeds. Instead, the *consensus and innovation* approach has been applied to the (convex) DC-OPF problem by Hug and Kar et al. [13,14], but the chosen distributed algorithm only provides approximate solutions even in the considered convex scenario.

The kind of approach we follow is alternative to the main trend in the literature, in the sense that we do not consider any convex relaxation and work directly on the

Correspondence: erseghe@dei.unipd.it
Università di Padova, Dipartimento di Ingegneria dell'Informazione, via G. Gradenigo 6/b, Padova, Italy

non-convex OPF problem. In this way, we can guarantee adherence to the original problem and develop an algorithm which is capable of identifying local minima. This idea was originally exploited in [15] where a distributed algorithm based upon ADMM was proposed. This algorithm provided undeniable evidence of the goodness of the intuition but had two major drawbacks. First, optimization for speed was cumbersome and required *centralized* coordination. Second, no guarantee on convergence was available, and in fact the algorithm often failed to converge. Although the convergence failure did not practically prevent the algorithm output for being usable, convergence is an issue that practically limits the algorithm speed.

In this paper, we wish to solve the above cited issues. To simplify system parameters and improve convergence speed, we remap the distributed problem in such a way to reveal the network *power flow*. In the ADMM formulation, the power flow variables are adequately weighted in order to force the algorithm to solve an approximate linear problem in the power flow variables in the first iterations (similarly to what happens with DC-OPF). The approximation is progressively abandoned in later iterations. This corresponds to the practical intuition that a linear power flow exchange problem provides a solution which is close to the optimum (some preliminary results on this aspect were recently presented at an international conference [16]). We also modify the plain ADMM algorithm and reinterpret it as a non-convex augmented Lagrangian method (see the work of Martinez and Birgin et al. [17,18]) where penalty parameters are constantly updated (increased) to always guarantee convergence. More specifically, a global convergence guarantee is available under the assumption that local solvers are efficient, in the sense that they can guarantee the identification of a (feasible) local minimum. This might not be an easy task in general, but it is a reasonable assumption when the number of local variables is controlled. Furthermore, a certificate of convergence to a local optimum is available when penalty parameters are bounded. The kind of coordination involved in this process is only local and therefore defines a fully distributed algorithm.

The rest of this paper is organized as follows. First, the reference OPF problem is presented and put in a networked form readily usable for obtaining a distributed algorithm. Then the distributed approach is discussed in abstract form and its convergence properties proved. Application to the specific OPF problem is then detailed, and the proposed distributed algorithm is finally tested in meaningful scenarios.

The OPF problem

We first introduce the OPF problem in its natural (centralized) formulation.

Standard formulation

Consider an electrical network of N nodes at steady state, where V_i, P_i, and Q_i represent, respectively, the local complex voltage, and the node's active and reactive powers. Assume that, at node i, a *local* cost is associated to active power production through a cost function $f_i(P_i)$. Assume that the electrical neighbors of node i are identified through the *neighbors* set $\mathcal{N}_i$, and that the line admittance $Y_{i,j}$, $j \in \mathcal{N}_i$, is known for each physical connection. Then the standard OPF problem has the form

$$\begin{aligned} &\min \sum_{i\in\mathcal{N}} f_i(P_i) \\ &\text{w.r.t. } V_i \in \mathbb{C}, P_i, Q_i \in \mathbb{R}, i \in \mathcal{N} \\ &\text{s.t. } P_i + jQ_i = V_i \sum_{j\in\mathcal{N}_i} Y_{i,j}^* V_j^* \\ &\quad \underline{V}_i \le |V_i| \le \overline{V}_i \\ &\quad \underline{P}_i \le P_i \le \overline{P}_i, \\ &\quad \underline{Q}_i \le Q_i \le \overline{Q}_i \end{aligned} \tag{1}$$

where $\mathcal{N} = \{1, \ldots, N\}$ is the nodes set. The first constraint in (1) refers to power flow equations (i.e., Kirchoff's laws). The remaining constraints are voltage and power constraint limitations, with $\underline{V}_i$, $\overline{V}_i$, $\underline{P}_i$, $\overline{P}_i$, $\underline{Q}_i$, $\overline{Q}_i$ local upper and lower bounds.

For the ease of simplicity here we refer to a basic OPF problem, but additional constraints can be easily added to (1), e.g., power flow constraints on specific lines. Constraints referred to resources such as storage systems and renewable generators (dispatchable or not dispatchable) can be included by suitably selecting the cost factor f_i, by introducing proper corrections to the cost function, or by inserting a time variable. Discrete variables can be also included in the problem formulation (e.g., the tap changing of the transformers, or the cost to turn on/off a generator), in which case a mixed-integer programming solver will be needed. The results that follow are valid for all the above generalizations.

Region-based formulation

We now wish to fully capture the network relations in (1), in such a way to be used in a distributed implementation. The idea is to *partition* the network in R regions, where the sets $\mathcal{R}_k$, $k = 1, \ldots, R$, identify nodes belonging to region k. We have

$$\mathcal{N} = \bigcup_{k=1}^{R} \mathcal{R}_k\,, \qquad \mathcal{R}_k \cap \mathcal{R}_h = \emptyset, \forall k \neq h\,. \tag{2}$$

Because of power flow equations in (1), the voltages of interest in region k are those belonging to set

$$\mathcal{V}_k = \bigcup_{i \in \mathcal{R}_k} \mathcal{N}_i \tag{3}$$

where $\mathcal{N}_i$ identify the neighbors of node i. Note that set $\mathcal{V}_k$ includes set $\mathcal{R}_k$ as a subset, as well as all those nodes which belong to neighbor regions and which have a direct connection (edge) with one of the nodes of $\mathcal{R}_k$. Accordingly, we identify the *local* voltage vectors $\boldsymbol{x}_k$ with entries $x_{k,\ell}$ by

$$\boldsymbol{x}_k = [x_{k,\ell}]_{\ell \in \mathcal{V}_k}, \qquad x_{k,\ell} = V_\ell \tag{4}$$

and the corresponding constraint region

$$\begin{aligned} \mathcal{X}_k = \Big\{ & \underline{V}_\ell \le |x_{k,\ell}| \le \overline{V}_\ell, \forall \ell \in \mathcal{V}_k, \\ & \underline{P}_i \le P_i \le \overline{P}_i, \\ & \underline{Q}_i \le Q_i \le \overline{Q}_i, \\ & P_i + jQ_i = x_{k,i} \sum_{j \in \mathcal{N}_i} Y^*_{i,j} x^*_{k,j}, \forall i \in \mathcal{R}_k \Big\} \end{aligned} \tag{5}$$

collecting voltage constraints, active and reactive power constraints, and power flow constraints, and to which we may add any additional constraint of interest. Regions $\mathcal{X}_k$ are deliberately chosen to be compact (closed and bounded) in order to strengthen later derivations and results.

Hence, a region-based equivalent formalization for (1) corresponds to the non-convex problem

$$\begin{aligned} &\min \sum_{k \in \mathcal{R}} F_k(\boldsymbol{x}_k) \\ &\text{w.r.t. } \boldsymbol{x}_k \in \mathcal{X}_k, k \in \mathcal{R} \\ &\text{s.t. } x_{k,\ell} = x_{h,\ell}, \ \forall \ell \in \mathcal{V}_k \cap \mathcal{V}_h, k, h \in \mathcal{R} \end{aligned} \tag{6}$$

where $\mathcal{R} = \{1, \dots, R\}$, function

$$F_k(\boldsymbol{x}_k) = \sum_{\ell \in \mathcal{R}_k} f_\ell(P_\ell) \tag{7}$$

collects local cost functions, and where the constraint is forcing equivalence between duplicated (voltage) variables in vectors $\boldsymbol{x}_k$.

Capturing the power flow

The formalization given in (6), although correct, is somehow unsatisfactory in terms of the slow convergence speed involved with its distributed implementation, and in terms of the difficulty in optimizing its system parameters (see [15]). The key point is that we are not using any electrical intuition that could help the distributed processing. The intuition we use is illustrated in Figure 1.

The idea with Figure 1 is the following. Consider two neighboring regions k, and h, and edge (i,j) connecting the two regions, i.e., with $i \in \mathcal{R}_k$ and $j \in \mathcal{R}_h$. It also is $\{i,j\} \subset \mathcal{V}_k$ and $\subset \mathcal{V}_h$. Then, equivalence between the local variables can be written in the form

$$\begin{aligned} x_{k,i} &= x_{h,i} \\ x_{k,j} &= x_{h,j} \end{aligned} \tag{8}$$

which is equivalent to the constraint in (6). However, equivalence can be also written in the form

$$\begin{aligned} x_{k,i} - x_{k,j} &= x_{h,i} - x_{h,j} \\ x_{k,i} + x_{k,j} &= x_{h,i} + x_{h,j} \end{aligned} \tag{9}$$

where the first equivalence captures the *power flow*, since the power flowing through line (i,j) is of the form $Z_{i,j}|V_i - V_j|^2$, i.e., it only depends on voltage differences as from the first of (9).

The corresponding formulation for the OPF problem can then be compactly written by using sets

$$\mathcal{O}_k = \Big\{ (i,j) \Big| i \in \mathcal{R}_k, j \in \mathcal{N}_i \cap (\mathcal{V}_k \backslash \mathcal{R}_k) \Big\} \tag{10}$$

collecting in region k those edges connecting a node of $\mathcal{R}_k$ to a node in a neighbor region. By further introducing two auxiliary variables $\boldsymbol{z}^-$ and $\boldsymbol{z}^+$ belonging to the linear spaces

$$\begin{aligned} \mathcal{Z}^- &= \{\boldsymbol{z}^- | z^-_{i,j} = -z^-_{j,i}, \ \forall (i,j) \in \mathcal{O}_k, k \in \mathcal{R}\} \\ \mathcal{Z}^+ &= \{\boldsymbol{z}^+ | z^+_{i,j} = z^+_{j,i}, \ \forall (i,j) \in \mathcal{O}_k, k \in \mathcal{R}\} \end{aligned} \tag{11}$$

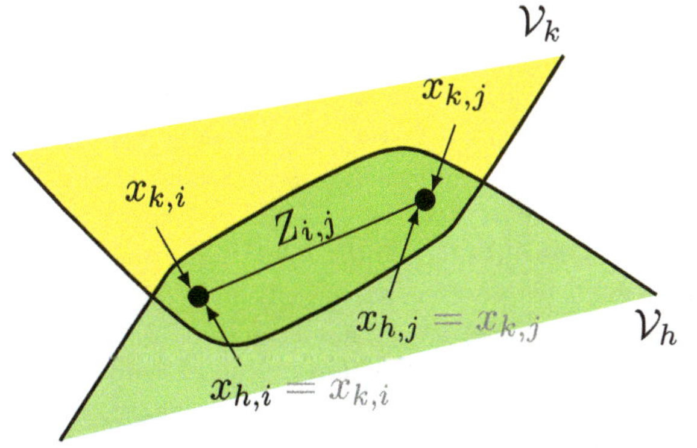

Figure 1 A way to capture the power flow on edge (i,j) with $i \in \mathcal{R}_k$ and $j \in \mathcal{R}_h$.

the OPF problem becomes

$$\begin{aligned} &\min \sum_{k\in\mathcal{R}} F_k(\boldsymbol{x}_k) \\ &\text{w.r.t. } \boldsymbol{x}_k \in \mathcal{X}_k, k \in \mathcal{R} \\ &\quad \boldsymbol{z}^- \in \mathcal{Z}^-, \boldsymbol{z}^+ \in \mathcal{Z}^+ \\ &\text{s.t. } \rho\,(x_{k,i} - x_{k,j}) = z_{i,j}^-, \\ &\quad \zeta\,(x_{k,i} + x_{k,j}) = z_{i,j}^+, \forall (i,j) \in \mathcal{O}_k, k \in \mathcal{R} \end{aligned} \tag{12}$$

where two positive constants ρ, ζ are used to differently weigh the power flow constraint on $\boldsymbol{z}^-$ (providing convergence on an approximate linear problem on power flow variables) from the full equivalence constraint on $\boldsymbol{z}^+$. The linear constraints in (12) can be also expressed in the compact matrix notation

$$\boldsymbol{z}_k = \begin{bmatrix} z_{i,j}^- \\ z_{i,j}^+ \end{bmatrix}_{(i,j)\in\mathcal{O}_k} = \boldsymbol{A}_k \boldsymbol{x}_k \tag{13}$$

where $\boldsymbol{A}_k$ is a sparse matrix of size $2|\mathcal{O}_k| \times |\mathcal{V}_k|$. In the typical case of large regions having a few connections with neighbors it is $|\mathcal{O}_k| \ll |\mathcal{V}_k|$.

The distributed approach

We now introduce the distributed algorithm in a general and abstract form, in order to assess its properties and capture its structure with a compact notation.

Reference optimization problem

The kind of problem we wish to solve in (12) is a non-convex problem of the form

$$\begin{aligned} &\min F(\boldsymbol{x}) \\ &\text{w.r.t. } \boldsymbol{x} \in \mathcal{X}, \boldsymbol{z} \in \mathcal{Z} \\ &\text{s.t. } \boldsymbol{A}\boldsymbol{x} = \boldsymbol{z} \end{aligned} \tag{14}$$

where $\boldsymbol{x} = [\boldsymbol{x}_k]_{k\in\mathcal{R}}$ collects all variables, $\boldsymbol{z} = [\boldsymbol{z}_k]_{k\in\mathcal{R}}$ collects all auxiliary variables, $F(\boldsymbol{x}) = \sum_{k\in\mathcal{R}} F_k(\boldsymbol{x}_k)$ is separable, $\mathcal{X} = \mathcal{X}_1 \times \ldots \times \mathcal{X}_R$ is a Cartesian product, set $\mathcal{Z} = \mathcal{Z}^- \times \mathcal{Z}^+$ is a linear space with associated projector $\boldsymbol{L}_{\mathcal{Z}}$, and $\boldsymbol{A} = \text{diag}(\boldsymbol{A}_1, \ldots, \boldsymbol{A}_R)$ has a block diagonal form. The results given in the following further consider $\mathcal{X}$ bounded (as we already assumed), and $F(\boldsymbol{x})$ continuous. We finally assume that (14) has a solution.

The smoothness of functions involved with the OPF problem ensure that a one-to-one relation exists between local minima of problem (14) and the corresponding Karush Kuhn Tucker (KKT) conditions. We have (e.g., see [19])

Theorem 1. (KKT stationary points)
The KKT stationary point conditions associated with the primal problem (14) are given by

$$\begin{aligned} &\boldsymbol{0} \in \partial F(\boldsymbol{x}) + \partial \eta_{\mathcal{X}}(\boldsymbol{x}) + \boldsymbol{A}^T \boldsymbol{\lambda} \\ &\boldsymbol{A}\boldsymbol{x} = \boldsymbol{z} \\ &\quad \boldsymbol{x} \in \mathcal{X}\,,\ \boldsymbol{z} \in \mathcal{Z}\,,\ \boldsymbol{\lambda} \perp \mathcal{Z} \end{aligned} \tag{15}$$

where ∂ is the proximal sub-gradient operator, and where $\eta_{\mathcal{A}}$ is the indicator function of set $\mathcal{A}$, with $\eta_{\mathcal{A}}(\boldsymbol{a}) = 0$ if $\boldsymbol{a} \in \mathcal{A}$ and $+\infty$ if $\boldsymbol{a} \notin \mathcal{A}$. Conditions (15) identify the local minima of (14). □

Augmented Lagrangian formalization

No global minimum ensurance is given in the present context, since the Lagrangian associated with problem (14) may suffer of a primal-dual gap. A remedy in this respect is to use a Powell Hestenes Rockafellar (PHR) *augmented* Lagrangian formulation. The *augmented* Lagrangian associated with problem (14) can be written in the form

$$\begin{aligned} L(\boldsymbol{x}, \boldsymbol{z}, \boldsymbol{\lambda}, \boldsymbol{\epsilon}) = F(\boldsymbol{x}) &+ \eta_{\mathcal{X}}(\boldsymbol{x}) + \eta_{\mathcal{Z}}(\boldsymbol{z}) \\ &+ \boldsymbol{\lambda}^T(\boldsymbol{A}\boldsymbol{x} - \boldsymbol{z}) + \frac{1}{2}\|\boldsymbol{A}\boldsymbol{x} - \boldsymbol{z}\|_{\boldsymbol{\epsilon}}^2 \end{aligned} \tag{16}$$

where $\|\boldsymbol{x}\|_{\boldsymbol{\epsilon}}^2 = \boldsymbol{x}^T \text{diag}(\boldsymbol{\epsilon})\boldsymbol{x}$ is a scaled norm, and where the entries of $\boldsymbol{\epsilon}$ are strictly positive. In (16), the couple $(\boldsymbol{x}, \boldsymbol{z})$ plays the role of primal variables, while $(\boldsymbol{\lambda}, \boldsymbol{\epsilon})$ play the role of dual variables (Lagrange multipliers). The *dual* function associated with (16) is

$$D(\boldsymbol{\lambda}, \boldsymbol{\epsilon}) = \min_{\boldsymbol{x},\boldsymbol{z}} L(\boldsymbol{x}, \boldsymbol{z}, \boldsymbol{\lambda}, \boldsymbol{\epsilon})\ . \tag{17}$$

The PHR augmented Lagrangian of (16) is well defined, in the sense that it ensures the typical properties of ordinary Lagrangians of convex functions, i.e., the zero duality gap property and the applicability of a saddle point theorem. The result is given in ([20], Theorem 11.59). Incidentally, we are using a vector of weighting factors $\boldsymbol{\epsilon}$ instead of a unique multiplication by scalar factor ϵ. This, however, does not modify derivation nor the final result.

Theorem 2. (Rockafellar-Wets)
1. Zero duality gap *Let $(\boldsymbol{x}^*, \boldsymbol{z}^*)$ be a solution to the primal problem (14), and let $(\boldsymbol{\lambda}^*, \boldsymbol{\epsilon}^*)$ be any maximizer of the dual function (17). The corresponding duality gap is zero, that is, we have*

$$F(\boldsymbol{x}^*) = D(\boldsymbol{\lambda}^*, \boldsymbol{\epsilon}^*)\ . \tag{18}$$

2. *Saddle point* *The solutions in 1 identify a saddle point of PHR augmented Lagrangian (16), that is*

$$\begin{aligned}(\boldsymbol{x}^*, \boldsymbol{z}^*) &\in \operatorname*{argmin}_{\boldsymbol{x},\boldsymbol{z}} L(\boldsymbol{x}, \boldsymbol{z}, \boldsymbol{\lambda}^*, \boldsymbol{\epsilon}^*)\\ (\boldsymbol{\lambda}^*, \boldsymbol{\epsilon}^*) &\in \operatorname*{argmax}_{\boldsymbol{\lambda},\boldsymbol{\epsilon}\geq 0} L(\boldsymbol{x}^*, \boldsymbol{z}^*, \boldsymbol{\lambda}, \boldsymbol{\epsilon})\,.\end{aligned} \tag{19}$$

Conversely, any saddle point (19) identifies a primal and dual solution, as from 1. □

In this context, the search for an optimum point can be turned into the search for a saddle point of the PHR augmented Lagrangian, which is in general more effective in terms of efficiency and speed. However, since only a local optimization point may be available for the first of (19) (because of non-convexity), then only *local* saddle points can be practically identified. It is then interesting to observe the following result, which is a straightforward consequence of the fact that local minima/maxima conditions of (19) correspond to KKT stationary point conditions (15), as the reader can easily verify.

Theorem 3. *There exists a one-to-one correspondence between local minima of the original problem (14), KKT stationary points (15), and local saddle points of the PHR augmented Lagrangian in (19).* □

As a consequence, the search for local minima can be mapped into a search for local saddle points of the augmented Lagrangian.

Alternating direction search for a local saddle point

The search for a local saddle point can be dealt with by using the method of [17] (see also [18]). In our context, the method can be mapped into an alternating direction algorithm of the form

$$\begin{aligned}\boldsymbol{x}_{t+1} &\in \arg\min_{\boldsymbol{x}\in\mathcal{X}} L(\boldsymbol{x}, \boldsymbol{z}_t, \boldsymbol{\lambda}_t, \boldsymbol{\epsilon}_t)\\ \boldsymbol{z}_{t+1} &\in \arg\min_{\boldsymbol{z}\in\mathcal{Z}} L(\boldsymbol{x}_{t+1}, \boldsymbol{z}, \boldsymbol{\lambda}_t, \boldsymbol{\epsilon}_t)\\ \boldsymbol{\lambda}_{t+1} &= \boldsymbol{\lambda}_t + \boldsymbol{E}_t(\boldsymbol{A}\boldsymbol{x}_{t+1} - \boldsymbol{z}_{t+1})\end{aligned} \tag{20}$$

where $\boldsymbol{E}_t = \operatorname{diag}(\boldsymbol{\epsilon}_t)$, and where $\boldsymbol{\epsilon}_t$ is suitably updated at each cycle by guaranteeing $\boldsymbol{\epsilon}_{t+1} \geq \boldsymbol{\epsilon}_t$. Note that, differently from [17], and similarly to what we have in ADMM, an independent update is used for $\boldsymbol{x}_t$ and $\boldsymbol{z}_t$. In turn, differently from ADMM, the weighting parameters $\boldsymbol{\epsilon}_t$ are updated in order to ensure convergence of the process in a non convex scenario.

Throughout the process, we assume that the commutation property

$$\boldsymbol{L}_{\mathcal{Z}}\boldsymbol{E}_t = \boldsymbol{E}_t\boldsymbol{L}_{\mathcal{Z}} \tag{21}$$

holds, which corresponds to the request

$$\epsilon_{k,i,j} = \epsilon_{h,j,i}\,, \qquad (i,j) \in \mathcal{O}_k, j \in \mathcal{R}_h, k, h \in \mathcal{R}\,. \tag{22}$$

We also assume that

$$\boldsymbol{\lambda}_0 \perp \mathcal{Z}\,. \tag{23}$$

These are light hypotheses guaranteeing that (20) simplifies to updates

$$\begin{aligned}\boldsymbol{x}_{t+1} &\in \arg\min_{\boldsymbol{x}\in\mathcal{X}} F(\boldsymbol{x}) + \frac{1}{2}\|\boldsymbol{A}\boldsymbol{x} - (\boldsymbol{z}_t - \boldsymbol{E}_t^{-1}\boldsymbol{\lambda}_t)\|_{\boldsymbol{\epsilon}_t}^2\\ \boldsymbol{z}_{t+1} &= \boldsymbol{L}_{\mathcal{Z}}\boldsymbol{A}\boldsymbol{x}_{t+1}\\ \boldsymbol{\lambda}_{t+1} &= \boldsymbol{\lambda}_t + \boldsymbol{E}_t(\boldsymbol{A}\boldsymbol{x}_{t+1} - \boldsymbol{z}_{t+1})\end{aligned} \tag{24}$$

and we also have

$$\boldsymbol{z}_{t+1} \in \mathcal{Z}\,, \qquad \boldsymbol{\lambda}_{t+1} \perp \mathcal{Z} \tag{25}$$

so that the third line in KKT conditions (15) is satisfied throughout the iterative process. Note that the update of $\boldsymbol{x}_t$ in the first of (24) corresponds to the parallel of a number of local updates because F is separable, and $\mathcal{X}$ is a Cartesian product. In addition, since the full minimum for the first of (24) may be not available, we relax the result by assuming that a local minimum is achieved and that the target function in this local minimum $\boldsymbol{x}_{t+1}$ is smaller than or equal to the function value in $\boldsymbol{x}_t$. Therefore, a reliability assumption on the local solver is required. Although this might be in general a strong request (e.g., see [21]), especially when the local constraints identify a very small feasibility region, we expect it to be reasonably met when the number of local variables is not too large (i.e., for small regions).

Interestingly, given the fact that $\mathcal{X}$ is bounded, then both sequences $\{\boldsymbol{x}_t\}$ and $\{\boldsymbol{z}_t\}$ are bounded. This may not be the case for $\{\boldsymbol{\lambda}_t\}$, but it is convenient to force this property by assuming

$$\boldsymbol{\lambda}_{t+1} = \mathcal{P}[\,\boldsymbol{\lambda}_t + \boldsymbol{E}_t(\boldsymbol{A}\boldsymbol{x}_{t+1} - \boldsymbol{z}_{t+1})] \tag{26}$$

with $\mathcal{P}[\boldsymbol{\lambda}] = \max(\boldsymbol{\lambda}_{\min}, \min(\boldsymbol{\lambda}, \boldsymbol{\lambda}_{\max}))$ a projection onto a compact box. The reason for this action will become clearer later on in the proof of Theorem 5.

Concerning penalty parameters $\boldsymbol{\epsilon}_t$, in the centralized fashion of [17] the update criterion on $\boldsymbol{\epsilon}_t$ is of the form

$$\boldsymbol{\epsilon}_{t+1} = \begin{cases}\boldsymbol{\epsilon}_t & \text{if}\,\Gamma_{t+1} \leq \theta\,\Gamma_t\\ \tau\boldsymbol{\epsilon}_t & \text{otherwise}\end{cases} \tag{27}$$

with constants $0 < \theta < 1$ and $\tau > 1$, and with

$$\Gamma_t = \|\boldsymbol{A}\boldsymbol{x}_t - \boldsymbol{z}_t\|_\infty \tag{28}$$

a measure of the primal gap (in infinity norm), in such a way to increase the penalty only if the primal gap is not decreasing sufficiently. The criterion can be also made local. The approach we propose is the following. We first check the primal gap decrease in region k via

$$\check{\epsilon}_{k,t+1} = \begin{cases} \|\epsilon_{k,t}\|_\infty \mathbf{1} & \text{if } \Gamma_{k,t+1} \leq \theta\, \Gamma_{k,t} \\ \tau \|\epsilon_{k,t}\|_\infty \mathbf{1} & \text{otherwise} \end{cases} \tag{29}$$

with $\mathbf{1}$ the all-ones vector, and with

$$\Gamma_{k,t} = \|A_k x_{k,t} - z_{k,t}\|_\infty \tag{30}$$

the local gap. We then select the smallest $\epsilon_{t+1} \geq \check{\epsilon}_{t+1}$ satisfying (29), which in our context implies

$$\epsilon_{k,i,j,t+1} = \max\left(\check{\epsilon}_{k,i,j,t+1}, \check{\epsilon}_{h,j,i,t+1}\right) \tag{31}$$

where $(i,j) \in \mathcal{O}_k, j \in \mathcal{R}_h, k, h \in \mathcal{R}$. This approach only requires local message exchanges. With this definition, the update is such that if one value of $\epsilon_{k,t}$ grows to ∞, then all the values in the network do so, as it is for the centralized counterpart (27).

The proposed solution is summarized in Algorithm 1.

Algorithm 1: Alternating direction search method

1. Update variable x_{t+1} using the first of (24). When a global minimum guarantee is not available, a local minimum must be identified, with the guarantee that the target function is decreased with respect to its value at x_t.
2. Update auxiliary variables $z_{t+1} = L_{\mathcal{Z}} A x_{t+1}$.
3. Update the Lagrange multipliers λ_{t+1} using (26).
4. Update the penalty parameters ϵ_{t+1} using (27) or (29)-(31).

Convergence guarantees

The important characteristic of Algorithm 1 is that, in the given scenario, it provides a distributed solution. The main difference with the inspiring technique of [17] lays in the use of an alternating search with respect to x and z (versus the joint minimum search on (x, z)), this being the key point for obtaining a *distributed* algorithm. Nevertheless, the algorithm always converges (despite the non-convex scenario), and convergence guarantees essentially equivalent to those of [17] can be derived.

We separately treat the case where the penalty constant parameters are bounded and the case where they are unbounded. For bounded parameters we have the following result.

Theorem 4. ***(Bounded penalties)***
Consider Algorithm 1, and assume that the sequence of penalty parameters $\{\epsilon_t\}$ is bounded. We have:

1. *Sequences $\{z_t\}$ and $\{\lambda_t\}$ converge to finite values, z^* and λ^*, respectively.*
2. *There exists a finite limit point (accumulation point) for the sequence $\{x_t\}$, and if $A^T A$ is invertible then sequence $\{x_t\}$ is further guaranteed to converge to a finite value x^*.*
3. *The triplets (x^*, z^*, λ^*), with x^* any limit point of $\{x_t\}$, satisfy the KKT conditions of (15), hence all limit points x^* identify a local minimum to the original problem. Even more, in the limit $t \to \infty$ any triplet (x_t, z_t, λ_t) satisfies the KKT stationarity conditions, i.e., identifies a local minimum and satisfies the constraint $Ax_t = z_t$.* □

Proof of Theorem 4. Consider that the sequence of penalty parameters $\{\epsilon_t\}$ is bounded, to have $\epsilon_t = \epsilon_\infty$ for $t \geq t_0$. For both (27) and (29), we have that $\Gamma_{t+1} \leq \theta\, \Gamma_t$ for $t > t_0$, and therefore λ_t is bounded and converges to a finite value λ_∞ (also in case the projection (26) is limiting the value to its maximum).

Now, by exploiting equivalence $z_t = L_{\mathcal{Z}} A x_t$, we rewrite the update of x_t in (24) in the form

$$\begin{aligned} x_{t+1} \in \underset{x \in \mathcal{X}}{\operatorname{argmin}}\ & F(x) + \frac{1}{2}\|(I - L_{\mathcal{Z}})Ax\|^2_{\epsilon_t} \\ & + \frac{1}{2}\|L_{\mathcal{Z}} A(x - x_t)\|^2_{\epsilon_t} + \lambda_t^T A x\,. \end{aligned} \tag{32}$$

By then using the shorthand notation

$$\begin{aligned} g_t &= F(x_t) + \eta_{\mathcal{X}}(x_t) + \frac{1}{2}\|Ax_t - z_t\|^2_{\epsilon_\infty} + \lambda_\infty^T A x_t \\ \zeta_t &= (\lambda_t - \lambda_\infty)^T A (x_t - x_{t+1})\,, \end{aligned}$$

and $\Delta g_t = g_{t+1} - g_t$, from (32) we have

$$\Delta g_t + \frac{1}{2}\|z_{t+1} - z_t\|^2_{\epsilon_\infty} \leq \zeta_t \leq |\zeta_t|\,, \quad t > t_0 \tag{33}$$

which implies $\Delta g_t \leq |\zeta_t|$ for $t > t_0$. By noting that $\|A(x_t - x_{t+1})\|$ is bounded because $\mathcal{X}$ is assumed bounded, and by recalling that $\lim_{t\to\infty} \lambda_t = \lambda_\infty$, then it also is $\lim_{t\to\infty} |\zeta_t| = 0$. This is sufficient to guarantee that Δg_t converges to 0 for $t \to \infty$, which can be proved by contradiction. Specifically, if Δg_t does not converge to 0 then there exists an infinite sequence for which $|\Delta g_t| \geq \epsilon > 0$. Moreover, since $\Delta g_t \leq |\zeta_t|$, where the right value can be made arbitrarily small for large t, there also exists an infinite sequence for which $\Delta g_t \leq -\epsilon$. By denoting the sequence as $\mathcal{S}_\epsilon \subset (t_0, \infty)$, this would imply

$$g_\infty - g_{t_0} = \sum_{t \notin \mathcal{S}_\epsilon} \Delta g_t + \sum_{t \in \mathcal{S}_\epsilon} \Delta g_t \leq \sum_{t \notin \mathcal{S}_\epsilon} |\zeta_t| - \sum_{t \in \mathcal{S}_\epsilon} \epsilon \,.$$

Since $|\zeta_t|$ is guaranteed to be exponentially decreasing because of the assumption $\Gamma_{t+1} \leq \theta\, \Gamma_t$, the above implies $g_\infty = -\infty$, hence a contradiction. Therefore, g_t converges to a finite value, and, as a consequence of (33), the weighted norm $\|z_{t+1} - z_t\|^2_{\epsilon_\infty}$ converges to 0, i.e., z_t converges to a finite value too. These results justify points 1 and 2.

To conclude with point 3, since x_{t+1} is assumed a local minimum, from (32) we also have, for $t > t_0$,

$$\begin{aligned} \mathbf{0} \in\ & \partial F(x_{t+1}) + \partial \eta_{\mathcal{X}}(x_{t+1}) + A^T \lambda_\infty \\ & + A^T E_\infty (I - L_{\mathcal{Z}}) A x_{t+1} \\ & + A^T E_\infty (z_{t+1} - z_t) + A^T (\lambda_t - \lambda_\infty) \end{aligned}$$

and since the values on the second and third lines tend to $\mathbf{0}$ in the limit, then in the limit, the KKT stationary point conditions (15) are satisfied.

As a consequence, bounded penalty parameters guarantee a convergence of the algorithm to a KKT stationary point, i.e., they imply the identification of a local minimum. Note that the result is sufficiently strong also in the case where $A^T A$ is not invertible (see second part of point 3). This is an important property since the invertibility of $A^T A$ is only ensured for a single-node regions choice $\mathcal{R}_k = \{k\}$.

The result for unbounded parameters assumes that the ill conditioning associated with very large/infinite values is adequately solved, e.g., by locally normalizing the minimization in (32) by the maximum penalty value $\|\epsilon_{k,t}\|_\infty$. We have

Theorem 5 (Unbounded penalties). *Consider Algorithm 1, and assume that the sequence of penalty parameters $\{\epsilon_t\}$ is unbounded. We have:*

1. *Sequence $\{z_t\}$ converges to a finite value, z^*.*
2. *There exists a finite limit point for the sequence $\{x_t\}$, and if $A^T A$ is invertible then sequence $\{x_t\}$ is ensured to converge to a finite value x^*.* □

Proof. *The results in the proof of Theorem 4 can be applied by suitably (locally) normalizing parameters. The kind of replacements we use are*

$$\epsilon_t \implies \tilde{\epsilon}_t = \left[\frac{\epsilon_{k,t}}{\|\epsilon_{k,t}\|_\infty} \right]_{k=1,\ldots,R}$$

$$F(x) \implies \tilde{F}(x) = \sum_{k=1}^{R} \frac{F_k(x_k)}{\|\epsilon_{k,t}\|_\infty}$$

$$\lambda_t \implies \tilde{\lambda}_t = \left[\frac{\lambda_{k,t}}{\|\epsilon_{k,t}\|_\infty} \right]_{k=1,\ldots,R}$$

which have the characteristic of providing bounded quantities. For both (27) and (29), all entries $\epsilon_{k,t}$ are diverging by construction, hence $\tilde{\lambda}_t$ is ensured to converge to $\mathbf{0}$ in the limit. Convergence is also guaranteed to be exponential, because of the presence of parameter $\tau > 1$ in the update of penalty parameters. These properties are fundamental and are ensured by use of projection (26). Furthermore, $\tilde{\epsilon}_t$ is guaranteed to converge to the all ones vector $\mathbf{1}$. By then investigating the counterparts to g_t and ζ_t, namely,

$$\tilde{g}_t = \tilde{F}(x_t) + \eta_{\mathcal{X}}(x_t) + \frac{1}{2}\|Ax_t - z_t\|^2_{\tilde{\epsilon}_t}$$

$$\tilde{\zeta}_t = \tilde{\lambda}_t^T A (x_t - x_{t+1})$$

we still verify that properties $\lim_{t\to\infty} |\tilde{\zeta}_t| = 0$ and

$$\Delta \tilde{g}_t + \frac{1}{2}\|z_{t+1} - z_t\|^2_{\tilde{\epsilon}_t} \leq \tilde{\zeta}_t \leq |\tilde{\zeta}_t| \tag{34}$$

hold, and we also have that $\Delta\tilde{g}_t$ converges to 0. Hence $\tilde{g}_t$ converges to a finite value, so that there exist limit points for the sequence $\{x_t\}$. From (34) we also find that z_t converges to a finite value. This proves the theorem.

Note that Theorem 5, although being able to prove convergence of both sequences $\{x_t\}$ and $\{z_t\}$, cannot guarantee that the limit solution is feasible, i.e., it satisfies $Ax_t = z_t$. As a matter of fact, in the limit, the minimization in (32) assumes the (approximate) form

$$x_{t+1} \in \operatorname*{argmax}_{x \in \mathcal{X}} \|(I - L_{\mathcal{Z}})Ax\|^2 + \|L_{\mathcal{Z}} A(x - x_t)\|^2 \tag{35}$$

which corresponds to an iterative algorithm for performing a projection of x onto the feasible space $\mathcal{X} \cap \{x | Ax = L_{\mathcal{Z}} Ax\}$, and in this context, the contribution $\|L_{\mathcal{Z}} A(x - x_t)\|^2$ plays the role of a proximity operator, forcing vicinity to the solution available from the previous step. Therefore, if the algorithm used to solve the local problem (32) is sufficiently powerful, then convergence to a feasible point is also ensured in the limit. This is the case, in practice, only for moderately non-convex scenarios.

The distributed OPF algorithm

The distributed OPF algorithm that we obtain by applying Algorithm 1 to problem (12) is summarized in Algorithm 2. The local penalty parameters update (29)-(31) is used.

Algorithm 2: Distributed OPF processing in region k (t denotes the iteration number)

1 **for** $t = 0$ **to** ∞ **do**
2   **if** $t = 0$ **then**
3   Initialize local voltages $\boldsymbol{x}_{k,0}$
4   **else**
5   Update local voltages $\boldsymbol{x}_{k,t}$ via

$$\boldsymbol{x}_{k,t} \in \underset{\boldsymbol{x}_k \in \mathcal{X}_k}{\operatorname{argmin}} F_k(\boldsymbol{x}_k) + \frac{1}{2}\boldsymbol{x}_k^T \boldsymbol{D}_{k,t}\boldsymbol{x}_k + \boldsymbol{y}_{k,t}^T \boldsymbol{x}_k$$

    where

6

$$\boldsymbol{D}_{k,t} = \boldsymbol{A}_k^T \operatorname{diag}(\boldsymbol{\epsilon}_{k,t-1})\boldsymbol{A}_k$$

$$\boldsymbol{y}_{k,t} = \boldsymbol{A}_k^T (\operatorname{diag}(\boldsymbol{\epsilon}_{k,t-1})\boldsymbol{z}_{k,t-1} - \boldsymbol{\lambda}_{k,t-1})$$

7   A local minimum must be identified, with the guarantee that the target function is decreased with respect to its value at $\boldsymbol{x}_{k,t-1}$.
8   **end if**
9   Prepare messages $\boldsymbol{m}_{k,t} = \boldsymbol{A}_k \boldsymbol{x}_{k,t}$
10   $\Rightarrow$ Broadcast messages $\boldsymbol{m}_{k,t}$ to neighbor regions
11   $\Leftarrow$ Receive messages $\boldsymbol{m}_{h,t}$ from neighbor regions h
12   Update auxiliary variables via
13

$$z^-_{k,i,j,t} = \frac{1}{2}(m^+_{k,i,j,t} - m^-_{h,j,i,t})$$

$$z^+_{k,i,j,t} = \frac{1}{2}(m^+_{k,i,j,t} + m^+_{h,j,i,t})$$

14   **if** $t = 0$ **then**
15   Initialize Lagrange multipliers $\boldsymbol{\lambda}_{k,0}$. If no a priori information is available, then set them to $\boldsymbol{0}$.
16   Initialize the local gap $\Gamma_{k,0} = \infty$
17   Initialize penalty parameters $\boldsymbol{\epsilon}_{k,0} = \boldsymbol{1}\xi$
18   **else**
19   Update Lagrange multipliers via

$$\boldsymbol{\lambda}_{k,t} = \mathcal{P}\left[\boldsymbol{\lambda}_{k,t-1} + \operatorname{diag}(\boldsymbol{\epsilon}_{k,t-1})(\boldsymbol{A}_k\boldsymbol{x}_{k,t} - \boldsymbol{z}_{k,t})\right]$$

    where $\mathcal{P}$ is a projection onto box $[\boldsymbol{\lambda}_{\min}, \boldsymbol{\lambda}_{\max}]$.
20   Update the local gap $\Gamma_{k,t} = \|\boldsymbol{A}_k\boldsymbol{x}_{k,t} - \boldsymbol{z}_{k,t}\|_\infty$
21   Locally update penalty parameters

$$\boldsymbol{\epsilon}_{k,t} = \begin{cases} \|\boldsymbol{\epsilon}_{k,t-1}\|_\infty \boldsymbol{1} & \text{if } \Gamma_{k,t} \le \theta\, \Gamma_{k,t-1} \\ \tau \|\boldsymbol{\epsilon}_{k,t-1}\|_\infty \boldsymbol{1} & \text{otherwise} \end{cases}$$

22     $\Rightarrow$ Broadcast $\epsilon_{k,i,j,t}$ to neighbor regions
23     $\Leftarrow$ Receive $\epsilon_{h,j,i,t}$ from neighbor region h
24   Correct local penalty parameters via

$$\epsilon_{k,i,j,t+1} = \max\left(\epsilon_{k,i,j,t+1}, \epsilon_{h,j,i,t+1}\right)$$

25   **end if**
26 **end for**

Note that two local message exchanges (denoted with arrows) are required in lines 10 to 11 and lines 22 to 23 to exchange, respectively, the updated values $\boldsymbol{x}_{k,t}$ (in order to update auxiliary variables) and the temptative penalty parameters updates $\check{\boldsymbol{\epsilon}}_{k,t}$ (in order to make sure that the final update satisfies (21)). In principle, a single message exchange could be obtained by postponing the penalty parameters correction of line 24 after the auxiliary variable update in line 13, at the cost of some sub optimality in performance.

Overall, the local processing effort of Algorithm 2 is light. The algorithm complexity is determined by the update of $\boldsymbol{x}_t$ in line 5, which corresponds to a region-based optimization problem, and which can be efficiently solved by state-of-the-art methods, e.g., interior point methods (IPMs). The remaining actions require a limited effort, especially in the standard case where a few connections are active with neighboring regions and auxiliary vectors are short (i.e., $|\mathcal{O}_k| \ll |\mathcal{V}_k|$).

We finally underline that five key parameters are used in Algorithm 2, and these need to be accurately set for good performance. We have:

1. Weighting constants ρ and ζ (they define matrices $\boldsymbol{A}_k$, see (12)-(13)). They should be chosen in such a way that $\rho \gg \zeta > 0$, in order to force the algorithm towards an approximate linear solution on power flow variables.
2. Initialization value for penalty parameters ξ. It should be set to a small value to guarantee a good algorithm outcome even when starting from a point very far from the optimum.
3. Penalty parameters update constants $0 < \theta < 1$ and $\tau > 1$. In order to avoid a rapid increasing behavior on penalty parameters, the constants should be set to values close to 1.

Performance assessment

The algorithm performance is tested using three different scenarios, namely: 1) the wide area network IEEE Power System Test Case Archive [22]; 2) the IEEE PES Distribution Test Feeder [23,24]; 3) a microgrid topology generated according to the model proposed in [25]. The networks in Scenarios 2) and 3) have a tree topology, while Scenario 1) involves networks with many loops where algorithm convergence may be an issue. All chosen scenarios are moderate sized networks, with moderate non-convexities, which constitute the applicability field of the proposed algorithm. Applicability to more complex networks with more severe non-convexities and a high number of loops (e.g., the Polish system models) requires use of some additional (quasi centralized) coordination between entities, and will be the subject of future investigation.

Description of the scenarios

A power losses minimization problem under voltage and power constraints is considered (i.e., $f_i(P_i) = P_i$), and the following settings are used in the various scenarios:

1. Networks sizes $N = 30, 57, 118$, and 300 are used. Constraints and load requests are set as from the MATPOWER distribution [26].
2. The $N = 123$ nodes network is used in single-phase fashion. The chosen settings are inspired by [6]. Load requests are set as given in the dataset, and generating capabilities ranges are added in the form $|Q_{G,i}| \le 1.2|Q_{L,i}|$, and $0 \le P_{G,i} \le 30$ kW, where the subscript L stands for *load* and G for *generation*. Voltage regulation is applied with $0.94 \le |V_i| \le 1.06$.
3. A unique network is selected with $N = 120$. The network is generated as four joint small-world graphs with 30 nodes (to limit the depth of the graph) and rewiring probability $p = 0.4$ (see also details in [16]). Lines lengths have an exponential distribution with parameter $\mu = 65.86$ m and a minimum distance set to 10 m. The impedance value is chosen $2.9400 + j0.0861$ Ω/km (class 1, 10 mm^2 cables). Load requests are randomly generated with an uniform distribution in [0, 3] kW, and with a uniform $\cos\phi$ with $\phi \in [-\frac{\pi}{8}, \frac{\pi}{8}]$. 20% of the nodes are given generation capabilities, randomly distributed in [0, 10] kW for active power and [−20, 20] kVAr for reactive power. Voltage regulation is applied in the range $0.9 \le |V_i| \le 1.1$.

Region partitioning

Region partitioning is a fundamental aspect for ensuring a good performance. Ideally, compact regions with very few outer connections guarantee limited complexity, accuracy of the solution, and controlled computational time. In the considered scenarios, region partitioning is chosen in such a way that a unique generator is available in each region, and the region further includes those loads which are electrically closer (in terms of line impedance) to the generator. Since this corresponds to an excessively fine partitioning in Scenario 2), for the IEEE feeder, the region choice is made in such a way that a local controller is placed at each network bifurcation point, and the associated region corresponds to all those nodes which are electrically closer to it (in terms of line impedance).

Simulation tools

The local optimization problem (see line 5 of Algorithm 2, or see the first of (24)) is solved by using IPOPT [27], an efficient IPM solver which allows a MatLab interface. Although a true optimality guarantee is not available, IPM methods are known to perform very well for OPF kind of problems. MUMPS linear solver is used within IPOPT, and the *warm start* option is used in such a way to start the local minimization process using the solution available from the previous iteration (this reduces computational times). The code is run on a MacBook Air and is written in MatLab [28].

Convergence test in the considered scenarios

A test on the behavior of Algorithm 2 in the three different scenarios using the parameters of Table 1 is illustrated in Figure 2. The starting point is chosen to be the all-ones vector $x_{k,0} = \mathbf{1}$, and Lagrange multipliers are initially set to zero, $\boldsymbol{\lambda}_{k,0} = \mathbf{0}$. This corresponds to the unavailability of any *a priori* information on both position and Lagrange multipliers and is therefore a worst case scenario. Iterations are stopped (and convergence is declared) when the primal gap $\|\boldsymbol{A}\boldsymbol{x}_t - \boldsymbol{z}_t\|_\infty$ (infinity norm) reaches 10^{-4}. The maximum values for Lagrange multipliers are set to $\boldsymbol{\lambda}_{\max} = 10^3 \cdot \mathbf{1}, \boldsymbol{\lambda}_{\min} = -10^3 \cdot \mathbf{1}$.

For the three scenarios considered, Figure 2 shows in the first column the voltages V_i (amplitude and phase diagram) at convergence, together with the active voltage constraints. Observe that all voltage limitations are met.

The second column of Figure 2 shows the behavior of the primal gap in norm 2 and norm ∞ as a function of the iteration number t. Although the curves are not strictly decreasing, they are clearly diminishing to zero-gap value. The penalty parameters update, illustrated in the third column of Figure 2, shows the ability of (29)-(31) of keeping a small gap between maximum and minimum values of $\boldsymbol{\epsilon}_t$. The fact that the parameters are always increasing is due to the sub optimality of the distributed criterion with respect to the centralized criterion (27) which would be more effective in limiting the increase of penalty parameters. Nevertheless, the algorithm converges to points very close to the optimum (see Table 1) despite the very badly chosen initial point. In this respect, the local IPM solvers are fully capable of resolving the limit problem (35) and hence guarantee convergence to a feasible point. Note that the slower convergence is experienced with Scenario 2), i.e., the IEEE feeder with $N = 123$. This is due to the fact that this is the network with highest depth due to its radial structure. This makes the distributed process particularly challenging since agreement must be obtained between regions that are very far one from the other.

Finally, in the fourth column of Figure 2, we provide the locally determined reactive power regulation ($Q_{G,i}$ stands for reactive power at generators), which show a converging behavior in accordance with the fact that the primal gap is vanishing. A perfectly equivalent behavior is found for active powers (but this is not shown in figure).

Performance evaluation

A more in-depth performance measure for the tests of Figure 2 is given in Table 1, where the distributed

Table 1 Performance starting from a remote point

	Network					
	IEEE 30	**IEEE 57**	**IEEE 118**	**IEEE 300**	**IEEE feeder 123**	**radial 120**
IPOPT						
Generated power P_G	190.80 MW	1.26 GW	4.25 GW	23.74 GW	3.53 MW	169.81 kW
Number of iterations	12	10	20	33	17	22
Processing time	0.17 s	0.29 s	1.56 s	20.97 s	1.55 s	1.45 s
Algorithm 2						
Number of regions R	5	7	54	69	24	25
ρ	3	3	5	5	10	10
ζ	1/3	1/3	1/5	1/5	1/10	1/10
ξ	3	3	10	10	30	30
θ	0.99	0.99	0.99	0.99	0.99	0.99
τ	1.02	1.02	1.02	1.02	1.02	1.02
Generated power P_G	191.07 MW	1.26 GW	4.26 GW	23.79 GW	3.55 MW	170.78 kW
Gap	0.14%	0.002%	0.25%	0.23%	0.42%	0.57%
Number of iterations	110	144	186	216	246	85
Processing time (aggregate)	8.93 s	20.03 s	119.18 s	214.88 s	110.22 s	32.28 s
Max processing time per region	2.08 s	6.42 s	3.25 s	12.16 s	10.36 s	5.18 s
Average processing time per region	1.79 s	2.86 s	2.21 s	3.11 s	4.59 s	1.29 s

approach of Algorithm 2 is compared with the performance of a centralized IPOPT solver.

Note that the performance gap with respect to a central solver is always below a 1% error, which is an impressive performance considering that we are dealing with a worst case situation, and that we are approaching the problem in distributed form with a severe network partitioning. As a matter of fact, the outstanding performance of IPMs is mainly due to their central coordination capabilities (e.g., see [15]). Incidentally, we observed that the performance of Algorithm 2 is almost independent of the chosen settings. As a consequence, the performance gap in Table 1 coincides with the ultimate accuracy that could be achieved after thousands of iterations for every studied case.

By inspecting the references, the reader can further appreciate the substantial improvement with respect to the performance of the ADMM-based algorithm of [15], and the sensibly improved network size and partitioning performance with respect to the preliminary algorithm version of [16].

Processing times

Some information on the processing times involved with Algorithm 2 is given in both Table 1 and Figure 3.

Figure 3 shows, for the six networks under consideration, the maximum local processing time and the aggregate processing time per iteration. These are almost constant throughout the iterative process, evidencing the fact that the processing time is approximately linear in the number of iterations. From Table 1, we can further extract some information on the time needed per region (the *max processing time per region*), which is in a range between 2 and 13 s, the value being in agreement with the literature on distributed OPF (e.g., see [6]).

Observe that communication delays were not taken into account in Figure 3 and Table 1, and in fact these can be made negligible by choosing a suitable communication technique. High data rate communication standards with associated short packet lengths are to be preferred. This is the case, for example, of broadband power line communication techniques which can guarantee packet lengths of less than a millisecond [29] and which can be deployed in small area applications (e.g., in micro grids). WiMax is a wireless alternative in these scenarios. For wide area applications, instead, optical fiber communications (e.g., gigabit Ethernet) are an appropriate solution.

Conclusions

In this paper, we proposed a distributed algorithm for OPF regulation based upon a non-convex formulation. By suitably controlling penalty parameters, the algorithm was proven to always converge under a proper assumption on local solver reliability. A certificate of convergence to a local minimum is also available under the request that penalty factors are bounded. The algorithm was shown to provide a reliable performance also in a worst case situation where the search for the optimum is initialized

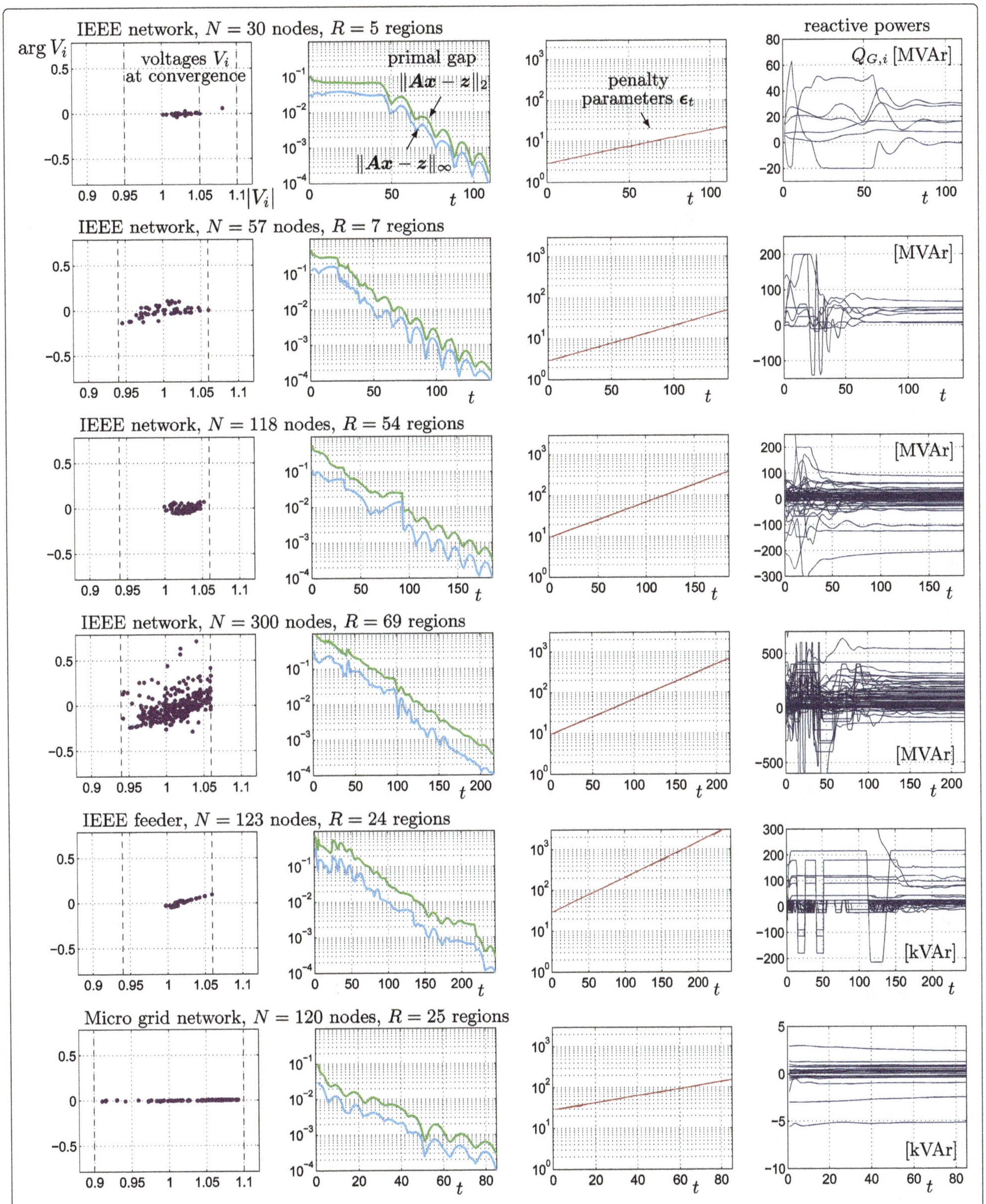

Figure 2 Performance of distributed OPF with IEEE and microgrid networks.

on a point very far from its final destination. The algorithm was proven to be efficient and fast and to be also robust with respect to a severe network partitioning. Its required computational effort was found to be of the order of state-of-the-art methods (using convex problem approximations to ease the convergence issue), with the

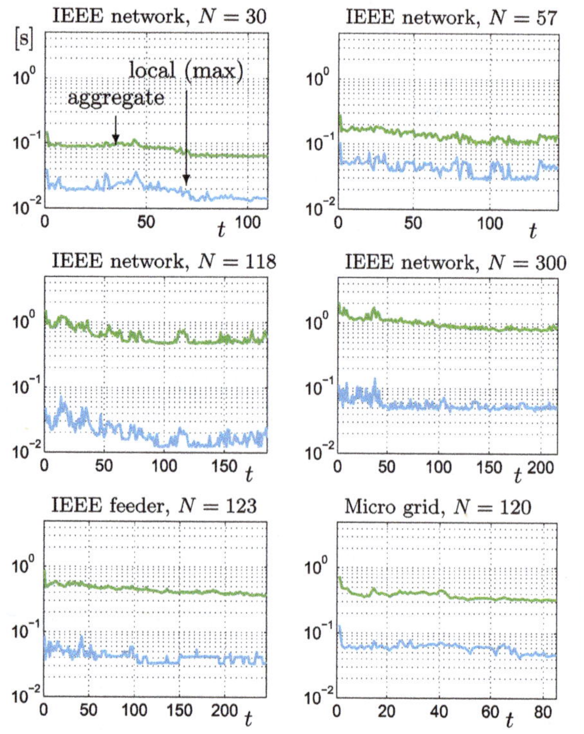

Figure 3 Local and aggregate processing times per iteration.

added value of allowing for a full adherence to the original problem since no (convex) approximation is used.

On the applicability side, the distributed algorithm is readily applicable on moderate time scales (tens of seconds) and on moderate sized networks (up to 300 nodes) for system optimization purposes, not concerning fast regulation (e.g., fault or protection issues require much faster time scales). In this scenario, the algorithm is also expected to be robust to packet losses, because of its alternating direction structure.

Applicability to larger network sizes, with many loops, and more severe non-convexities is instead out of the scope of the present work. As a matter of fact, the proposed alternating direction search allows distributing the processing burden, but might not find an agreement (or it might take too long) in harsh situations. To overcome these difficulties, two strategies can be jointly employed. On the one side, some criteria to determine the optimal region partition strategy should be identified. On the other side, some additional coordination between agents should be used, e.g., a proper distributed generalization of the techniques used in the work of Martinez and Birgin et al. [18] which could also be capable of closing the performance gap with respect to a centralized solver. Use of recent advances on ADMM accelerated methods and scaling techniques (e.g., see [30]) is also an interesting option but need to be suitably adapted to a non-convex context. These aspects are left for future investigations.

Competing interests

The author declares that he has no competing interests.

Authors' information

Tomaso Erseghe was born in 1972. He received the Laurea (M.Sc degree) and the Ph.D. in Telecommunication Engineering from the University of Padova, Italy in 1996 and 2002, respectively. Since 2003, he is Assistant Professor (Ricercatore) at the Department of Information Engineering, University of Padova. His current research interest is in the fields of distributed algorithms for telecommunications, and smart grids optimization. His research activity also covered the design of ultra-wideband transmission systems, properties and applications of the fractional Fourier transform, and spectral analysis of complex modulation formats.

References

1. J Lavaei, SH Low, Zero duality gap in optimal power flow problem. IEEE Trans. Power Syst. **27**(1), 92–107 (2012)
2. S Sojoudi, J Lavaei, in *IEEE Conference on Decision and Control (CDC)*. On the exactness of semidefinite relaxation for nonlinear optimization over graphs: Part I (Florence, Italy, 2013), pp. 1043–1050
3. S Sojoudi, J Lavaei, in *IEEE Conference on Decision and Control (CDC)*. On the exactness of semidefinite relaxation for nonlinear optimization over graphs: part II (Florence, Italy, 2013), pp. 1051–1057
4. R Madani, S Sojoudi, J Lavaei, Convex relaxation for optimal power flow problem: mesh networks. IEEE Trans. Power Syst. **30**(1), 199–211 (2015)
5. AYS Lam, B Zhang, DN Tse, in *IEEE 51st Annual Conference on Decision and Control (CDC 2012)*. Distributed algorithms for optimal power flow problem (Maui, HI, 2012), pp. 430–437
6. B Zhang, AYS Lam, A Dominguez-Garcia, DN Tse, An optimal and distributed method for voltage regulation in power distribution systems. To appear in IEEE Trans. Power Syst.
7. E Dall'Anese, H Zhu, GB Giannakis, Distributed optimal power flow for smart microgrids. IEEE Trans. Smart Grid. **4**(3), 1464–1475 (2013)
8. E Dall'Anese, SV Dhople, BB Johnson, GB Giannakis, Decentralized optimal dispatch of photovoltaic inverters in residential distribution systems. IEEE Trans. Energy Conv. **29**(4), 957–967 (2014)
9. D Gayme, U Topcu, Optimal power flow with large-scale storage integration. IEEE Trans. Power Syst. **28**(2), 709–717 (2013)
10. T Erseghe, S Tomasin, Power flow optimization for smart microgrids by SDP relaxation on linear networks. IEEE Trans. Smart Grid. **4**(2), 751–762 (2013)
11. P Šulc, S Backhaus, M Chertkov, Optimal distributed control of reactive power via the alternating direction method of multipliers. IEEE Trans. Energy Conversion. **29**(4), 968–977 (2014)
12. S Magnusson, PC Weeraddana, C Fischione, A distributed approach for the optimal power flow problem based on ADMM and sequential convex approximations. To appear in IEEE Trans. on Control of Network Systems
13. S Kar, G Hug, in *Power and Energy Society General Meeting, 2012 IEEE*. Distributed robust economic dispatch in power systems: a consensus + innovations approach (San Diego, CA, 2012), pp. 1–8
14. J Mohammadi, S Kar, G Hug, Distributed approach for DC optimal power flow calculations. arXiv (2014). arxiv.org/abs/1410.4236
15. T Erseghe, Distributed optimal power flow using ADMM. IEEE Trans. Power Syst. **29**(5), 2370–2380 (2014)
16. T Erseghe, in *IEEE International Conference on Smart Grid Communications, 2014*. A distributed algorithm for fast optimal power flow regulation in smart grids (Venice, Italy, 2014)
17. R Andreani, EG Birgin, JM Martínez, ML Schuverdt, On augmented lagrangian methods with general lower-level constraints. SIAM J. Optimization. **18**(4), 1286–1309 (2007)
18. E Birgin, J Martínez, *Practical Augmented Lagrangian Methods for Constrained Optimization*. (Society for Industrial and Applied Mathematics, Philadelphia, PA, 2014)
19. OL Mangasarian, *Nonlinear Programming*, vol. 10. (Society for Industrial and Applied Mathematics, Philadelphia, 1994)

20. RT Rockafellar, R Wets, in *Fundamental Principles of Mathematical Sciences*. Variational analysis, vol. 317 (Springer Berlin, 1998)
21. A Castillo, RP O'Neill, Computational performance of solution techniques applied to the ACOPF. Federal Energy Regulatory Commission, Optimal Power Flow Paper. **5** (2013)
22. RD Christie, Power Systems Test Case Archive. www.ee.washington.edu/research/pstca
23. WH Kersting, in *IEEE Power Engineering Society Winter Meeting, 2001*. Radial distribution test feeders, vol. 2 (Columbus, OH, 2001), pp. 908–912
24. Group, D.T.F.W.: Distribution test feeders. ewh.ieee.org/soc/pes/dsacom/testfeeders 2010
25. GA Pagani, M Aiello, Power grid network evolutions for local energy trading. arXiv (2012). arxiv.org/abs/1201.0962
26. RD Zimmerman, CE Murillo-Sánchez, RJ Thomas, Matpower: Steady-state operations, planning, and analysis tools for power systems research and education. Power Systems, IEEE Trans. **26**(1), 12–19 (2011)
27. A Wächter, LT Biegler, On the implementation of an interior-point filter line-search algorithm for large-scale nonlinear programming. Math. Program. **106**, 25–57 (2006)
28. MATLAB, *Version 7.13.0.564 (R2011b)*. (The MathWorks Inc., Natick, Massachusetts, 2011)
29. AR Di Fazio, T Erseghe, E Ghiani, M Murroni, P Siano, F Silvestro, Integration of renewable energy sources, energy storage systems, and electrical vehicles with smart power distribution networks. J. Ambient Intell. Humanized Comput. **4**(6), 663–671 (2013)
30. T Goldstein, B O'Donoghue, S Setzer, R Baraniuk, Fast alternating direction optimization methods. SIAM J. Imaging Sci. **7**(3), 1588–1623 (2014)

13

A novel information transferring approach for the classification of remote sensing images

Jianqiang Gao*, Lizhong Xu, Jie Shen, Fengchen Huang and Feng Xu

Abstract

Traditional remote sensing images classification methods focused on using a large amount of labeled target data to train an efficient classification model. However, these approaches were generally based on the target data without considering a host of auxiliary data or the additional information of auxiliary data. If the valuable information from auxiliary data could be successfully transferred to the target data, the performance of the classification model would be improved. In addition, from the perspective of practical application, these valuable information from auxiliary data should be fully used. Therefore, in this paper, based on the transfer learning idea, we proposed a novel information transferring approach to improve the remote sensing images classification performance. The main rationale of this approach is that first, the information of the same areas associated with each pixel is modeled as the intra-class set, and the information of different areas associated with each pixel is modeled as the inter-class set, and then the obtained texture feature information of each area from auxiliary is transferred to the target data set such that the inter-class set is separated and intra-class set is gathered as far as possible. Experiments show that the proposed approach is effective and feasible.

Keywords: Transfer learning; Image classification; Texture feature information; Support vector machine (SVM)

1 Introduction

Remote sensing images classification is a complex process that may be affected by many factors, such as the availability of high-quality images, proper classification method, and the analytical ability of scientists. For a particular problem, it is often difficult to identify the best classifier due to the lack of a guideline for selection and the availability of suitable classification approaches to band. Therefore, many researchers proposed all kinds of algorithms to address the remote sensing images classification problems. In [1], the authors built textural information model that use spatial information, and then proposed a wavelet-based multi-scale strategy to characterize local texture, taking the physical nature of the data into account, then the extracted textural information was used as new feature to build a texture kernel and the final kernel was the weighted sum of a kernel made with the spectral information and the texture kernel. In [2], the authors proposed applying kernels on a segmentation graph method. Fauvel et al. [3] proposed a spatial-spectral kernel-based approach with the spatial and spectral information were jointly used for the classification. A kernel-based block matrix decomposition approach for the classification of remotely sensed images was proposed by Gao et al. [4]. Tuia et al. [5] used active learning to adapt remote sensing image classifiers. Their goal is to select these pixels in an intelligent fashion that minimizes their number and maximizes their information content. Two strategies based on uncertainty and clustering of the data space are considered to perform active selection. In [6], Dos Santos J.A. et al. proposed a method for interactive classification of remote sensing images considering multiscale segmentation. Their aim is to improve the selection of training samples using the features from the most appropriate scales of representation. They use a boosting-based active learning strategy to select regions at various scales for user's relevance feed back. However, these approaches may ignore the auxiliary data of the remote sensing images. In other words, they do not take the auxiliary data into account in the classification model. In this paper, we aim to transfer the texture feature information from the auxiliary data to

*Correspondence: jianqianggaohh@126.com
College of Computer and Information Engineering, Hohai University, Xikang Road No.1, 210098 Nanjing, China

the target data to improve the classification performance of remote sensing images.

In the traditional classification learning framework, a classification task is to first train a classification model on a labeled training data. And then, the learned model is used to classify a test data set. Hence, under such a framework, the learning method relies on the availability of a large amount of labeled data. In practice, high-quality labeled data are often hard to come by, especially for learning tasks in a new region. Labeling data in a new region involves much human labor and is time-consuming, such as [5,6]. But, fortunately, some auxiliary data such as the texture information are easy to obtain. Therefore, it is reasonable to consider that how to make full use of the valuable texture information of some auxiliary data to improve the classification performance.

Recently, transfer learning [7] has become a popular machine learning method which utilizes auxiliary data for learning. Transfer learning is concerned with adapting knowledge acquired from one source domain to solving problems in another different but related target domain [8]. Generally speaking, traditional machine learning models assume that the training samples collected previously inherit the same feature and distribution as new, incoming data samples during operation [9]. However, in many real-world cases, this assumption does not always hold. In fact, in regard to data classification in non stationary environment, it is not unlikely that the training data set follows a different data distribution as compared with the actual incoming data samples during operation. Such as, in communication channels, discrete signals generated by a specific sequence from a source could be corrupted by Gaussian noise in the transmission process; so, the received signals could deviate from the signal sequence [10]. In this case, traditional machine learning models may not be able to perform well when dealing with the new data samples in the target domain. Hence, the ability of transfer learning would greatly improve the robustness of machine learning models by transferring and adapting knowledge learned from one domain to another related, but different domain. On the other hand, a large set of data samples from a particular task normally is required to train an effective machine learning model [11]. The main principle of transfer learning is that even though the data distributions in the source and target domains are different, some common knowledge across both domains can be adapted for learning [12].

Many researchers have proposed all kinds of methods to transfer learning information or knowledge from auxiliary data. In [13], authors proposed a TrAdaBoost transfer learning framework which constructed a high-quality classification model for target domain by a small number of labeled data and auxiliary data. In [14], authors proposed an extensional method called MultiSource-TrAdaBoost to extend the TrAdaBoost framework for solving multiple sources. In [15], authors proposed a matrix factorization framework to build two mapping matrices for the training images and the auxiliary text data. Based on the co-occurrence data, the correlative principle was introduced to transfer knowledge from text to images by Qi et al. [16]. The authors of reference [17] use an auxiliary data set to construct the pseudo text for each target image, and then, by exploiting the semantic structure of the pseudo text data, the visual features are mapped to the semantic space which respects the text structure. Generally speaking, these methods attempted to transfer information from a lot of auxiliary data to train a more effective model for target data. In our paper, we employ the texture feature information of auxiliary data set to build the similarity matrix for target data set, and then by exploiting the texture information structure of the similarity matrix, the valuable features are mapped to the spectral space and the textural space. At last, the original

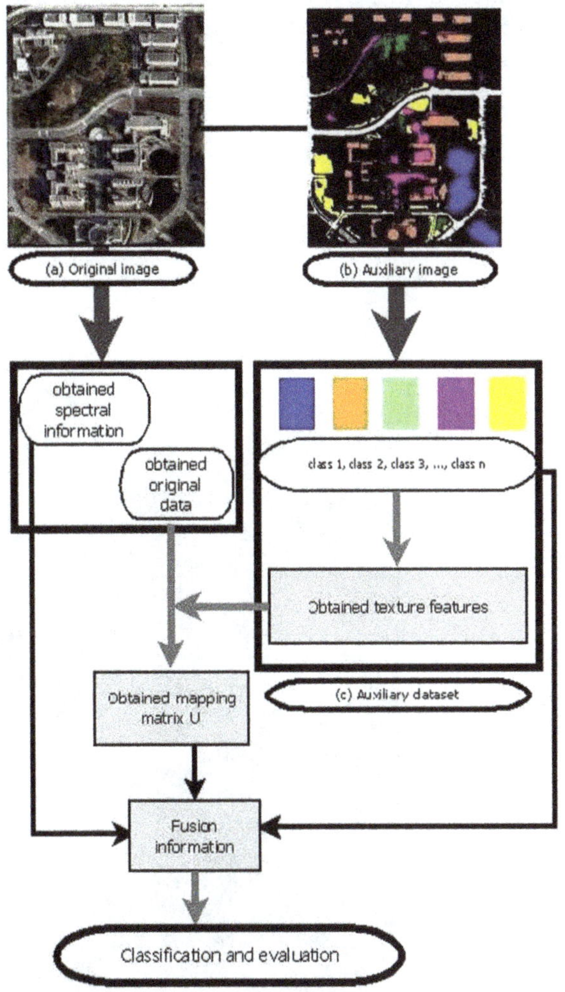

Figure 1 Flowchart of proposed approach.

Table 1 Information classes and training and test samples for PUD

Class No	Samples			
	Name	Train	Test	Auxiliary data
1	Asphalt	548	6304	300
2	Meadow	540	18146	300
3	Gravel	392	1815	300
4	Tree	524	2912	300
5	Metal sheet	265	1113	300
6	Bare soil	532	4572	300
7	Bitumen	375	981	300
8	Bricks	514	3364	300
9	Shadow	231	795	300
–	–	3921	40002	2700

spectral information is combined with texture information to improve the performance of classification model. In order to solve the shortcomings of scale sensitive and more time consuming, Zhang et al. [22] proposed a potential support vector machine (PSVM) algorithm, which uses a novel objective function to overcome the problem of scale sensitivity in SVM.

The remainder of this paper is organized as follows. Section 2 briefly reviews the formulations of relevant knowledge. In Section 3, the derivation process of the proposed method is described in detail. The effectiveness of the proposed method is demonstrated in Section 4 by experiments on remote sensing images. Finally, Section 5 concludes this paper.

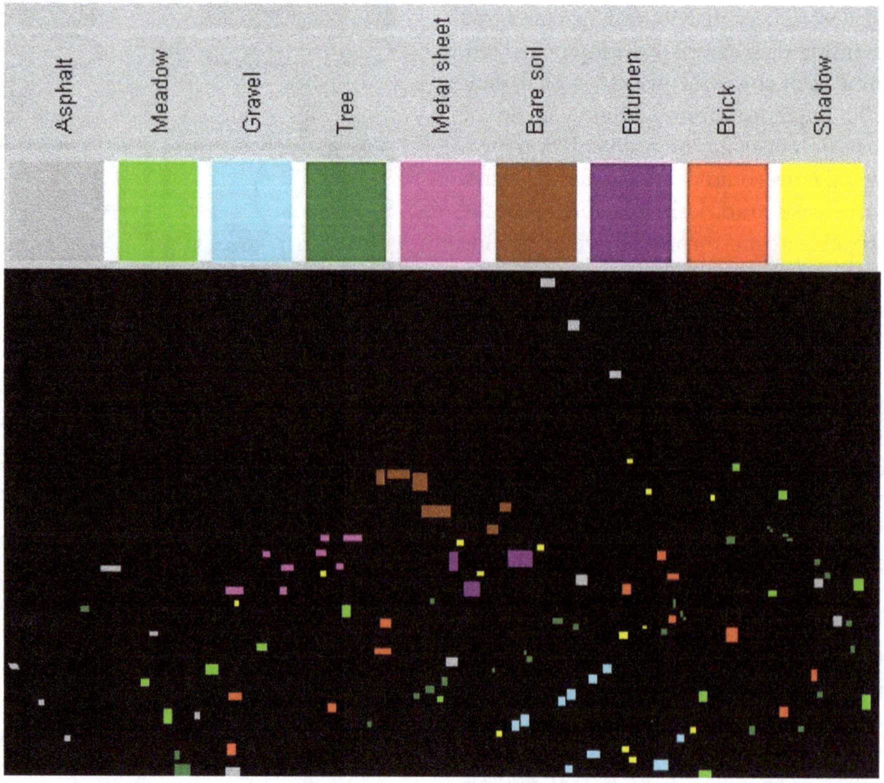

Figure 2 The distribution of training data for PUD.

2 Relevant knowledge

2.1 Transferring knowledge of feature representations

A new case of clustering problems, known as self-taught clustering, was proposed by Dai et al. [18]. Self-taught clustering (STC) is an instance of unsupervised transfer learning, which aims at clustering a small collection of unlabeled data in the target domain with the help of a large amount of unlabeled data in the source domain. STC tries to learn a common feature space across domains, which helps in clustering in the target domain. The objective function of STC is shown as follows:

$$J(\widetilde{X}_T,\widetilde{X}_S,\widetilde{Z}) = I(X_T,Z) - I(\widetilde{X}_T,\widetilde{Z}) + \lambda\left[I(X_S,Z) - I(\widetilde{X}_S,\widetilde{Z})\right], \quad (1)$$

where X_S and X_T are the source and target domain data, respectively. Z is a shared feature space by X_S and X_T, and $I(\cdot,\cdot)$ is the mutual information between two random variables. Suppose that there exist three clustering functions $C_{X_T}: X_T \rightarrow \widetilde{X}_T$, $C_{X_S}: X_S \rightarrow \widetilde{X}_S$, and $C_Z: Z \rightarrow \widetilde{Z}$, where $\widetilde{X}_T, \widetilde{X}_S$, and $\widetilde{Z}$ are corresponding clusters of X_T, X_S, and Z, respectively. The aim of STC is to learn $\widetilde{X}_T$ by solving the optimization problem (1):

$$\arg\min_{\widetilde{X}_T,\widetilde{X}_S,\widetilde{Z}} J(\widetilde{X}_T,\widetilde{X}_S,\widetilde{Z}). \quad (2)$$

An iterative algorithm for solving the optimization function (2) was given in [18].

2.2 Fisher linear discriminant analysis (FLDA)

The main goal of FLDA is to perform dimension reduction while preserving as much information as possible. Linear discriminant analysis aims to find the optimal transformation matrix such that the class structure of the original high-dimensional space is preserved in the low-dimensional space. But in hyperspectral remote sensing images classification problem, generally dimension of the feature vectors is very high with respect to the number of feature vectors. In this subsection, we briefly review the two-dimension Fisher discriminant analysis (2DFLDA) method by Kong et al. [19] proposed to handle the reduce dimensional problem. The main content can be summarized as follows:

Let c be the number of classes, N_i be the number of selected samples from ith class, N be the number of total selected samples from each class, A_j^i be the jth image from ith class, and m_i be the mean image of ith class. $N = \sum_{i=1}^{c} N_i$, $m_i = \frac{1}{N}\sum_{j=1}^{N_i} A_j^i$, $(i = 1, \cdots, c)$. The optimal projection matrix $G = [g_1, g_2, \cdots, g_l]$ can be found in 2DFLDA. Where l is at most $\min(c-1, N)$. We can obtain the optimal projection matrix by maximizing the following criterion:

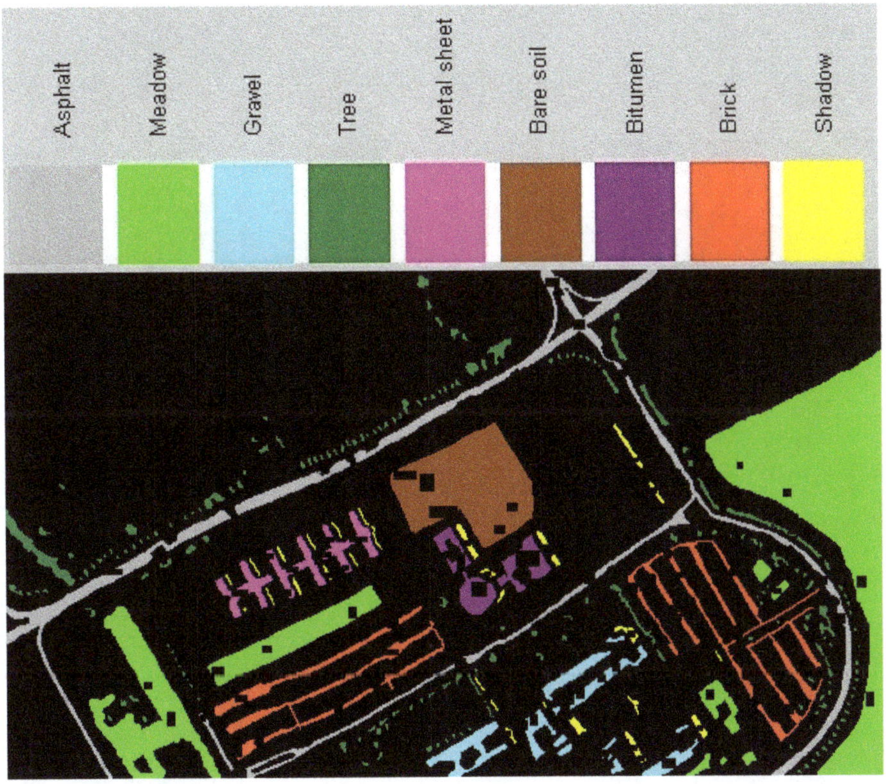

Figure 3 The distribution of test data for PUD.

$$J(G) = \frac{G^T S_b G}{G^T S_w G}, \quad (3)$$

where S_b and S_w are the inter-class and intra-class scatter matrices, respectively. $S_b = \sum_{i=1}^{c} (m_i - m_0)^T (m_i - m_0)$, $S_w = \sum_{i=1}^{c} \sum_{j=1}^{N_i} \left(A_j^i - m_i\right)^T \left(A_j^i - m_i\right)$. $m_0 = \frac{1}{c} \sum_{i=1}^{c} m_i$ is the global mean image of all classes.

3 Learning for information transferring

In this section, based on gray level co-occurrence matrix (GLCM), we first obtain the texture feature information of an image. In addition, the feature matrix of auxiliary data for remote sensing images can be obtained, as described in the following. According to Equation 3, compute matrices S_b and S_w, and solve the optimal projection matrix G. Let λ_i, $(i = 1, 2, \cdots, l)$ be the absolute values of the diagonal elements of the G corresponds to matrix. The value of k is determined such that E is at least some fixed percentage of the whole energy of the image. In our following experiments, we choose $E = 99.99\%$:

$$\frac{\sum_{i=1}^{k} \lambda_i}{\sum_{i=1}^{l} \lambda_i} \geqslant E, \quad (4)$$

Figure 1 shows a block diagram of our simple system. In the next, we will introduce the proposed approach which can be summarized as follows.

Table 2 Confusion matrices, κ and time (s) of PUD

Class no.	1	2	3	4	5	6	7	8	9	UA (%)
Spectral space										
1	5,244	37	130	22	19	16	368	460	8	83.19
2	0	12,230	0	2,223	0	3,675	0	18	0	67.40
3	29	8	1,194	0	0	3	1	580	0	65.79
4	0	36	0	2,858	1	17	0	0	0	98.15
5	0	1	2	2	1,105	0	0	0	3	99.28
6	5	118	0	40	99	4,216	0	24	0	92.21
7	96	0	1	0	0	0	872	12	0	88.89
8	31	19	185	3	0	31	8	3,087	0	91.77
9	21	0	7	0	0	0	0	0	767	96.48
PA(%)	96.65	97.69	78.60	55.52	90.28	52.98	69.82	73.83	98.59	
OA(%) = 78.93										
κ = 0.7340										
t=1.26s										
Fusion space										
1	5,286	5	4	41	3	1	403	547	14	83.85
2	0	18,145	0	1	0	0	0	0	0	99.99
3	0	0	1,782	0	0	33	0	0	0	98.18
4	0	0	0	2,911	1	0	0	0	0	99.97
5	0	0	0	4	1,109	0	0	0	0	99.64
6	0	0	11	26	33	4,502	0	0	0	98.47
7	95	0	0	0	0	0	872	14	0	88.89
8	32	0	0	0	0	0	8	3,324	0	98.81
9	21	0	0	0	0	0	0	3	771	96.98
PA(%)	97.28	99.97	99.17	97.59	96.77	99.25	67.97	85.49	98.22	
OA(%) = 96.75										
κ = 0.9562										
t=0.51s										

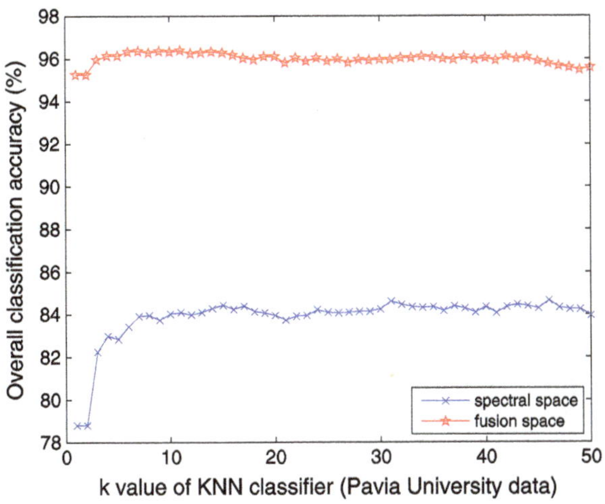

Figure 4 The overall classification accuracy of PUD with KNN.

3.1 Notations

In this paper, we consider two data sets. One is the target data set (*viz.* original image) which only includes spectral information. The other is the auxiliary data set which consists of texture information (Please consult Figure 1). Both the two data sets include c classes. Let $\mathbb{R}^k$ and $\mathbb{R}^m$ be the spectral information and texture information feature spaces. And without loss of generality, we use $S^{(t)}$ and $S^{(a)}$ to represent target data set and auxiliary data set, respectively. Denote the feature matrix of target data set as $\mathbf{X}^{(t)} \in \mathbb{R}^{k \times n^{(t)}}$, the feature matrix of spectral information of auxiliary data set as $\mathbf{X}^{(a)} \in \mathbb{R}^{k \times n^{(a)}}$, and the texture feature information matrix in auxiliary data set as $\mathbf{T}^{(a)} \in \mathbb{R}^{m \times n^{(a)}}$. For target data set, we assume that each sample corresponds to particular auxiliary information. We use $S^{(t)}$ to represent the target data as below Equation 5:

$$S^{(t)} = \left\{ \left(\mathbf{x}_i^{(t)}, \widehat{\mathbf{f}}_i^{(t)}, y_i^{(t)} \right) \Big| 1 \leqslant i \leqslant n^{(t)} \right\}, \tag{5}$$

where $\mathbf{x}_i^{(t)} \in \mathbb{R}^k$ is the column vector of $\mathbf{X}^{(t)}$, $\widehat{\mathbf{f}}_i^{(t)} \in \mathbb{R}^m$ is the feature vector of the pseudo texture feature information of target data, and $y_i^{(t)} \in \{1, 2, \cdots, c\}$ is the class label of target data. Similarly, we use $S^{(a)}$ to represent the auxiliary data as below Equation 6:

$$S^{(a)} = \left\{ \left(\mathbf{x}_j^{(a)}, \mathbf{f}_j^{(a)}, y_j^{(a)} \right) \Big| 1 \leqslant j \leqslant n^{(a)} \right\}, \tag{6}$$

where $\mathbf{x}_j^{(a)} \in \mathbb{R}^k$ is the column vector of $\mathbf{X}^{(a)}$, $\mathbf{f}_j^{(a)} \in \mathbb{R}^m$ is the feature vector of the texture feature information of target data, and $y_j^{(a)} \in \{1, 2, \cdots, c\}$ is the class label of auxiliary data. In addition, we use C_w and C_b to represent the relationship of $\mathbf{x}_i^{(t)}$ and $\mathbf{x}_j^{(t)}$ as follows in Equations 7 and 8, respectively:

$$C^{(w)} = \left\{ \left(\mathbf{x}_i^{(t)}, \mathbf{x}_j^{(t)} \right) \Big| y_i^{(t)} = y_j^{(t)} \right\}, \tag{7}$$

$$C^{(b)} = \left\{ \left(\mathbf{x}_i^{(t)}, \mathbf{x}_j^{(t)} \right) \Big| y_i^{(t)} \neq y_j^{(t)} \right\}. \tag{8}$$

Equation 7 shows that $\mathbf{x}_i^{(t)}$ and $\mathbf{x}_j^{(t)}$ are in the same class in target data set. And then, Equation 8 shows that $\mathbf{x}_i^{(t)}$ and $\mathbf{x}_j^{(t)}$ are in the different classes in target data set.

3.2 Construct the similarity matrix of $S^{(t)}$ and $S^{(a)}$

As we all know, there are same spectrum and texture information for the same region (or field). Therefore, the similarity matrix with very important information for target data set is constructed based on the similarities between samples in $S^{(t)}$ and $S^{(a)}$. For the sample $x_i^{(t)}$ in $S^{(t)}$, the most similar sample in $S^{(a)}$ is defined as Equation 9:

$$f\left(\mathbf{x}_j^{(a)}\right) = \min_{\mathbf{x}_j^{(a)}} d\left(\mathbf{x}_i^{(t)}, \mathbf{x}_j^{(a)}\right), \forall j, y_i^{(t)} = y_j^{(a)}, \left(j = 1, 2, \cdots, n^{(a)}\right), \tag{9}$$

where $d(\cdot, \cdot)$ is the Euclidean distance in $\mathbb{R}^k$. The Equation 9 shows that the auxiliary data set corresponding to $\mathbf{x}_j^{(a)}$ can approximately reflect the similarity relationship of

Table 3 Information classes and training and test samples for HUD

Class No	Samples			
	Name	Train	Test	Auxiliary data
1	Road	488	64,650	100
2	Roof	242	56,560	100
3	Tree	266	24,765	100
4	Bare soil	230	36,650	100
5	Water	182	36,600	100
6	Shadow	304	21,220	100
–	–	8,220	240,445	600

$\mathbf{x}_i^{(t)}$. So, we can obtain the $\widehat{\mathbf{f}}_i^{(t)} = \mathbf{x}_j^{(a)}$. In the following steps, we will obtain the similarity matrix of intra-class $\mathbf{W}_w$ and the similarity matrix of inter-class $\mathbf{W}_b$ by computing Equations 10 and 11:

$$\mathbf{W}_w = \begin{cases} w_{ij}^{(w)} = d\left(\widehat{\mathbf{f}}_i^{(t)}, \mathbf{f}_j^{(a)}\right), \; y_i^{(t)} = y_j^{(t)} \\ 0, \qquad\qquad\qquad\qquad \text{otherwise} \end{cases}, \tag{10}$$

$$\mathbf{W}_b = \begin{cases} w_{ij}^{(b)} = d\left(\widehat{\mathbf{f}}_i^{(t)}, \mathbf{f}_j^{(a)}\right), \; y_i^{(t)} \neq y_j^{(t)} \\ 0, \qquad\qquad\qquad\qquad \text{otherwise} \end{cases}. \tag{11}$$

In Equations 10 and 11, $w_{ij}^{(w)}$ and $w_{ij}^{(b)}$ are the elements of $\mathbf{W}_w$ and $\mathbf{W}_b$, respectively. $d(\cdot,\cdot)$ is the Euclidean distance between two feature vectors with very important texture information. For $\mathbf{W}_w$ and $\mathbf{W}_b$, in order to simplify the calculation, we have done the approximate calculation. The specific steps are as follows:

Firstly, we build feature matrices of similarity matrix $S^{(t)}$ by using $\widehat{\mathbf{f}}_i^{(t)}$ $(i = 1, 2, \cdots, n^{(t)})$ and auxiliary data set matrix $S^{(a)}$ by using $\mathbf{f}_j^{(a)}$ $(j = 1, 2, \cdots, n^{(a)})$, respectively. *Viz:*

$$S_{\widehat{\mathbf{f}}}^{(t)} = \left[\widehat{\mathbf{f}}_1^{(t)}, \widehat{\mathbf{f}}_2^{(t)}, \cdots, \widehat{\mathbf{f}}_{n^{(t)}}^{(t)}\right], \tag{12}$$

$$S_{\mathbf{f}}^{(a)} = \left[\mathbf{f}_1^{(a)}, \mathbf{f}_2^{(a)}, \cdots, \mathbf{f}_{n^{(a)}}^{(a)}\right]. \tag{13}$$

Figure 5 The airborne remote sensing digital image of HUD.

Secondly, we build the similarity matrix of intra-class $\mathbf{W}_w$ and the similarity matrix of inter-class $\mathbf{W}_b$ by using the feature vector of each sample of $S_{\hat{\mathbf{f}}}^{(t)}$ and $S_{\mathbf{f}}^{(a)}$. At last, the $\mathbf{W}_w$ and $\mathbf{W}_b$ as Equations 14 and 15.

$$\mathbf{W}_w = \sum_{i=1}^{n^{(t)}} \sum_{j=1}^{n^{(a)}} (S_{\hat{\mathbf{f}}_i}^{(t)} - S_{\mathbf{f}_j}^{(a)})^T \left(S_{\hat{\mathbf{f}}_i}^{(t)} - S_{\mathbf{f}_j}^{(a)}\right), \quad (14)$$

$$\mathbf{W}_b = \sum_{i=1}^{n^{(t)}} \sum_{j=1}^{n^{(a)}} \left(\overline{S}_{\hat{\mathbf{f}}_i}^{(t)} - \overline{S}_{\mathbf{f}_j}^{(a)}\right)^T \left(\overline{S}_{\hat{\mathbf{f}}_i}^{(t)} - \overline{S}_{\mathbf{f}_j}^{(a)}\right), \quad (15)$$

where $\overline{S}_{\hat{\mathbf{f}}_i}^{(t)}$ is the ith row mean value of $S_{\hat{\mathbf{f}}}^{(t)}$ and $\overline{S}_{\mathbf{f}_j}^{(a)}$ is the jth row mean value of $S_{\mathbf{f}}^{(a)}$.

3.3 Information transferring of auxiliary data

In this paper, our goal is to learn an optimal linear mapping matrix $\mathbf{U} \in \mathbb{R}^{k \times m}$ which project the texture information from auxiliary data set to the target data set. That is because the texture information of an image is very important; meanwhile, it can enhance the image detail by introducing the texture information of auxiliary data. We formulate the regularization framework for information transferring of auxiliary data as follows:

$$\min_{\mathbf{U}} F(\mathbf{U}) = \left\| \mathbf{U}^T \mathbf{X}^{(a)} - \mathbf{T}^{(a)} \right\|_F^2 + \Omega(\mathbf{U}), \quad (16)$$

where $|| \cdot ||_F^2$ is the Frobenius norm, and $\Omega(\cdot)$ is the regularization constraint on $S^{(t)}$. In this framework, we project the texture feature information in $S^{(a)}$ from the auxiliary data set space to the target data set space. Meanwhile, the constraint on $S^{(t)}$ is taken into account. In this paper, we define $\Omega(\cdot)$ as follows:

$$\Omega(\mathbf{U}) = \alpha \Psi_w(\mathbf{U}) - (1 - \alpha)\Psi_b(\mathbf{U}), \quad (17)$$

where Ψ_w is the similarity constraints on C_w, Ψ_b is the diversity constraints on C_b, $\alpha (0 < \alpha < 1)$ is regularization parameter for balancing the tradeoff between

Table 4 Confusion matrices, κ and time (s) of HUD

Class no.	1	2	3	4	5	6	UA (%)
Spectral space							
1	44,609	19,748	0	188	0	105	69.00
2	16,283	37,422	250	777	4	1824	66.16
3	184	1,160	22,429	11	662	319	90.57
4	189	4,576	52	31,832	0	1	86.85
5	33	4	47	0	36,484	32	99.68
6	176	3041	1,165	2	4,358	12,478	58.80
PA(%)	72.57	56.74	93.68	97.02	87.90	84.55	
OA(%) = 77.05							
κ = 0.7145							
t=18s							
Fusion space							
1	64,634	16	0	0	0	0	99.98
2	0	56,560	0	0	0	0	100.0
3	0	638	22,843	0	662	622	92.24
4	0	1,250	0	35,400	0	0	96.59
5	0	33	46	0	36,485	36	99.69
6	0	267	1,457	0	4,366	15,130	71.30
PA(%)	100.0	96.25	93.83	100.0	87.89	95.83	
OA(%) = 96.09							
κ = 0.9515							
t=8s							

within-class and between-class constraints. Specifically, Ψ_w is formulated as follows:

$$\Psi_w(\mathbf{U}) = \sum_{\left(\mathbf{x}_i^t,\mathbf{x}_j^t\right)\in C_w} w_{ij}^{(w)} \left\|\mathbf{U}^T\mathbf{x}_i^{(t)} - \mathbf{U}^T\mathbf{x}_j^{(t)}\right\|_F^2 = tr\left(\mathbf{U}^T\mathbf{X}^{(t)}\mathbf{P}_w\left(\mathbf{X}^{(t)}\right)^T\mathbf{U}\right) \tag{18}$$

where $\mathbf{P}_w = \mathbf{I} - \mathbf{D}_w^{-\frac{1}{2}}\mathbf{W}_w\mathbf{D}_w^{-\frac{1}{2}}$ the normalized Laplacian matrix, $\mathbf{I}$ is a unit matrix, and $\mathbf{D}_w = diag\left(\mathbf{W}_w \cdot \mathbf{1}\right)$ is a weight matrix whose diagonal elements are $\mathbf{D}_w^{ii} = \sum_{j=1}^{n^{(t)}} w_{ij}^{(w)}$, and $tr(\cdot)$ denotes the trace function. Similarly, Ψ_b can be formulated as Equation 19

$$\Psi_b(\mathbf{U}) = \sum_{\left(\mathbf{x}_i^t,\mathbf{x}_j^t\right)\in C_b} w_{ij}^{(b)} ||\mathbf{U}^T\mathbf{x}_i^{(t)} - \mathbf{U}^T\mathbf{x}_j^{(t)}||_F^2 = tr\left(\mathbf{U}^T\mathbf{X}^{(t)}\mathbf{P}_b\left(\mathbf{X}^{(t)}\right)^T\mathbf{U}\right) \tag{19}$$

where $\mathbf{P}_b = \mathbf{I} - \mathbf{D}_b^{-\frac{1}{2}}\mathbf{W}_b\mathbf{D}_b^{-\frac{1}{2}}$ the normalized Laplacian matrix, $\mathbf{D}_b = diag\left(\mathbf{W}_b \cdot \mathbf{1}\right)$ is a weight matrix whose diagonal elements are $\mathbf{D}_b^{ii} = \sum_{j=1}^{n^{(t)}} w_{ij}^{(b)}$.

Through the above analysis, the objective function in Equation 16 can be rewritten as follows:

$$\begin{aligned}
\min_{\mathbf{U}} F(\mathbf{U}) &= \left\|\mathbf{U}^T\mathbf{X}^{(a)} - T^{(a)}\right\|_F^2 + \Omega(\mathbf{U}) \\
&= \left\|\mathbf{U}^T\mathbf{X}^{(a)} - T^{(a)}\right\|_F^2 + \alpha\Psi_w(\mathbf{U}) - (1-\alpha)\Psi_b(\mathbf{U}) \\
&= \left\|\mathbf{U}^T\mathbf{X}^{(a)} - T^{(a)}\right\|_F^2 + \alpha tr\left(\mathbf{U}^T\mathbf{X}^{(t)}\mathbf{P}_w\left(\mathbf{X}^{(t)}\right)^T\mathbf{U}\right) \\
&\quad - (1-\alpha)tr\left(\mathbf{U}^T\mathbf{X}^{(t)}\mathbf{P}_b\left(\mathbf{X}^{(t)}\right)^T\mathbf{U}\right).
\end{aligned} \tag{20}$$

It is obvious that the above optimization is a convex problem, which can be achieved using existing convex optimization packages, such as **fminunc** and **fmincon** functions [20], **SeDuMi** [21]. The detailed description of the overall pseudo algorithm process is given in Algorithm 1.

Algorithm 1 Proposed approach

Input: Datasets $S^{(t)}$, $S^{(a)}$, regularization parameter α.
First stage:
Step 1. The value of k is determined by using Equations 3 and 4;
Step 2. Construct three matrices $\mathbf{X}^{(t)}$, $\mathbf{X}^{(a)}$ and $\mathbf{T}^{(a)}$, meanwhile, calculate matrices $\mathbf{P}_w$ and $\mathbf{P}_b$;
Step 3. Set initialize $\mathbf{U}_0 = \mathbf{0}$;
Step 4. According to the gradient descent method, solve the $\mathbf{U}$ by using optimization packages;
Step 5. Until convergence or maximum iteration number achieves;
Output: The mapping matrix $\mathbf{U}$.
Second stage:
Step 1. For training and test datasets, the valuable texture features are mapped to the spectral space, then combined original spectral space. Viz. $[\mathbf{U} * \mathbf{T}^{(a)}; \mathbf{X}^{(a)}]$ and $[\mathbf{U} * \mathbf{T}^{(s)}; \mathbf{X}^{(t)}]$, where, $\mathbf{T}^{(s)}$ is pseudo texture matrix of test dataset.
Step 2. The effective classification model can be obtained by using SVM or K-Nearest Neighbor (KNN) classifier to train.

4 Experimental results and analysis

In this section, we demonstrate the effectiveness of the proposed approach on remote sensing images classification tasks. The available data set, namely Pavia University data set and Hohai University data set, are used for experiments. In order to evaluate the efficiency of proposed method, the Gaussian radial basis kernel function is employed in our experiment as Equation 21. And then, the penalty term C and the width of kernel g are need to be tuned. In addition, the two parameters were set using five-fold cross validation strategy. Each original data set was scaled between [-1, 1] by using a per band range stretching method.

$$k_\sigma(x_i, x_j) = \exp\left(-\frac{\left\|x_i - x_j\right\|^2}{2\sigma^2}\right) = \exp\left(-g \cdot \left\|x_i - x_j\right\|^2\right), \tag{21}$$

4.1 Pavia University data set (PUD)

Pavia dataset is around the Engineering School at the University of Pavia. It is 610 × 340 pixels. The spatial resolution is 1.3 m per pixel. Twelve channels have been removed due to noise. The remaining 103 spectral channels are processed. Nine classes of interest are considered: asphalt, meadow, gravel, tree, metal sheet, bare soil, bitumen, bricks, and shadow. The training and test sets for each class are given in Table 1.

In our experiments, the product's accuracy (PA) and the user's accuracy (UA) are defined as Equations 22 and 23, respectively:

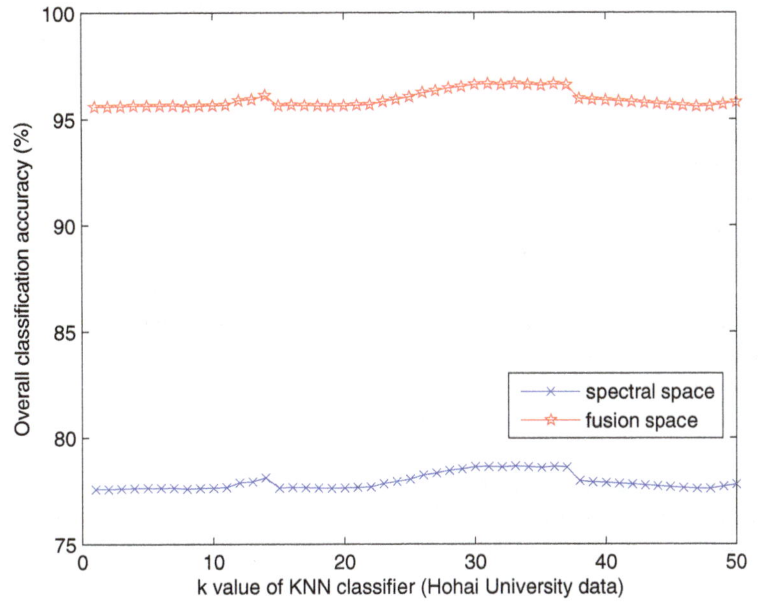

Figure 6 The overall classification accuracy of HUD with KNN.

$$PA_i = \frac{x_{i,i}}{x_{+i}}, \quad (22)$$

$$UA_i = \frac{x_{i,i}}{x_{i+}}, \quad (23)$$

where $x_{i,i}$ is the value on the major diagonal of the ith row in the confusion matrix, x_{i+} is the total number of the ith row, and x_{+i} is the total number of the ith column. To measure the agreement between the classification and the reference data, we compute the kappa coefficient (κ) based on the following equation, where N is the number of total pixels.

$$\kappa = \frac{\left[N\sum_{i=1}^{k} x_{i,i} - \sum_{i=1}^{k}(x_{i+} \times x_{+i})\right]}{\left[N^2 - \sum_{i=1}^{k}(x_{i+} \times x_{+i})\right]}. \quad (24)$$

The distribution of training data and test data are listed in Figures 2 and 3, respectively. All the algorithms are tested in MATLAB (2010b) running on a PC with Intel Core 2 Celeron (2.40 GHz) with 2 GB of RAM. The two parameters C and g (from 2^{-10} to 2^{10}, the step is $2^{0.5}$) are determined by fivefold cross-validation strategy. According to the experiments, we can see that $C = 64$, $g = 8$ is the best choice in spectral space, while $C = 16$, $g = 16$ is the best choice in fusion space. In addition, we also found that the trained model is more efficient in fusion space than only in spectral space by using SVM. The confusion matrices of PUD were shown in Table 2.

According to Table 2, we found that the proposed approach with the fusion space gives better results as compared to the spectral space applied on PUD. In addition, the proposed method gives more overall accuracy (OA) (96.63%) and kappa value (0.9621) as compared to the method original spectral space. So, the proposed method can improve the overall classification accuracy and kappa value. In addition, it is worth noting that the elapsed time of proposed method is less than the original method.

Table 5 OA (%), κ and time (s) for SVM, PSVM, Mbsvd, Mbqrcp and proposed method with SVM

	PUD			HUD		
Method	**OA**	**κ**	**Time**	**OA**	**κ**	**Time**
SVM [22]	78.93	0.7340	1.26	77.05	0.7145	18
PSVM [22]	95.36	0.9466	0.91	91.88	0.9345	11
Mbsvd [4]	95.91	0.9489	1.01	92.03	0.9366	28
Mbqrcp [4]	96.60	0.9548	0.80	94.74	0.9411	10
Proposed	96.75	0.9562	0.51	96.09	0.9515	8

Figure 4 will show the overall classification accuracy by using KNN classifier. According to Figure 4, we can see that the proposed method gives the best results.

4.2 Hohai University data set

The data is the airborne remote sensing digital orthophoto map images acquired in February 2012, at the location of Jiangning campus of Hohai University, Nanjing city, Jiangsu province, P.R. China. This data set is at a spatial resolution of 0.5 m, and the size of image is $1,400 \times 1,024$ pixels. In this data set, we only considered six classes such as road, roof, tree, bare soil, water, and shadow to characterize this area. The class definitions and the number of samples for each experiment is listed in Table 3.

The airborne remote sensing digital image of Hohai University data set (HUD) is shown in Figure 5. According to the experiments, we can see that $C = 2^{0.5} \approx 1.4142$, and $g = 2^{7.5} \approx 181.0193$ is the best choice in fusion space for HUD.

Table 4 shows the confusion matrices, *kappa* values, and elapsed time obtained for different space.

According to Table 4, the OA in classification accuracy obtained by proposed approach (96.09%) was much higher than that obtained by original method (77.05%). In addition, it is worth noting that the elapsed time of proposed algorithm is less than original algorithms. Meanwhile, we also found that in the fusion space, the result is better as compared to in the spectral space in terms of the accuracy of each class classification and κ.

Figure 6 will display the overall classification accuracy by using KNN classifier.

From Figure 6, we can obtain that the proposed approach gives a better result with respect to the OA by using KNN classifier. In order to demonstrate the effectiveness of the proposed approach on remote sensing image classification task, the comparison with other techniques proposed in the literature is implemented in the following experiment. Table 5 gives the overall accuracy and kappa value of different data sets. The best results are reported in Table 5 according to different approaches.

From Table 5, we found that the proposed method shows better performance as compared to other approaches in terms of OA, κ, and running time. This is because the valuable texture information is employed in classification process. Hence, in the classification phase, the classification performance is improved.

5 Conclusions

In this paper, we proposed an information transferring approach to enhance remote sensing images classification performance. The main idea of the proposed method is that the texture feature information of auxiliary data set is transferred to the target data set, and then, the classification model is trained by using SVM or KNN classifier. And finally, experimental results show our approach is feasible.

In addition, the authors realize that more work must be done to improve the classification results in the further. Such as, how to choose a suitable method in classification tasks for remote sensing images. In addition, how to avoid negative transfer is an important open issue that is attracting more and more attention in the future. Of course, in this paper, how to determine the parameter λ, how to transfer other valuable information. This will be an interesting open issue.

Competing interests

The authors declare that they have no competing interests.

Acknowledgements

This work is supported partly by the National Natural Science Foundation of PR China (No. 61271386) and by the Graduates' Research Innovation Program of Higher Education of Jiangsu Province of PR China (No. CXZZ13-0239), and the Industrialization Project of Universities in Jiangsu Province PR China (No. JH10-9).

References

1. G Mercier, F Girard-Ardhuin, Partially supervised oil-slick detection by SAR imagery using kernel expansion. IEEE Trans. Geosci. Remote Sensing. **44**(10), 2839-2846 (2006)
2. Z Harchaoui, F Bach, in *IEEE Conference on Computer Vision and Pattern Recognition (CVPR)*. Image classification with segmentation graph kernels, (2007), pp. 1–8
3. M Fauvel, J Chanussot, JA Benediktsson, A spatial-spectral kernel-based approach for the classification of remote-sensing images. Pattern Recognit. **45**(1), 381–392 (2012)
4. J Gao, L Xu, A Shi, F Huang, A kernel-based block matrix decomposition approach for the classification of remotely sensed images. Appl. Math. Comput. **228**, 531–545 (2014)
5. D Tuiaa, E Pasollib, WJ Emeryc, Using active learning to adapt remote sensing image classifiers. Remote Sensing Environ. **115**(9), 2232–2242 (2011)
6. JA Dos Santos, PH Gosselin, PF Sylvie, RDS Torres, AX Falcao, Interactive multiscale classification of high-resolution remote sensing images. Selected Topics Appl. Earth Observations Remote Sensing, IEEE J. **99**, 1–15 (2013)
7. SJ Pan, Q Yang, A survey on transfer learning. Knowledge and Data Engineering. IEEE Trans. **22**(10), 1345–1359 (2010)
8. G Boutsioukis, I Partalas, I Vlahavas, in *Recent advances in reinforcement learning, 9th European Workshop EWRL*. Transfer learning in multi-agent reinforcement learning domains (Athens, Greece, 2011)
9. B Kocer, A Arslan, Genetic transfer learning. Expert Syst. Appl. **37**, 6997–7002 (2010)
10. DI Ostry, Synthesis of accurate fractional Gaussian noise by filtering. IEEE Trans. Inf. Theory. **52**(4), 1609–1623 (2006)
11. Z Xu, S Sun, in *Neural information processing, 19th International Conference ICONIP*. Multi-source transfer learning with multiview Adaboost (Doha, Qatar, 2012)
12. S Yang, M Lin, C Hou, C Zhang, Y Wu, A general framework for transfer sparse subspace learning. Neural Comput. Appl. **21**(7), 1801–1817 (2012)
13. Y Yao, G Doretto, in *Computer Vision and Pattern Recognition (CVPR), 2010 IEEE Conference on. IEEE*. Boosting for transfer learning with multiple sources, (2010), pp. 1855–1862
14. W Dai, Y Chen, G Xue, Q Yang, Y Yu, in *Proceedings of Advances in Neural Information Processing Systems (NIPS)*. Translated learning: transfer learning across different feature spaces, (2008), pp. 353–360
15. Y Zhu, Y Chen, Z Lu, *et al*, in *Special Track on AI and the Web, associated with The Twenty-Fourth AAAI Conference on Artificial Intelligence (AAAI)*. Heterogeneous Transfer Learning for Image Classification, (2011)

16. GJ Qi, C Aggarwal, T Huang, in *Proceedings of the 20th international conference on World wide web*. Towards semantic knowledge propagation from text corpus to web images (ACM, 2011), pp. 297–306
17. Y Wei, Y Zhao, Z Zhu, Y Xiao, in *Intelligent Information Hiding and Multimedia Signal Processing (IIH-MSP) 2012 Eighth International Conference on*. Knowledge transferring for Image Classification (IEEE, 2012), pp. 347–350
18. W Dai, Q Yang, G Xue, Y Yu, in *Proc. 25th Int'l Conf. Machine Learning*. Self-Taught Clustering, (2008), pp. 200–207
19. H Kong, EK Teoh, JG Wang, R Venkateswarlu, Two dimensional fisher discriminant analysis: Forget about small sample size problem. Proc. IEEE Intern. Conf. Acoustics Speech, Signal Process. **2**, 761–764 (2005)
20. MathWorks (2013). [Online]. Available: http://www.mathworks.com
21. JF Sturm, Using SeDuMi 1.02, a MATLAB toolbox for optimization over symmetric cones. Optimization Methods Softw. **11**(1–4), 625–653 (1999)
22. R Zhang, J Ma, An improved SVM method P-SVM for classification of remotely sensed data. Int. J. Remote Sensing. **29**(20), 6029–6036 (2008)

14

Performance analysis of α-β-γ tracking filters using position and velocity measurements

Kenshi Saho* and Masao Masugi

Abstract

This paper examines the performance of two position-velocity-measured (PVM) α-β-γ tracking filters. The first estimates the target acceleration using the measured velocity, and the second, which is proposed for the first time in this paper, estimates acceleration using the measured position. To quantify the performance of these PVM α-β-γ filters, we analytically derive steady-state errors that assume that the target is moving with constant acceleration or jerk. With these performance indices, the optimal gains of the PVM α-β-γ filters are determined using a minimum-variance filter criterion. The performance of each filter under these optimal gains is then analyzed and compared. Numerical analyses clarify the performance of the PVM α-β-γ filters and verify that their accuracy is better than that of the general position-only-measured α-β-γ filter, even when the variance in velocity measurement noise is comparatively large. We identify the conditions under which the proposed PVM α-β-γ filter outperforms the general α-β-γ filter for different ratios of noise variance in the velocity and position measurements. Finally, numerical simulations verify the effectiveness of the PVM α-β-γ filters for a realistic maneuvering target.

Keywords: α-β-γ filter; Moving target tracking; Position and velocity measurements; Steady-state error; Optimal gains; Minimum variance filter criterion

Introduction

Remote monitoring systems embedded in robots and vehicles require the capability to accurately track moving objects. Tracking filters, such as Kalman filters, extended Kalman filters (EKFs), and particle filters, are commonly used for this purpose [1-5]. These can accurately track movement based on adaptive filtering, which minimizes the error in the predicted position based on dynamical and measurement models. However, these techniques have a relatively heavy computational load, and in some cases their use is impractical. Moreover, their design is conducted empirically, because it is difficult to evaluate the validity of the design parameters (i.e., the process noise) [6,7].

One effective approach that does not suffer from these problems is known as an α-β-γ filter. These are simple tracking filters that assume constant acceleration during the sampling interval [8]. Because of their small computational load, they have been employed in various tracking systems [9-12]. Moreover, there are only three design parameters (the α, β, and γ gains), from which the performance indices can be analytically calculated. Consequently, it is simpler to design an appropriate α-β-γ filter than to construct other tracking filters (e.g., the Kalman filter). Many researchers have studied the analytical performance and design methodology of optimal gains in the α-β-γ filter by assuming simple and practical motion models [8,13-17]. Based on these fundamental studies, recent work has investigated effective gain-setting algorithms for various maneuvering targets [18,19]. Simple gain-setting algorithms have enabled the effectiveness of α-β-γ filters to be verified in various real-world applications, such as motor position control [9] and human fall detection [10].

Traditionally, tracking filter techniques have been applied to radar, sonar, and global positioning systems that measure position only [6]. However, various sensing systems that can accurately measure velocity have recently been developed thanks to technical advances in various sensors and sensor networks, such as the micro-Doppler radar network [20,21]. Consequently, the application of tracking filters to such sensing systems has become an

*Correspondence: saho@fc.ritsumei.ac.jp
Department of Electronic and Computer Engineering, Ritsumeikan University, 1-1-1 Noji-Higashi, Kusatsu, Shiga 525-8577, Japan

important area of research [22-25]. We can expect measured velocities to improve the accuracy of tracking compared with trackers that use position measurements alone. However, when the reliability of the velocity measurements is low, the tracking accuracy may deteriorate. Thus, the relationship between tracking accuracy and measurement noise is very important for the implementation of α-β-γ filters using both position and velocity measurements. Although position-velocity-measured (PVM) tracking filters have been investigated [23-28], the number of such studies is quite small compared with those on general tracking filters that measure only position. Additionally, most studies on PVM tracking filters use Kalman or particle filters. Several applications of PVM α-β-γ filters have been reported [29,30], but these studies do not investigate the filters' theoretical performance, meaning the tracking system parameters are designed empirically. Thus, their analytical properties have not been adequately investigated.

This paper analyzes PVM α-β-γ filters and compares their performance with that of a general α-β-γ filter. For a fair comparison with the general α-β-γ filter, the number of filter gains is fixed to three. As a result, two PVM α-β-γ filters are considered, one of which is being proposed for the first time in this paper. We analytically derive filter performance indices for PVM α-β-γ filters. The derived performance indices are then calculated using the gains determined by a minimum-variance (MV) filter criterion [31], which is the optimal gain design for the general α-β-γ filter. A performance evaluation using numerical analyses and simulations verifies the relationships between measurement noise, filter gains, and filter performance. Moreover, we show that the accuracy of the proposed PVM α-β-γ filter is better than that of the general α-β-γ filter, even when the error in velocity measurements is relatively large.

General α-β-γ filter using position-only measurements

In this section, we summarize the definition and performance of the general α-β-γ filter, which uses position measurements alone. We also review some design methods for filter gains.

The α-β-γ filter predicts the position, velocity, and acceleration of a moving target based on a constant acceleration model using three filter gains [8,13]. This filter iterates prediction and smoothing processes. The prediction process is expressed by the following equations:

$$x_{pk} = x_{sk-1} + Tv_{sk-1} + \left(T^2/2\right) a_{sk-1}, \tag{1}$$

$$v_{pk} = v_{sk-1} + Ta_{sk-1}, \tag{2}$$

$$a_{pk} = a_{sk-1}, \tag{3}$$

where x_{sk} is the smoothed target position at time kT, T is the sampling interval, x_{pk} is the predicted target position, v_{sk} is the smoothed target velocity, v_{pk} is the predicted target velocity, a_{sk} is the smoothed target acceleration, and a_{pk} is the predicted target acceleration. The smoothing process is expressed as follows:

$$x_{sk} = x_{pk} + \alpha(x_{ok} - x_{pk}), \tag{4}$$

$$v_{sk} = v_{pk} + (\beta/T)(x_{ok} - x_{pk}), \tag{5}$$

$$a_{sk} = a_{pk} + \left(\gamma/T^2\right)(x_{ok} - x_{pk}), \tag{6}$$

where x_{ok} is the measured target position, and α, β, and γ are filter gains. The definition of the α-β-γ filter does not include process noise [8,31].

Filter performance indices

To evaluate the tracking performance of the α-β-γ filters, the two steady-state error performance indices can be derived from (1) to (6) [6,8,13,14]. These indices are more effective in evaluating the steady-state tracking accuracy than the error covariance matrix in the Kalman filter equation, which is the usual performance indicator for tracking filters. This is because the error covariance matrix overrates the variance in the errors that is caused by measurement noise, as verified by Ekstrand (see Section 9.8 of [6]). In addition, the relationship between basic properties such as the filter bandwidth and the error covariance matrix is not sufficiently clarified [6,7]. Thus, the indices that are explained in the following subsections are useful when designing α-β-γ filters.

Steady-state error for a target under constant acceleration (smoothing performance index)

An important function of the tracking filter is the reduction of random errors caused by measurement noise. One index of this performance is the steady-state error of a target under constant acceleration considering sensor noise. We assume that x_{ok} contains noise with variance B_x, and that the target moves with constant acceleration. The variance of the predicted target position in the steady-state is calculated using B_x and filter gains as [8,13]:

$$\begin{aligned}\sigma_p^2 &= E\left[\left(x_{pk} - x_{tk}\right)^2\right] \\ &= \frac{8\beta^2 + \alpha(4-2\alpha-\beta)(2\alpha\beta - \gamma(2-\alpha))}{(2-\alpha)(4-2\alpha-\beta)(2\alpha\beta-\gamma(2-\alpha))}B_x,\end{aligned} \tag{7}$$

where x_{tk} is the true target position and $E[\ ]$ indicates the mean. Note that the mean error $E[x_{pk} - x_{tk}]$ is zero, because the assumed target motion is the same as the motion model of the α-β-γ filter (constant acceleration target). We call σ_p^2 the smoothing performance index.

Steady-state error for a target with constant jerk (tracking performance index)
The tracking filter is required to track complicated motion including jerks. In the α-β-γ filter, steady-state bias error occurs when tracking a target moving with constant jerk, because the filter is based on a constant acceleration model. This error is an index of the tracking performance. When $x_{ok} = J(kT)^3/6$ (J is the constant jerk) and the measurement errors are not considered, the steady-state predicted error is expressed as [14]:

$$e_{\rm fin} = \lim_{k\to\infty} (x_{ok} - x_{pk}) = JT^3/\gamma. \tag{8}$$

We call $e_{\rm fin}$ the tracking performance index. The smaller these tracking/smoothing performance indices, the better the tracking filter. However, there is a trade-off between $e_{\rm fin}$ and $\sigma_{\rm p}^2$, and this is a very important consideration in the design of tracking filters [6].

Here, we discuss the design of the α-β-γ filter and compare it with the Kalman filter design. Table 1 summarizes the design parameters, performance indices, and gain calculation method for these filters [7,31,32]. (Note that details of the gain design methods of the α-β-γ filter are explained in the next subsection.) As shown in (7) and (8), the above indices can be directly calculated using the filter gains that we have designed. In contrast, for the Kalman filter, we must design the covariance matrix of the process noise. However, the relationship between this and the performance index (error covariance matrix) has not been rigorously established [7,32]. Moreover, the error covariance matrix gives a misleading evaluation of tracking filter performance, as mentioned earlier in this subsection. Therefore, the design of appropriate process noise is conducted empirically and/or by Monte Carlo simulations (see Section 6 of [6]). Consequently, it is simpler to design an appropriate α-β-γ filter than to construct a Kalman filter or EKF [7,23,26-28].

Gain design methods

Various approaches can be used to determine appropriate gains for the α-β-γ filter. The main approach is to derive gains from the Kalman filter equations, because the α-β-γ filter can be considered as the steady-state Kalman filter [15-17]. However, it is difficult to select appropriate process noise for the motion model, for the same reason as the difficulties in designing a Kalman filter mentioned in the previous subsection. In addition, the performance of an α-β-γ filter derived from the Kalman filter is not optimal when evaluated using the performance indices expressed in (7) and (8) [31].

To avoid these problems, the MV filter criterion has been proposed [14,31]. This criterion determines the gains by minimizing the smoothing performance index $\sigma_{\rm p}^2$ under the condition that the tracking performance index $e_{\rm fin}$ is constant [31]. As shown in (8), the tracking performance index depends only on γ. Thus, for the general α-β-γ filter, the optimal gains with the MV filter criterion are determined by:

$$\begin{aligned} &\arg\min_{\alpha,\beta}\ \sigma_{\rm p}^2 \\ &\text{sub. to}\quad \gamma = \text{const.} \end{aligned} \tag{9}$$

As shown in this equation, the MV filter criterion does not require the process noise of the motion model, unlike the Kalman filter-based approach [15-17]. In [14], it was reported that the performance (evaluated using the tracking/smoothing performance indices of (7) and (8)) is better than that of other α-β-γ filters derived from the Kalman filter equations. Thus, this paper uses the MV filter criterion to determine the optimal gains.

PVM α-β-γ filters

As described in the 'Introduction' section, the performance of PVM tracking filters has not been fully investigated. Hence, we focus on PVM α-β-γ filters. In this section, we derive the smoothing and tracking performance indices ($\sigma_{\rm p}^2$ and $e_{\rm fin}$) for PVM α-β-γ filters. As mentioned above, we ensure a fair comparison with the general α-β-γ filter by fixing the number of gains to three. We can define two types of PVM α-β-γ filter. The first has been used in several tracking systems that measure both position and velocity [29,30]. However, its performance indices have not been derived, and thus the gain determination has so far been conducted empirically. The second type is a new PVM α-β-γ filter that is being proposed for the first time in this paper. The aim of this new filter is to achieve accurate tracking, even when the noise in the velocity measurements is comparatively large. The performance indices $\sigma_{\rm p}^2$ and $e_{\rm fin}$ are derived analytically for each PVM α-β-γ filter.

Table 1 Summary of the design and gain calculation of the α-β-γ and Kalman filters

	Design parameter	Performance index	Gain calculation
α-β-γ filter	Gain γ (or $\Gamma_{\text{A-V}}$ of (29))	Tracking/smoothing performance indices	Based on relationship between gains derived using the Kalman filter equation or MV filter criterion
Kalman filter	Covariance matrix of process noise	Covariance matrix of errors	Adaptively calculated with Riccati equation

Acceleration smoothed by measured velocity (A-V)-type PVM α-β-γ filter

Using the measured velocity v_{ok}, several researchers have used a PVM α-β-γ filter with the smoothing process [29,30]:

$$x_{sk} = x_{pk} + \alpha(x_{ok} - x_{pk}), \quad (10)$$
$$v_{sk} = v_{pk} + \beta(v_{ok} - v_{pk}), \quad (11)$$
$$a_{sk} = a_{pk} + (\gamma/T)(v_{ok} - v_{pk}), \quad (12)$$

and a prediction process that is the same as that in the general α-β-γ filter (expressed in (1) to (3)). Compared with the general α-β-γ filter, the second terms of (5) and (6) have been changed to use the measured velocity. Equation (11) shows that the smoothed velocity can be estimated using the measured velocity. This is the natural expansion of the general α-β-γ filter considering the velocity measurements. Additionally, as shown in (12), the smoothed acceleration is also estimated using the measured velocity. We call this PVM α-β-γ filter the acceleration smoothed by measured velocity (A-V) filter.

The performance indices of the A-V filter are derived from (1) to (3) and (10) to (12). For simplicity, we assume that the noise in the position and velocity measurements is uncorrelated. The smoothing performance index is then derived as

$$\sigma^2_{\mathrm{p,\,A\text{-}V}} = \frac{\alpha}{2-\alpha}B_{\mathrm{x}} + \frac{f_1(\alpha,\beta,\gamma)}{f_2(\alpha,\beta,\gamma)}T^2 B_{\mathrm{v}}, \quad (13)$$

where B_{v} is the variance of the noise in v_{ok}, and

$$\begin{aligned} f_1(\alpha,\beta,\gamma) = {} & \alpha^2(\beta-1)\left(4\beta^2 - 2\beta\gamma - \gamma^2 + 4\gamma\right) \\ & + \alpha\left(6\beta^2\gamma - 4\beta^3 + 8\beta^2 + 3\beta\gamma^2 - 16\beta\gamma - 2\gamma^2 1 + 8\gamma\right) \\ & - 4\beta^2\gamma + 2\beta\gamma(4-\gamma), \end{aligned} \quad (14)$$

$$\begin{aligned} f_2(\alpha,\beta,\gamma) = {} & 2\alpha\beta(2-\alpha)(4-2\beta-\gamma) \\ & \times\left(\alpha^2 + \alpha\beta + \gamma - \alpha^2\beta - \alpha\beta\right). \end{aligned} \quad (15)$$

The derivation of (13) is given in the Appendix. Then, the tracking performance index can be derived as

$$e_{\mathrm{fin,\,A\text{-}V}} = \frac{12 - 6\beta - \gamma}{12\alpha\gamma} J T^3. \quad (16)$$

Again, details of the derivation are given in the Appendix.

Acceleration smoothed by measured position (A-P)-type PVM α-β-γ filter

As it uses the measured velocity, we expect the A-V filter to realize better tracking accuracy than the general α-β-γ filter. However, the performance of the A-V filter deteriorates when the variance B_{v} is large. To reduce this deterioration, we consider another PVM α-β-γ filter whose smoothing process is expressed as follows:

$$x_{sk} = x_{pk} + \alpha(x_{ok} - x_{pk}), \quad (17)$$
$$v_{sk} = v_{pk} + \beta(v_{ok} - v_{pk}), \quad (18)$$
$$a_{sk} = a_{pk} + \left(\gamma/T^2\right)(x_{ok} - x_{pk}), \quad (19)$$

and whose prediction process is the same as in the general α-β-γ filter (i.e., (1) to (3)). The difference from the A-V filter is that the smoothed acceleration is estimated using the measured position, i.e., (6) in the general α-β-γ filter. We call this new PVM α-β-γ filter the acceleration smoothed by measured position (A-P) filter. It appears that the performance of the A-P filter is better than that of the A-V filter when B_{v} is relatively large. In contrast, the A-V filter appears to outperform the A-P filter when B_{v} is relatively small. Moreover, when B_{v} is relatively large, it is unclear whether the performance of the A-P filter or the general α-β-γ filter is better. In the next section, these cases are investigated and clarified with theoretical analyses.

We can derive the smoothing performance index for the A-P filter as:

$$\sigma^2_{\mathrm{p,\,A\text{-}P}} = \frac{g_1(\alpha,\beta,\gamma)B_{\mathrm{x}} + g_2(\alpha,\beta)T^2 B_{\mathrm{v}}}{g_3(\alpha,\beta,\gamma)}, \quad (20)$$

where

$$\begin{aligned} g_1(\alpha,\beta,\gamma) = {} & 8\alpha^3\beta(2-\beta)(\beta-1) \\ & + 2\alpha^2\left(\beta^3\gamma + 4\beta^3 - 3\beta^2\gamma - 8\beta^2 + 6\beta\gamma - 4\gamma\right) \\ & + \alpha\gamma\left(2\beta^3 + \beta^2\gamma + 4\beta^2 - \beta\gamma - 24\beta + 16\right) \\ & - 4\beta\gamma(2-\beta)^2, \end{aligned} \quad (21)$$

$$g_2(\alpha,\beta) = 8\beta^2(\alpha + \beta - \alpha\beta - 2), \quad (22)$$

$$\begin{aligned} g_3(\alpha,\beta,\gamma) = {} & (16 - 8\beta - \beta\gamma - 8\alpha + 4\alpha\beta) \\ & \cdot\left(2\alpha^2\beta^2 - 2\alpha^2\beta - 2\alpha\beta^2 + \alpha\beta\gamma - \alpha\gamma\right. \\ & \left. -2\beta\gamma + 2\gamma\right), \end{aligned} \quad (23)$$

and the tracking performance index is

$$e_{\mathrm{fin,\,A\text{-}P}} = JT^3/\gamma. \quad (24)$$

Note that the tracking performance index is the same as in the general α-β-γ filter, as shown in (8). The derivation of these performance indices is given in the Appendix.

Performance analysis and comparison

In this section, we compare the performance of the A-V filter, A-P filter, and general MV (GMV) α-β-γ filter (which measures position only). The optimal gains are calculated with the MV filter criterion [31], and performance is analyzed using the derived tracking/smoothing performance indices and the calculated gains. The relationship between measurement noise (B_x and B_v) and filter performance is clarified for various gain settings.

Optimal gain calculation with MV filter criterion

First, we calculate the optimal gains of the A-V filter. Under the MV filter criterion, we assume that the tracking performance index is constant. With (16), the tracking performance index depends on

$$C_{\text{A-V}} = \frac{12 - 6\beta - \gamma}{12\alpha\gamma}. \tag{25}$$

Thus, $C_{\text{A-V}}$ is constant in the MV filter criterion. Solving this for γ, we obtain

$$\gamma = \frac{6(2 - \beta)}{12\alpha C_{\text{A-V}} + 1}. \tag{26}$$

Substituting (26) into (13) gives the smoothing performance index $\sigma^2_{\text{p, A-V}}(\alpha, \beta, C_{\text{A-V}})$, which is used to calculate the optimal gains for constant $C_{\text{A-V}}$. Then, we determine the optimal α and β for each $C_{\text{A-V}}$ by:

$$\begin{aligned} &\arg\min_{\alpha,\beta} \ \sigma^2_{\text{p, A-V}}(\alpha, \beta, C_{\text{A-V}}) \\ &\text{sub. to} \quad C_{\text{A-V}} = \text{const.} \end{aligned} \tag{27}$$

Next, we consider the optimal gain calculation of the A-P filter. As shown in (24), the tracking performance index of the A-P filter depends only on γ. Consequently, γ is constant when $e_{\text{fin, A-P}}$ is constant. Thus, we determine the optimal α and β for each γ by:

$$\begin{aligned} &\arg\min_{\alpha,\beta} \ \sigma^2_{\text{p, A-P}} \\ &\text{sub. to} \quad \gamma = \text{const.} \end{aligned} \tag{28}$$

We now give the gain calculation results using (27) and (28) and compare these with the gains from the GMV filter. First, to simplify the discussion, we define the following two parameters.

- The reciprocal of $C_{\text{A-V}}$ is defined as

$$\Gamma_{\text{A-V}} = 1/C_{\text{A-V}}. \tag{29}$$

  With (8), (16), and (24), $\Gamma_{\text{A-V}}$ corresponds to γ in the A-P and GMV filters.
- The ratio of the two variances of measurement noise is defined as

$$R_v = T^2 B_v / B_x. \tag{30}$$

  The smoothing performance of the PVM α-β-γ filters depends on this ratio, as we can see from (13) and (20). The relationship between R_v and the performance indices is very important for the design of tracking filters that use both the measured position and velocity.

Figure 1 shows the gain calculations for the PVM and GMV filters with $R_v = 1/2$ for each value of γ or $\Gamma_{\text{A-V}}$. Here, we have used the gradient descent technique to minimize (27) and (28); the complexity (using big O notation) of this operation is $O(n^2)$ [33]. The mean calculation time to determine each (α, β, γ) is 56.3 s using an Intel CORE i7-4600U CPU@2.10 GHz 2.70 GHz. This time is acceptable, because the gain calculation is conducted in the filter design process before its application in a tracking system. As shown in Figure 1, the value of β in the A-V filter is relatively large compared with that in the GMV filter. This is because β is the gain for velocity smoothing, and the measurement accuracy of the velocity is better than that of the position. For the same reason, the value of β (α) is larger (smaller) in the A-P filter than in the other filters. These examples indicate that the gains of the PVM filters depend on the relationship between the accuracy of the position and velocity measurements.

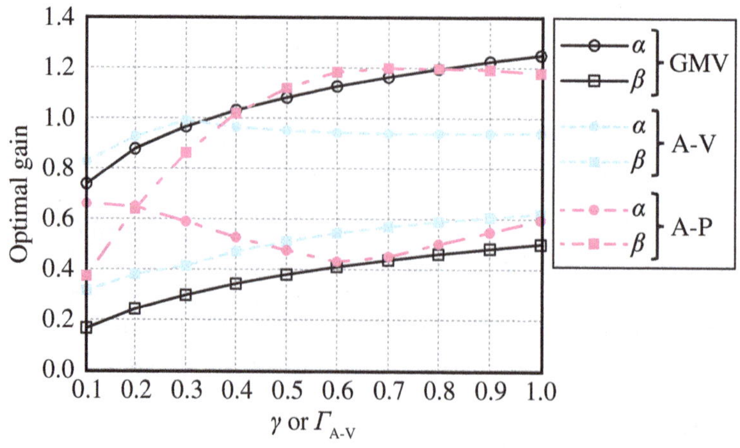

Figure 1 Calculation results for the optimal gains when $R_v = 1/2$.

Analysis results and discussion

Using the calculated optimal gains, we conduct performance analyses of the A-V, A-P, and GMV filters. The smoothing performance indices of these filters are calculated using (7), (13), and (20) under the assumption that the tracking performance indices are constant (i.e., $\Gamma_{\text{A-V}}$ is constant for the A-V filter and γ is constant for the other filters). We assume that the sampling interval T and the variance of the measured position error B_x are normalized to 1.

Figure 2 shows the smoothing performance indices as a function of γ or $\Gamma_{\text{A-V}}$ for $R_v = 1/2$ and 7. For relatively small R_v, shown in Figure 2a, the PVM α-β-γ filters outperform the GMV filter, especially for large values of γ or $\Gamma_{\text{A-V}}$. In this case, the A-V filter realizes the best performance. This is because accurately measured velocities improve the performance of both the smoothing and tracking. Moreover, for larger values of R_v, shown in Figure 2b, the performance deterioration in the proposed A-P filter is small compared with that in the A-V filter. This is because the smoothed acceleration in the A-P filter is calculated using the measured position. The smoothing performance of the A-P filter is better than that of the GMV filter when $\gamma \geq 0.6$ for $R_v = 7$. This result implies that the proposed A-P filter can realize better performance than the GMV filter, even when the noise in the velocity measurements is large. Figure 3 shows the smoothing performance indices as a function of R_v for γ and $\Gamma_{\text{A-V}}$ values of 0.9. As shown in this figure, for $R_v = 10$ (i.e., the noise variance in the velocity measurements is ten times as large as that in the position measurements), the A-P filter achieves better smoothing performance than the GMV filter. In contrast, the performance of the A-V filter deteriorates when R_v is comparatively large.

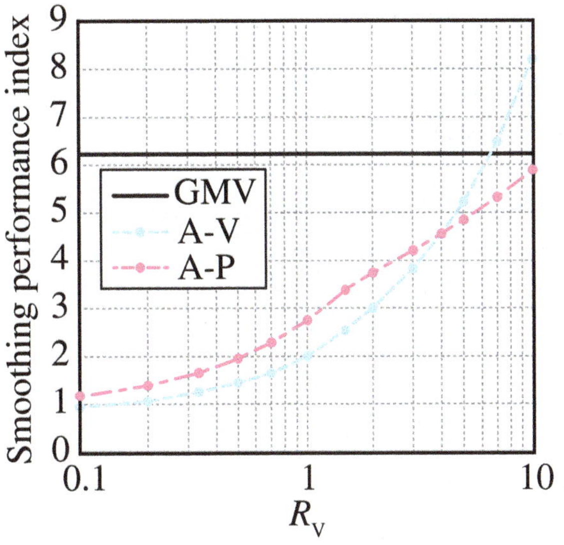

Figure 3 Relationship between smoothing performance indices and the ratio R_v when $\Gamma_{A-V} = \gamma = 0.9$.

Table 2 summarizes the properties of the α-β-γ filters considered in this paper. This table indicates that the A-V filter realizes accurate tracking for small R_v, whereas the proposed A-P filter realizes better accuracy than the other filters for relatively large R_v. Additionally, when

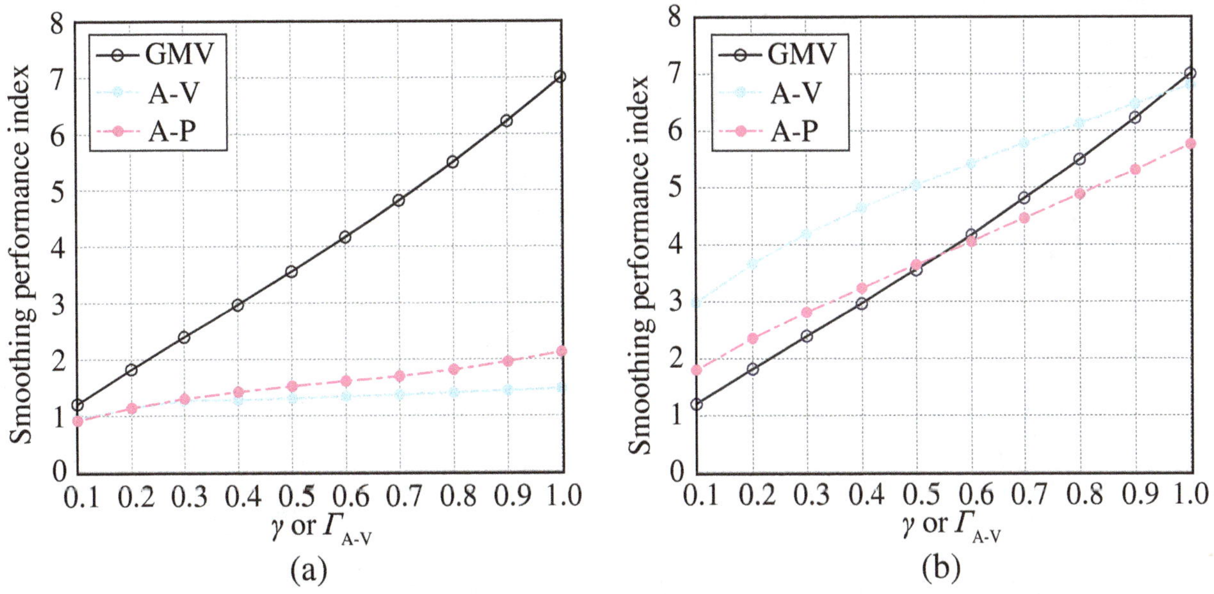

Figure 2 Relationship between smoothing performance indices and Γ_{A-V} or γ. **(a)** $R_v = 1/2$, **(b)** $R_v = 7$.

Table 2 Summary of the properties of the α-β-γ filters considered in this paper

	GMV filter	A-V filter	A-P filter[a]
Measurement parameter	**Position**	**Position and velocity**	
Prediction process		(1) to (3)	
Smoothing process	(4) to (6)	(10) to (12)	(17) to (19)[a]
Smoothing performance index	(7)	(13)[a]	(20)[a]
Tracking performance index	(8)	(16)[a]	(8)[a]
Suitable when R_v is	very large[a]	small[a]	large[a]

[a]Indicates novel results in this paper.

R_v becomes large, the GMV filter realizes the best performance, which suggests that we should not use the measured velocity in this case.

Cramér–Rao bound evaluation

This section calculates the fundamental performance limitation of the PVM and conventional tracking problems using a Cramér–Rao bound (CRB) evaluation. Moreover, we evaluate the tracking accuracy of the PVM filters using Monte Carlo simulations and compare this with the CRBs.

The CRBs in the position estimation are calculated by the Riccati-like recursion used in [34]. The CRB is the lower bound of the covariance of the state estimation, which is expressed as

$$E\left[\left(\hat{\boldsymbol{x}}_k - \boldsymbol{x}_{tk}\right)\left(\hat{\boldsymbol{x}}_k - \boldsymbol{x}_{tk}\right)^T\right] \geq \boldsymbol{J}_k^{-1} = \boldsymbol{P}_k, \tag{31}$$

where $\hat{\boldsymbol{x}}_k$ is the target state estimate based on all measurements collected up to and including time kT, the target state is composed of the position, velocity, and acceleration in the form $(x_k, v_k, a_k)^T$, $\boldsymbol{J}_k$ is the filtering information matrix defined in [35], and $\boldsymbol{P}_k$ is the CRB. When we do not use the process noise, the recursive formula for $\boldsymbol{J}_k$ can be expressed as [36]:

$$\boldsymbol{J}_{k+1} = \left(\boldsymbol{F}^{-1}\right)^T \boldsymbol{J}_k \boldsymbol{F}^{-1} + \boldsymbol{H}^T \boldsymbol{R}^{-1} \boldsymbol{H}, \tag{32}$$

where $\boldsymbol{F}$ is the state transition matrix, $\boldsymbol{H}$ is the observation matrix, and $\boldsymbol{R}$ is the covariance matrix of measurement noise. In the PVM tracking problem, these are expressed as [26]:

$$\boldsymbol{F} = \begin{pmatrix} 1 & T & T^2/2 \\ 0 & 1 & T \\ 0 & 0 & 1 \end{pmatrix}, \tag{33}$$

$$\boldsymbol{H} = \begin{pmatrix} 1 & 0 & 0 \\ 0 & 1 & 0 \end{pmatrix}, \tag{34}$$

$$\boldsymbol{R} = \begin{pmatrix} B_x & 0 \\ 0 & B_v \end{pmatrix}. \tag{35}$$

A detailed explanation is provided in [34].

First, we calculate and compare the CRBs of the general position-only-measured and PVM tracking problems. In the position-only-measured tracking problem, $\boldsymbol{H}$ and $\boldsymbol{R}$ are expressed as:

$$\boldsymbol{H} = \begin{pmatrix} 1 & 0 & 0 \end{pmatrix}, \tag{36}$$

$$\boldsymbol{R} = \begin{pmatrix} B_x \end{pmatrix}, \tag{37}$$

and $\boldsymbol{F}$ is same as (33). We set $B_x = 1$ and $T = 1$. Figure 4a shows the calculated CRBs of the position estimation. As shown in this figure, the performance limitation of the PVM tracking problem is less than that of the position-only-measured tracking problem. The limitation of the PVM tracking problem with $R_v = 0.1$ is small even for relatively large k. When $R_v = 10$, the CRB is almost the same as that for the position-only-measured tracking problem

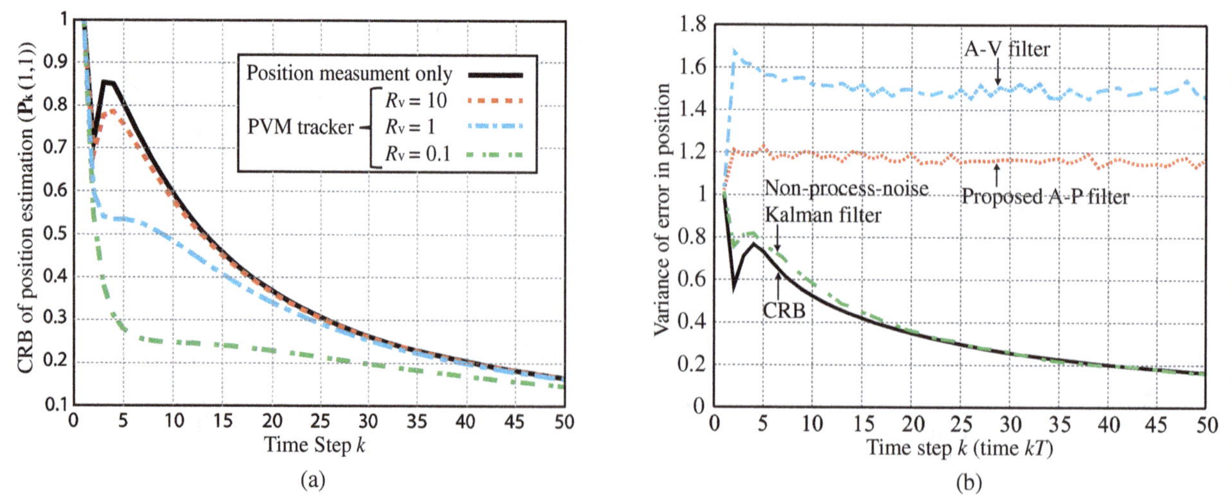

Figure 4 CRB evaluation results. **(a)** CRBs of PVM and position-only-measured tracking problems, **(b)** Monte Carlo simulation results of the PVM tracking filters when $R_v = 10$.

at approximately $k > 10$. However, no deterioration in the CRBs has occurred.

Next, we compare the CRBs with the performance of the PVM α-β-γ filters calculated using Monte Carlo simulations. We set the number of Monte Carlo simulations to 10,000, the initial state to $(0, 0.5, 0.005)^T$, $B_x = 1$, and $T = 1$. For reference, the simulation results for the non-process-noise Kalman filter [32] are also presented. Figure 4b shows the CRB and the Monte Carlo simulation results for the A-V and A-P filters for γ $(\Gamma_{\text{A-V}}) = 0.1$ and the non-process-noise Kalman filters where $R_v = 10$. As shown in this figure, the accuracy of the PVM α-β-γ filters is worse than that of the non-process-noise Kalman filter whose performance is close to the CRB. This is because the α-β-γ filter uses fixed gains, unlike the Kalman filter. However, the proposed A-P filter produces a smaller difference between the CRBs and error variances than the A-V filter. Additionally, the computational load of the proposed filter is smaller than that of the Kalman filter, as we shall discuss later.

Simulation assuming radar tracking of a maneuvering target

Finally, we use numerical simulations to investigate the performance of each filter for a realistic maneuvering target. In this subsection, we simulate the Doppler radar tracking [20,21,29] of a maneuvering target and compare the tracking errors given by the three filters considered in this paper and an EKF [1,2]. Figure 5 shows the simulation scenario. Figure 5a,b shows the true target motion and the radar position, respectively. Two-dimensional (2D) tracking of the point target is assumed, and the received radar signals are calculated using ray-tracing, as in [29]. We assume there are two Doppler radars located at $(x, y) = (0, 0)$ and $(0.5\text{ m}, 0)$. The sampling interval T is 1 ms, and the transmitting signal is an ultrawide-band pulse with a center frequency of 26.4 GHz and bandwidth of 2 GHz. The radars measure the position using ranging results and the velocity using the Doppler shift [29]. White Gaussian noise is added to the ranging and Doppler shift estimations to control R_v. Figure 5c shows the true target position at each time.

We now describe the composition of the tracking filters. For 2D tracking, the α-β-γ filter is composed as follows along each axis:

$$\begin{pmatrix} x_{sk} \\ v_{xsk} \\ a_{xsk} \\ y_{sk} \\ v_{ysk} \\ a_{ysk} \end{pmatrix} = \begin{pmatrix} x_{pk} \\ v_{xpk} \\ a_{xpk} \\ y_{pk} \\ v_{ypk} \\ a_{ypk} \end{pmatrix} + \boldsymbol{K} \begin{pmatrix} x_{ok} - x_{pk} \\ v_{xok} - v_{xpk} \\ y_{ok} - y_{pk} \\ v_{yok} - v_{ypk} \end{pmatrix}, \tag{38}$$

$$\begin{pmatrix} x_{pk} \\ v_{xpk} \\ a_{xpk} \\ y_{pk} \\ v_{ypk} \\ a_{ypk} \end{pmatrix} = \begin{pmatrix} 1 & T & T^2/2 & 0 & 0 & 0 \\ 0 & 1 & T & 0 & 0 & 0 \\ 0 & 0 & 1 & 0 & 0 & 0 \\ 0 & 0 & 0 & 1 & T & T^2/2 \\ 0 & 0 & 0 & 0 & 1 & T \\ 0 & 0 & 0 & 0 & 0 & 1 \end{pmatrix} \begin{pmatrix} x_{sk-1} \\ v_{xsk-1} \\ a_{xsk-1} \\ y_{sk-1} \\ v_{ysk-1} \\ a_{ysk-1} \end{pmatrix}, \tag{39}$$

where $\boldsymbol{K}$ is the gain matrix, x_k, v_{xk}, and a_{xk} denote position, velocity, and acceleration along the x-axis, y_k, v_{yk}, and a_{yk} denote position, velocity, and acceleration along the y-axis, and subscripts 's', 'p', and 'o' denote 'smoothed', 'predicted', and 'observed (measured)', respectively. In the

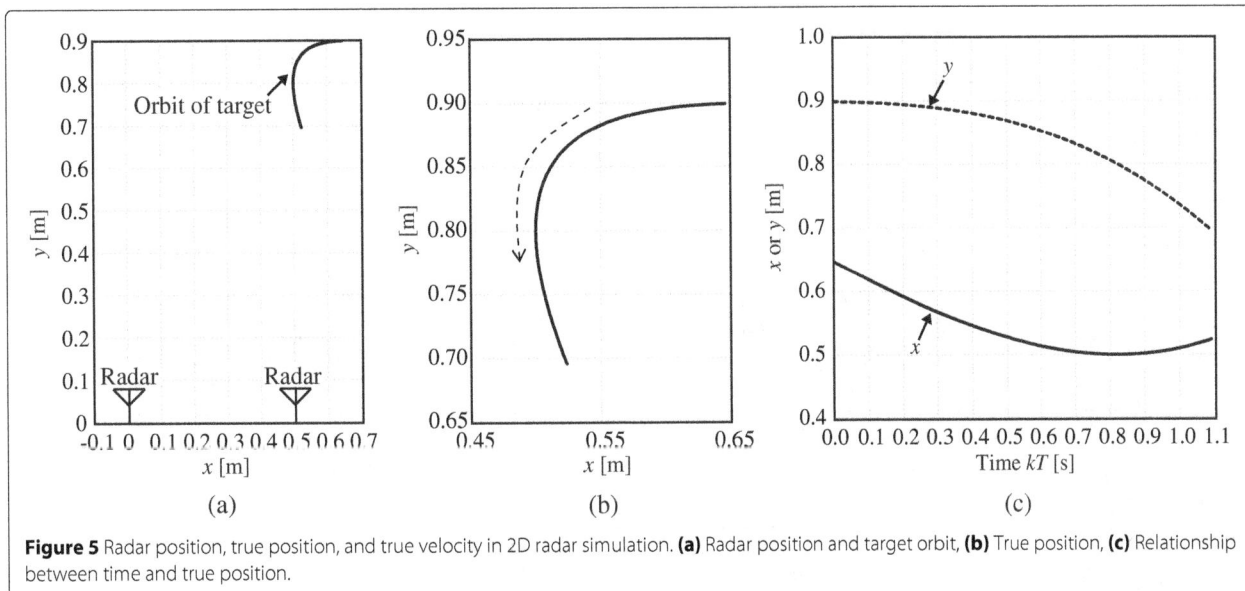

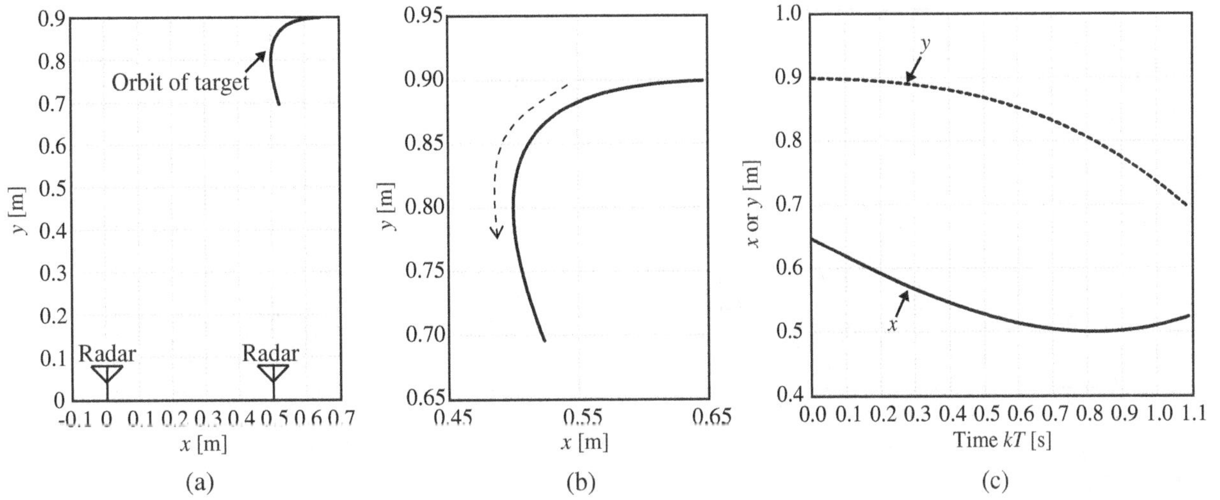

Figure 5 Radar position, true position, and true velocity in 2D radar simulation. **(a)** Radar position and target orbit, **(b)** True position, **(c)** Relationship between time and true position.

A-V filter tracking, $\boldsymbol{K}$ is expressed as:

$$\boldsymbol{K}_{\textbf{A-V}} = \begin{pmatrix} \alpha & 0 & 0 & 0 \\ 0 & \beta & 0 & 0 \\ 0 & \gamma/T & 0 & 0 \\ 0 & 0 & \alpha & 0 \\ 0 & 0 & 0 & \beta \\ 0 & 0 & 0 & \gamma/T \end{pmatrix}. \tag{40}$$

In the A-P filter tracking, $\boldsymbol{K}$ is expressed as:

$$\boldsymbol{K}_{\textbf{A-P}} = \begin{pmatrix} \alpha & 0 & 0 & 0 \\ 0 & \beta & 0 & 0 \\ \gamma/T^2 & 0 & 0 & 0 \\ 0 & 0 & \alpha & 0 \\ 0 & 0 & 0 & \beta \\ 0 & 0 & \gamma/T^2 & 0 \end{pmatrix}. \tag{41}$$

Additionally, in the EKF tracking, $\boldsymbol{K}$ is the Kalman gain matrix calculated by the Kalman filter equations, and a nonlinear measurement model is used [1]. The EKF considers the correlation between the $x - y$ axes, unlike the α-β-γ filters. The process noise is taken to be the zero-mean random-acceleration noise given in [2], and we empirically set this to realize errors that are as small as possible.

Figure 6 shows the results of the 2D radar tracking. Figure 6a,b shows the position prediction error $\sqrt{(x_{pk} - x_{tk})^2 + (y_{pk} - y_{tk})^2}$ (where y_{tk} is the true y) for γ and $\Gamma_{\text{A-V}}$ values of 0.2 and 0.8 with the GMV and PVM α-β-γ filters when the mean R_v is 0.426. The gains of the GMV and PVM α-β-γ filters are calculated according to this mean R_v. As for the previous analyses and simulations, the accuracy of the PVM α-β-γ filters is somewhat better than that of the GMV filters when the gains are relatively large. In both cases, the EKF realizes the best performance. This is because it considers the correlated noise of the axes and has four times as many gains as

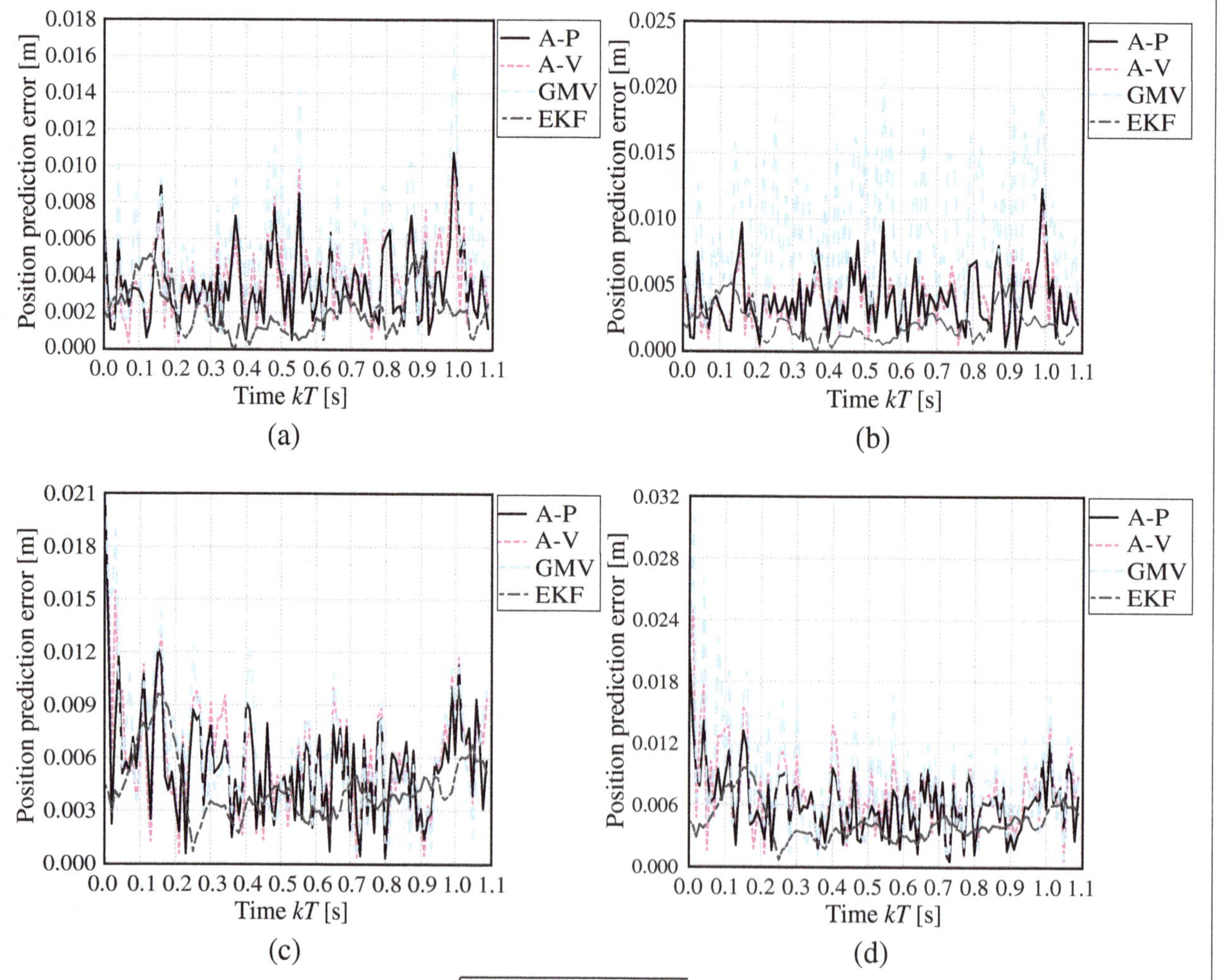

Figure 6 2D radar simulation results (prediction error $\sqrt{(x_{pk} - x_{tk})^2 + (y_{pk} - y_{tk})^2}$). **(a)** $R_v = 0.426$ and $\Gamma_{A-V} = \gamma = 0.2$, **(b)** $R_v = 0.426$ and $\Gamma_{A-V} = \gamma = 0.8$, **(c)** $R_v = 6.67$ and $\Gamma_{A-V} = \gamma = 0.8$, **(d)** $R_v = 6.67$ and $\Gamma_{A-V} = \gamma = 0.8$.

the α-β-γ filters. Moreover, these gains change adaptively. However, the PVM α-β-γ filters realize relatively good accuracy with fixed gains and a small computational load. Table 3 shows the required number of addition, multiplication, and inversion operations of matrices for each time step k for the EKF and PVM α-β-γ filter. As shown in this table, the computational load of the PVM α-β-γ filter is smaller than that of the EKF.

Next, we present results for when the velocity measurement noise is large compared with the position measurement noise. Figure 6c,d shows the position prediction error for γ and $\Gamma_{\text{A-V}}$ values of 0.2 and 0.8 with the GMV and PVM α-β-γ filters when the mean R_v is 6.67. The EKF realizes the best performance in both cases, as for the previous scenario. When $\gamma = \Gamma_{\text{A-V}} = 0.2$, the difference between the three α-β-γ filters is slight, as suggested by our theoretical analysis. When $\gamma = \Gamma_{\text{A-V}} = 0.8$, the proposed A-P filter realizes slightly better accuracy than the other two α-β-γ filters. Table 4 lists the mean error of the results shown in Figure 6. From this table, we can see that the mean accuracy of the EKF is better than that of the PVM α-β-γ filters in all cases. However, the accuracy of the PVM α-β-γ filters is sufficiently high. As shown in Figure 6, the positioning errors of the PVM α-β-γ filters are almost smaller than the wavelength of the radar signal (which is approximately 0.0114 m, corresponding to 26.4 GHz). For various remote sensing applications using radar, sonar, and laser, an accuracy of better than the wavelength is often expected. Thus, the above results indicate that the accuracy of the PVM filters is sufficient for various remote sensing applications. In contrast, the errors in the GMV filter are often larger than the wavelength. The mean error of the A-V filter is 0.418 times that of the GMV filter when $R_v = 0.426$ and $\gamma = \Gamma_{\text{A-V}} = 0.8$. Moreover, the A-P filter even realizes better accuracy when $R_v = 6.67$. These results indicate that the PVM filters enable accurate tracking with simple calculations and few gains when the velocity measurement noise is relatively small, even for 2D radar tracking applications.

Conclusions

In this paper, we have examined the performance of two PVM α-β-γ filters: the A-V filter and the newly proposed A-P filter. The A-V filter estimates smoothed acceleration using the measured velocity, whereas the proposed A-P filter uses the measured position. We analytically derived the tracking and smoothing performance indices of each filter. Based on these performance indices, we calculated the optimal gains of the PVM α-β-γ filters with the MV filter criterion. The performance of the A-V and A-P filters was investigated in terms of the calculated gains, and we compared the output with that from the GMV filter. Numerical analyses verified that the A-V filter realizes better accuracy than the GMV filter when the ratio R_v is relatively small. Moreover, the proposed A-P filter achieved better performance when both R_v and the gain γ were comparatively large. The proposed A-P filter achieved the best performance for $R_v = 7$ and $\gamma \geq 0.6$. In particular, even for $R_v = 10$, which means that the variance of noise in the velocity measurements is ten times that in the position measurements, the A-P filter was more accurate than the GMV filter when $\gamma = 0.9$. Finally, numerical simulations verified the effectiveness of the A-V and A-P filters for a realistic 2D radar application. Moreover, the simulation results matched those from numerical analyses using the derived performance indices. Thus, the performance analyses presented in this paper will be useful for the design of actual tracking systems using position and velocity measurements. One limitation of the current study is our assumption that there are three filter gains. The relaxation of this assumption is an important area of future work that will enable the realization of more accurate tracking filters.

Table 3 Required number of matrix operations for the tracking filtering in each time step

	Addition	Multiplication	Inversion
PVM α-β-γ filter	1	2	0
EKF	5	11	1

Table 4 Mean of the predicted errors in 2D radar simulations (units: mm)

R_v	$\Gamma_{\text{A-V}}$ or γ	A-P filter	A-V filter	GMV filter	EKF
0.426	0.2	3.49	3.82	5.29	2.07
	0.8	3.93	3.91	9.35	2.07
6.67	0.2	5.48	5.91	6.21	4.28
	0.8	6.07	7.09	8.40	4.28

Appendix

Derivation of (13)

The true position of a target under constant acceleration is expressed as

$$x_{tk} = x_{tk-1} + Tv_{tk-1} + \left(T^2/2\right) a_{tk-1}, \tag{42}$$

where v_t and a_t are the true velocity and acceleration. With (1) and (42), the variance of the predicted position errors is

$$\begin{aligned}
\sigma_p^2 &= E\left[\left(x_{pk} - x_{tk}\right)^2\right] \\
&= E\left[\left(x_{sk-1} - x_{tk-1}\right)^2\right] + T^2 E\left[\left(v_{sk-1} - v_{tk-1}\right)^2\right] \\
&\quad + \left(T^4/4\right) E\left[\left(a_{sk-1} - a_{tk-1}\right)^2\right] \\
&\quad + 2TE\left[\left(x_{sk-1} - x_{tk-1}\right)\left(v_{sk-1} - v_{tk-1}\right)\right] \\
&\quad + T^2 E\left[\left(x_{sk-1} - x_{tk-1}\right)\left(a_{sk-1} - a_{tk-1}\right)\right] \\
&\quad + T^3 E\left[\left(v_{sk-1} - v_{tk-1}\right)\left(a_{sk-1} - a_{tk-1}\right)\right].
\end{aligned} \tag{43}$$

Because we assume a steady state, the variances and covariances in (43) do not depend on k. Consequently, we can define these variances and covariances as:

$$\sigma_{\mathrm{sx}}^2 = E\left[(x_{sk} - x_{tk})^2\right] = E\left[\left(x_{sk-1} - x_{tk-1}\right)^2\right], \quad (44)$$

$$\sigma_{\mathrm{sv}}^2 = E\left[(v_{sk} - v_{tk})^2\right] = E\left[\left(v_{sk-1} - v_{tk-1}\right)^2\right], \quad (45)$$

$$\sigma_{\mathrm{sa}}^2 = E\left[(a_{sk} - a_{tk})^2\right] = E\left[\left(a_{sk-1} - a_{tk-1}\right)^2\right], \quad (46)$$

$$\begin{aligned}\sigma_{\mathrm{sxv}}^2 &= E\left[(x_{sk} - x_{tk})(v_{sk} - v_{tk})\right] \\ &= E\left[\left(x_{sk-1} - x_{tk-1}\right)\left(v_{sk-1} - v_{tk-1}\right)\right],\end{aligned} \quad (47)$$

$$\begin{aligned}\sigma_{\mathrm{sxa}}^2 &= E\left[(x_{sk} - x_{tk})(a_{sk} - a_{tk})\right] \\ &= E\left[\left(x_{sk-1} - x_{tk-1}\right)\left(a_{sk-1} - a_{tk-1}\right)\right],\end{aligned} \quad (48)$$

$$\begin{aligned}\sigma_{\mathrm{sva}}^2 &= E\left[(v_{sk} - v_{tk})(a_{sk} - a_{tk})\right] \\ &= E\left[\left(v_{sk-1} - v_{tk-1}\right)\left(a_{sk-1} - a_{tk-1}\right)\right].\end{aligned} \quad (49)$$

Substituting (44) to (49) into (43), we have

$$\sigma_{\mathrm{p}}^2 = \sigma_{\mathrm{sx}}^2 + T^2\sigma_{\mathrm{sv}}^2 + \left(T^4/4\right)\sigma_{\mathrm{sa}}^2 + 2T\sigma_{\mathrm{sxv}}^2 + T^2\sigma_{\mathrm{sxa}}^2 + T^3\sigma_{\mathrm{sva}}^2. \quad (50)$$

The variances and covariances in this equation are derived as functions of the filter gains and the variances of measurement noise. With (1) and (10), we have

$$x_{sk} = (1-\alpha)\left(x_{sk-1} + Tv_{sk-1} + \left(T^2/2\right)a_{sk-1}\right) + \alpha x_{ok}. \quad (51)$$

We can rewrite (42) as

$$x_{tk} = (1-\alpha)\left(x_{tk-1} + Tv_{tk-1} + \left(T^2/2\right)a_{tk-1}\right) + \alpha x_{tk}. \quad (52)$$

Using (51) and (52), the smoothing error is expressed as

$$\begin{aligned}x_{sk} - x_{tk} = {} & (1-\alpha)\left\{(x_{sk-1} - x_{tk-1}) + T(v_{sk-1} - v_{tk-1})\right. \\ & \left. + \left(T^2/2\right)(a_{sk-1} - a_{tk-1})\right\} \\ & + \alpha(x_{ok} - x_{tk}).\end{aligned} \quad (53)$$

Thus, the variance of this error is calculated as

$$\begin{aligned}\sigma_{\mathrm{sx}}^2 &= E\left[(x_{sk} - x_{tk})^2\right] \\ &= (1-\alpha)^2\left\{E\left[\left(x_{sk-1} - x_{tk-1}\right)^2\right] + T^2E\left[\left(v_{sk-1} - v_{tk-1}\right)^2\right]\right. \\ &\quad + \left(T^4/4\right)E\left[\left(a_{sk-1} - a_{tk-1}\right)^2\right] \\ &\quad + 2TE\left[\left(x_{sk-1} - x_{tk-1}\right)\left(v_{sk-1} - v_{tk-1}\right)\right] \\ &\quad + T^2E\left[\left(x_{sk-1} - x_{tk-1}\right)\left(a_{sk-1} - a_{tk-1}\right)\right] \\ &\quad \left. + T^3E\left[\left(v_{sk-1} - v_{tk-1}\right)\left(a_{sk-1} - a_{tk-1}\right)\right]\right\} \\ &\quad + \alpha^2E\left[(x_{ok} - x_{tk})^2\right] \\ &\quad + 2\alpha(1-\alpha)\left\{E\left[\left(x_{sk-1} - x_{tk-1}\right)(x_{ok} - x_{tk})\right]\right. \\ &\quad + TE\left[\left(v_{sk-1} - v_{tk-1}\right)(x_{ok} - x_{tk})\right] \\ &\quad \left. + \left(T^2/2\right)E\left[\left(a_{sk-1} - a_{tk-1}\right)(x_{ok} - x_{tk})\right]\right\}.\end{aligned} \quad (54)$$

Here,

$$E\left[(x_{ok} - x_{tk})^2\right] = B_{\mathrm{x}}. \quad (55)$$

The following relations are satisfied because of the steady-state assumption and because the smoothed parameters are a linear combination of the measured parameters:

$$E\left[\left(x_{sk-1} - x_{tk-1}\right)(x_{ok} - x_{tk})\right] = 0, \quad (56)$$

$$E\left[\left(v_{sk-1} - v_{tk-1}\right)(x_{ok} - x_{tk})\right] = 0, \quad (57)$$

$$E\left[\left(a_{sk-1} - a_{tk-1}\right)(x_{ok} - x_{tk})\right] = 0. \quad (58)$$

Substituting (44) to (49) and (55) to (58) into (54), we obtain

$$\begin{aligned}\sigma_{\mathrm{sx}}^2 = {} & (1-\alpha)^2\left(\sigma_{\mathrm{sx}}^2 + T^2\sigma_{\mathrm{sv}}^2 + \left(T^4/4\right)\sigma_{\mathrm{sa}}^2 + 2T\sigma_{\mathrm{sxv}}^2\right. \\ & \left. + T^2\sigma_{\mathrm{sxa}}^2 + T^3\sigma_{\mathrm{sva}}^2\right) + \alpha^2B_{\mathrm{x}}.\end{aligned} \quad (59)$$

This can be simplified to

$$\begin{aligned}& \alpha(2-\alpha)\sigma_{\mathrm{sx}}^2 - (1-\alpha)^2\left(T^2\sigma_{\mathrm{sv}}^2 + \left(T^4/4\right)\sigma_{\mathrm{sa}}^2 + 2T\sigma_{\mathrm{sxv}}^2\right. \\ & \left. + T^2\sigma_{\mathrm{sxa}}^2 + T^3\sigma_{\mathrm{sva}}^2\right) = \alpha^2B_{\mathrm{x}}.\end{aligned} \quad (60)$$

In the same way, other variances and covariances are calculated using (1) to (3) and (10) to (12) as follows:

$$\begin{aligned}\sigma_{\mathrm{sv}}^2 &= E\left[(v_{sk} - v_{tk})^2\right] \\ &= (1-\beta)^2\left(\sigma_{\mathrm{sv}}^2 + T^2\sigma_{\mathrm{sa}}^2 + 2T\sigma_{\mathrm{sva}}^2\right) + \beta^2B_{\mathrm{v}},\end{aligned} \quad (61)$$

$$\begin{aligned}\sigma_{\mathrm{sa}}^2 &= E\left[(a_{sk} - a_{tk})^2\right] \\ &= \sigma_{\mathrm{sa}}^2 + \left(\gamma^2/T^2\right)B_{\mathrm{v}} + \left(\gamma^2/T^2\right)\left(\sigma_{\mathrm{sv}}^2 + T^2\sigma_{\mathrm{sa}}^2 + 2T\sigma_{\mathrm{sva}}^2\right) \\ &\quad - (2\gamma/T)\left(\sigma_{\mathrm{sva}}^2 + \sigma_{\mathrm{sa}}^2\right)\end{aligned} \quad (62)$$

$$\begin{aligned}\sigma_{\mathrm{sxv}}^2 &= E\left[(x_{sk} - x_{tk})(v_{sk} - v_{tk})\right] \\ &= (1-\alpha)(1-\beta)\left(\sigma_{\mathrm{sxv}}^2 + T\sigma_{\mathrm{sxa}}^2 + T\sigma_{\mathrm{sv}}^2 + T^2\sigma_{\mathrm{sva}}^2\right. \\ &\quad \left. + (T^2/2)\sigma_{\mathrm{sva}}^2 + (T^3/2)\sigma_{\mathrm{sa}}^2\right)\end{aligned} \quad (63)$$

$$\begin{aligned}\sigma_{\mathrm{sxa}}^2 &= E\left[(x_{sk} - x_{tk})(a_{sk} - a_{tk})\right] \\ &= (1-\alpha)\left(\sigma_{\mathrm{sxa}}^2 + T\sigma_{\mathrm{sva}}^2 + \left(T^2/2\right)\sigma_{\mathrm{sa}}^2\right) \\ &\quad - (g(1-a)/T)\left(\sigma_{\mathrm{sxv}}^2 + T\sigma_{\mathrm{sxa}}^2 + T\sigma_{\mathrm{sv}}^2\right. \\ &\quad \left. + T^2\sigma_{\mathrm{sva}}^2 + \left(T^2/2\right)\sigma_{\mathrm{sva}}^2 + \left(T^3/2\right)\sigma_{\mathrm{sa}}^2\right),\end{aligned} \quad (64)$$

$$\begin{aligned}\sigma_{\mathrm{sva}}^2 &= E\left[(v_{sk} - v_{tk})(a_{sk} - a_{tk})\right] \\ &= (1-\beta)(1-\gamma)\left(\sigma_{\mathrm{sva}}^2 + T\sigma_{\mathrm{sa}}^2\right) - (\gamma(1-\beta)/T) \\ &\quad \times\left(\sigma_{\mathrm{sv}}^2 + T\sigma_{\mathrm{sva}}^2\right) + (\beta\gamma/T)B_{\mathrm{v}},\end{aligned} \quad (65)$$

where $$E\left[(v_{ok} - v_{tk})^2\right] = B_{\mathrm{v}}, \quad (66)$$

and the following is satisfied because we assume that the measurement position and velocity noise are uncorrelated:

$$E\left[(x_{ok} - x_{tk})(v_{ok} - v_{tk})\right] = 0. \quad (67)$$

Equations (61) to (65) can be simplified to:

$$\beta(2-\beta)\sigma_{sv}^2 - (1-\beta)^2\left(T^2\sigma_{sa}^2 + 2T\sigma_{sva}^2\right) = \beta^2 B_v, \quad (68)$$

$$\begin{aligned}&\gamma(2-\gamma)\sigma_{sa}^2 - \left(\gamma^2/T^2\right)\sigma_{sv}^2 - (2\gamma(1-\gamma)/T)\\ &\sigma_{sva}^2 = \left(\gamma^2/T^2\right)B_v,\end{aligned} \quad (69)$$

$$\begin{aligned}&(\alpha+\beta-\alpha\beta)\,\sigma_{sxv}^2 - (1-\alpha)(1-\beta)\\ &\left(T\sigma_{sv}^2 + \left(T^3/2\right)\sigma_{sa}^2 + T\sigma_{sxa}^2 + \left(3T^2/2\right)\sigma_{sva}^2\right) = 0,\end{aligned} \quad (70)$$

$$\begin{aligned}&(\alpha+\gamma-\alpha\gamma)\,\sigma_{sxa}^2 + (\gamma(1-\alpha)/T)\left(T\sigma_{sv}^2 + \sigma_{sxv}^2\right)\\ &-\left((\alpha-1)(\gamma-1)\,T^2/2\right)\sigma_{sa}^2\\ &-\left((\alpha-1)(3\gamma-2)\,T/2\right)\sigma_{sva}^2 = 0,\end{aligned} \quad (71)$$

$$\begin{aligned}&(\beta+2\gamma-2\beta\gamma)\,\sigma_{sva}^2 + \gamma(1-\beta)\sigma_{sv}^2\\ &-(1-\beta)(1-\gamma)T\sigma_{sa}^2 = (\beta\gamma/T)B_v.\end{aligned} \quad (72)$$

Solving the linear system involving (59) and (68) to (72), we obtain:

$$\sigma_{sx}^2 = \frac{\alpha}{2-\alpha}B_x + \frac{f_1(\alpha,\beta,\gamma)(1-\alpha)^2}{f_2(\alpha,\beta,\gamma)}T^2 B_v, \quad (73)$$

$$\sigma_{sv}^2 = \frac{2\beta^2 + 2\gamma - 3\beta\gamma}{\beta(4-2\beta-\gamma)}B_v, \quad (74)$$

$$\sigma_{sa}^2 = \frac{2\gamma^2}{\beta(4-2\beta-\gamma)}\frac{B_v}{T^2}, \quad (75)$$

$$\sigma_{sxv}^2 = \frac{(1-\alpha)(1-\beta)\left(4\alpha\beta^2 - \alpha\gamma^2 - 4\alpha\beta\gamma + 4\alpha\gamma + 4\beta\gamma\right)}{f_2(\alpha,\beta,\gamma)/\alpha}TB_v, \quad (76)$$

$$\sigma_{sxa}^2 = \frac{\gamma(\alpha-1)\left(4\alpha\beta^2 - 4\alpha\beta - \gamma^2 + 4\gamma\right)}{f_2(\alpha,\beta,\gamma)/\alpha}B_v, \quad (77)$$

$$\sigma_{sva}^2 = \frac{\gamma(2\beta-\gamma)}{\beta(4-2\beta-\gamma)}\frac{B_v}{T}, \quad (78)$$

where $f_1(\alpha,\beta,\gamma)$ and $f_2(\alpha,\beta,\gamma)$ are expressed as (14) and (15). Substituting (73) to (78) into (50), we arrive at (13).

Derivation of (16)

We first derive the relationship between the measured signals (x_{ok} and v_{ok}) and the predicted position x_{pk} in the z-domain and then obtain the tracking performance index using the final value theorem. Applying a z-transform to (1) to (3) and (10) to (12), we obtain:

$$X_p(z) = X_s(z)/z + TV_s(z)/z + \left(T^2/2\right)A_s(z)/z, \quad (79)$$

$$V_p(z) = V_s(z)/z + TA_s(z)/z, \quad (80)$$

$$A_p(z) = A_s(z)/z, \quad (81)$$

$$X_s(z) = X_p(z) + \alpha(X_o(z) - X_p(z)), \quad (82)$$

$$V_s(z) = V_p(z) + \beta(V_o(z) - V_p(z)), \quad (83)$$

$$A_s(z) = A_p(z) + (\gamma/T)(V_o(z) - V_p(z)). \quad (84)$$

Substituting (84) into (81), we have

$$A_p(z) = \frac{\gamma}{z-1}\cdot\frac{V_o(z) - V_p(z)}{T}. \quad (85)$$

Substituting (83) into (80) gives

$$(z+\beta-1)V_p(z) = \beta V_o(z) + zTA_p(z). \quad (86)$$

Substituting (85) into (86), the relationship between the predicted and measured velocities is calculated as

$$V_p(z) = \frac{(\beta+\gamma)z - \beta}{z^2 + (\beta+\gamma-2)z - \beta + 1}V_o(z). \quad (87)$$

Substituting (87) into (85), the relationship between the predicted acceleration and the measured velocities is calculated as

$$A_p(z) = \frac{\gamma(z-1)}{z^2 + (\beta+\gamma-2)z - \beta + 1}\frac{V_o(z)}{T}. \quad (88)$$

Substituting (82) to (84), (87), and (88) into (79), the relationship between the predicted position and the measured position and velocity is written as

$$\begin{aligned}X_p(z) = &\frac{\alpha}{z+\alpha-1}X_o(z)\\ &+\frac{z((2\beta+\gamma)z - 2\beta+\gamma)}{2(z+\alpha-1)(z^2 + (\beta+\gamma-2)z - \beta + 1)}TV_o(z).\end{aligned} \quad (89)$$

Thus, the z-transform of the error $x_{ok} - x_{pk}$ is expressed as

$$\begin{aligned}E_p(z) = &\frac{z-1}{z+\alpha-1}X_o(z)\\ &-\frac{z((2\beta+\gamma)z - 2\beta+\gamma)}{2(z+\alpha-1)(z^2 + (\beta+\gamma-2)z - \beta + 1)}TV_o(z).\end{aligned} \quad (90)$$

Here, the measured position and velocity of a target with constant jerk J are:

$$x_{ok} = J(kT)^3/6, \quad (91)$$

$$v_{ok} = J(kT)^2/2, \quad (92)$$

and their z-transforms are:

$$X_o(z) = \frac{z\left(z^2 + 4z + 1\right)}{6(z-1)^4}JT^3, \quad (93)$$

$$V_o(z) = \frac{z(z+1)}{2(z-1)^3}JT^2. \quad (94)$$

Substituting (93) and (94) into (90), we have

$$E_p(z) = \frac{z\left(2z^2 + (8-4\beta-\gamma)z + 2 - 2\beta\right)}{12(z-1)(z+\alpha-1)\left(z^2 + (\beta+\gamma-2)z - \beta + 1\right)}JT^3. \quad (95)$$

With the final value theorem $\lim_{z\to 1}(z-1)E_p(z)$, we have (16).

Derivation of (20)

Using the same procedure as for the A-V filter, the linear system with respect to the variances and covariances of the smoothing parameters is calculated using (1) to (3) and (17) to (19) as:

$$\alpha(2-\alpha)\sigma_{sx}^2 - (1-\alpha)^2\left(T^2\sigma_{sv}^2 + (T^4/4)\,\sigma_{sa}^2 + 2T\sigma_{sxv}^2 + T^2\sigma_{sxa}^2 + T^3\sigma_{sva}^2\right) = \alpha^2 B_x, \tag{96}$$

$$\beta(2-\beta)\sigma_{sv}^2 - (1-\beta)^2\left(T^2\sigma_{sa}^2 + 2T\sigma_{sva}^2\right) = \beta^2 B_v, \tag{97}$$

$$(\gamma(4-\gamma)/4)\,\sigma_{sa}^2 - \left(\gamma - 2/T^4\right)\left(\sigma_{sx}^2 + T^2\sigma_{sv}^2 + 2T\sigma_{sxv}^2\right) + \gamma(2-\gamma)\left(\sigma_{sxa}^2/T^2 + \sigma_{sva}^2/T\right) = \left(\gamma^2/T^4\right)B_x, \tag{98}$$

$$(\alpha+\beta-\alpha\beta)\,\sigma_{sxv}^2 - (1-\alpha)(1-\beta)\left(T\sigma_{sv}^2 + (T^3/2)\,\sigma_{sa}^2 + T\sigma_{sxa}^2 + (3T^2/2)\sigma_{sva}^2\right)\sigma_{sxv}^2 = 0, \tag{99}$$

$$(\alpha+\gamma-\alpha\gamma)\,\sigma_{sxa}^2 - (\alpha-1)(\gamma-1)T\sigma_{sva}^2 - ((\alpha-1)(\gamma-2)/4)\,T^2\sigma_{sa}^2 + \gamma(1-\alpha)(\sigma_{sx}^2/T^2 + \sigma_{sv}^2 + 2\sigma_{sxv}^2/T) = \left(\alpha\gamma/T^2\right)B_x, \tag{100}$$

$$(2\beta+3\gamma-3\beta\gamma)\,\sigma_{sva}^2/2 + \gamma(1-\beta) \times \left(\sigma_{sxv}^2/T^2 + \sigma_{sv}^2/T + \sigma_{sxa}^2/T^2\right) - (\beta-1)(\gamma-2)T\sigma_{sa}^2/2 = 0. \tag{101}$$

Solving the linear system involving (96) to (101), we obtain:

$$\sigma_{sx}^2 = \frac{g_{1x}(\alpha,\beta,\gamma)B_x + (1-\alpha)^2 g_2(\alpha,\beta)T^2 B_v}{g_3(\alpha,\beta,\gamma)}, \tag{102}$$

$$\sigma_{sv}^2 = \frac{g_{1v}(\alpha,\beta,\gamma)B_x/T^2 + g_{2v}(\alpha,\beta,\gamma)B_v}{(2-\beta)g_3(\alpha,\beta,\gamma)}, \tag{103}$$

$$\sigma_{sa}^2 = \frac{g_{1a}(\alpha,\beta,\gamma)B_x/T^4 + g_{2a}(\alpha,\beta,\gamma)B_v/T^2}{(2-\beta)g_3(\alpha,\beta,\gamma)}, \tag{104}$$

$$\sigma_{sxv}^2 = \frac{g_{1xv}(\alpha,\beta,\gamma)B_x/T + g_{2xv}(\alpha,\beta,\gamma)TB_v}{g_3(\alpha,\beta,\gamma)}, \tag{105}$$

$$\sigma_{sxa}^2 = \frac{g_{1xa}(\alpha,\beta,\gamma)B_x/T^2 + 2(1-\alpha)(2-\beta)\gamma^2 g_2(\alpha,\beta)B_v}{\gamma(2-\beta)g_3(\alpha,\beta,\gamma)}, \tag{106}$$

$$\sigma_{sva}^2 = \frac{g_{1va}(\alpha,\beta,\gamma)B_x/T^3 + g_{2va}(\alpha,\beta,\gamma)B_v/T}{(2-\beta)g_3(\alpha,\beta,\gamma)}, \tag{107}$$

where $g_2(\alpha,\beta)$ and $g_3(\alpha,\beta,\gamma)$ are expressed as (22) and (23), and

$$\begin{aligned} g_{1x}(\alpha,\beta,\gamma) &= 8\alpha^3\beta(2-\beta)(\beta-1) - 2\alpha^2\left(3\beta^3\gamma - 4\beta^3 - 9\beta^2\gamma + 8\beta^2 + 2\beta\gamma + 4\gamma\right) \\ &\quad + \alpha\gamma\left(10\beta^3 + \beta^2\gamma - 28\beta^2 - \beta\gamma + 8\beta + 16\right) \\ &\quad - 4\beta\gamma(2-\beta)^2, \end{aligned} \tag{108}$$

$$g_{1v}(\alpha,\beta,\gamma) = 8\gamma^2(1-\beta)^2(2-\beta)(\alpha+\beta-\alpha\beta-2), \tag{109}$$

$$\begin{aligned} g_{2v}(\alpha,\beta,\gamma) &= 8\alpha^3\beta^2(2-\beta)(1-\beta) + 2\alpha^2\beta^2(\beta-2) \\ &\quad \times (3\beta\gamma - 12\beta - 3\gamma + 8) \\ &\quad - \alpha\beta^2\left(22\beta^2\gamma - 16\beta^2 + \beta\gamma^2 - 64\beta\gamma + 32\beta - \gamma^2 + 40\gamma\right) \\ &\quad + 2\beta^2\gamma(\beta-1)(8\beta+\gamma-16), \end{aligned} \tag{110}$$

$$g_{1a}(\alpha,\beta,\gamma) = 4\beta\gamma^2(\beta-2)\left(2\alpha\beta^2 - 6\alpha\beta - 2\beta^2 - \beta\gamma + 4\alpha + 4\beta + \gamma\right), \tag{111}$$

$$g_{2a}(\alpha,\beta,\gamma) = 4\beta^2\gamma(\alpha-2)\left(2\alpha\beta^2 - 6\alpha\beta - 2\beta^2 - \beta\gamma + 4\alpha + 4\beta + \gamma\right), \tag{112}$$

$$g_{1xv}(\alpha,\beta,\gamma) = 2\beta\gamma(\alpha-1)(\beta-2)(\beta-1)(4\alpha+\gamma), \tag{113}$$

$$g_{2xv}(\alpha,\beta,\gamma) = 2\beta^2(\alpha-1)(\alpha-2)(\beta-1)(4\alpha+\gamma), \tag{114}$$

$$\begin{aligned} g_{1xa}(\alpha,\beta,\gamma) &= \gamma^2(2-\beta)\left(8\alpha^2\beta^3 - 24\alpha^2\beta^2 + 16\alpha^2\beta + 2\alpha\beta^3\gamma - 8\alpha\beta^3 - 6\alpha\beta^2\gamma + 16\alpha\beta^2 \right. \\ &\quad - 4\alpha\beta\gamma + 8\alpha\gamma - 2\beta^3\gamma - \beta^2\gamma^2 + \beta\gamma^2 \\ &\quad \left. + 16\beta\gamma - 16\gamma\right), \end{aligned} \tag{115}$$

$$g_{1va}(\alpha,\beta,\gamma) = 2\beta\gamma^2(2-\beta)(\beta-1)(4\alpha\beta - 8\alpha - 4\beta - \gamma + 8), \tag{116}$$

$$g_{2va}(\alpha,\beta,\gamma) = 2\beta^2\gamma(2-\alpha)(\beta-1)(4\alpha\beta - 8\alpha - 4\beta - \gamma + 8). \tag{117}$$

Substituting (102) to (107) into (50), we arrive at (20).

Derivation of (24)

The derivation process is the same as for the A-V filter. By applying a z-transform to (1) to (3) and (17) to (19) and their simplified forms, the predicted parameters in the z-domain are derived as:

$$A_p(z) = \frac{\gamma z}{z-1} \cdot \frac{X_o(z) - X_p(z)}{T^2}, \quad (118)$$

$$V_p(z) = \frac{\gamma z^2}{(z-1)(z+\beta-1)} \cdot \frac{X_o(z) - X_p(z)}{T} + \frac{\beta}{z+\beta-1} V_o(z), \quad (119)$$

$$X_p(z) = \frac{(2\alpha+\gamma)X_o(z) + 2\beta T V_o(z) + 2(1-\beta)T V_p(z) + T^2 A_p(z)}{2z + 2\alpha + \gamma - 2}. \quad (120)$$

Substituting (118) and (119) into (120), the relationship between the predicted position and the measured position and velocity is expressed as

$$X_p(z) = \frac{h_1(z)}{h_2(z)} X_o(z) + \frac{2\beta z(z-1)}{h_2(z)} T V_o(z), \quad (121)$$

where

$$h_1(z) = \gamma z^3 + (2\alpha - \beta\gamma + 2\gamma)z^2 + (2\alpha\beta - 4\alpha + \beta\gamma - 2\gamma)z - 2\alpha\beta - \beta\gamma + 2\alpha + \gamma, \quad (122)$$

$$h_2(z) = (\gamma+2)z^3 + (2\alpha + 2\beta + 2\gamma - \beta\gamma - 6)z^2 + (2\alpha\beta + \beta\gamma - 4\alpha - 4\beta - 2\gamma + 6)z - 2\alpha\beta - \beta\gamma + 2\alpha + 2\beta + \gamma - 2. \quad (123)$$

From (121), the z-transform of the predicted error is

$$E_p(z) = X_o(z) - X_p(z) = \frac{2(z-1)^2(z+\beta-1)}{h_2(z)} X_o(z) - \frac{2\beta z(z-1)}{h_2(z)} T V_o(z). \quad (124)$$

Substituting (93) and (94) into (124), the error for a target with constant jerk is given by

$$E_p(z) = \frac{z\left(z^2 - 2\beta z + 4z - \beta + 1\right)}{3(z-1)h_2(z)} J T^3. \quad (125)$$

Applying the final value theorem to (125), we have (24).

Abbreviations

PVM: position-velocity-measured; MV: minimum-variance; A-V filter: acceleration smoothed by measured velocity-type PVM α-β-γ filter; A-P filter: acceleration smoothed by measured position-type PVM α-β-γ filter; GMV filter: general minimum-variance α-β-γ filter; CRB: Cramér–Rao bound; EKF: extended Kalman filter.

Competing interests

The authors declare that they have no competing interests.

Acknowledgements

This work was supported in part by the Ministry of Internal Affairs and Communications of Japan and JSPS KAKENHI Grant Number 26880023.

References

1. MJ Jahromi, HK Bizaki, Target tracking in MIMO radar systems using velocity vector. J. Inf. Sys. Telecommun. **2**, 150–158 (2014)
2. K Dae-Bong, H Sun-Mog, Multiple-target tracking and track management for an FMCW radar network. EURASIP J. Adv. Sig. Proc. **2013**, 159 (2013)
3. H Cheng, Z Tao, Z Chao, Accurate three-dimensional tracking method in bistatic forward scatter radar. EURASIP J. Adv. Sig. Proc. **2013**, 66 (2013)
4. H Niknejad, A Takeuchi, S Mita, D McAllester, On-road multivehicle tracking using deformable object model and particle filter with improved likelihood estimation. IEEE Trans. Intel. Transport. Sys. **13**, 748–758 (2012)
5. M Daun, F Ehlers, Tracking algorithms for multistatic sonar systems. EURASIP J. Adv. Sig. Proc. **2010**, 461538 (2010)
6. B Ekstrand, Some aspects on filter design for target tracking. J. Control Sci. Eng., 870890 (2012)
7. Y Bar-Shalom, XR Li, T Kirubarajan, *Estimation with Applications to Tracking and Navigation*. (Wiley-Interscience, New York City, USA, 2001)
8. D Tenne, T Singh, Characterizing performance of α-β-γ filters. IEEE Trans. Aero. Elec. Sys. **38**, 1072–1087 (2002)
9. NH Khin, YF Che, SML Eileen, WX Liang, Alpha beta gamma filter for cascaded PID motor position control. Procedia Eng. **41**, 244–250 (2012)
10. YS Lee, HJ Lee, in *Proc. of Int. Conf. Advanced Communication Technology 2009 (ICACT2009)*. Multiple object tracking for fall detection in real-time surveillance system (IEEE Phoenix Park, 2009), pp. 2308–2312
11. Y Wang, Feature point correspondence between consecutive frames based on genetic algorithm. Int. J. Robot. Automation. **21**, 35–38 (2006)
12. K Daniilidis, C Krauss, M Hansen, G Sommer, Real-time tracking of moving objects with an active camera. Real-Time Imaging. **4**, 3–20 (1998)
13. Y Kosuge, M Ito, T Okada, S Mano, Steady-state errors of an α-β-γ filter for radar tracking. Electron. Commun. Japan (Part III: Fundamental Electronic Sci.) **85**, 65–79 (2002)
14. Y Kosuge, M Ito, in *Proc. of the 40th SICE Annual Conf.* A necessary and sufficient condition for the stability of an α-β-γ filter (The Society of Instrument and Control Engineers Nagoya, 2001), pp. 7–12
15. CC Arcasoy, G Ouyang, Analytical solution of α-β-γ tracking filter with a noisy jerk as correlated target maneuver model. IEEE Trans. Aero. Elec. Sys. **33**, 347–353 (1997)
16. JJ Sudano, The α-β-γ tracking filter with a noisy jerk as the maneuver model. IEEE Trans. Aero. Elec. Sys. **29**, 578–580 (1993)
17. PR Kalata, The tracking index: A generalized parameter for α-β and α-β-γ target trackers. IEEE Trans. Aero. Elec. Sys. **AES-20**, 174–182 (1984)
18. W Chun-Mu, C Ching-Kao, C Tung-Te, A new EP-based α-β-γ-δ filter for target tracking. Math. Comput. Simul. **81**, 1785–1794 (2011)
19. D Mohammed, K Mokhtar, O Abdelaziz, M Abdelkrim, A new IMM algorithm using fixed coefficients filters (fastIMM). Int. J. of Electron. Commun. (AEÜ). **64**, 1123–1127 (2009)
20. R Kozma, L Wang, K Iftekharuddin, E McCracken, M Khan, K Islam, SR Bhurtel, RM Demirer, A radar-enabled collaborative sensor network integrating COTS technology for surveillance and tracking. Sensors. **12**, 1336–1351 (2012)
21. JH Lim, A Terzis, I-J Wang, in *Proc. of 2010 IEEE 35th Conf. Local Computer Networks (LCN)*. Tracking a non-cooperative mobile target using low-power pulsed Doppler radars (IEEE Denver, CO, 2010), pp. 913–920
22. YJ Hong, KD Yong, BS Hwan, S vladimir, Joint initialization and tracking of multiple moving objects using Doppler information. IEEE Trans. Sig. Proc. **59**, 3447–3452 (2011)
23. BR Geetha, KV Ramachandra, A three state Kalman filter with range and range-rate measurements. Int. J. Comput. Appl. **3**, 85–101 (2013)
24. X Zhu, J Hong, W Cui, in *Proc. of 4th IEEE Conf. on Industrial Electronics and Applications 2009*. Study on radar data processing algorithm with improved Kalman filter (IEEE Xi'an, 2009), pp. 3826–3829
25. K Jonghyuk, S Salah, 6DoF SLAM aided GNSS/INS navigation in GNSS denied and unknown environments. J. Global Pos. Sys. **4**, 120–128 (2005)
26. KV Ramachandra, BR Mohan, BR Geetha, A three-state Kalman tracker using position and rate measurements. IEEE Trans. Aero. Elec. Sys. **29**, 215–222 (1993)
27. RJ Fitzgerald, Simple tracking filters: Position and velocity measurements. IEEE Trans. Aero. Elec. Sys. **AES-18**, 531–537 (1982)

28. FR Castella, Tracking accuracies with position and rate measurements. IEEE Trans. Aero. Elec. Sys. **AES-17**, 433–437 (1980)
29. H Yamazaki, K Saho, T Sato, in *Proc. of 10th International Conference on Space, Aeronautical and Navigational Electronics*. Accurate shape estimation method for multiple moving targets with UWB Doppler radar interferometers (IEICE Hanoi, 2013), pp. 7–12
30. C Zheng, *Tracking vehicular motion-position using V2V communication*. (Master's thesis, the University of Waterloo, 2010)
31. Y Kosuge, M Ito, Evaluating an α-β filter in terms of increasing a track update-sampling rate and improving measurement accuracy. Electron. Commun. in Japan (Part I: Communications). **86**, 10–20 (2003)
32. Y Kosuge, in *Proc. of SICE Annual Conference, 2008*. Non-process-noise tracking filter using a constant velocity model (The Society of Instrument and Control Engineers Tokyo, 2008), pp. 2670–2674
33. P Baldi, Gradient descent learning algorithm overview: A general dynamical systems perspective. IEEE Trans. Neural Netw. **6**, 182–195 (1995)
34. B Ristic, A Farina, M Hernandez, Cramér-Rao lower bound for tracking multiple targets. IEE Proc. Radar Sonar Navig. **151**, 129–134 (2004)
35. P Tichavsky, CH Muravchik, A Nehorai, Posterior Cramér-Rao bounds for discrete-time nonlinear filtering. IEEE Trans. Sig. Process. **46**, 1386–1396 (1998)
36. B Ristic, A Farina, D Benvenuti, MS Arulampalam, Performance bounds and comparison of nonlinear filters for tracking a ballistic object on reentry. IEE Proc. Radar Sonar Navig. **150**, 65–70 (2003)

Sparse signal recovery with unknown signal sparsity

Wenhui Xiong*†, Jin Cao† and Shaoqian Li

Abstract

In this paper, we proposed a detection-based orthogonal match pursuit (DOMP) algorithm for compressive sensing. Unlike the conventional greedy algorithm, our proposed algorithm does not rely on the priori knowledge of the signal sparsity, which may not be known for some application, e.g., sparse multipath channel estimation. The DOMP runs binary hypothesis on the residual vector of OMP at each iteration, and it stops iteration when there is no signal component in the residual vector. Numerical experiments show the effectiveness of the estimation of signal sparsity as well as the signal recovery of our proposed algorithm.

Keywords: Sparsity; GLRT; OMP; Compressive sensing

1 Introduction

Compressive sensing (CS) [1,2], a framework to solve the under-determined system, has drawn great research attention in recent years. The CS problem can be modeled as finding the sparse solution of $\boldsymbol{h}$ for equation

$$\boldsymbol{y} = \boldsymbol{X}\boldsymbol{h} + \boldsymbol{n}, \tag{1}$$

where the observation $\boldsymbol{y} \in \boldsymbol{R}^{m\times 1}$ is obtained by using the sensing matrix $\boldsymbol{X} \in \boldsymbol{R}^{m\times n}$ to measure the k-sparse signal $\boldsymbol{h} \in \boldsymbol{R}^{n\times 1}$. In CS framework, the sensing matrix $\boldsymbol{X}$ in (1), is a 'fat' matrix, i.e., $m < n$.

To find the sparse solution of $\boldsymbol{h}$, i.e., recover the sparse signal, one can adopt either the convex relaxation based method, e.g., basis pursuit (BP) [3] or greedy algorithms, e.g., orthogonal matching pursuit (OMP) [4], regularized OMP (ROMP) [5], StOMP [6], etc. The greedy algorithm is often used for its low computational complexity and easy to implement. To implement the greedy algorithm, one needs to know the priori information on the signal's sparsity k. For example, in OMP and its variant, e.g., ROMP, the signal sparsity k must be specified so that the computation stops after k iterations. Other greedy algorithm such as subspace pursuit (SP) [7] also needs to know the value of k so that exact k candidate atoms could be selected at each iteration.

*Correspondence: whxiong@uestc.edu.cn
†Equal contributors
National Key Laboratory of Communication, University of Electronic Science and Technology of China, Chengdu 611731, China

The multipath channels, e.g., underwater acoustic (UWA) channel in sonar system [8] and Rayleigh fading channel in wireless communication [9], can be modeled as FIR filter. Those channels can be viewed as sparse signals according to experimental data [10,11]. Thus, CS can be applied for channel estimation. In [12], the authors shown that the CS approach achieves better estimation performance than the conventional methods.

In reality, the number of the channel taps, i.e., the signal sparsity, is usually unknown. Therefore, the greedy algorithms cannot be applied directly. In [13], the authors proposed the sparsity adaptive matching pursuit (SAMP) which does not need the signal sparsity information. In SAMP, the threshold is still needed to stop the iteration, and the performance of SAMP is sensitive to the threshold selection. In [14], stopping rules, i.e., the residual $\boldsymbol{r}_t$ at tth iteration meets $\|\boldsymbol{r}_t\|_{\ell_2} < \|\boldsymbol{n}\|_{\ell_2}$, for OMP under noise provides theoretical guarantee for sparse signal recovery.

In this paper, we proposed the detection-based orthogonal match pursuit (DOMP) algorithm which systematically provides the stop threshold based on the signal detection criteria. This is a more general threshold finding approach for stopping the OMP than the threshold proposed in [14]. Since the proposed DOMP is able to recover sparse signal without sparsity, it can be applied to the sparse channel estimation.

The rest of this paper is organized as follows. The analysis of residual vector of OMP is shown in Section 2. Section 3 discusses the hypothesis test on the residual vector. In Section 4, the threshold determined by given false alarm probability (P_{FA}) is discussed. The efficiency of the proposed stopping criteria is shown by numerical experiments in Section 5.

2 Analysis of residual vector in OMP

In this section, we show the property of the residual vector of OMP which motivates us to apply signal detection technique to determine the stopping criteria of OMP. In this study, we assume the sensing matrix $\boldsymbol{X}$ satisfies the RIP condition, i.e., $\delta_{k+1} < \frac{1}{\sqrt{k+1}}$, which guarantees the perfect recovery without noise perturbation [15].

The OMP can be viewed as the successive interference cancellation method, i.e., at each iteration the strongest signal component is subtracted from the residual vector. We denote the residual vector at the tth iteration by $\boldsymbol{r}_t$, the support of signal at the tth iteration by S_t, the sub-matrix formed by the columns of $\boldsymbol{X}$ according to the support S_t by $\boldsymbol{X}_{S_t}$, and the rest of the matrix $\boldsymbol{X}$ by $\boldsymbol{X}_{\bar{S}_t}$.

At the tth iteration, the column index i of $\boldsymbol{X}_{\bar{S}_t}$ which has the highest correlation with the residual vector $\boldsymbol{r}_t$ is added to the support set, i.e., $S_t = S_{t-1} \bigcup i$. After updating the signal support, the residual vector is updated by projecting $\boldsymbol{y}$ onto the null-space of $\boldsymbol{X}_{S_t}$, i.e.,

$$\boldsymbol{r}_t = \boldsymbol{P}_t^{\perp} \boldsymbol{y} \tag{2}$$

where $\boldsymbol{P}_t^{\perp} = \boldsymbol{I} - \boldsymbol{P}_t$ is orthogonal projector onto null space of $\boldsymbol{X}_{S_t}$ and $\boldsymbol{P}_t = \boldsymbol{X}_t \left(\boldsymbol{X}_t^T \boldsymbol{X}_t\right)^{-1} \boldsymbol{X}_t^T \in R^{m \times m}$. Thus, the residual vector $\boldsymbol{r}_t$ after t iterations can be expressed as

$$\boldsymbol{r}_t = \boldsymbol{P}_t^{\perp} \boldsymbol{X} \boldsymbol{h} + \boldsymbol{P}_t^{\perp} \boldsymbol{n}, \tag{3}$$

We denote the support of the k-sparse signal $\boldsymbol{h}$ by supp(h), i.e., supp(h) $:= \{i \in \{1, 2, \ldots, n\} | h(i) \neq 0\}$, where $h(i)$ is the ith element of vector $\boldsymbol{h}$. When the support obtained via the iteration is the supper-set of the actual support of the signal, i.e., supp($\boldsymbol{h}$) $\subset S_t$, there is no signal component in the residual vector r_t.

Thus, we can adopt the signal detection method to test whether the signal component exists in the residual vector after each iteration. Since one entry of $\boldsymbol{h}$, indexed by largest column correlation with $\boldsymbol{X}_{\bar{S}_t}$, is set to zero at the tth iteration, we can define the signal component after t iteration in the residual r_t as

$$\boldsymbol{h}_t := \begin{cases} \boldsymbol{h}_t(i) = 0, & i \in S_t, \\ \boldsymbol{h}_t(i) = \boldsymbol{h}(i), & \text{others}, \end{cases}$$

Then, (3) is equivalent to

$$\boldsymbol{r}_t = \boldsymbol{P}_t^{\perp} \boldsymbol{X} \boldsymbol{h}_t + \boldsymbol{P}_t^{\perp} \boldsymbol{n}.$$

According to the definition of RIP [16], for real signal $\boldsymbol{h}$, $\boldsymbol{X}$ obeys

$$(1 - \delta_k) \|\boldsymbol{h}\|_{\ell_2}^2 \leq \|\boldsymbol{X}_{S_t} \boldsymbol{h}\|_{\ell_2}^2 \leq (1 + \delta_k) \|\boldsymbol{h}\|_{\ell_2}^2, \tag{4}$$

for all subsets S_t with $\|S_t\|_{l_0} < k$. Since we assumed that $\delta_{k+1} < \frac{1}{\sqrt{k+1}}$, the sensing matrix $\boldsymbol{X}$ meets the RIP condition with $\delta_k < \frac{1}{\sqrt{k-1+1}} \leq 1$, i.e., $\delta_k < 1$. Therefore, $\|\boldsymbol{X}_{S_t} \boldsymbol{h}\|_{\ell_2}^2 \geq (1 - \delta_k) \|\boldsymbol{h}\|_{\ell_2}^2 > 0$, for any $\boldsymbol{h} \neq \boldsymbol{0}$. In other words, equation $\boldsymbol{X}_{S_t} \boldsymbol{h} = 0$ has no nonzero solution, or any t columns of $\boldsymbol{X}$ are linearly independent. We have $rank(\boldsymbol{P}_t) = t$. Since $rank\,(\boldsymbol{P}_t) + rank\left(\boldsymbol{P}_t^{\perp}\right) = m$, $\boldsymbol{P}_t^{\perp}$ is not a row full rank matrix, and vector $\boldsymbol{P}_t^{\perp} \boldsymbol{y}$ is of degenerated multivariate normal distribution. To derive the distribution of the residual vector $\boldsymbol{r}_t$, the residual is further projected onto a subspace formed by taking any $m - t$ rows from $\boldsymbol{P}_t^{\perp}$. Since any $m - t$ rows of $\boldsymbol{P}_t^{\perp}$ are linearly independent, i.e., $rank(\boldsymbol{P}_t) = m - t$, the sub-matrix formed by these rows is of full row rank.

We denote the projection matrix by $\boldsymbol{P}_{m-t} = \boldsymbol{M}_{m-t} \boldsymbol{P}_t^{\perp}$, where $\boldsymbol{M}_{m-t}$ is a matrix that takes $m - t$ rows from other matrix. For example, $\boldsymbol{M}_3$ can be

$$\boldsymbol{M}_3 = \begin{bmatrix} 1\ 0\ 0\ 0 \cdots 0 \\ 0\ 1\ 0\ 0 \cdots 0 \\ 0\ 0\ 1\ 0 \cdots 0 \end{bmatrix}, \tag{5}$$

where m is the number of measurements, and t is the iteration times. $\boldsymbol{P}_{m-t} = \boldsymbol{M}_{m-t} \boldsymbol{P}_t^{\perp}$ is the sub-matrix formed by the $m - t$ rows of $\boldsymbol{P}_{m-t}^{\perp}$. $\boldsymbol{P}_{m-t}$ projects r_t onto a subspace of rank $m - t$, that is $\boldsymbol{z}_t = \boldsymbol{P}_{m-t} \cdot \boldsymbol{r}_t$. Since any $m - t$ rows of $\boldsymbol{P}_t^{\perp}$ are linearly independent, and other t rows can be linearly represented by these $m - t$ rows, any $\boldsymbol{M}_{m-t}$ with full row rank projects the residual vector $\boldsymbol{r}_t$ onto the identical subspace. Thus, we take any $m - t$ rows from $\boldsymbol{P}_t^{\perp}$ for the further projection.

Define $\boldsymbol{C}_{m-t} := \boldsymbol{P}_{m-t} \boldsymbol{P}_{m-t}^T$. If there is only noise in the residual vector, that is, $\boldsymbol{z}_t = \boldsymbol{P}_{m-t} \boldsymbol{n}$, then the projected residual $\boldsymbol{z}_t$ follows

$$\boldsymbol{z}_t \sim \mathcal{N}\left(0, \sigma^2 \boldsymbol{C}_{m-t}\right). \tag{6}$$

If the residual vector consists the signal component and noise, i.e., $\boldsymbol{r}_t = \boldsymbol{P}_{m-t}\left(\boldsymbol{X} \boldsymbol{h}_t + \boldsymbol{n}\right)$, the distribution of z_t is

$$\boldsymbol{z}_t \sim \mathcal{N}\left(0, \left(\theta_t + \sigma^2\right) \boldsymbol{C}_{m-t}\right). \tag{7}$$

where $\theta_t = \|\boldsymbol{h}_t\|_{\ell_2}$ is an unknown parameter.

3 Hypothesis test on residual vector

With the PDF of the residual vector known, we can form the binary hypothesis test on whether there are signal components in the residual vector after t iterations,

$$\begin{aligned} H_0 &: \boldsymbol{z}_t = \boldsymbol{P}_{m-t} \cdot \boldsymbol{n} \\ H_1 &: \boldsymbol{z}_t = \boldsymbol{P}_{m-t} \cdot \left(\boldsymbol{X} \boldsymbol{h}_t + \boldsymbol{n}\right). \end{aligned} \tag{8}$$

If H_0 is decided, the iteration stops. Since one entry of signal $\boldsymbol{h}_t$ is set to zero at each iteration, $\|\boldsymbol{h}_t\|_{\ell_2}$ decreases after each iteration, and it needs to be estimated.

According to (6) and (7), the PDF of the residual vector under H_0 and H_1 are respectively given by

$$p(\boldsymbol{z}_t; H_0) = \frac{1}{(2\pi)^{\frac{m-t}{2}} \det^{1/2}(\boldsymbol{C}_{H_0})} \cdot \exp\left[-\frac{1}{2}\boldsymbol{z}_t^T \boldsymbol{C}_{H_0}^{-1} \boldsymbol{z}_t\right], \tag{9}$$

$$p(\boldsymbol{z}_t; \theta_t, H_1) = \frac{1}{(2\pi)^{\frac{m-t}{2}} \det^{1/2}(\boldsymbol{C}_{H_1})} \cdot \exp\left[-\frac{1}{2}\boldsymbol{z}_t^T \boldsymbol{C}_{H_1}^{-1} \boldsymbol{z}_t\right], \tag{10}$$

where $\boldsymbol{C}_{H_0} = \sigma^2 \boldsymbol{C}_{m-t}$, $\boldsymbol{C}_{H_1} = (\theta_t + \sigma^2) \boldsymbol{C}_{m-t}$. Then, the binary hypothesis test can be conducted using the generalized likelihood ratio test (GLRT) [17]. H_1 is decided if the following inequality holds

$$\begin{aligned} \ln L(\boldsymbol{z}_t) &= \ln \frac{p\left(\boldsymbol{z}_t; \hat{\theta}_t, H_1\right)}{p(\boldsymbol{z}_t; H_0)} \\ &= \frac{1}{2}\left(\frac{1}{\sigma^2} - \frac{1}{\hat{\theta}_t + \sigma^2}\right) \boldsymbol{z}_t^T \boldsymbol{C}_{m-t}^{-1} \boldsymbol{z}_t. \\ &\quad + \frac{m-t}{2} \ln \frac{\sigma^2}{\hat{\theta}_t + \sigma^2} \\ &> \gamma', \end{aligned} \tag{11}$$

where $\hat{\theta}_t$ is the maximum likelihood estimation (MLE) of θ_t at each iteration,

$$\hat{\theta}_t = \begin{cases} \frac{\boldsymbol{z}_t^T \boldsymbol{C}_{m-t}^{-1} \boldsymbol{z}_t}{m-t} - \sigma^2, & \frac{\boldsymbol{z}_t^T \boldsymbol{C}_{m-t}^{-1} \boldsymbol{z}_t}{m-t} - \sigma^2 > 0 \\ 0, & \frac{\boldsymbol{z}_t^T \boldsymbol{C}_{m-t}^{-1} \boldsymbol{z}_t}{m-t} - \sigma^2 \le 0 \end{cases} \tag{12}$$

When $\hat{\theta}_t = 0$, i.e., no signal component exists, the iteration stops; otherwise, the $\hat{\theta}_t$ is plugged into (11) for further test. After plugging the $\hat{\theta}_t$ and simplification, the test statistics is given by

$$\frac{m-t}{2}\left[\frac{\boldsymbol{z}_t^T \boldsymbol{C}_{m-t}^{-1} \boldsymbol{z}_t}{\sigma^2(m-t)} - \ln\left(\frac{\boldsymbol{z}_t^T \boldsymbol{C}_{m-t}^{-1} \boldsymbol{z}_t}{\sigma^2(m-t)}\right) - 1\right] > \gamma'. \tag{13}$$

Since the function $g(x) = x - \ln x - 1$ in (13) is a monotonically increasing function of x for $x > 1$, and its inverse function g^{-1} exists for $x > 1$, (13) can be rewritten as

$$\frac{m-t}{2} g\left(\frac{\boldsymbol{z}_t^T \boldsymbol{C}_{m-t}^{-1} \boldsymbol{z}_t}{\sigma^2(m-t)}\right) > \gamma', \tag{14}$$

In (14), $\frac{\boldsymbol{z}_t^T \boldsymbol{C}_{m-t}^{-1} \boldsymbol{z}_t}{(m-t)} - \sigma^2 > 0$, we have $\frac{\boldsymbol{z}_t^T \boldsymbol{C}_{m-t}^{-1} \boldsymbol{z}_t}{\sigma^2(m-t)} > 1$. Therefore, (14) is simplified as

$$\frac{\boldsymbol{z}_t^T \boldsymbol{C}_{m-t}^{-1} \boldsymbol{z}_t}{\sigma^2(m-t)} > g^{-1}\left(\frac{2\gamma'}{m-t}\right) = \gamma''. \tag{15}$$

Finally, we obtain the detector $T(\boldsymbol{z}_t) = \boldsymbol{z}_t^T \boldsymbol{C}_{m-t}^{-1} \boldsymbol{z}_t$, and choose H_1 if

$$T(\boldsymbol{z}_t) = \boldsymbol{z}_t^T \boldsymbol{C}_{m-t}^{-1} \boldsymbol{z}_t > \sigma^2 (m-t)\gamma'' = \gamma_t. \tag{16}$$

In other words, when $T(\boldsymbol{z}_t)$ is greater than the threshold γ_t, signal component remains in residual vector, and iteration should be continued.

4 Threshold selection

The threshold selection is crucial in the binary hypothesis test. We use the constant false alarm (CFA) criteria to determine the value of threshold γ_t. Recall that the detector is in the quadratic form of $T = \boldsymbol{v}^T \boldsymbol{B} \boldsymbol{v}$, where $\boldsymbol{B}$ is a symmetric $n \times n$ matrix and $\boldsymbol{v}$ is an $n \times 1$ vector following $\mathcal{N}(0, \mathcal{C})$. With $\boldsymbol{B} = \boldsymbol{C}^{-1}$, we know that T follows the chi-square distribution with n degrees of freedom. Thus, we have

$$\frac{T(\boldsymbol{z}_t)}{\sigma^2} = \frac{\boldsymbol{z}_t^T}{\sigma} \boldsymbol{C}_{m-t}^{-1} \frac{\boldsymbol{z}_t}{\sigma} \sim \chi_{m-t}^2, \quad H_0,$$

$$\frac{T(\boldsymbol{z}_t)}{\theta_t + \sigma^2} = \frac{\boldsymbol{z}_t^T}{\sqrt{\theta_t + \sigma^2}} \boldsymbol{C}_{m-t}^{-1} \frac{\boldsymbol{z}_t}{\sqrt{\theta_t + \sigma^2}} \sim \chi_{m-t}^2, \quad H_1.$$

Therefore, false alarm probability and detect probability are given as

$$P_{FA} = P\{T(\boldsymbol{z}_t) > \gamma_t; H_0\} = Q_{\chi_{m-t}^2}\left(\frac{\gamma_t}{\sigma^2}\right), \tag{17}$$

$$P_D = P\{T(\boldsymbol{z}_t) > \gamma_t; H_1\} = Q_{\chi_{m-t}^2}\left(\frac{\gamma_t}{\theta_t + \sigma^2}\right), \tag{18}$$

where $Q_{\chi_\nu^2}(a)$ is the right-tail probability of Chi-Square χ_ν^2 function given by

$$Q_{\chi_\nu^2}(a) = \begin{cases} 2Q(\sqrt{a}), & \nu = 1 \\ 2Q(\sqrt{a}) + f(a), & \nu > 1, \nu \text{ is odd} \\ \exp\left(-\frac{1}{2}a\right) \sum_{k=0}^{\frac{\nu}{2}-1} \frac{\left(\frac{a}{2}\right)^k}{k!}, & \nu \text{ is even} \end{cases} \tag{19}$$

where $f(a) = \frac{\exp(-\frac{1}{2}a)}{\sqrt{\pi}} \sum_{k=1}^{\frac{\nu-1}{2}} \frac{(k-1)!(2a)^{k-\frac{1}{2}}}{(2k-1)!}$ and $Q(\sqrt{a}) = \int_{\sqrt{a}}^{\infty} \frac{1}{\sqrt{2\pi}} \exp\left(-\frac{1}{2}t^2\right) dt$.

The stopping threshold γ_t shown in (17) can be calculated using numerical method [17]. Our proposed DOMP is shown in Algorithm 1.

Algorithm 1 DOMP

Input: X, y, σ^2, P_{FA}

$r_0 \leftarrow y$

$S_0 \leftarrow \emptyset$

$t \leftarrow 1$

repeat

$u_t \leftarrow X^T r_{t-1}$

$S_t \leftarrow S_{t-1} \cup \arg\max_{i\in\{1,2,\dots,n\}} |u_t(i)|$

$r_t \leftarrow P^{\perp}_{S_t} y$

$z_t \leftarrow M_{m-t} r_t$ {M_{m-t} is a matrix selecting $m-t$ entries from r_t}

$t \leftarrow t+1$

until $T(z_t) \leq \gamma_t$ {$T(z_t)$ and γ_t need to be calculated for each iteration}

Output: $\hat{h} = \begin{cases} \hat{h}_{S_t} = \arg\min_z \|y - X_{S_t} z\|_{\ell_2} \\ \hat{h}_{\bar{S}_t} = 0 \end{cases}, \hat{k} = t-1,$

$\hat{S} = S_t$ {output the recovered signal, estimated sparsity and recovered support of the signal}

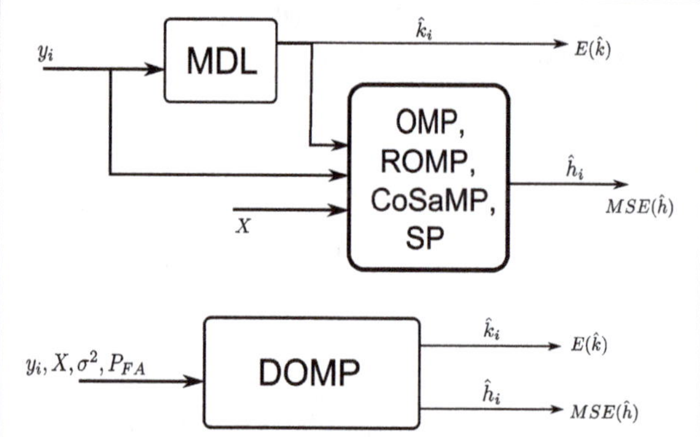

Figure 1 Pfa versus Pd of the signal component detection in the residual, SNR = 3 dB and SNR = 5 dB.

5 Numerical results

In this section, we present the numerical results of proposed DOMP algorithm. To evaluate the performance of DOMP, we define the mean square error of the estimated vector by

$$\mathrm{MSE}(\hat{h}) = \frac{1}{N}\sum_{i=1}^{N} \|\hat{h}_i - h\|^2_{\ell_2}. \tag{20}$$

where $\hat{h}_i$ is the recovered h of the ith experiment, and N is the number of experiments. N is set to be 5,000 in all our numerical experiments.

The detector $T(z_t)$ of DOMP checks whether there is signal in residual for each iteration. First, we show that the detection performance of the $T(z_t)$ on residuals for each iteration. In this test, the sensing matrix is a 128 × 256 Gaussian matrix whose elements follow i.i.d. Gaussian distribution of $\mathcal{N}(0,1)$. A 3-sparse signal, whose nonzero elements are all ones, is sensed. For each P_{FA}, we perform 1,000 trials. The residual at the ith iteration is denoted as r_i, and the curves of logarithmic scaled (dB) P_{FA} versus P_D at each iteration for different SNRs are shown in Figure 1. The detection probabilities of signal components are high for the first two iterations (when there exists signal components in the residual), and the detection probabilities are low after three iterations (when the residual has no signal component) for P_{FA} between -30 and $-10dB$. In other words, P_{FA} about $0.001-0.1$ provides good tradeoff between P_{FA} and P_D.

We then compare the performance of the support recovery rate and the MSE of the recovered signal using 1) OMP with sparsity k known; 2) OMP with unknown sparsity with stopping rule of $\|r_t\|_{\ell_2} < \|n\|_{\ell_2}$ as proposed in [14]; 3) DOMP with different false alarm probabilities, i.e., $P_{FA} = 0.05$, $P_{FA} = 0.01$, and $P_{FA} = 0.001$. The sensing matrix is Gaussian matrix whose elements follow i.i.d. Gaussian distribution $\mathcal{N}(0,1)$. The nonzero elements of the 256-dimensional signal are set to one. In Figures 2 and 3, the performance of these methods are shown as number of measurements (dimension of y) increases, while the sparsity of the signal is set to be 4 for SNR = 5 dB. The results shown that the OMP with sparsity k known has the best performance followed by DOMP, and OMP with stopping rule, $\|r_t\|_{\ell_2} < \|n\|_{\ell_2}$. For DOMP with different P_{FA}, the successful support recovery rate increases for lower P_{FA} as the number of measurements increases, e.g., DOMP with $P_{FA} = 0.01$ outperforms DOMP with $P_{FA} = 0.05$ when the number of measurements is greater than 60. Note in Figure 3, we can observe the crossovers of DOMP with different P_{FA} as the dimension of y increases. This is due to the fact that detection probability is a

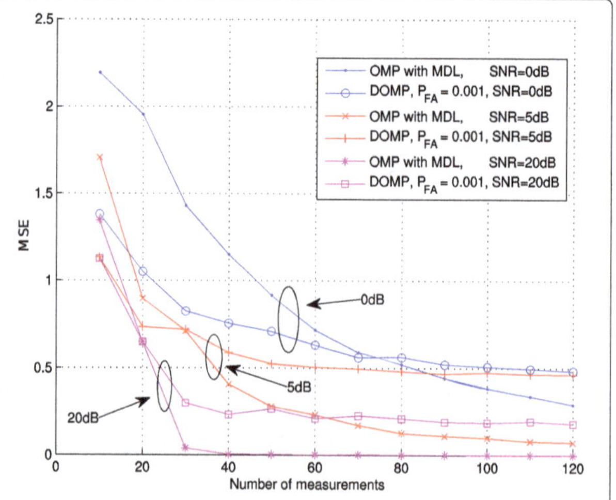

Figure 2 The MSE of the recovered signal using DOMP and OMP with/without sparsity information as number of measurements (dimension of y) increases, SNR = 5 dB.

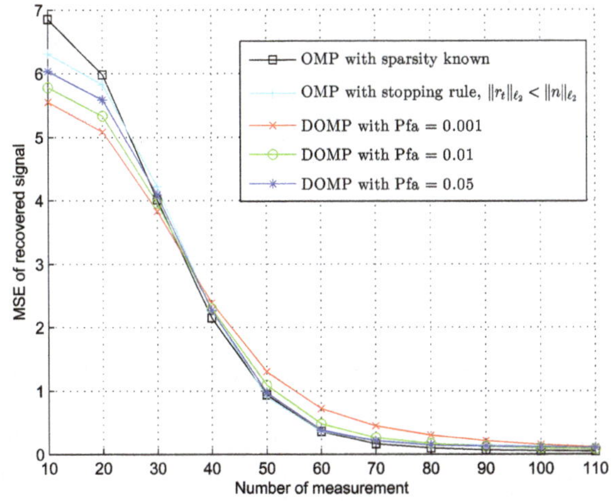

Figure 3 The percentage of successfully recovered support of signal using DOMP and OMP with/without sparsity information as number of measurements (dimension of *y*) increases, SNR = 5 dB.

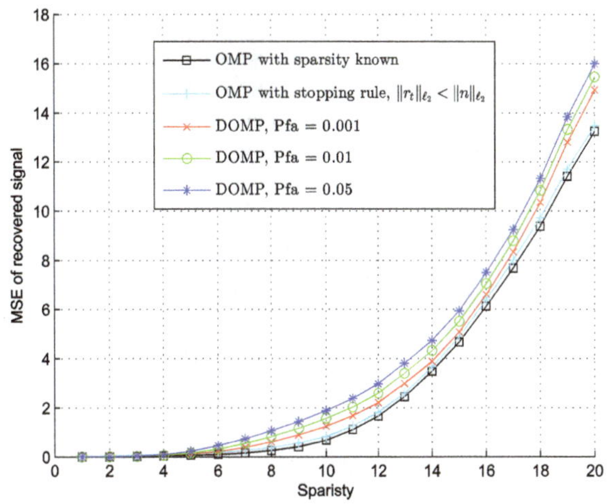

Figure 5 The MSE of recovered signal using DOMP and OMP with/without sparsity information as the sparsity increases, SNR = 5 dB.

increasing function both of the number of measurements and P_{FA}. With small number of measurements, the effect of lower less number of measurement is more dominant than the effect of lower P_{FA}. Thus, higher P_{FA} results better support recovery performance for DOMP when the number of measurements is small. As the number of measurement increases, the effect of the more measurement dominates, and the DOMP with lower P_{FA} performs better.

In Figures 4 and 5, we show the performance as sparsity of signal increase, while keep the number of measurements fixed to be 128. The figures show again that the OMP with known sparsity outperforms other methods. Our proposed DOMP outperforms the OMP with stopping rule, $\|\boldsymbol{r}_t\|_{\ell_2} < \|\boldsymbol{n}\|_{\ell_2}$. The DOMP with lower P_{FA} has higher support recovery rate and lower MSE.

It is worth noting that in reality, the sparsity information may not be known in prior. Thus, one may not be able to directly apply OMP. Minimum description length (MDL) criterion is often used, in this scenario, to estimate the sparsity of the signal [18], i.e., the eigenvalues of the sample covariance matrix R of the received signal $\boldsymbol{y}$, denoted by λ_i is used to estimate the signal sparsity as

$$\hat{k} = \underset{k \in \{1,2,\ldots,n\}}{\arg\min}\ \mathrm{MDL}(k), \tag{21}$$

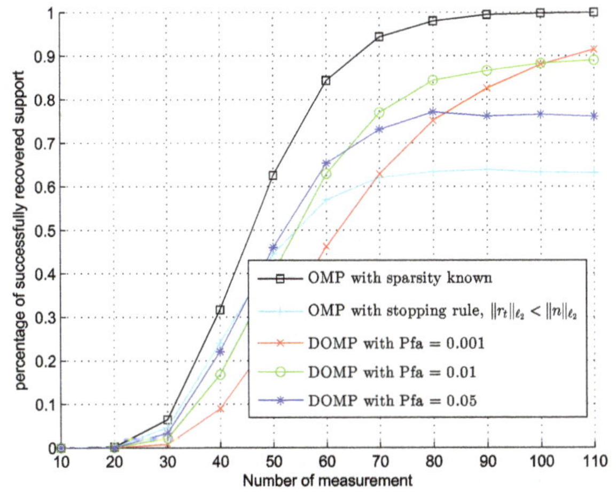

Figure 4 The percentage of successfully recovered support of signal using DOMP and OMP with/without sparsity information as the sparsity increases, SNR = 5 dB.

Table 1 The estimated sparsity of signal by DOMP and MDL

Sparsity *k*	DOMP		MDL	
	Mean	Std	Mean	Std
1	1.00	0	1.06	0.34
2	2.00	0	1.91	0.64
3	3.06	0.32	2.15	0.85
4	4.48	1.03	3.22	1.43
5	6.53	1.87	2.71	1.61
6	8.77	2.54	4.78	3.18
7	10.98	2.81	7.82	3.07
8	13.33	3.09	9.51	1.15
9	15.67	3.47	9.82	0.45
10	17.56	3.39	9.99	0.10

Sensing matrix is 128 × 256 random matrix whose entries are i.i.d Gaussian variables with mean 0 and variance 1. SNR = 5 dB, P_{FA} = 0.05. Trials = 1,000.

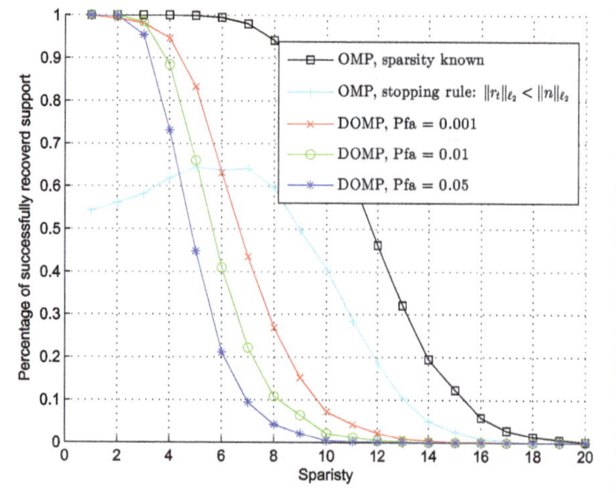

Figure 6 Block diagram for comparing DOMP and other greedy algorithms with MDL.

where MDL(k) is given by

$$\mathrm{MDL}(k) = -\log\left(\frac{\prod_{i=k+1}^{m} \lambda_i^{\frac{1}{(m-k)}}}{\frac{1}{(m-k)}\sum_{i=k+1}^{m}\lambda_i}\right)^{(m-k)n} + \frac{1}{2}k(2m-k)\log n. \quad (22)$$

We now compare the accuracy of estimation of the signal sparsity by DOMP and MDL. In this experiment, the signal dimension is set to be $n = 256$, and the sensing matrix $\boldsymbol{X}$ is a 128×256 Gaussian matrix whose entries are i.i.d Gaussian with mean zero and variance of one. The k-sparse signal $\boldsymbol{h}$ is generated by randomly setting k entries in $\boldsymbol{h}$ to be one and other entries of $\boldsymbol{h}$ to be zero. The experiment is conducted for SNR = 5 dB.

Table 2 The parameters of cost207 channel

	BUx6	RAx4
Sample Frequency (MHz)	18.4	18.4
Path delays (μs)	0.0 0.4 1.0 1.6 5.0 6.6	0.0 0.2 0.4 0.6
Average path gain (dB)	-3 0 -3 -5 -2 -4	0 -2 -10 -20
Support of $\boldsymbol{h}$	[1, 8, 19, 30, 93, 123]	[1, 5, 8, 12]
Dimension of $\boldsymbol{h}$	128	128

The estimated signal sparsity is shown in Table 1 for DOMP and MDL. We can observe that our proposed detection method gives accurate sparsity estimation for low sparsity signal. Actually, the estimated sparsity is the number of iteration for DOMP. Therefore, the average number of iterations for DOMP can be found in the Table 1, which actually matches the signal's sparsity for low sparsity case.

Adopting the scheme shown in Figure 6, we compare the performance of DOMP and other greedy pursuit algorithms, OMP, CoSaMP, ROMP, and SP with the signal sparsity estimated using MDL criterion. Similar with the previous experiment, we choose the sensing matrix to be the Gaussian matrix whose entries follow i.i.d Gaussian distributed of $\mathcal{N}(0,1)$. The support S of the signal is randomly selected, and the amplitude of the nonzero elements of the sparse signals $\boldsymbol{h}$ are drawn from standard Gaussian distribution. The noise $\boldsymbol{n}$ is a zero mean Gaussian noise. In Figure 7, the signal recovery MSE for different number of measurements (dimension of $\boldsymbol{y}$) is shown. In this figure, we can observe that the estimated error of DOMP is less than other greedy pursuit algorithms with MDL when the number of measurements is less than 40.

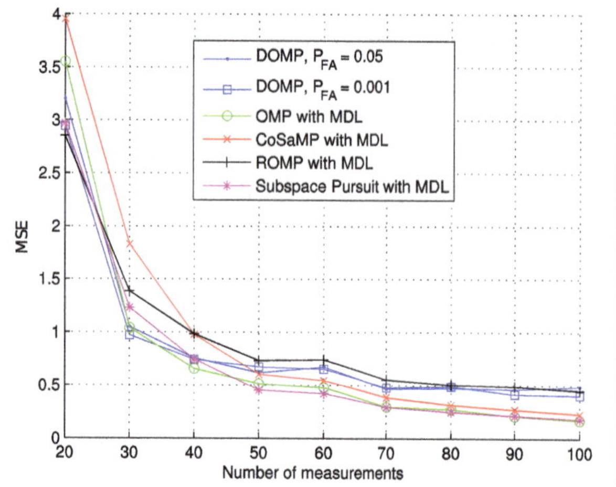

Figure 7 Performance of estimated error between DOMP and other greedy pursuits with MDL for different number of measurements (dimension of y) SNR = 0 dB.

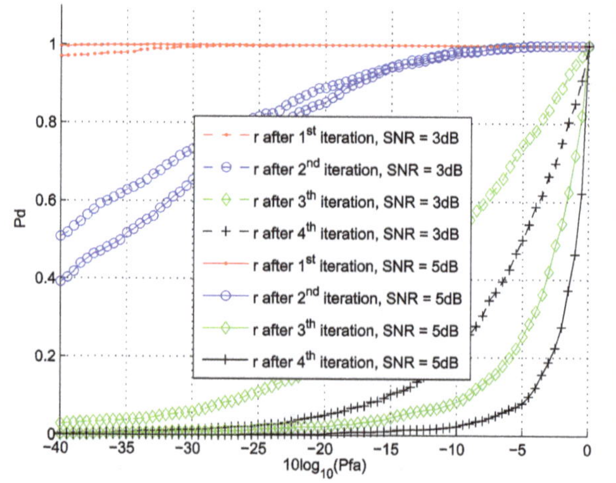

Figure 8 MSE of BUx6 channel estimation with DOMP and OMP with MDL.

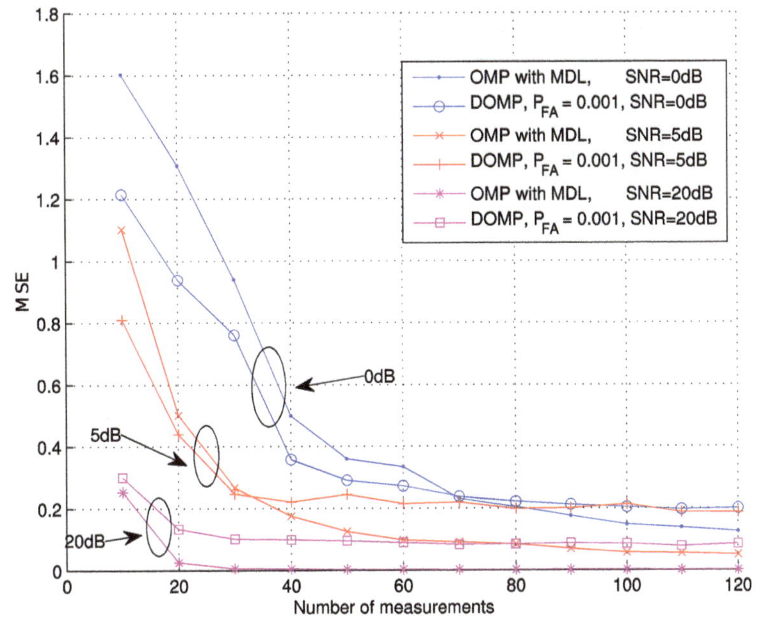

Figure 9 MSE of RAx4 channel estimation with DOMP and OMP with MDL.

One of the applications for DOMP is the channel estimation [12] since the number of channel taps are usually unknown. We compare the performances of estimating Rayleigh fading channel by MDL based OMP and DOMP following the same scheme shown in Figure 6. The Rayleigh fading channel $\boldsymbol{h}$ is given by cost207 model [19] with the parameters shown in Table 2. Since the sensing matrix $\boldsymbol{X}$ in wireless channel model $\boldsymbol{y} = \boldsymbol{X}\boldsymbol{h} + \boldsymbol{n}$ is a Toeplitz matrix constructed by the transmitted sequence $\boldsymbol{x}$ [20], we construct the sensing matrix by circular shift of the Gaussian vector whose elements are drawn from $\mathcal{N}(0, 1)$, which models the correlator output of spread spectrum signal. Figures 8 and 9 show the MSE of estimated BUx6 and RAx4 channel using DOMP and MDL based OMP, respectively. These two figures show that the error of estimation by DOMP is less than MDL-based OMP when the number of measurement m is less than 80.

6 Conclusions

In this paper, we proposed a detection-based OMP algorithm called DOMP. This method forms GLRT for each iteration to test if signal component exists in the residual vector. When no signal component exists, the algorithm stops the iteration. In this paper, we use OMP, a classical greedy algorithm, to apply this detection-based method. We envision that the detection-based method can be apply to other greedy algorithms for iteration stopping rules.

The numerical results show that the proposed DOMP outperforms the classical OMP algorithm without prior sparsity information at lower SNR or number of measurements. We use cost207 wireless channel estimation as an example to show the effectiveness of DOMP. The DOMP can be readily applied to other sparse recovery problems, e.g., underwater channel in sonar system and radar system, where the signal sparsity is unknown.

Competing interests

The authors declare that they have no competing interests.

Acknowledgements

This work was supported by the National Natural Science Foundation of China (Grant: 61101093, 61101090) and Fundamental Research Funds for the Central Universities (ZYGX2013J113).

References

1. DL Donoho, Compressed sensing. IEEE Trans. Inform. Theory. **52**(4), 1289–1306 (2006). doi:10.1109/TIT.2006.871582
2. EJ Candes, MB Wakin, An introduction to compressive sampling. IEEE Signal Process. Mag. **25**(2), 21–30 (2008). doi:10.1109/MSP.2007.914731
3. SS Chen, Donoho DL, MA Saunders, Atomic decomposition by basis pursuit. SIAM J. Sci. Comput. **20**(1), 33–61 (1998). doi:10.1137/S1064827596304010. Accessed 2013-12-19
4. JA Tropp, AC Gilbert, Signal recovery from random measurements via orthogonal matching pursuit. IEEE Trans. Inform. Theory. **53**(12), 4655–4666 (2007). doi:10.1109/TIT.2007.909108
5. D Needell, R Vershynin, Signal recovery from incomplete and inaccurate measurements via regularized orthogonal matching pursuit. IEEE J. Selected Topics Signal Process. **4**(2), 310–316 (2010). doi:10.1109/JSTSP.2010.2042412
6. DL Donoho, Y Tsaig, I Drori, J-L Starck, Sparse solution of underdetermined systems of linear equations by stagewise orthogonal matching pursuit. IEEE Trans. Inform. Theory. **58**(2), 1094–1121 (2012). doi:10.1109/TIT.2011.2173241
7. W Dai, O Milenkovic, Subspace pursuit for compressive sensing signal reconstruction. IEEE Trans. Inform. Theory. **55**(5), 2230–2249 (2009). doi:10.1109/TIT.2009.2016006

8. WC Knight, RG Pridham, SM Kay, Digital signal processing for sonar. Proc. IEEE. **69**(11), 1451–1506 (1981). doi:10.1109/PROC.1981.12186
9. D Tse, P Viswanath, *Fundamentals of wireless communication.* (Cambridge University Press, 2005), p. 04947
10. CR Berger, S Zhou, JC Preisig, P Willett, Sparse channel estimation for multicarrier underwater acoustic communication: from subspace methods to compressed sensing. IEEE Trans. Signal Process. **58**(3), 1708–1721 (2010). doi:10.1109/TSP.2009.2038424
11. AAM Saleh, RA Valenzuela, A statistical model for indoor multipath propagation. IEEE J. Selected Areas Commun. **5**(2), 128–137 (1987). doi:10.1109/JSAC.1987.1146527
12. WU Bajwa, J Haupt, AM Sayeed, R Nowak, Compressed channel sensing: a new approach to estimating sparse multipath channels. Proc. IEEE. **98**(6), 1058–1076 (2010). doi:10.1109/JPROC.2010.2042415
13. TT Do, L Gan, N Nguyen, TD Tran, in *Signals, Systems and Computers, 2008 42nd Asilomar Conference On.* Sparsity adaptive matching pursuit algorithm for practical compressed sensing, (2008), pp. 581–587. 00121
14. T. T Cai, L Wang, Orthogonal matching pursuit for sparse signal recovery with noise. IEEE Trans. Inform. Theory. **57**(7), 4680–4688 (2011). doi:10.1109/TIT.2011.2146090
15. J Wang, B Shim, On the recovery limit of sparse signals using orthogonal matching pursuit. IEEE Trans. Signal Process. **60**(9), 4973–4976 (2012). doi:10.1109/TSP.2012.2203124
16. EJ Candes, T Tao, Decoding by linear programming. IEEE Trans. Inform. Theory. **51**(12), 4203–4215 (2005). doi:10.1109/TIT.2005.858979
17. SM Kay, *Fundamentals of statistical signal processing volume 2: detection theory.* (Prentice Hall PTR, 1993)
18. M Wax, T Kailath, Detection of signals by information theoretic criteria. IEEE Trans. Acoust. Speech Signal Process. **33**(2), 387–392 (1985). doi:10.1109/TASSP.1985.1164557
19. M Failli, Digital land mobile radio communications COST 207. EC (1989)
20. J Haupt, WU Bajwa, G Raz, R Nowak, Toeplitz compressed sensing matrices with applications to sparse channel estimation. IEEE Trans. Inform. Theory. **56**(11), 5862–5875 (2010). doi:10.1109/TIT.2010.2070191

Permissions

All chapters in this book were first published in EURASIP-JASP, by Springer; hereby published with permission under the Creative Commons Attribution License or equivalent. Every chapter published in this book has been scrutinized by our experts. Their significance has been extensively debated. The topics covered herein carry significant findings which will fuel the growth of the discipline. They may even be implemented as practical applications or may be referred to as a beginning point for another development.

The contributors of this book come from diverse backgrounds, making this book a truly international effort. This book will bring forth new frontiers with its revolutionizing research information and detailed analysis of the nascent developments around the world.

We would like to thank all the contributing authors for lending their expertise to make the book truly unique. They have played a crucial role in the development of this book. Without their invaluable contributions this book wouldn't have been possible. They have made vital efforts to compile up to date information on the varied aspects of this subject to make this book a valuable addition to the collection of many professionals and students.

This book was conceptualized with the vision of imparting up-to-date information and advanced data in this field. To ensure the same, a matchless editorial board was set up. Every individual on the board went through rigorous rounds of assessment to prove their worth. After which they invested a large part of their time researching and compiling the most relevant data for our readers.

The editorial board has been involved in producing this book since its inception. They have spent rigorous hours researching and exploring the diverse topics which have resulted in the successful publishing of this book. They have passed on their knowledge of decades through this book. To expedite this challenging task, the publisher supported the team at every step. A small team of assistant editors was also appointed to further simplify the editing procedure and attain best results for the readers.

Apart from the editorial board, the designing team has also invested a significant amount of their time in understanding the subject and creating the most relevant covers. They scrutinized every image to scout for the most suitable representation of the subject and create an appropriate cover for the book.

The publishing team has been an ardent support to the editorial, designing and production team. Their endless efforts to recruit the best for this project, has resulted in the accomplishment of this book. They are a veteran in the field of academics and their pool of knowledge is as vast as their experience in printing. Their expertise and guidance has proved useful at every step. Their uncompromising quality standards have made this book an exceptional effort. Their encouragement from time to time has been an inspiration for everyone.

The publisher and the editorial board hope that this book will prove to be a valuable piece of knowledge for researchers, students, practitioners and scholars across the globe.

List of Contributors

Antonia Maria Masucci
ETIS/ENSEA - University of Cergy Pontoise - CNRS, 6 Avenue de Ponceau, 95014 Cergy, France
INRIA Paris-Rocquencourt, Le Chesnay Cedex, France

Elena Veronica Belmega
ETIS/ENSEA - University of Cergy Pontoise - CNRS, 6 Avenue de Ponceau, 95014 Cergy, France

Inbar Fijalkow
ETIS/ENSEA - University of Cergy Pontoise - CNRS, 6 Avenue de Ponceau, 95014 Cergy, France

Yu Wang
Department of Computer and Information engineering, Beijing Technology and Business University, Beijing, China

Yongsheng Zhao
Department of Mechanical Engineering, Yanshan University, Qinhuangdao City, Hebei Province, China

Yi Chen
Department of Computer and Information engineering, Beijing Technology and Business University, Beijing, China

Balázs Fodor
Technische Universität Braunschweig, Institute for Communications Technology, Schleinitzstr. 22, 38106, Braunschweig, Germany

Florian Pflug
Technische Universität Braunschweig, Institute for Communications Technology, Schleinitzstr. 22, 38106, Braunschweig, Germany

Tim Fingscheidt
Technische Universität Braunschweig, Institute for Communications Technology, Schleinitzstr. 22, 38106, Braunschweig, Germany

Xiumei Li
School of Information Science and Engineering, 58 Haishu Road, Hangzhou Normal University, 311121 Hangzhou, China

Guoan Bi
School of Electronic and Electrical Engineering, 50 Nanyang Ave., Nangyang Technological University, Singapore 639798, Singapore

José Antonio Cortés
Atmel Spain, Torre C2, Polígono Puerta Norte, A-23 Zaragoza, Spain

Alfredo Sanz
Atmel Spain, Torre C2, Polígono Puerta Norte, A-23 Zaragoza, Spain

Pedro Estopiñán
Atmel Spain, Torre C2, Polígono Puerta Norte, A-23 Zaragoza, Spain

José Ignacio García
Atmel Spain, Torre C2, Polígono Puerta Norte, A-23 Zaragoza, Spain

Carlo Muscas
Department of Electrical and Electronic Engineering, University of Cagliari, Piazza d'Armi 1, 09123 Cagliari, Italy

Marco Pau
Department of Electrical and Electronic Engineering, University of Cagliari, Piazza d'Armi 1, 09123 Cagliari, Italy

Paolo Attilio Pegoraro
Department of Electrical and Electronic Engineering, University of Cagliari, Piazza d'Armi 1, 09123 Cagliari, Italy

Sara Sulis
Department of Electrical and Electronic Engineering, University of Cagliari, Piazza d'Armi 1, 09123 Cagliari, Italy

José Carlos Palomares Salas
Research Group PAIDI-TIC-168: Computational Instrumentation and Industrial Electronics (ICEI), Av. Ramón Puyol S/N., E-11202 Algeciras-Cádiz, Spain
Area of Electronics, Polytechnic School of Engineering, University of Cádiz, Av. Ramón Puyol S/N., E-11202 Algeciras-Cádiz, Spain

Juan José González de la Rosa
Research Group PAIDI-TIC-168: Computational Instrumentation and Industrial Electronics (ICEI), Av. Ramón Puyol S/N., E-11202 Algeciras-Cádiz, Spain
Area of Electronics, Polytechnic School of Engineering, University of Cádiz, Av. Ramón Puyol S/N., E-11202 Algeciras-Cádiz, Spain

José María Sierra Fernández
Research Group PAIDI-TIC-168: Computational Instrumentation and Industrial Electronics (ICEI), Av. Ramón Puyol S/N., E-11202 Algeciras-Cádiz, Spain
Area of Electronics, Polytechnic School of Engineering, University of Cádiz, Av. Ramón Puyol S/N., E-11202 Algeciras-Cádiz, Spain

Agustín Agüera Pérez
Research Group PAIDI-TIC-168: Computational Instrumentation and Industrial Electronics (ICEI), Av. Ramón Puyol S/N., E-11202 Algeciras-Cádiz, Spain
Area of Electronics, Polytechnic School of Engineering, University of Cádiz, Av. Ramón Puyol S/N., E-11202 Algeciras-Cádiz, Spain

Alexei V Nikitin
Avatekh Inc., 901 Kentucky Street, Suite 303, Lawrence, KS 66044, USA
Dept. of Electrical and Computer Engineering, Kansas State University, Manhattan, KS 66506, USA

Ruslan L Davidchack
Dept. of Mathematics, University of Leicester, Leicester LE1 7RH, UK

Jeffrey E Smith
BAE Systems Technology Solutions, 6 New England Executive Park, Burlington, MA 01803, USA

Kuandong Gao
School of Communication and Information Engineering, University of Electronic Science and Technology of China, 2008, Road XiYuan, Chengdu, 611731 Sichuan, China

Hui Chen
School of Communication and Information Engineering, University of Electronic Science and Technology of China, 2008, Road XiYuan, Chengdu, 611731 Sichuan, China

Huaizong Shao
School of Communication and Information Engineering, University of Electronic Science and Technology of China, 2008, Road XiYuan, Chengdu, 611731 Sichuan, China

Jingye Cai
School of Communication and Information Engineering, University of Electronic Science and Technology of China, 2008, Road XiYuan, Chengdu, 611731 Sichuan, China

Wen-Qin Wang
School of Communication and Information Engineering, University of Electronic Science and Technology of China, 2008, Road XiYuan, Chengdu, 611731 Sichuan, China

Masahiro Iwahashi
Department of Electrical, Electronics and Information Engineering, Nagaoka University of Technology, 1603-1 Kamitomioka, Nagaoka, Niigata, Japan

Taichi Yoshida
Department of Electrical, Electronics and Information Engineering, Nagaoka University of Technology, 1603-1 Kamitomioka, Nagaoka, Niigata, Japan

Norrima Binti Mokhtar
Department of Electrical Engineering, University of Malaya, 50603 Kuala Lumpur, Malaysia

Hitoshi Kiya
Department of Information and Communication Systems, Faculty of System Design, Tokyo Metropolitan University, 6-6 Asahigaoka, Hino, Tokyo, Japan

Luisa Alfieri
Department of Electrical Engineering and Information Technology, University of Naples Federico II, Via Claudio, 21, 80125 Napoli, Italy

Tomaso Erseghe
Universita di Padova, Dipartimento di Ingegneria dell'Informazione, via G. Gradenigo 6/b, Padova, Italy

Jianqiang Gao
College of Computer and Information Engineering, Hohai University, Xikang Road No.1, 210098 Nanjing, China

Lizhong Xu
College of Computer and Information Engineering, Hohai University, Xikang Road No.1, 210098 Nanjing, China

Jie Shen
College of Computer and Information Engineering, Hohai University, Xikang Road No.1, 210098 Nanjing, China

Fengchen Huang
College of Computer and Information Engineering, Hohai University, Xikang Road No.1, 210098 Nanjing, China

Feng Xu
College of Computer and Information Engineering, Hohai University, Xikang Road No.1, 210098 Nanjing, China

Kenshi Saho
Department of Electronic and Computer Engineering, Ritsumeikan University, 1-1-1 Noji-Higashi, Kusatsu, Shiga 525-8577, Japan

Masao Masugi
Department of Electronic and Computer Engineering, Ritsumeikan University, 1-1-1 Noji-Higashi, Kusatsu, Shiga 525-8577, Japan

Wenhui Xiong
National Key Laboratory of Communication, University of Electronic Science and Technology of China, Chengdu 611731, China

Jin Cao
National Key Laboratory of Communication, University of Electronic Science and Technology of China, Chengdu 611731, China

Shaoqian Li
National Key Laboratory of Communication, University of Electronic Science and Technology of China, Chengdu 611731, China